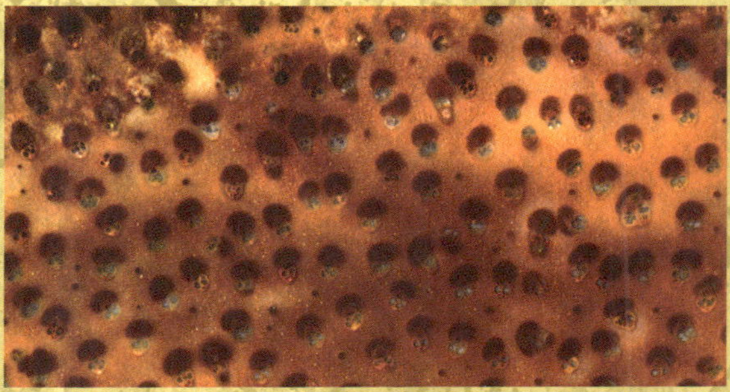

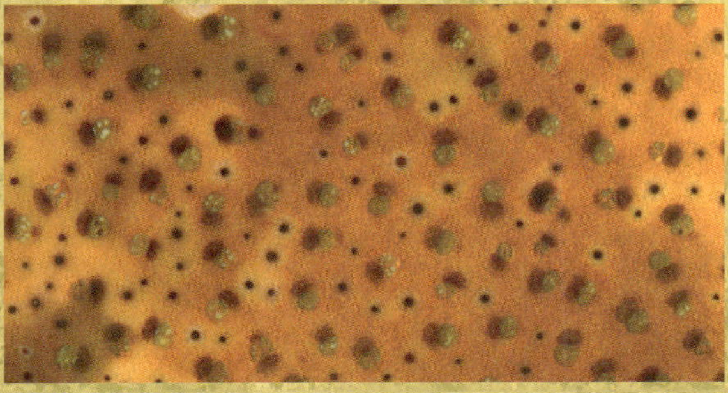

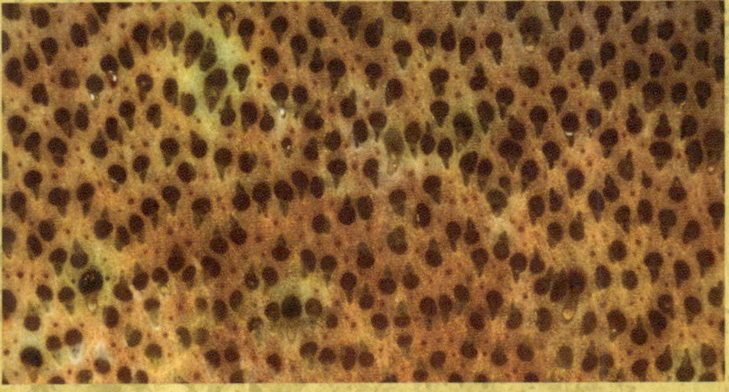

Mom & Dad,

Margo took me back to the Petrified Forest this year. I remembered our trip there on our way to Hawaii!. In *my* memory, you went there just because I loved rocks & minerals so much. I picked up a shard of petrified wood, talked with Dad about how it was formed & cherished it, and that visit all these years. I saw this book with all the amazing photos of the structure of petrified wood and had to get it for you. So... thank you for taking me there as a child, thank you for feeding & inspiring my love of the natural world, and thank you for being my parents. I love you!

Merry Christmas 2015

Love,

Ann

ANCIENT FORESTS

A CLOSER LOOK AT FOSSIL WOOD

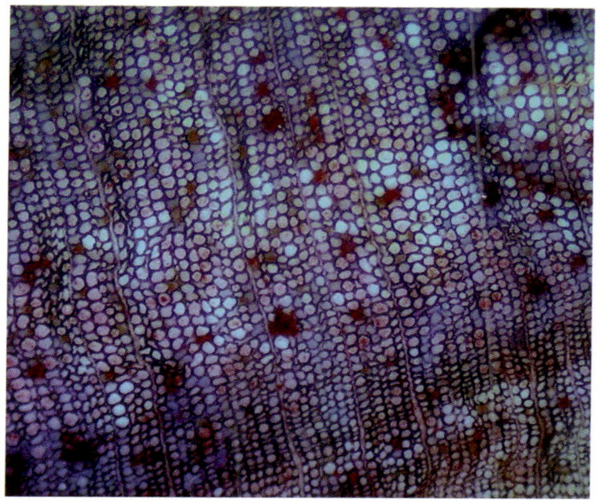

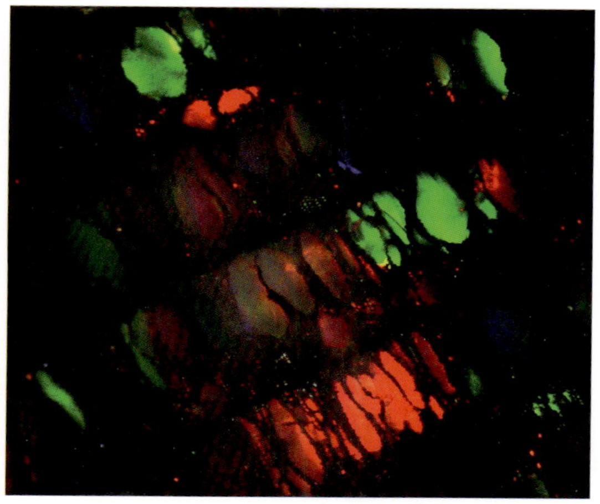

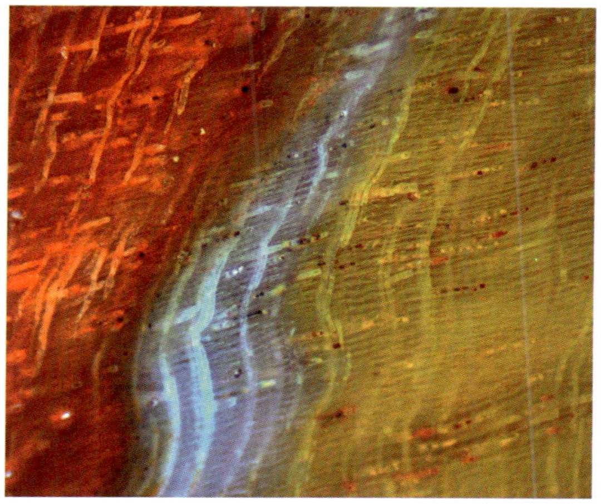

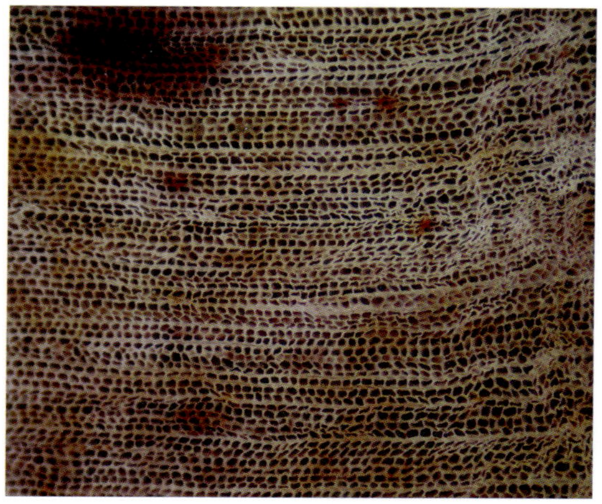

Frank J. Daniels and Richard D. Dayvault

PHOTOGRAPHY BY FRANK J. DANIELS

WESTERN COLORADO PUBLISHING COMPANY

First Edition

Western Colorado Publishing Company
2024 Freedom Court
Grand Junction, Colorado 81503-9522
(970) 242-5255
email: publisher@westerncoloradopublishing.com
website: www.westerncoloradopublishing.com

Design by Frank J. Daniels and Kitty Anderson Nicholason

The 100% acid-free paper used in this publication meets the minimum requirements of the American National Standard for Information Sciences – Permanence of Paper for Printed Library Materials.

Cataloging-in-Publication Data prepared by Western Colorado Publishing Company

Daniels, Frank J. 1949-
Dayvault, Richard D. 1948-
Ancient forests: a closer look at fossil wood/Frank J. Daniels, Richard D. Dayvault

Includes bibliographical references, glossary, and index.
ISBN-13: 978-0-9662938-1-4 hardcover
ISBN-10: 0-9662938-1-9 hardcover
1. Trees, Fossil. 2. Plants, Fossil. 3. Petrification.
4. Paleobotany. 5. Minerology. I. Title.
QE991 .D– 2006
561.16
Library of Congress Control Number: 2005938200

Printed and bound in Canada
Published in the United States of America

ACKNOWLEDGMENTS

Numerous individuals assisted in the preparation of this book by loaning specimens, providing information, and offering technical advice. Special thanks to two fossil wood collectors who provided guidance and inspiration and who now hunt perfect specimens in the Great Beyond: E. P. Akin of Shreveport, Louisiana, purveyor of the world's finest fossil *Palmoxylon* and a fine gentleman, and Shirts Quant of Prineville, Oregon, owner of the incredible agatized cone and a pioneer in Oregon fossil wood collecting.

Curators of three major museum collections offered access and friendly assistance with all necessary scheduling and details. For this we are eternally grateful. We extend our gratitude to Washington State University in Pullman, Washington, for help with ***The Lyle and Lela Jacklin Collection*** of fossil woods of the American Northwest. The assistance of Diane Johnson Cornelius, Dr. Gary D. Webster, and Dr. Peter B. Larson was much appreciated. Special thanks to the Rice Northwest Museum of Rocks and Minerals in Hillsboro, Oregon, and to Dennis and Mary Murphy for access to ***The Dennis and Mary Murphy Collection*** of fossil woods of the American West. We also especially thank Jim and Cathy Gray and staff for access to ***The Jim and Cathy Gray Collection*** of fossil woods of the American Southwest, located in Holbrook, Arizona.

Ed Rigel of Applegate Lapidary deserves special recognition for his efforts seeking out new and diverse fossil wood locations around the world. His provision of beautiful *Rhexoxylon* specimens from Zimbabwe to include in this book is much appreciated. George Mustoe of Western Washington University in Bellingham deserves our appreciation for providing technical reviews, discussions of supramolecular templating, and a number of fine electron microscope images. Ken Kemner and Max Boyanov of the Argonne National Laboratory analyzed specimens of green wood, confirming chromium and iron as coloring agents. Thanks are extended to Dr. Don Tidwell, professor emeritus at Brigham Young University, for providing guidance on many subjects and reviewing much of the manuscript for the book.

Many others went out of their way to provide information and specimens to photograph. Thank you to Russell and Adrienne Albers, Phil and Joni Andrist, Glenn Archer, Jose Arias, Bob Beck, Dan Beck, John Bennett, Ed Blankman, Danny Bittle, Robert Bockholt, Bill and Faye Branson, Gary Buss, Joe and Lucy Cota, Thomas and Michaela Davidsson, Norman Donpon, Judy Elkins, Bob Finnell, Captain N. A. Foshee, W. David Haggerty, Mike Hankins, Steven Hatch, Rudiger Hesse, Glen Hodson, Peter Huber, Ruth Jenkins, Marc Jobin, Edith Johnston, Ed "Chicken" Jones, Vince Jones, Ron Justman, Neil Korpusik, Hans Koser, Dean Lowdermilk, Adam Luna, Ray Lyman, Benoit Macray, Rod Miller, Kent Mower, Gene Mueller, John Needham, Charles Nicholas, Paul Obnich, Dan and Hope Rigel, Tom Robertson, Rick Rolater, Bill Rose, Larry Sanchez, Luis and Ana De Los Santos, Stephen Schuchman, Dr. William Shadish, Ernie Shirley, Russell Sims, Jane and Forrest Smith, Steve Speer, Joe Taubr, Michael Viney, Sterling and Josie Wade, Vint and Donna Wade, Jimmy Walker, Larry Walker, Betty Warrington, Bob Wurster, Joe Zash, and Tom Zienka. For sharing their knowledge and otherwise being of assistance, we would like to thank Greg Caffey, Mike Carlson, Dewey Dietz, Louis Goodman, Jan Herman, Bill Hood, Chuck Hutchinson, Dan Kile, Conrad Labandeira, Tom Michalski, Jim Mills, Terry Moxon, Robert Stoufer, Reed Waldron, and Walt Wright.

Jim Ross of Absolute Clarity & Calibration, LLC, provided technical assistance with the Meiji EMZ-5TR stereo microscope used to produce micro images of fossils. John Limbach of Biodisc Inc. provided photomicrographs of modern wood thin sections. Photographs of landscapes and large specimens were shot with a Pentax 6X7 medium format camera using 45 mm, 75 mm, 100 mm macro, 135 mm macro, and 200 mm lenses. Other photographs were taken with a Pentax ZX5n 35 mm camera using 50 mm and 100 mm macro lenses. Film used was Fuji Velvia 50. Micro images were taken with the Meiji EMZ-5TR stereo microscope using a Sony MVC-CD400 digital camera. Scanning electron microscope images were taken using a Tescan Vega operated at 10 kilovolts.

A note on photograph captions: When known to a reasonable degree of certainty, specimen photograph captions include the location, the geologic formation and age, the specimen owner, and the size measured to the nearest 0.5 centimeter across the widest dimension depicted.

We expect to offer supplements to this work in the future as new information and specimens become available. To register to be notified of future releases, please write or email the publisher. Our sister company, Western Colorado Trading Company, offers top quality specimens of fossil wood for sale to private collectors and museums. The Fossil Wood Institute was estabished to help preserve fossil wood collections for future generations to enjoy. To find out more information about The Fossil Wood Institute, see our website or contact us at fossilwoodinstitute@westerncoloradopublishing.com. If you are aware of any information that would be helpful for future works, please contact Frank J. Daniels though the publishing company.

Western Colorado Publishing Company
2024 Freedom Court
Grand Junction, CO 81503-9255
(970) 242-5255
publisher@westerncoloradopublishing.com

Western Colorado Trading Company
2024 Freedom Court
Grand Junction, CO 81503-9255
(970) 242-5255
tradingcompany@westerncoloradopublishing.com

Signed and numbered limited edition copies of photographs in this book and other photographs by Frank J. Daniels are available from Western Colorado Publishing Company. Contact photography@westerncoloradopublishing.com.

Website: **www.westerncoloradopublishing.com**

Our first book on the subject, ***Petrified Wood: The World of Fossilized Wood, Cones, Ferns, and Cycads***, ISBN 0966293800, is available directly from the publisher. Retail price is $75.00 US. It has been the number one best seller for books in the category of "Fossil Plants" for most of the past 7 years on Amazon.com.

CHOICE *Current Reviews for Academic Libraries*
"Petrified Wood is beautifully illustrated and serves to introduce the novice to the gamut of fossil wood types.... For the professional paleobotanist, geologist, and paleontologist, this illustrated survey of the major kinds of petrified woods from sites around the world can be quite useful.... A worthwhile addition to a general library collection."

FOSSIL NEWS
"This is a beautiful coffee table book that will delight both the eye and mind of any aficionado of paleobotany. You certainly couldn't find a better pictorial treatment of the subject anywhere."

Rock & Gem
"With 420 beautifully reproduced color photographs and easily read text, this book is both a coffee table beauty and a most interesting and useful text. An exciting new book ... with superb photographs."

FOREWORD

Elm-like fossil wood. Maryhill, Washington, USA.

My first book, *Petrified Wood: The World of Fossilized Wood, Cones, Ferns, and Cycads,* was published in 1998. *Petrified Wood* presents photographs of hundreds of beautiful fossil wood specimens from worldwide locations and discusses aspects of paleobotany, geology, stratigraphy, and mineralogy. *Ancient Forests: A Closer Look at Fossil Wood* results from additional questions and a desire for a more detailed examination of the subject. As with many areas of study, the closer you look, the more questions arise. While this book is somewhat far ranging, a few concepts guide our course. First is the wonder of beauty. Or is it the beauty of wonder? This is what originally drew my attention to fossil woods. Specimens shown in this book are either beautiful or have a fascinating story to tell. Beauty is more wonderful when rare and is especially so in the presence of perfection. "Beauty is truth, truth beauty, –that is all ye know on earth, and all ye need to know." [John Keats, 1819]

In some ways, the focus on beauty limits the usefulness of this book. By selecting predominately beautiful specimens, we ignore the vast majority of fossil wood. Countless fossil wood locations in the United States and around the world produce bland specimens that are of great interest to paleobotanists. For the most part, this book dismisses them. To me the trade-off is justified because this book relies on photography to present the story. Many fine books feature the science of fossil woods. To the extent *Ancient Forests* offers an inventory of wood from worldwide locations, it is a collection of well-silicified, attractive specimens. *Ancient Forests* also aims to assist with identification of fossil woods. A closer look at wood anatomy enriches our understanding of woods and their origins. The woods discussed for identification are often available to a collector or student. Many are from the western United States and others hail from distant, exotic locations around the globe. Our discussion primarily involves identification by observing transverse sections through low-power magnification.

Another fascinating aspect of the closer look deals with mineralization. To explain this complex chemical process, I enlisted Dick Dayvault, a field geologist with broad experience and an abiding interest in the subject. Dick explains how fossil wood mineralization is mainly the story of silicification. He also provides detailed information about cycads, limb casts, and short shoots from Utah.

Although we discuss and display fossil woods from worldwide locations, this book provides a particularly detailed view of fossil woods from the Western United States. From museum collections and privately held specimens to macro photography and micro imaging, this closer look leads us from fossil wood locations to microscopic views of cell structure and mineralization. We illustrate almost every point with photographs, through which we strive to offer information while expressing beauty.

Frank J. Daniels
Grand Junction, Colorado
May 2006

***Araucarioxylon*. Northern Arizona, USA.**

CONTENTS

INTRODUCTION

The study of plant fossils offers a window into forests of the past through which we can glimpse an intriguing former world. Fossils offer paleobotanists an opportunity to learn about ancient forests and flora while gathering evidence of plant evolution. While the study of all forms of plant fossils aids understanding, fossil woods are of particular importance. Petrifactions retain more anatomical information of the plant than do other forms of fossilization. Fossil wood often includes a stem with vascular tissue and, therefore, preserves abundant cellular information. Keep in mind the incredible rarity of these fossils in relation to all the trees that have graced the surface of this planet. Fossils are rarely the end product of a tree's existence. The overwhelming majority of plants simply die and rot, adding humus to the soil. Conditions necessary for petrifaction infrequently and sporadically present themselves to a tree. Even after fossilizing, many plant fossils disintegrate from exposure to the elements. Countless others reside deep in the earth beyond our reach. And as noted by Jay Ellis Ransom in his 1955 pamphlet, *Petrified Forest Trails*, "While millions of tons of petrified tree trunks lie scattered over the length and breadth of America, only a very small proportion of all that is available in the unburied state outside of our National Monuments can be considered ultra-high quality."

Fossil wood includes any form of preservation of prehistoric wood. All petrified wood is fossil wood, but not all fossil wood is petrified wood. Far more plant fossils take the form of compressions, impressions, molds, and casts than actually emplace silica and permineralize. Some form of silica composes most of the fossil wood depicted in this book. Silica replacement can take many forms, from chalcedony, jasper, and opal to sardonyx and beautiful banded agate. Silicas frequently replace wood structure, preserving it in a manner allowing cutting and fine polishing. Many other minerals that replace wood do not allow faithful preservation of anatomical structure.

Back to the subject of casts. A fossil wood mold is the cavity left by a buried piece of wood that then decays or otherwise dissolves, leaving a void in the earth. A cast forms by filling of the mold by mud, volcanic debris, sand, or mineral material that preserves the shape and texture of the mold and thereby the original log. Several well-known locations for collecting beautiful casts exist in the United States—these include Texas Springs in Nevada, the South Fork of the Crooked River in Oregon, Dubois and the Wiggins Fork River in Wyoming, Tom Miner Basin in Montana, and Yellow Cat in Utah. Some fossil wood perfectly preserves, cell by cell, every element of the original structure, and some is a true cast, preserving no portion of the original structure. Fossil wood comes in an infinite variety of forms, including every possible phase between perfectly preserved petrified wood and true casts. Some fossil wood has the appearance of a cast, but on closer examination is seen to retain an element of structure, almost as if it retains a ghost of the long-gone wood.

MACRO PHOTOGRAPHY ESSAY

Ratka, Hungary.

Vantage, Washington, USA.

Vantage, Washington, USA.

Sweet Home, Oregon, USA.

Owyhee Mountains, Oregon, USA.

Elko County, Nevada, USA.

THIS SERIES OF MACRO PHOTOGRAPHS, photographed with a 35 mm film camera using macro lenses, illustrates the range within fossil wood from extremely well-preserved wood structures to the complete absence of residual wood structures. The initial photographs depict a variety of fossil wood cell structures. As the series of photographs progresses, wood structures become less apparent, to the point where they entirely disappear and become non-wood mineralization. The final photographs are agate.

Yakima, Washington, USA.

Queensland, Australia.

Sweet Home, Oregon, USA.

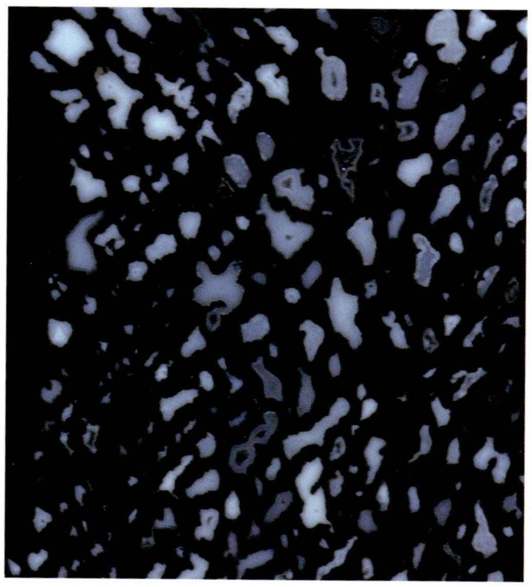
Sweet Home, Oregon, USA.

Henry Mountains, Utah, USA.

Henry Mountains, Utah, USA.

Henry Mountains, Utah, USA.

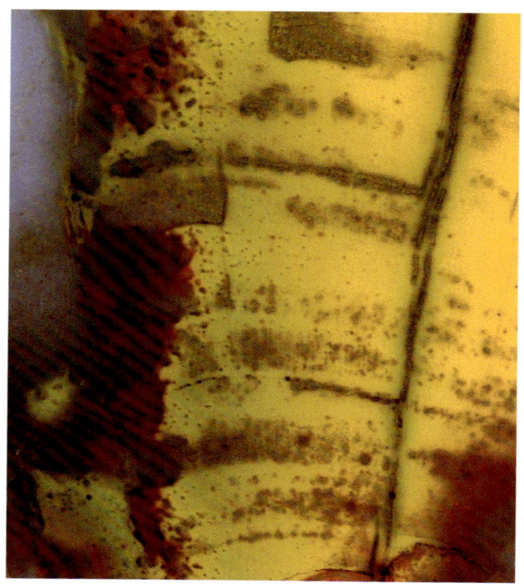
Henry Mountains, Utah, USA.

Henry Mountains, Utah, USA.

Henry Mountains, Utah, USA.

Central Oregon, USA.

Hampton Butte, Oregon, USA.

South Fork, Crooked River, Oregon, USA.

Douglas County, Colorado, USA.

Tasmania, Australia.

Zimbabwe.

Basalt, Nevada, USA.

Madagascar.

Madagascar.

Santa Cruz Province, Argentina.

Northern Arizona, USA.

Northern Arizona, USA.

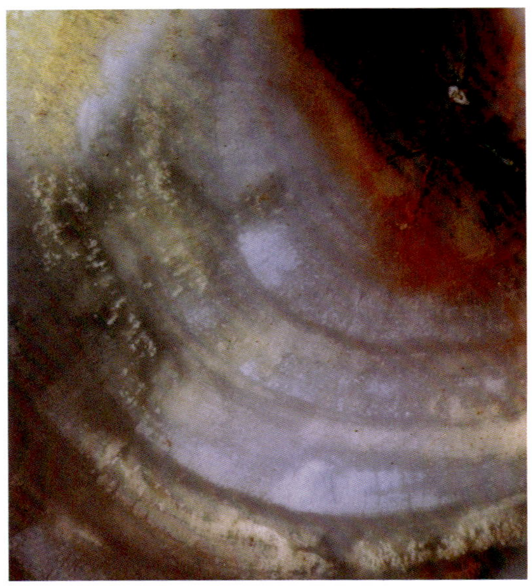
Northern Arizona, USA.

Northern Arizona, USA.

Northern Arizona, USA.

Northern Arizona, USA.

Northern Arizona, USA.

Northern Arizona, USA.

Northern Arizona, USA.

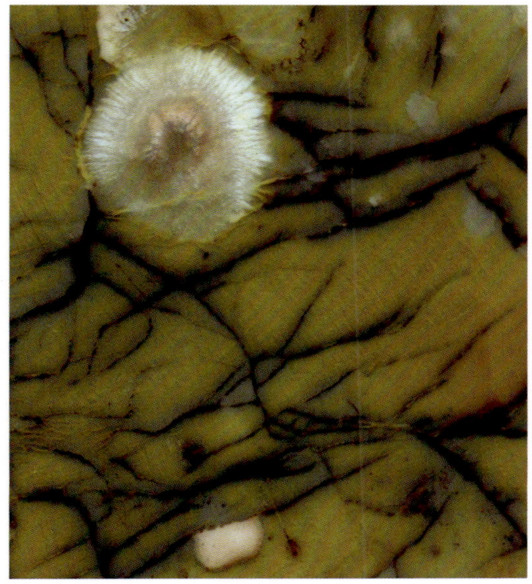
Henry Mountains, Utah, USA.

Little Humboldt River, Nevada, USA.

Little Humboldt River, Nevada, USA.

Hubbard Basin, Nevada, USA.

Hubbard Basin, Nevada, USA.

Madagascar.

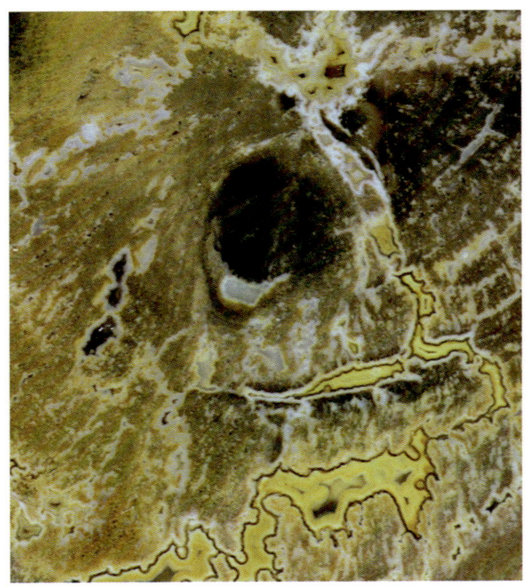
Madagascar.

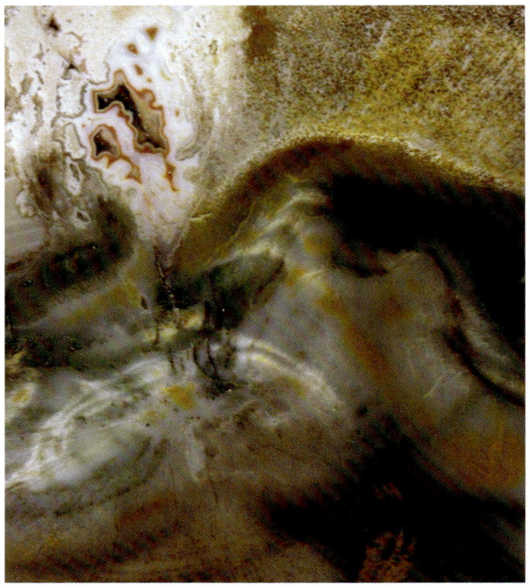
Madagascar.

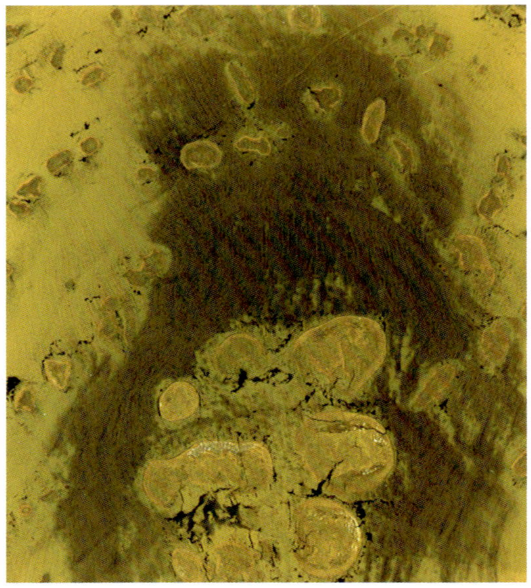
Northern Arizona, USA.

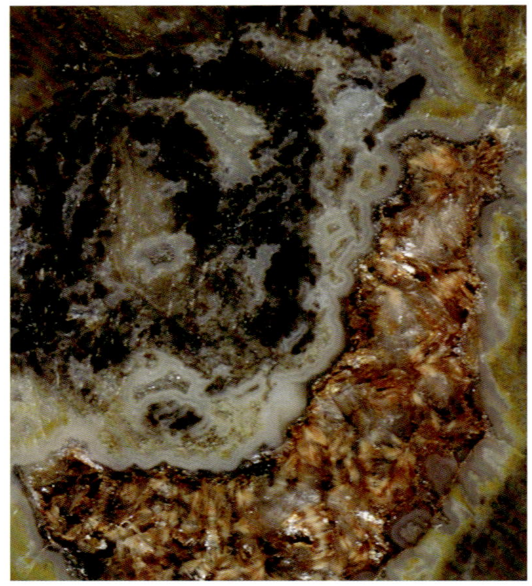
Northern Arizona, USA.

South Fork, Crooked River, Oregon, USA.

South Fork, Crooked River, Oregon, USA.

South Fork, Crooked River, Oregon, USA.

Sweet Home, Oregon, USA.

Sweet Home, Oregon, USA.

Tom Miner Basin, Montana, USA.

Tom Miner Basin, Montana, USA.

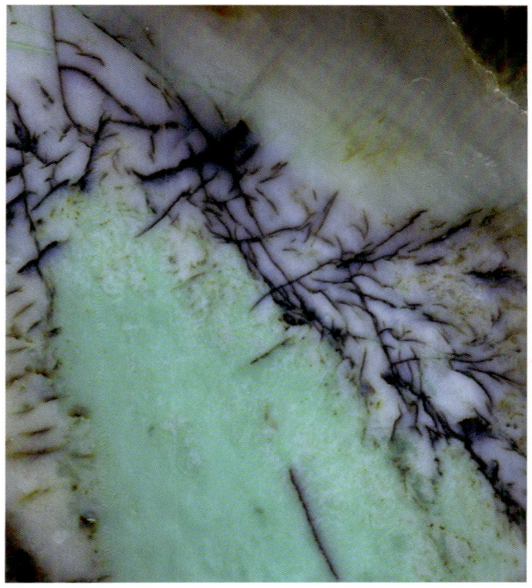
Java, Indonesia.

South Fork, Crooked River, Oregon, USA.

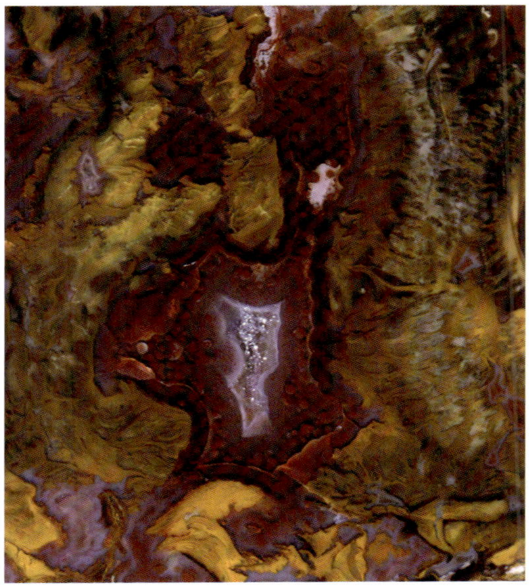
Eastern Utah, USA.

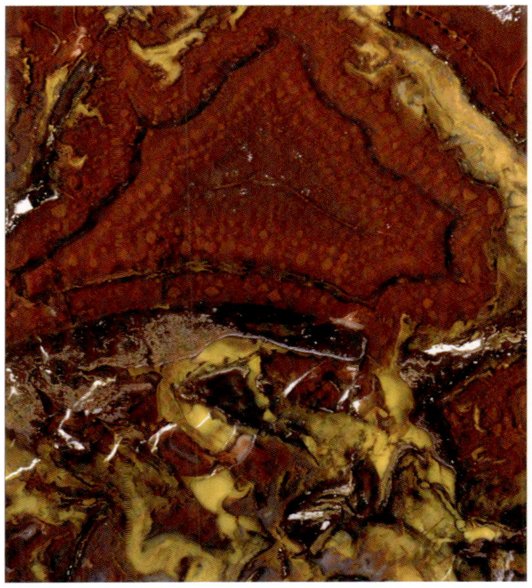
Eastern Utah, USA.

Sand Draw, Wyoming, USA.

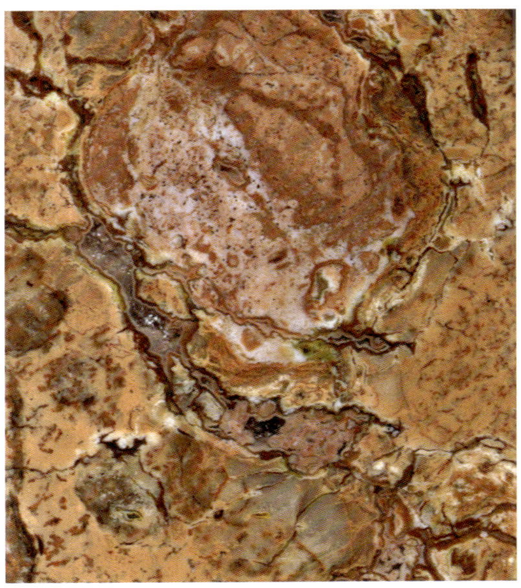
Madagascar.

Madagascar.

Rio Negro Province, Argentina.

Chubut Province, Argentina.

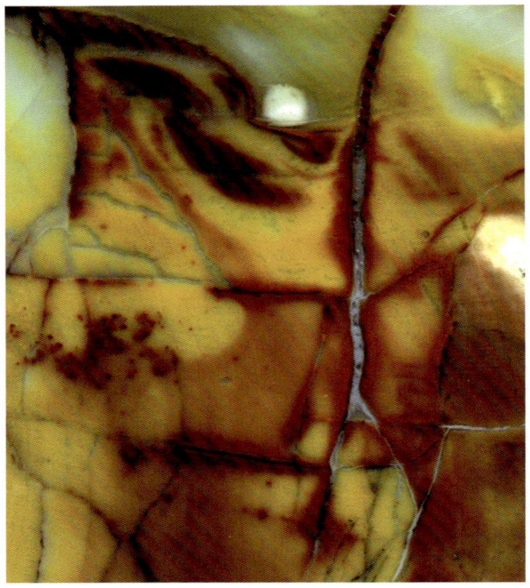
Henry Mountains, Utah, USA.

Henry Mountains, Utah, USA.

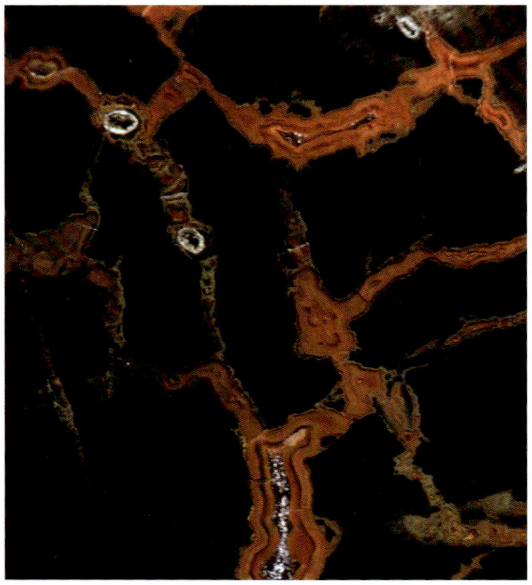
Henry Mountains, Utah, USA.

Moab, Utah, USA.

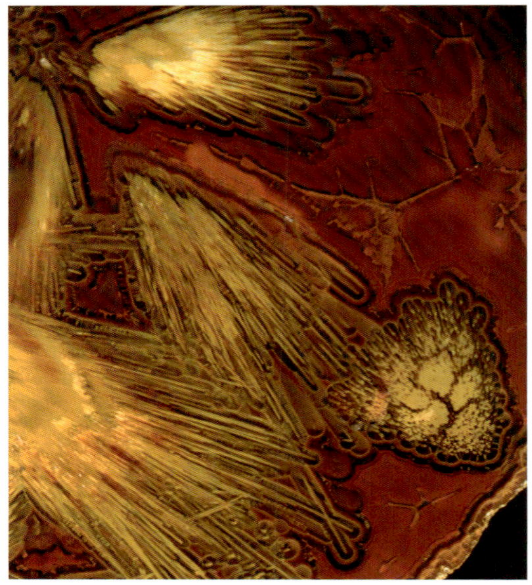
Moab, Utah, USA.

Moab, Utah, USA.

Yellow Cat, Utah, USA.

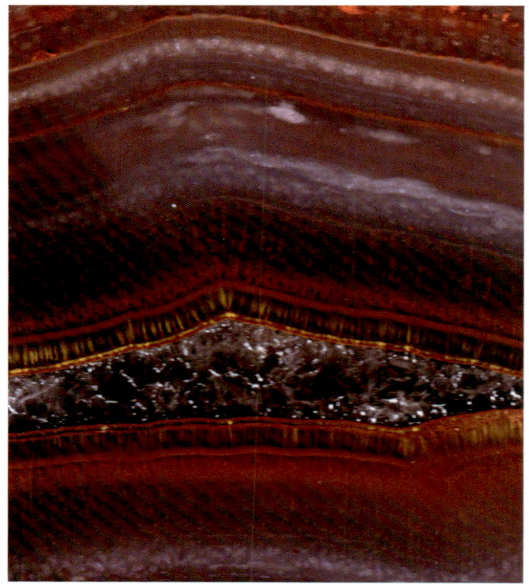
Yellow Cat, Utah, USA.

Moab, Utah, USA.

Yellow Cat, Utah, USA.

Moab, Utah, USA.

Dendritic agate. Davis Mountains, Texas.

Laguna agate. Chihuahua, Mexico.

Dryhead agate. Montana, USA.

Laguna agate. Chihuahua, Mexico.

Laguna agate. Chihuahua, Mexico.

Laguna agate. Chihuahua, Mexico.

MICRO IMAGING ESSAY

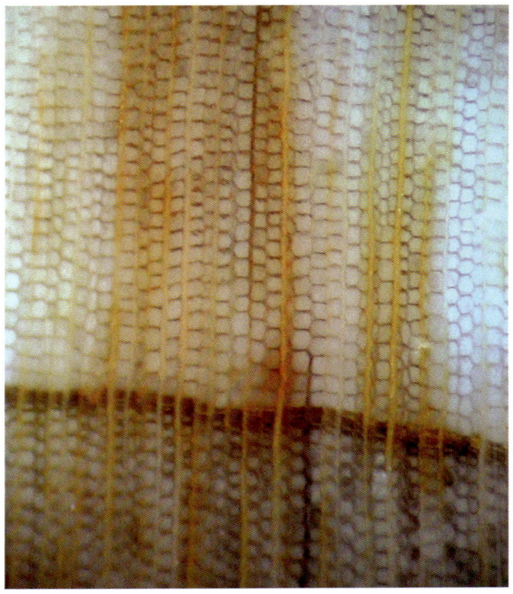
Vantage, Washington, USA.

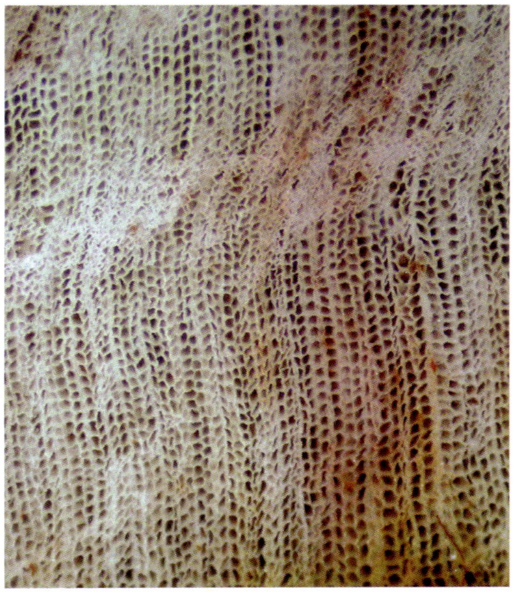
Grassy Mountain, Oregon, USA.

Vantage, Washington, USA.

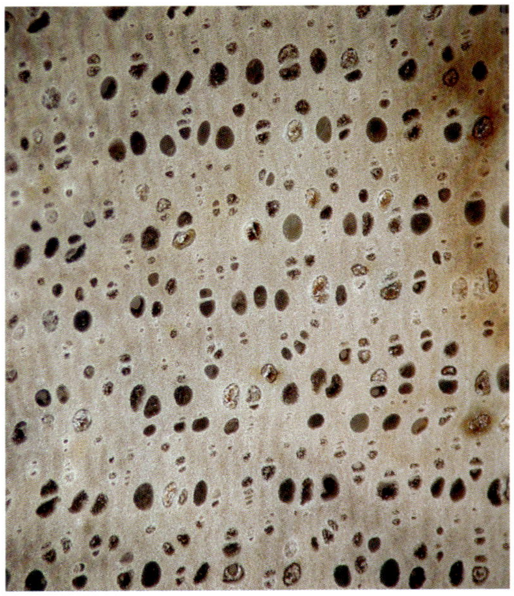
Sweet Home, Oregon, USA.

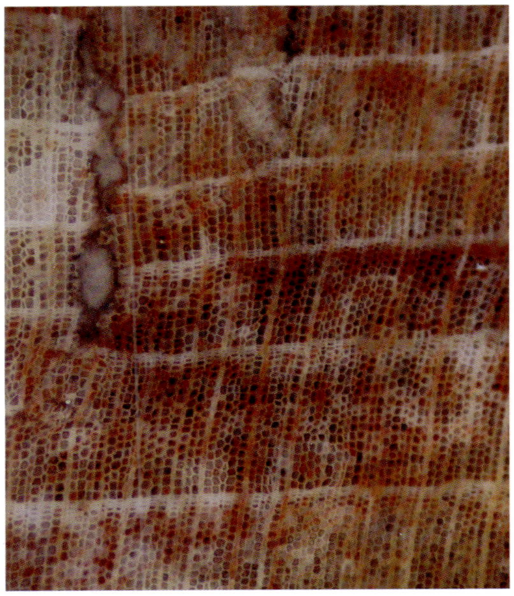
Kizilcahaman, Turkey.

Sweet Home, Oregon, USA.

THIS SERIES OF MICRO IMAGES, obtained though a microscope employing a digital camera, offers a closer look at types of fossil wood mineralization from extremely well-preserved wood structures to the complete absence of residual wood structures. The initial images present a variety of fossil wood cell structures. As the series of images progresses, wood structures become less apparent, to the point where they entirely disappear and become non-wood mineralization. The final images are agate.

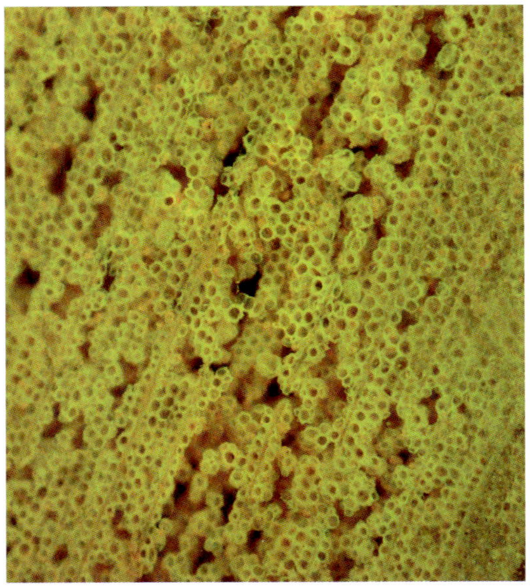
Henry Mountains, Utah, USA.

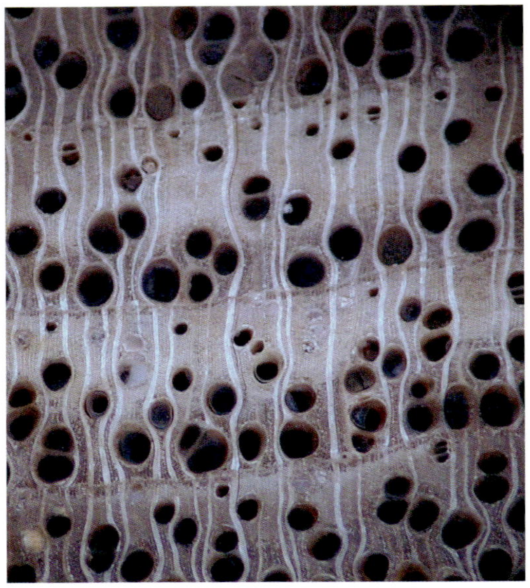
McDermitt, Oregon, USA.

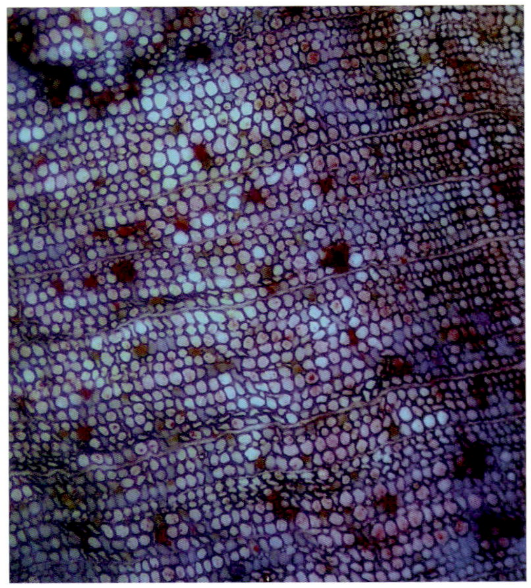
Madagascar.

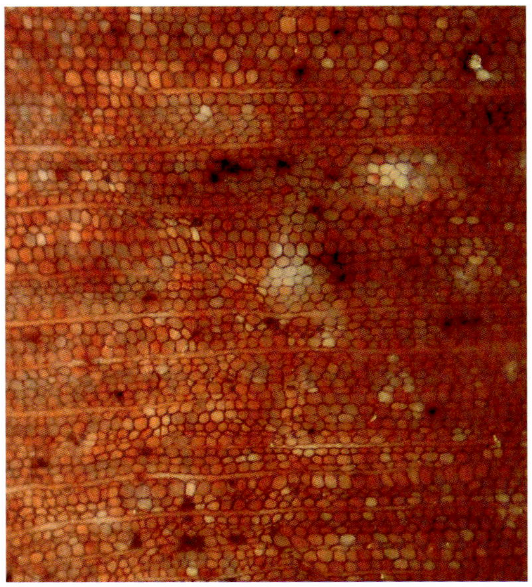
Madagascar.

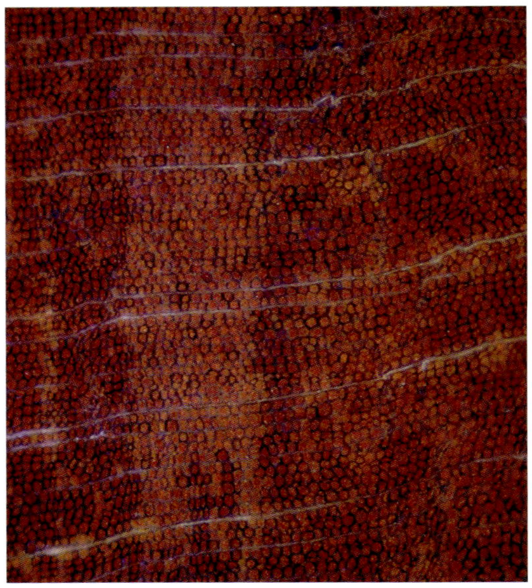
Northern Arizona, USA.

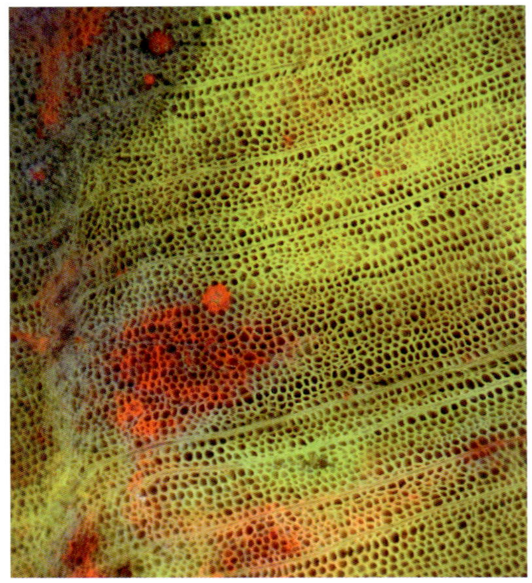
Northern Arizona, USA.

Northern Arizona, USA.

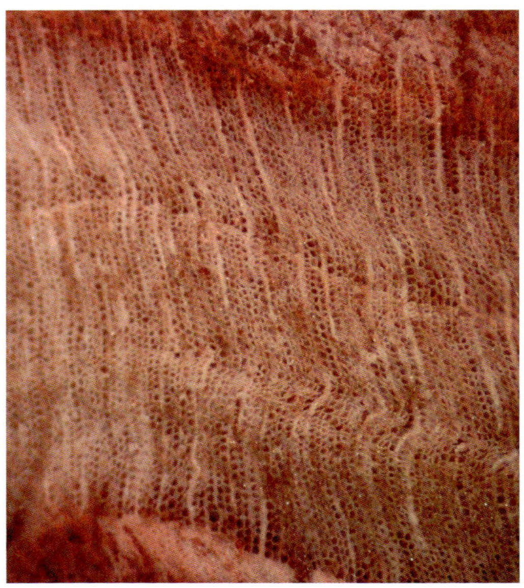
Goldfield, Nevada, USA.

Northern Arizona, USA.

Madagascar.

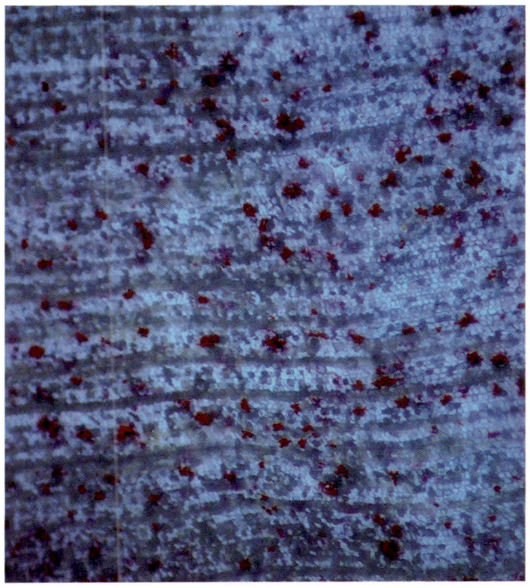

Madagascar.

Elko County, Nevada, USA.

Saddle Mountain, Washington, USA.

Yakima, Washington, USA.

Yakima, Washington, USA.

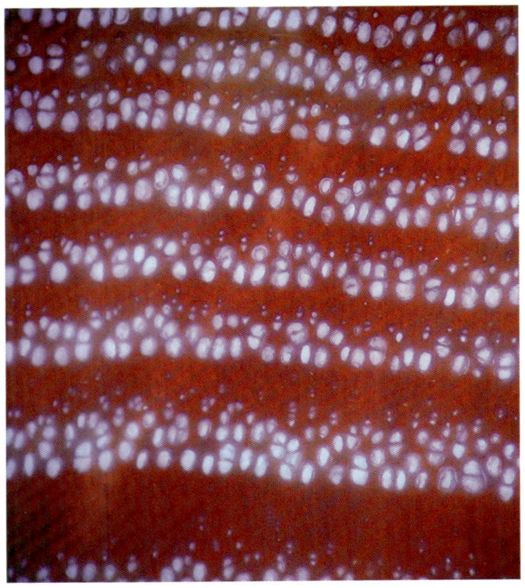

Succor Creek, Oregon/Idaho, USA.

Northern Arizona, USA.

Northern Arizona, USA.

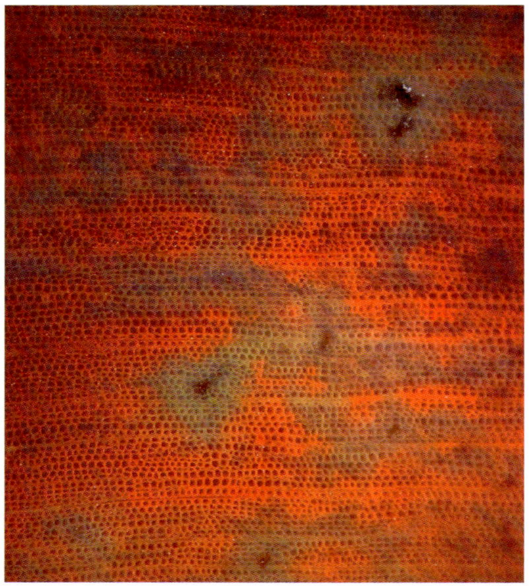
Northern Arizona, USA.

Circle Cliffs, Utah, USA.

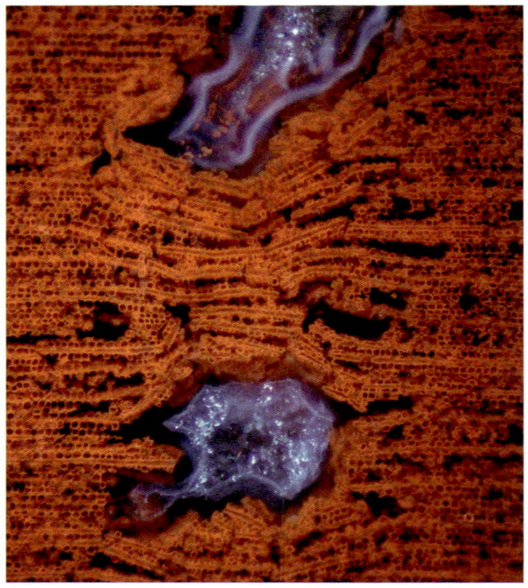
Henry Mountains, Utah, USA.

Henry Mountains, Utah, USA.

Douglas County, Colorado, USA.

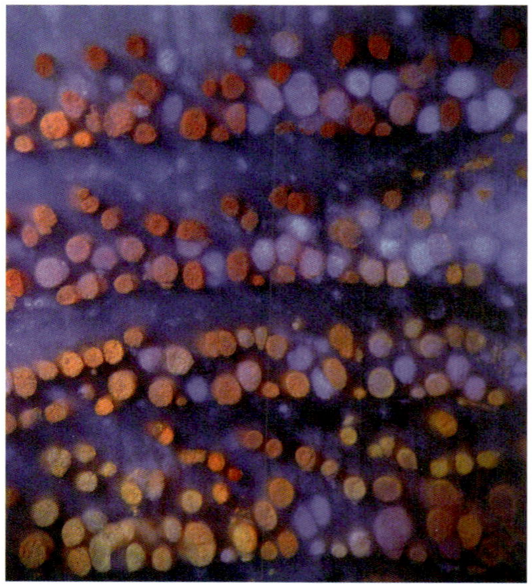
Grassy Mountain, Oregon, USA.

Grassy Mountain, Oregon, USA.

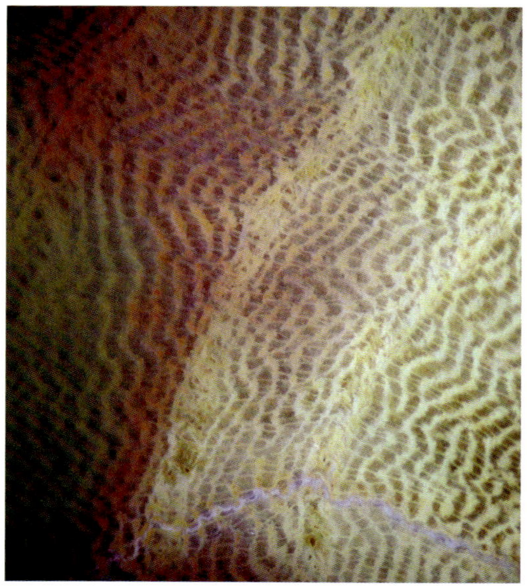
Hoodoo Basin, Oregon, USA.

Hubbard Basin, Nevada, USA.

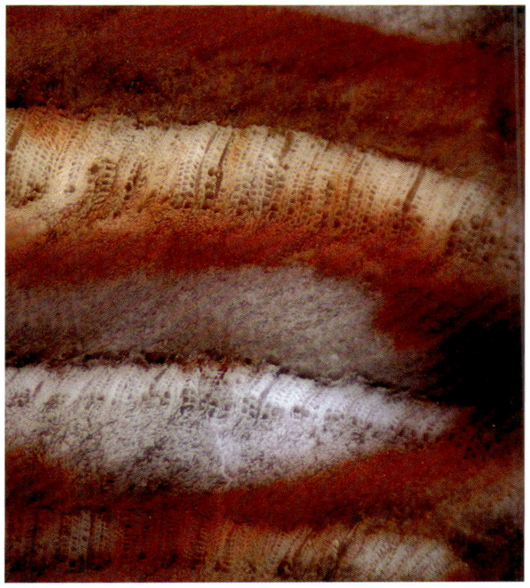

Elko County, Nevada, USA.

Grassy Mountain, Oregon, USA.

Grassy Mountain, Oregon, USA.

Northern Arizona, USA.

Northern Arizona, USA.

Northern Arizona, USA.

Northern Arizona, USA.

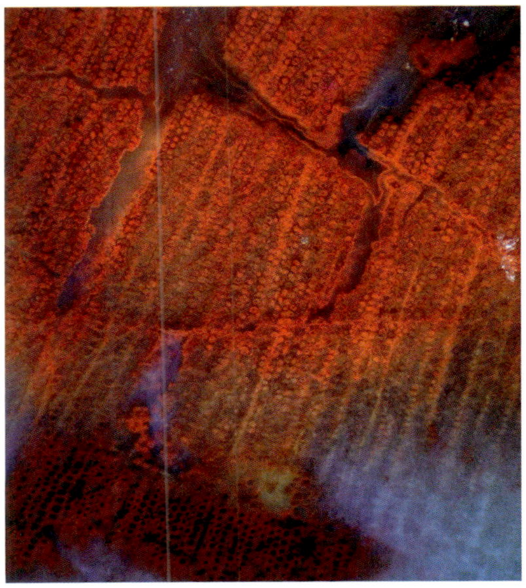

Henry Mountains, Utah, USA.

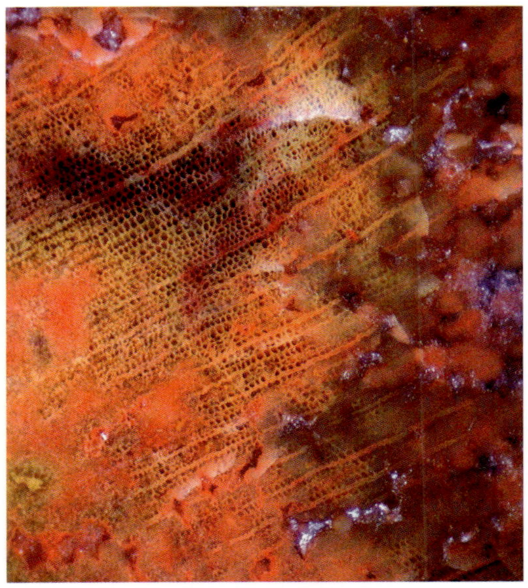

Henry Mountains, Utah, USA.

Goldfield, Nevada, USA.

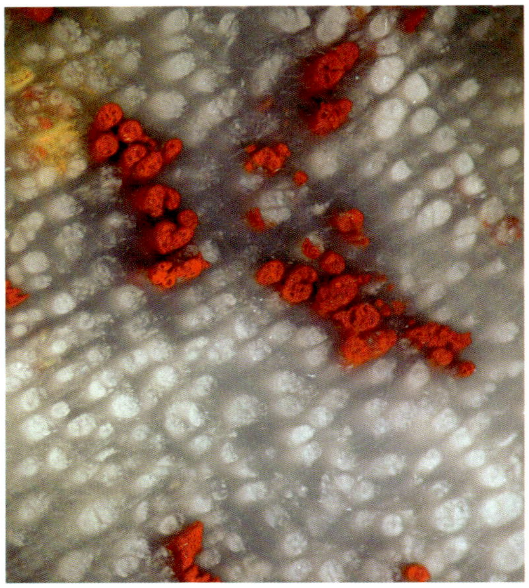

Grassy Mountain, Oregon, USA.

Owyhee Mountains, Oregon, USA.

Owyhee Mountains, Oregon, USA.

Northern Arizona, USA.

Northern Arizona, USA.

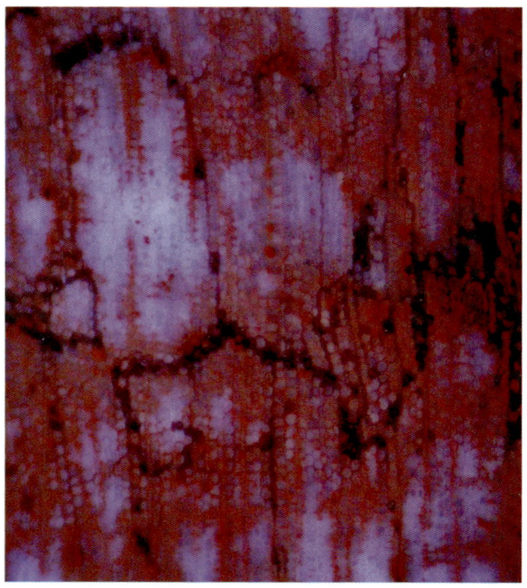

Northern Arizona, USA.

Northern Arizona, USA.

Northern Arizona, USA.

Northern Arizona, USA.

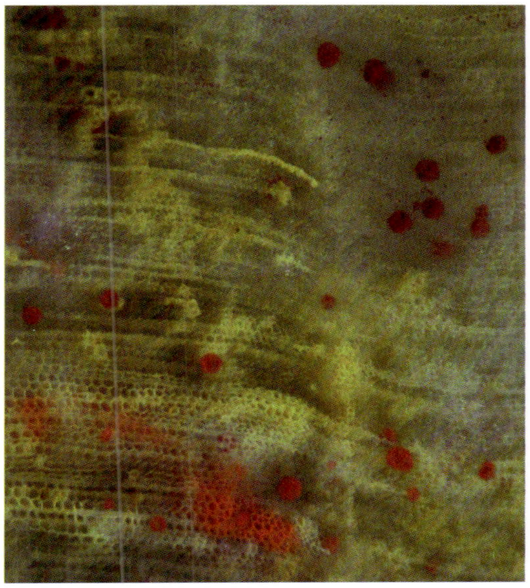

Northern Arizona, USA.

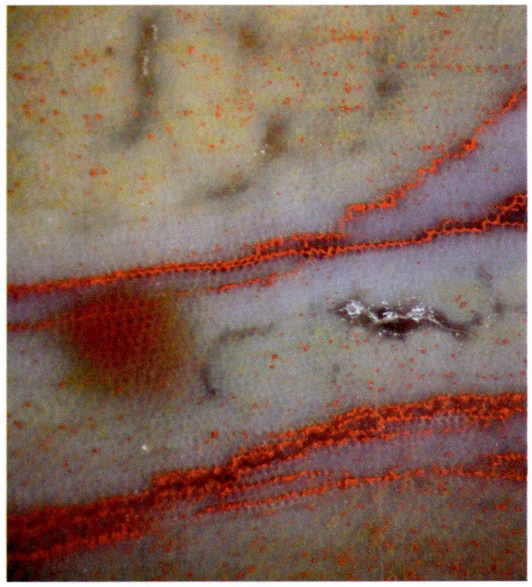

Northern Arizona, USA.

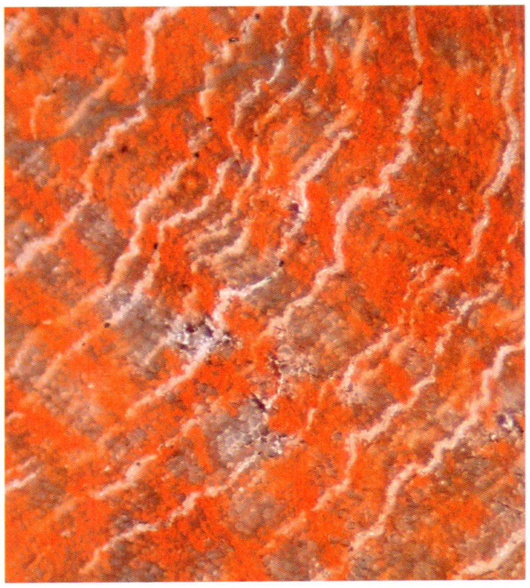

Northern Arizona, USA.

Northern Arizona, USA.

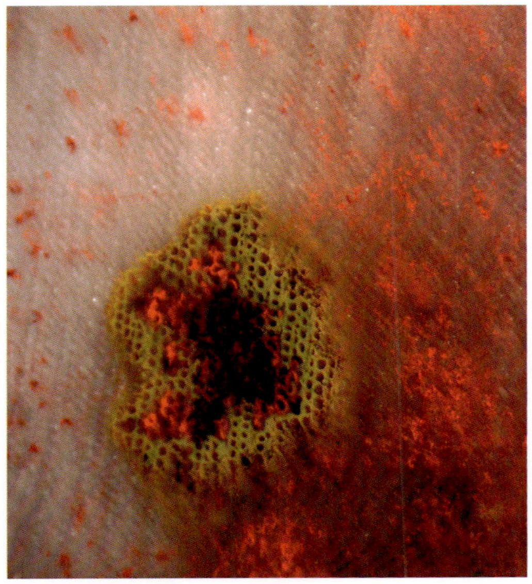

Northern Arizona, USA.

Northern Arizona, USA.

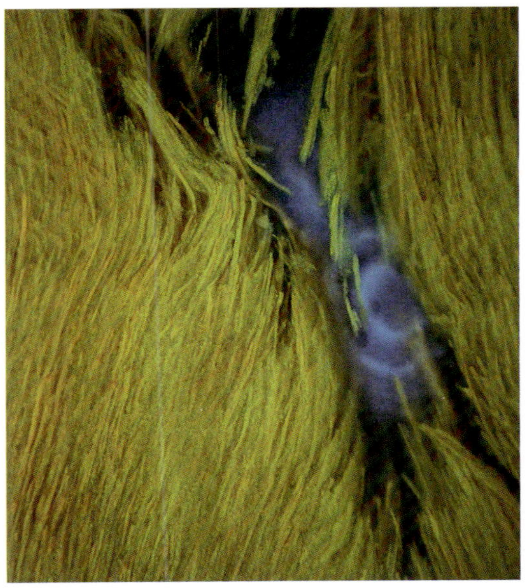

Northern Arizona, USA.

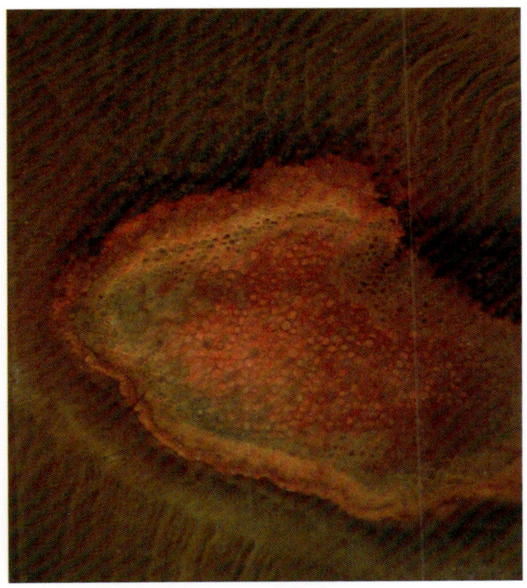

Hubbard Basin, Nevada, USA.

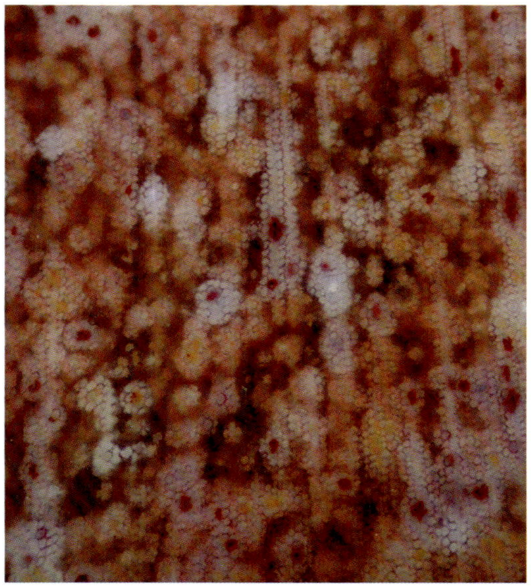

Madagascar.

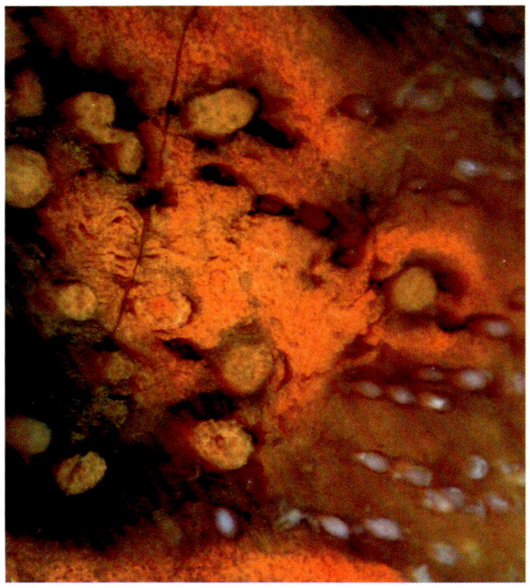

Douglas County, Colorado, USA.

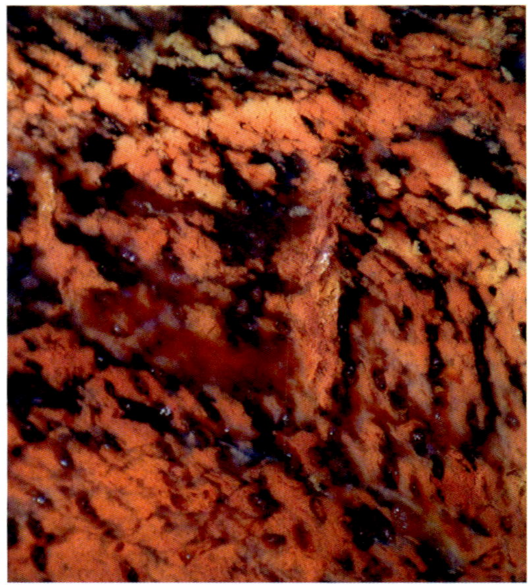

Douglas County, Colorado, USA.

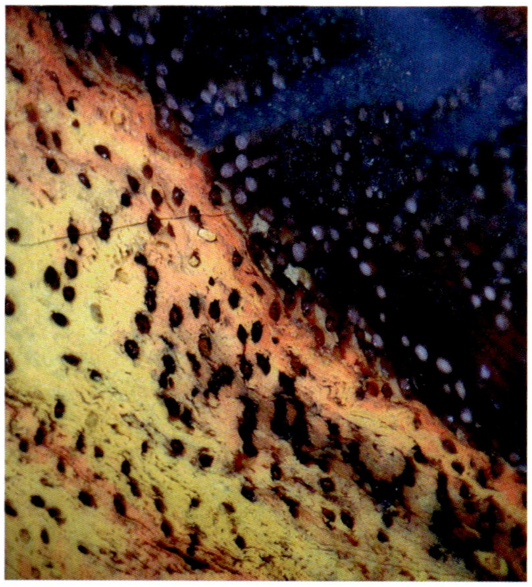

Douglas County, Colorado, USA.

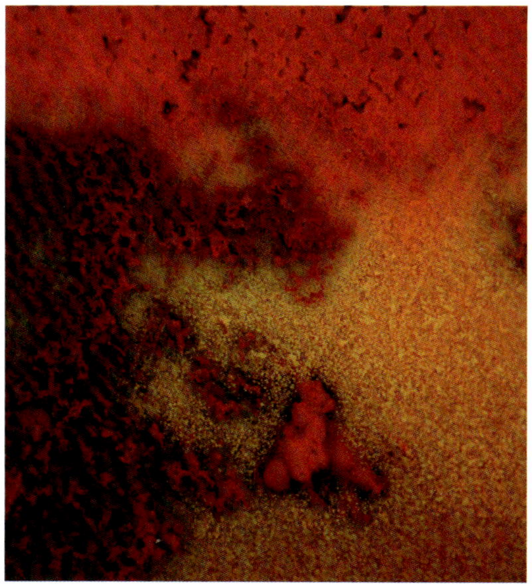

Henry Mountains, Utah, USA.

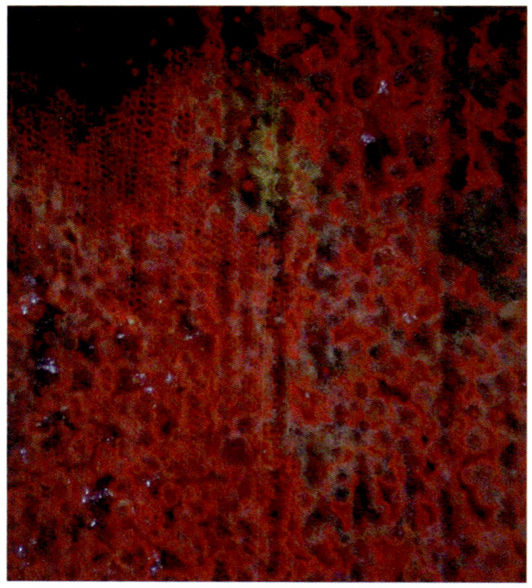

Henry Mountains, Utah, USA.

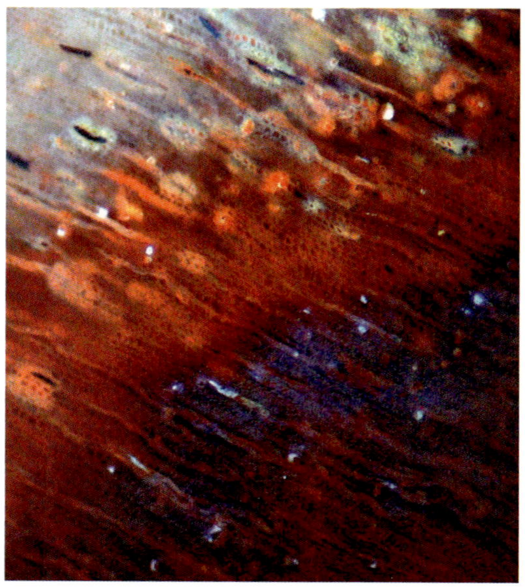

Henry Mountains, Utah, USA.

Henry Mountains, Utah, USA.

Henry Mountains, Utah, USA.

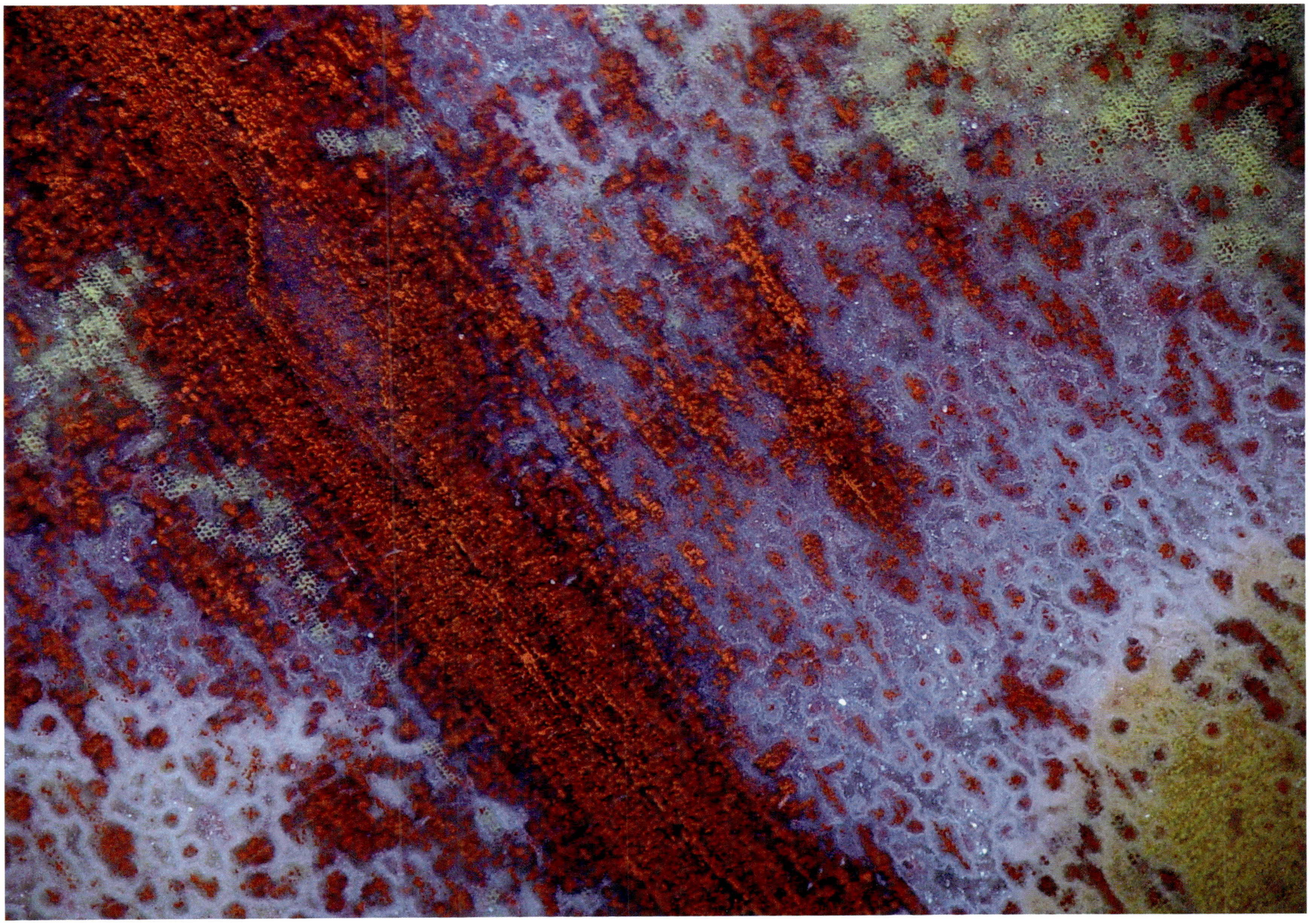

Henry Mountains, Utah, USA.

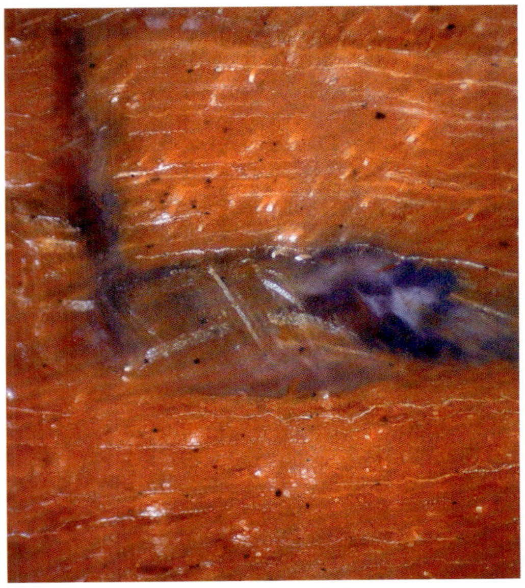

Henry Mountains, Utah, USA.

Henry Mountains, Utah, USA.

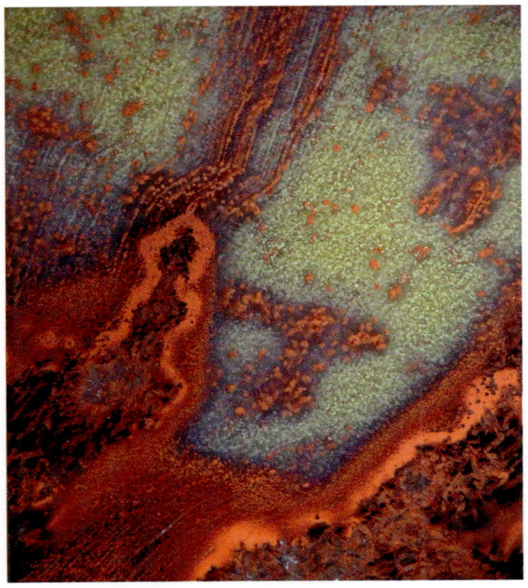

Henry Mountains, Utah, USA.

Henry Mountains, Utah, USA.

Hubbard Basin, Nevada, USA.

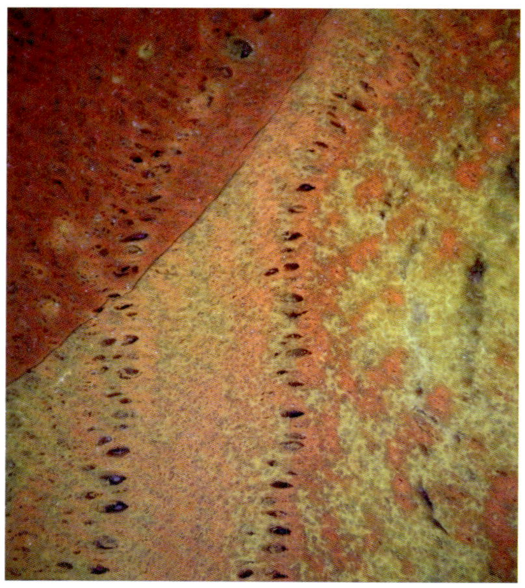

Basalt, Nevada, USA.

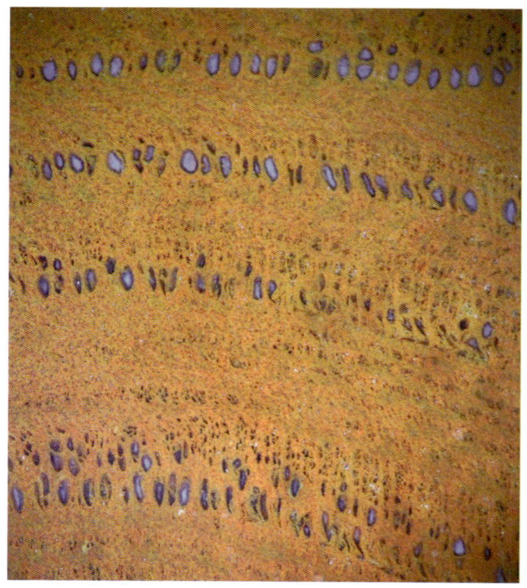

Basalt, Nevada, USA.

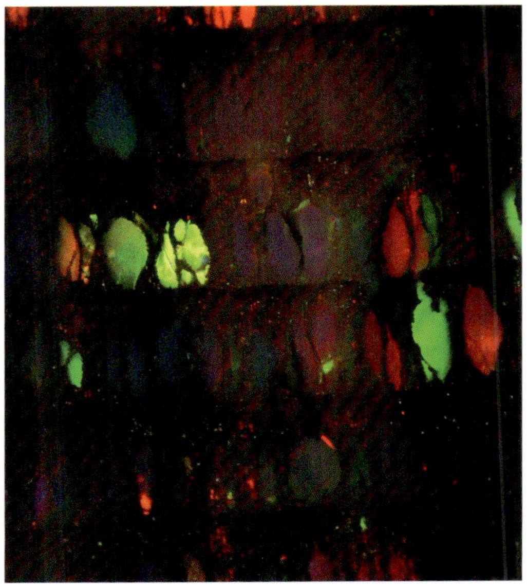
Virgin Valley, Nevada, USA.

Virgin Valley, Nevada, USA.

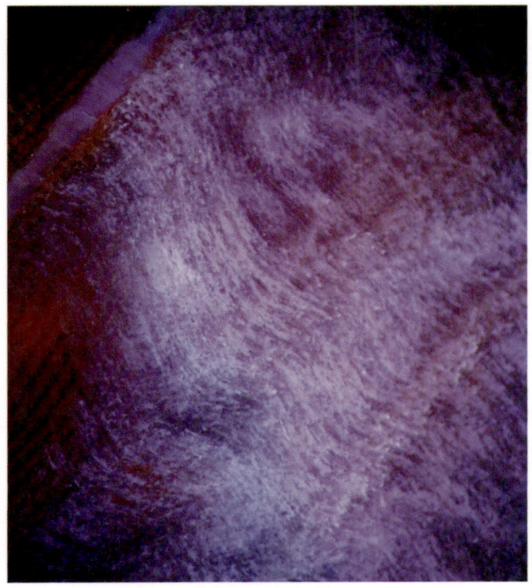
Owyhee Mountains, Oregon, USA.

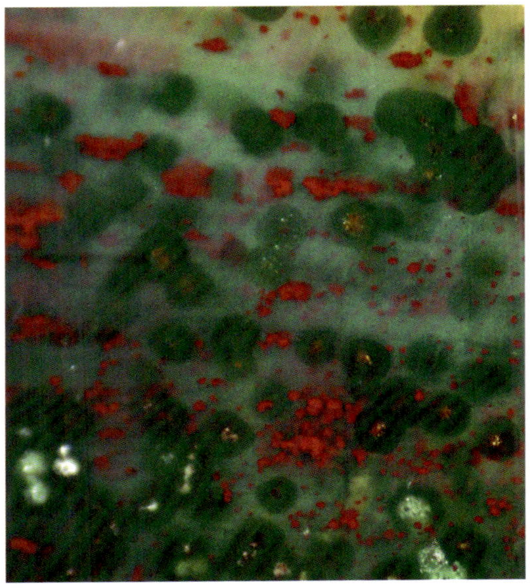
Hampton Butte, Oregon, USA.

Grassy Mountain, Oregon, USA.

Northern Arizona, USA.

Northern Arizona, USA.

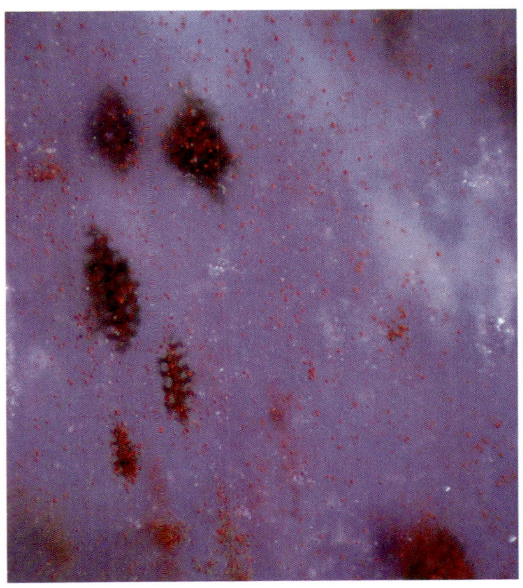
Northern Arizona, USA.

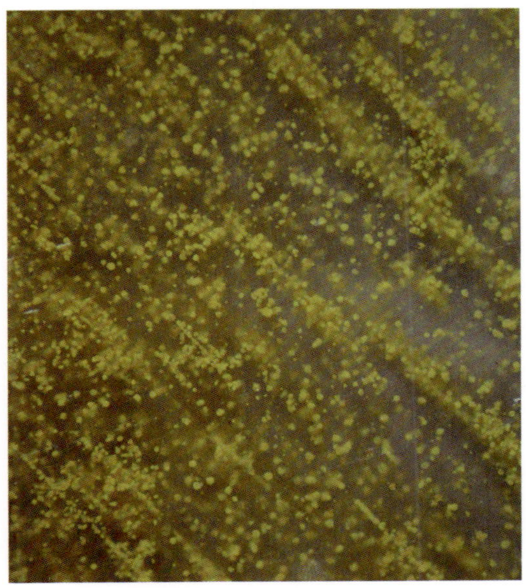
Northern Arizona, USA.

Northern Arizona, USA.

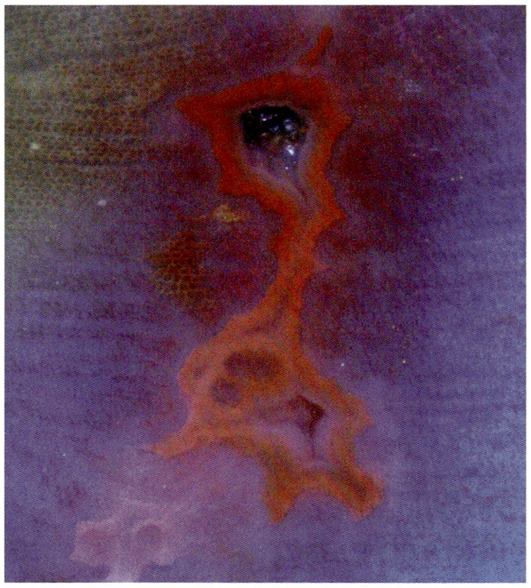
Northern Arizona, USA.

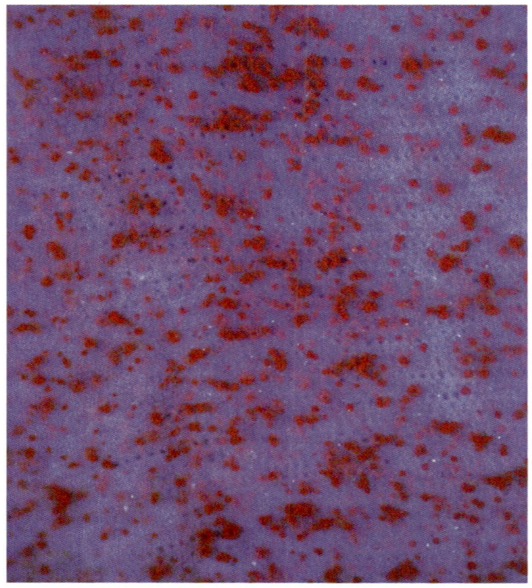
Northern Arizona, USA.

Northern Arizona, USA.

Northern Arizona, USA.

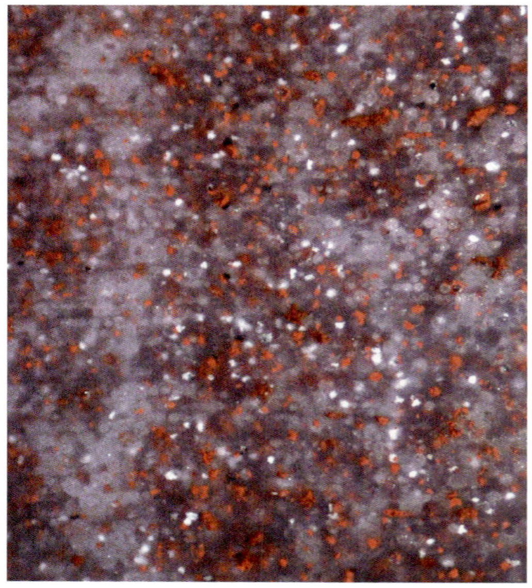
Northern Arizona, USA.

Northern Arizona, USA.

Northern Arizona, USA.

Northern Arizona, USA.

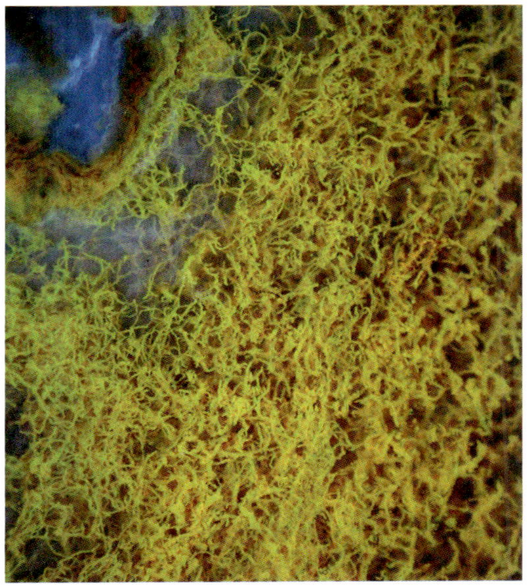
Northern Arizona, USA.

Northern Arizona, USA.

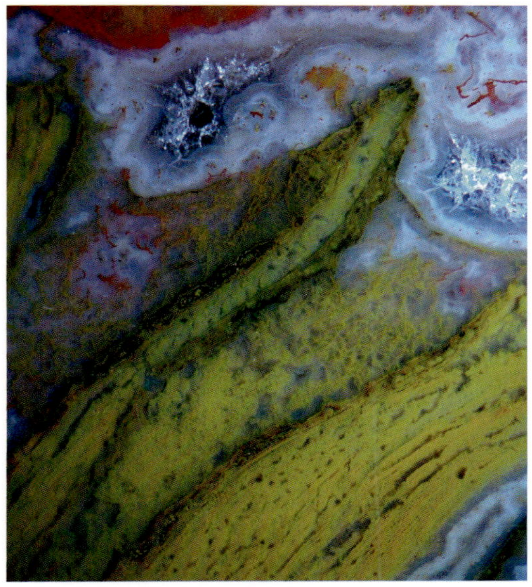
Northern Arizona, USA.

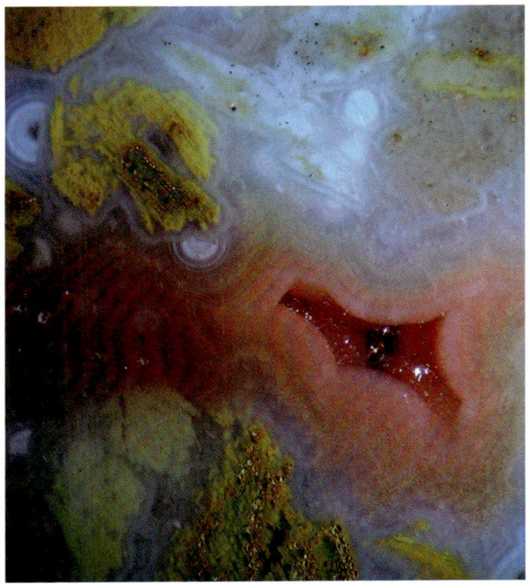
Northern Arizona, USA.

South Fork, Crooked River, Oregon, USA.

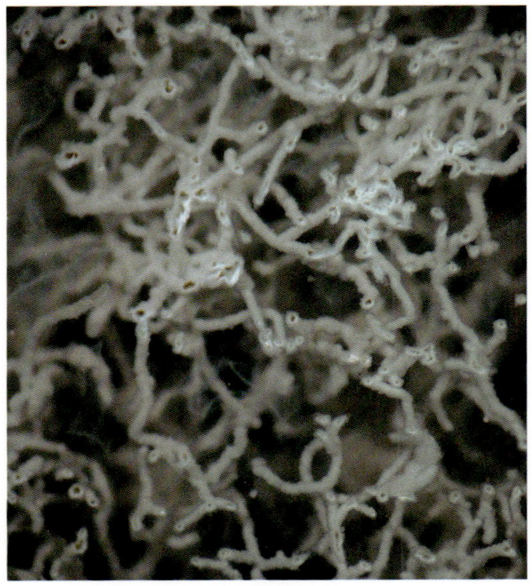
South Fork, Crooked River, Oregon, USA.

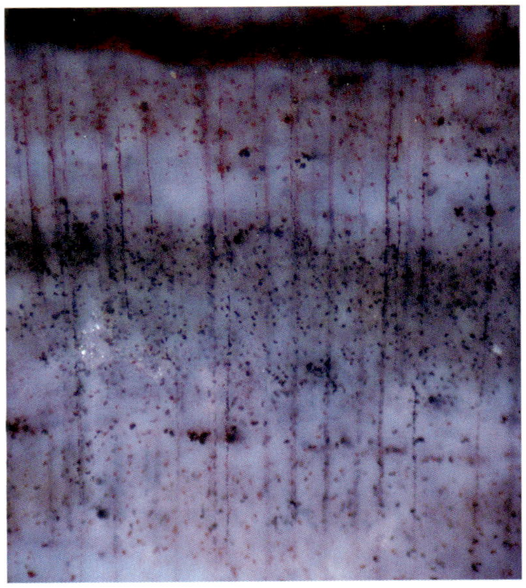
Hubbard Basin, Nevada, USA.

Hubbard Basin, Nevada, USA.

Hubbard Basin, Nevada, USA.

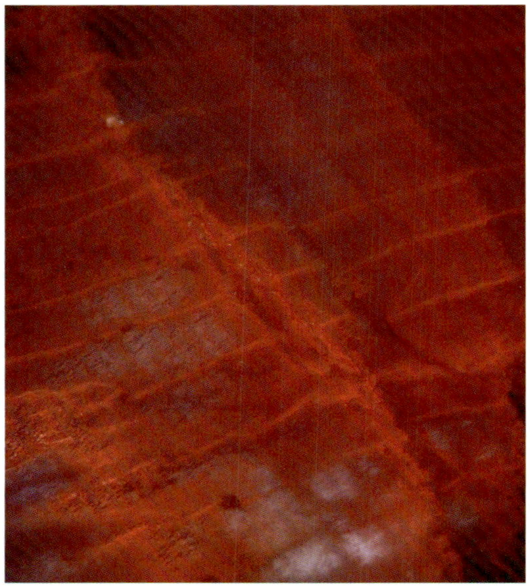

Elko County. Nevada, USA.

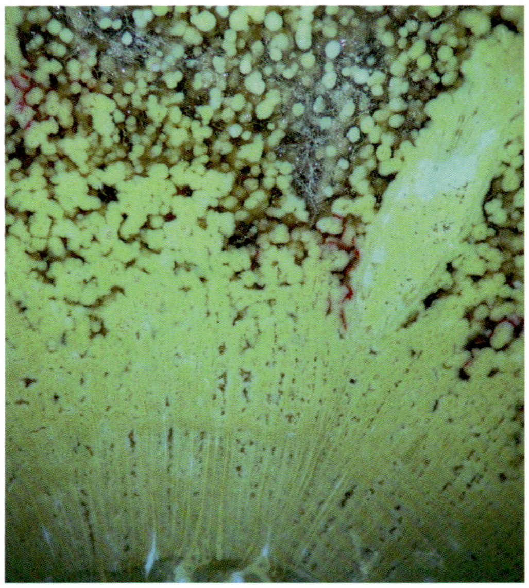

Madagascar.

Douglas County, Colorado, USA.

Douglas County, Colorado, USA.

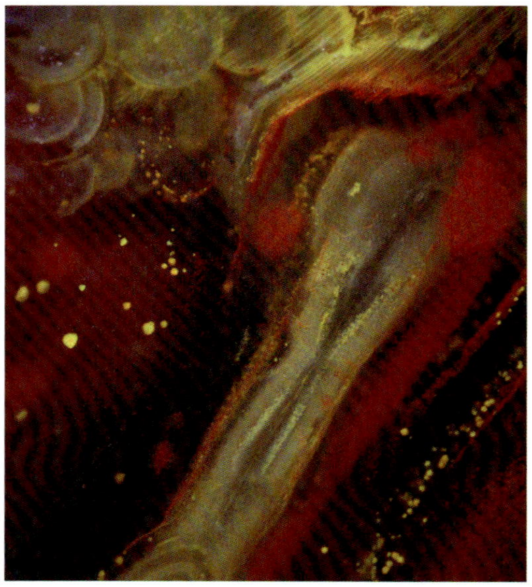

Eastern Utah, USA.

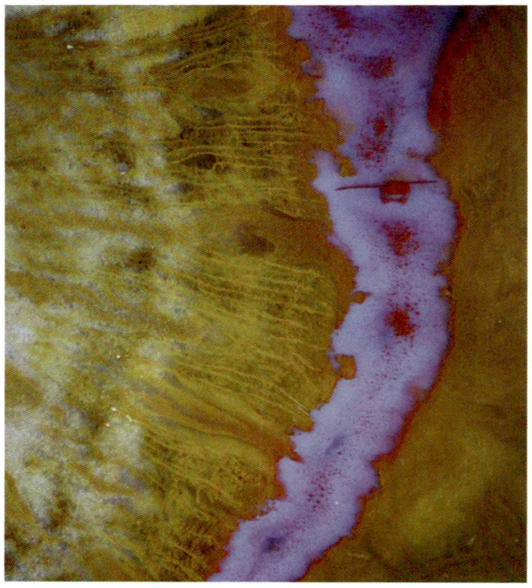

Eastern Utah, USA.

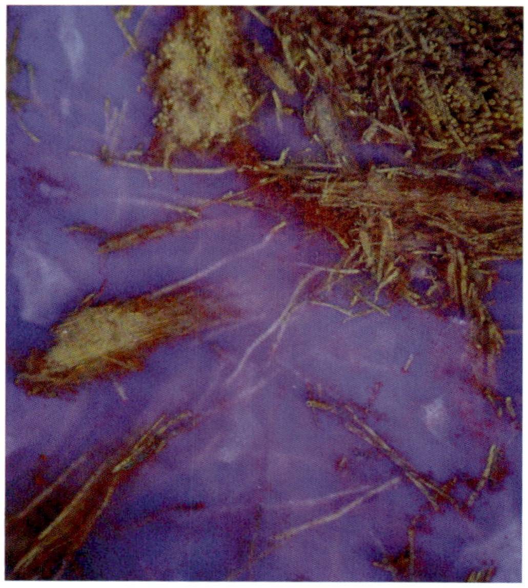

Eastern Utah, USA.

Circle Cliffs, Utah, USA.

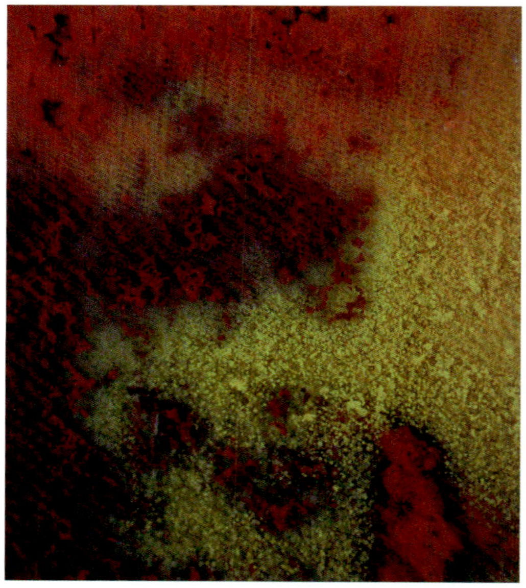

Henry Mountains, Utah, USA.

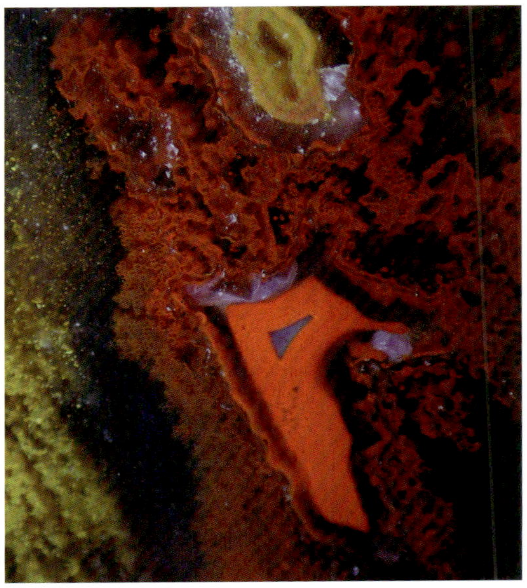

Henry Mountains, Utah, USA.

Henry Mountains, Utah, USA.

Henry Mountains, Utah, USA.

Henry Mountains, Utah, USA.

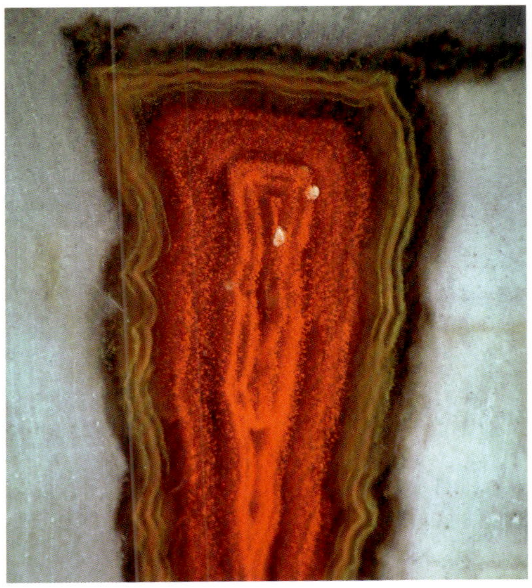

Henry Mountains, Utah, USA.

Henry Mountains, Utah, USA.

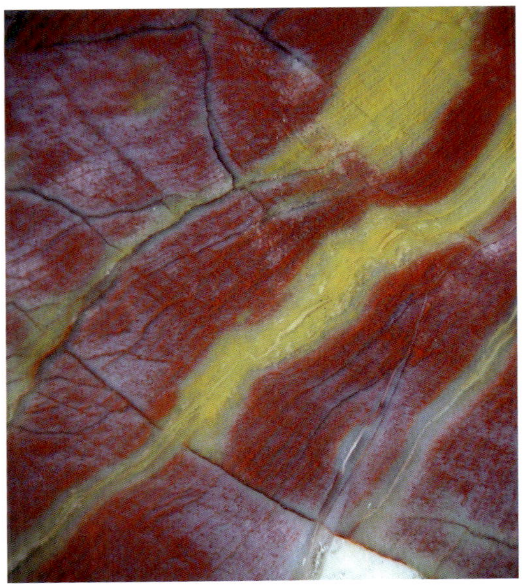

Henry Mountains, Utah, USA.

Chubut Province, Argentina.

Chubut Province, Argentina.

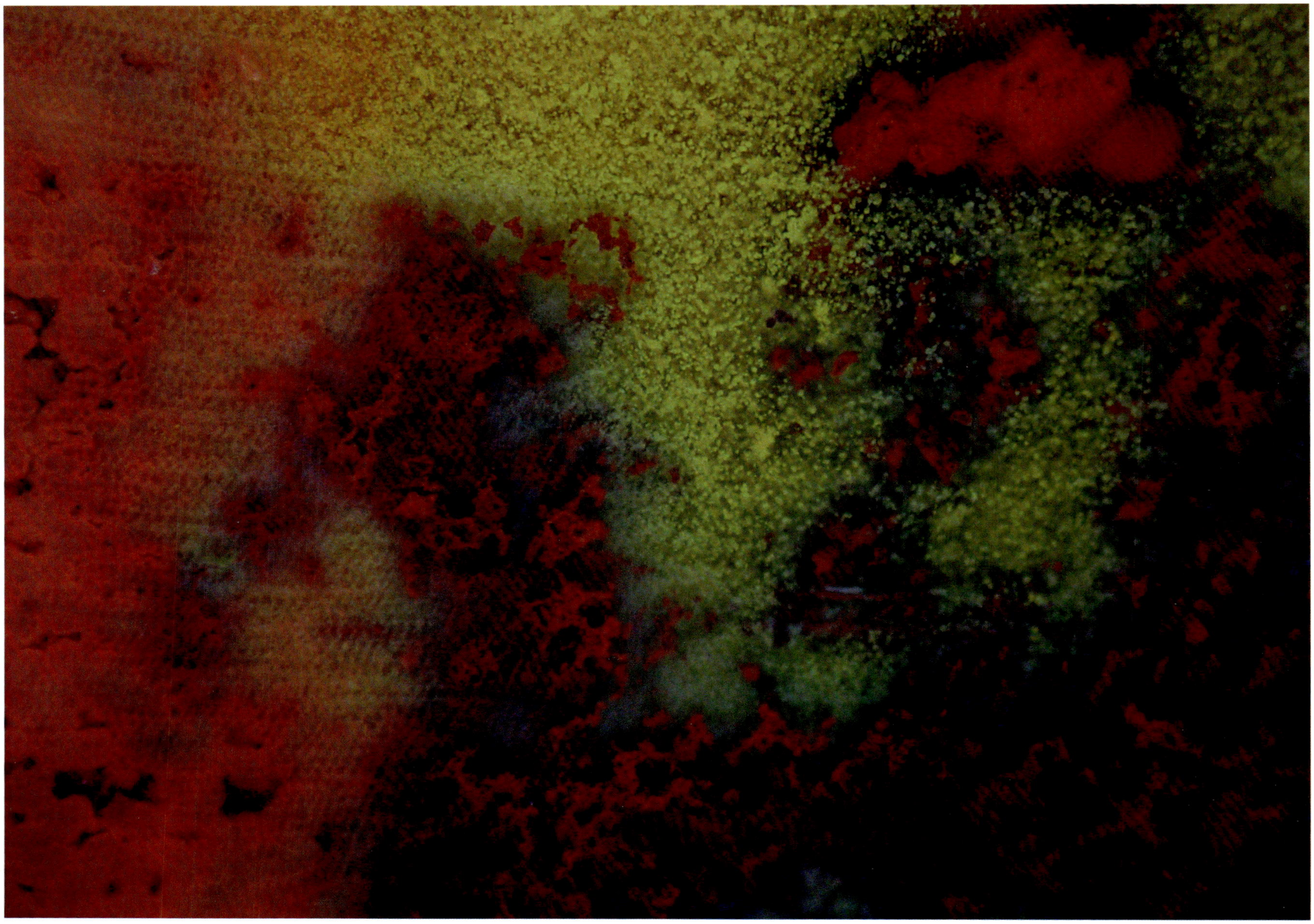
Henry Mountains, Utah, USA.

Chubut Province, Argentina.

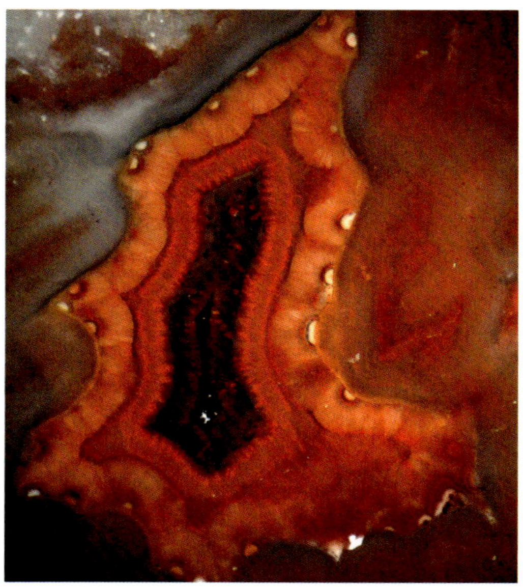
Rio Negro Province, Argentina.

Rio Negro Province, Argentina.

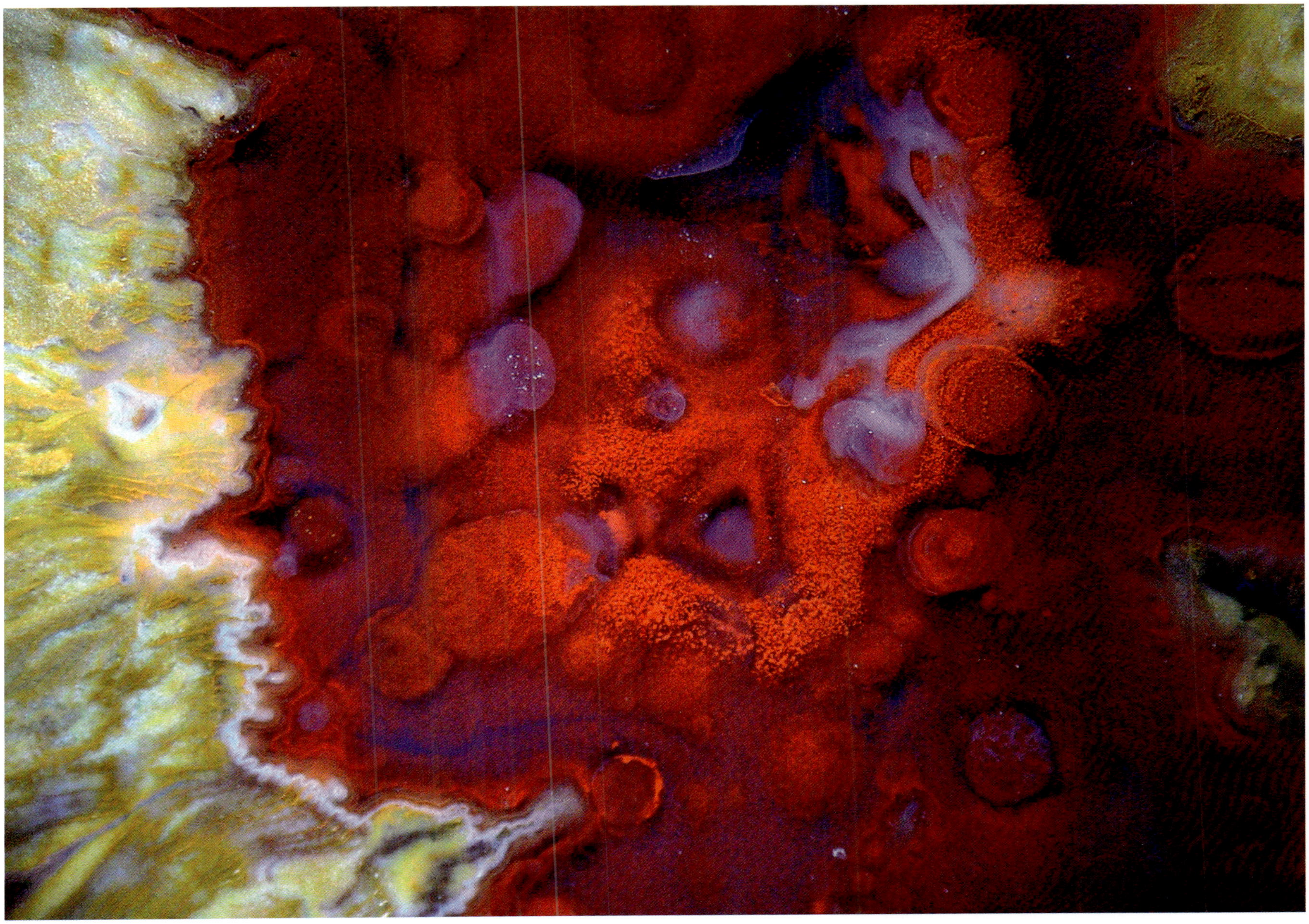
Eastern Utah, USA.

Rio Negro Province, Argentina.

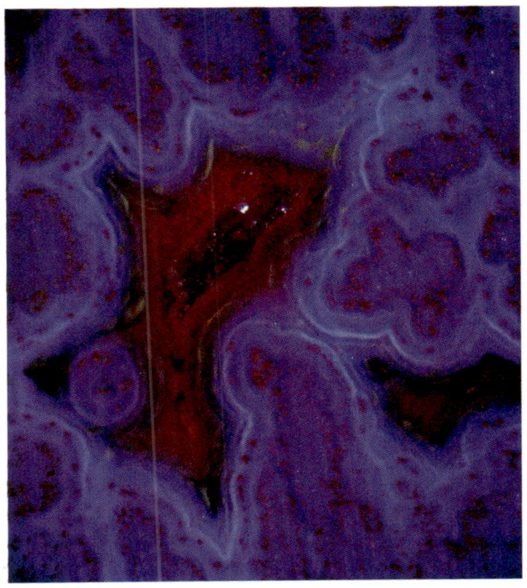
Northern Arizona, USA.

Northern Arizona, USA.

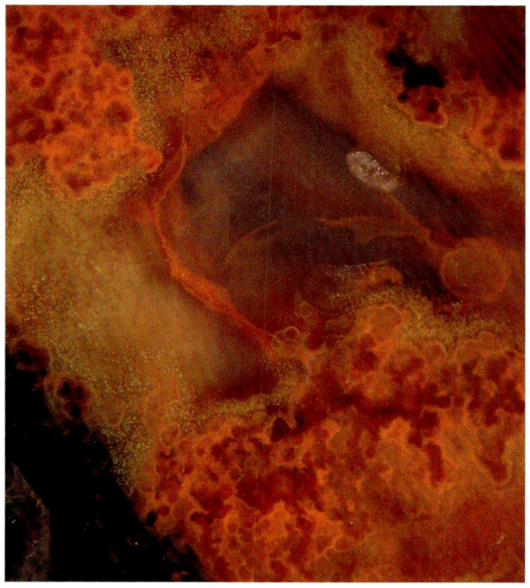
Northern Arizona, USA.

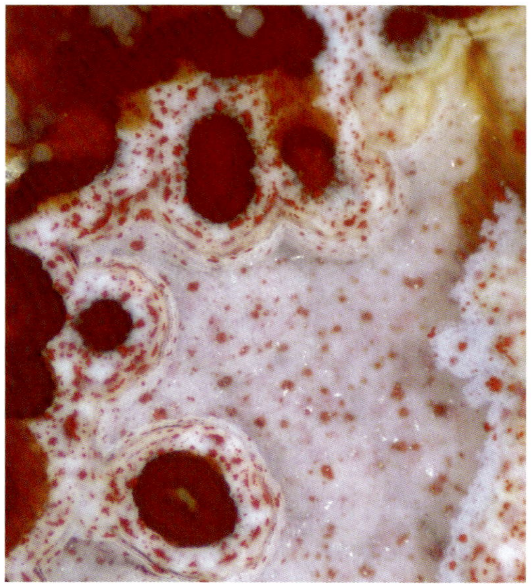
Northern Arizona, USA.

Hampton Butte, Oregon, USA.

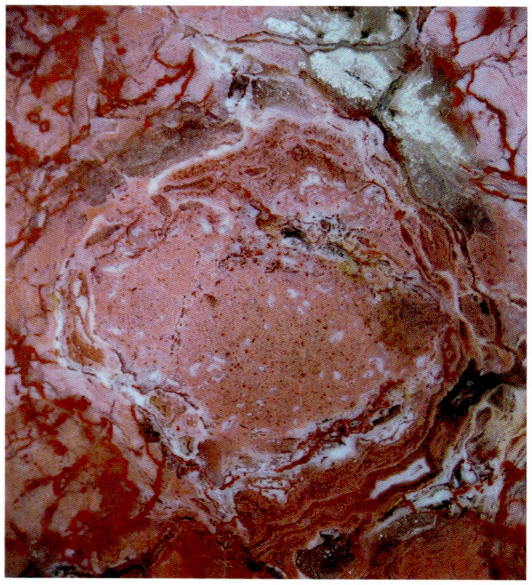
Madagascar.

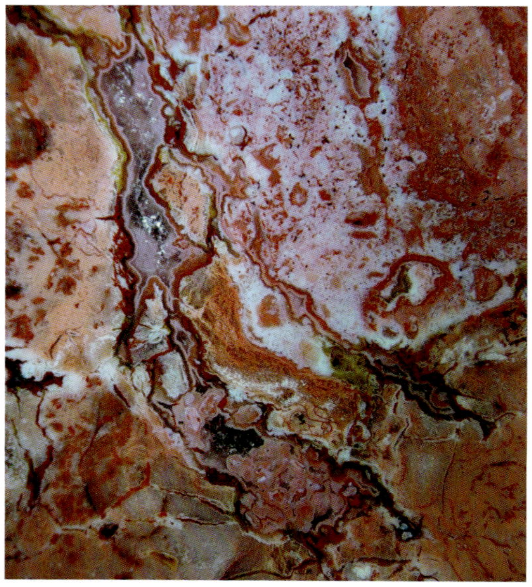
Madagascar.

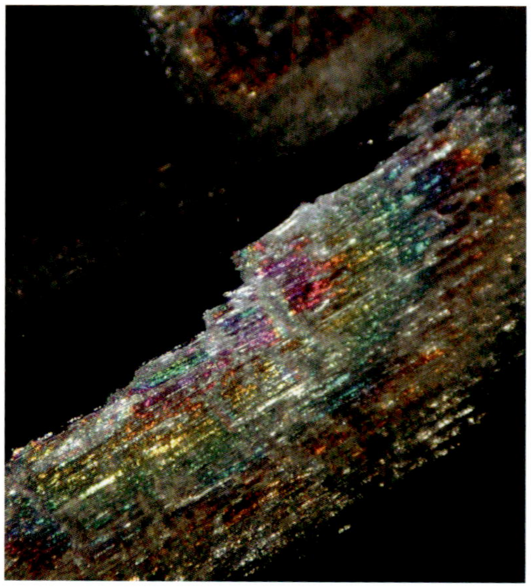
Budweis, Czech Republic.

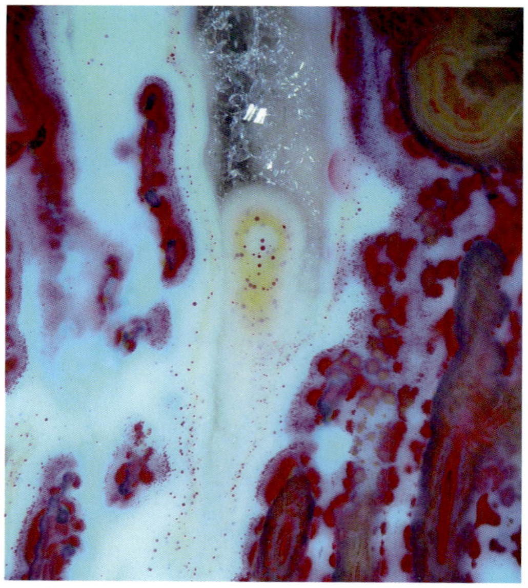
Madagascar.

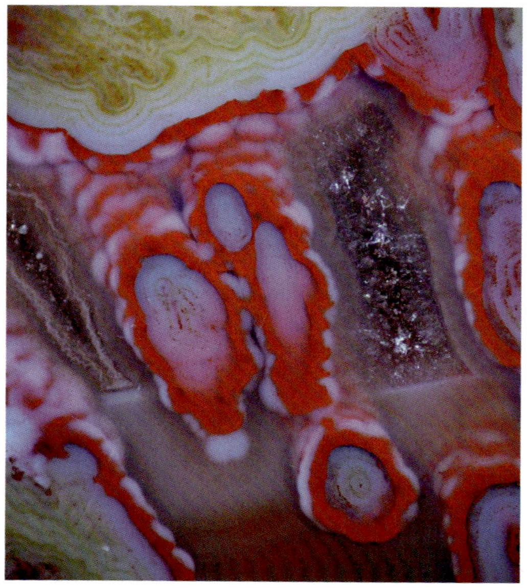
Madagascar.

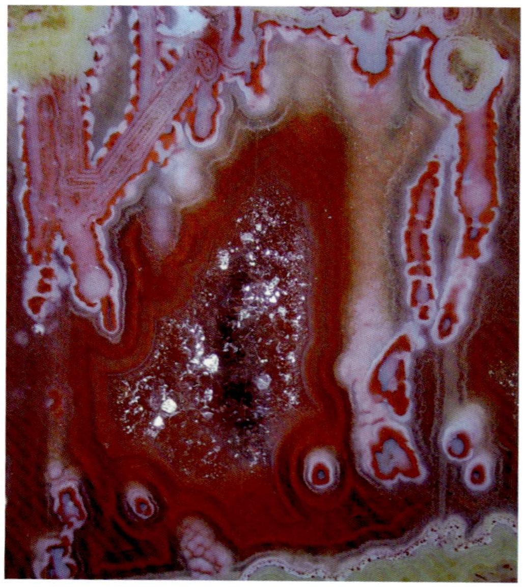
Madagascar.

Henry Mountains, Utah, USA.

Henry Mountains, Utah, USA.

Henry Mountains, Utah, USA.

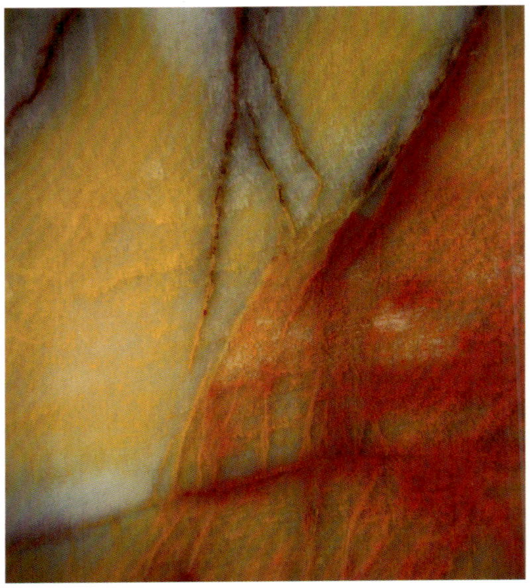
Henry Mountains, Utah, USA.

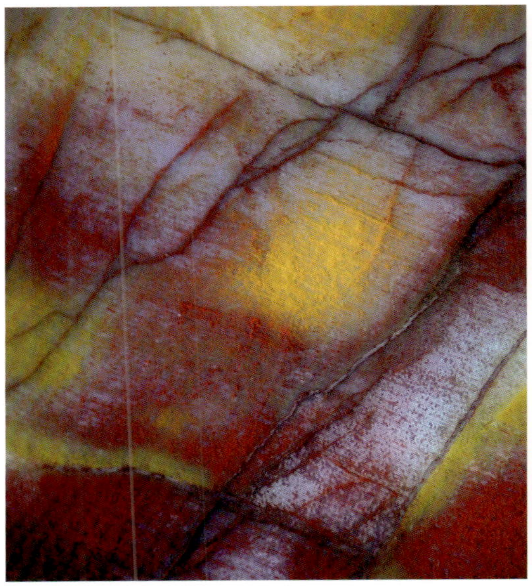
Henry Mountains, Utah, USA.

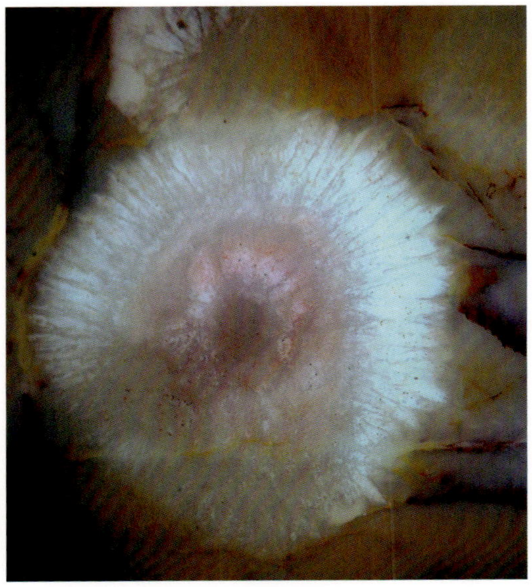
Henry Mountains, Utah, USA.

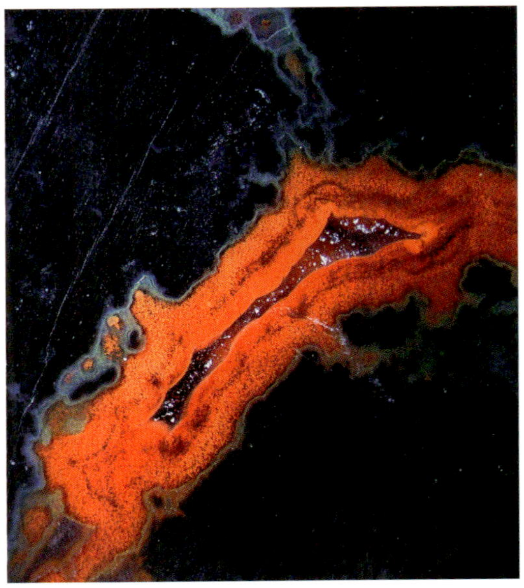
Henry Mountains, Utah, USA.

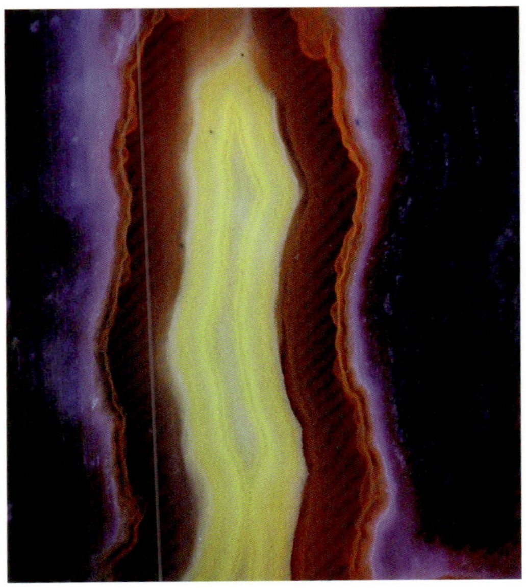
Henry Mountains, Utah, USA.

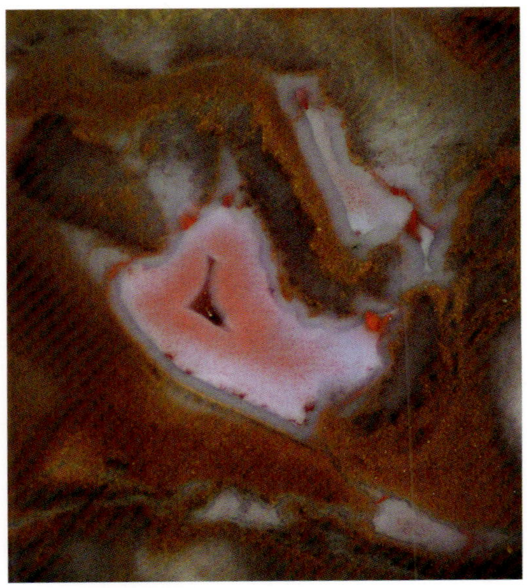
Moab, Utah, USA.

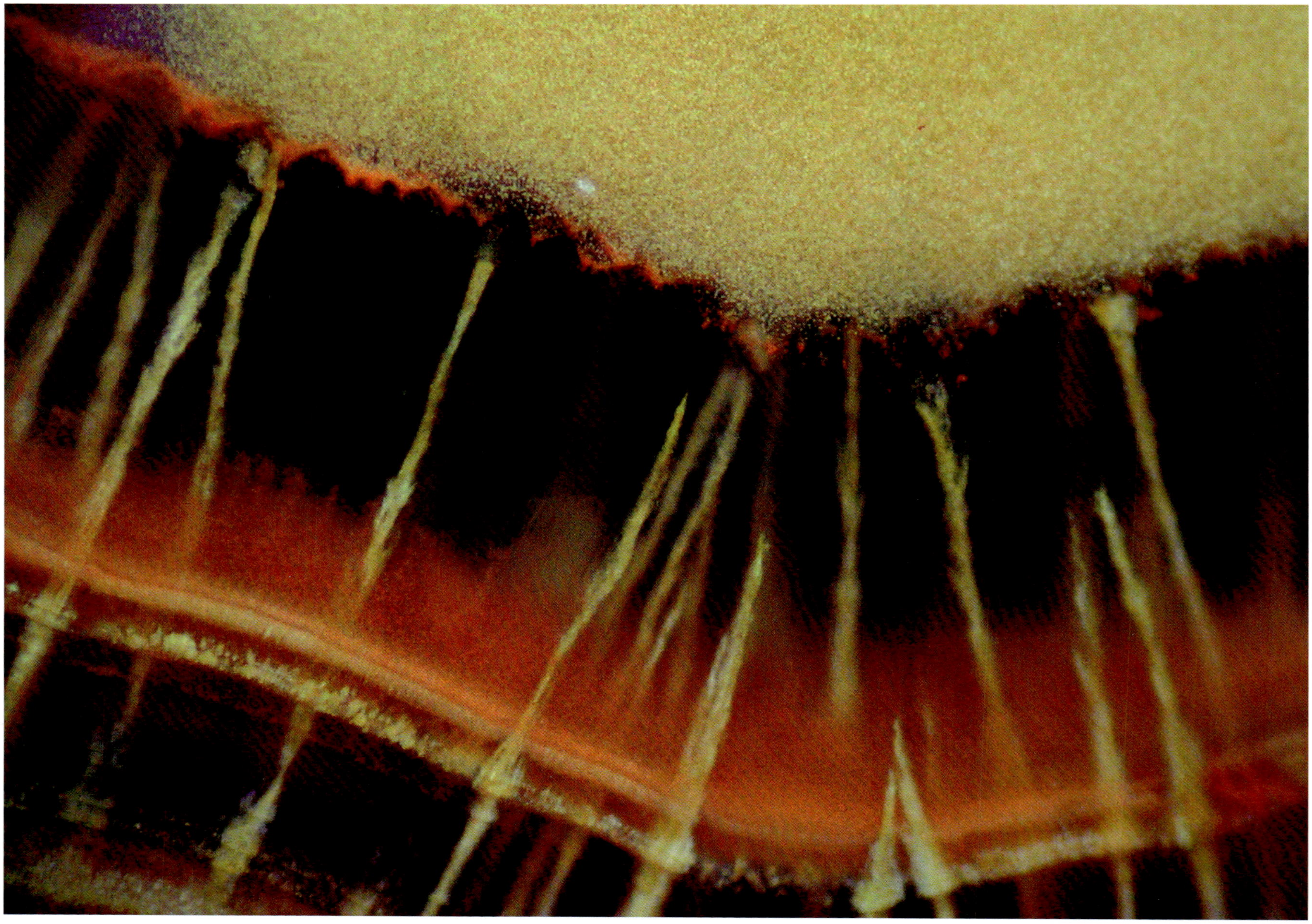

Moab, Utah, USA.

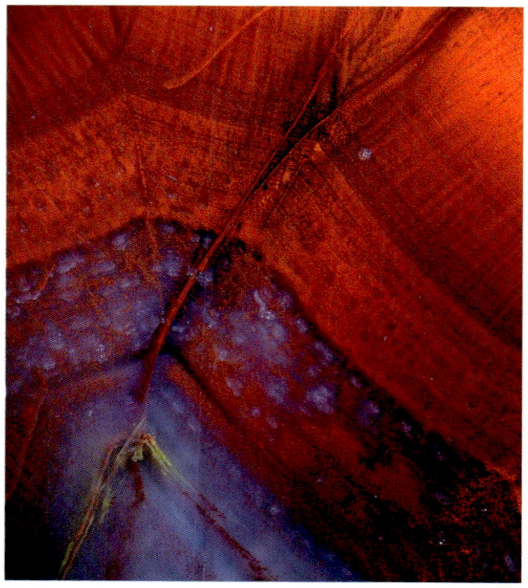

Yellow Cat, Utah, USA.

Moab, Utah, USA.

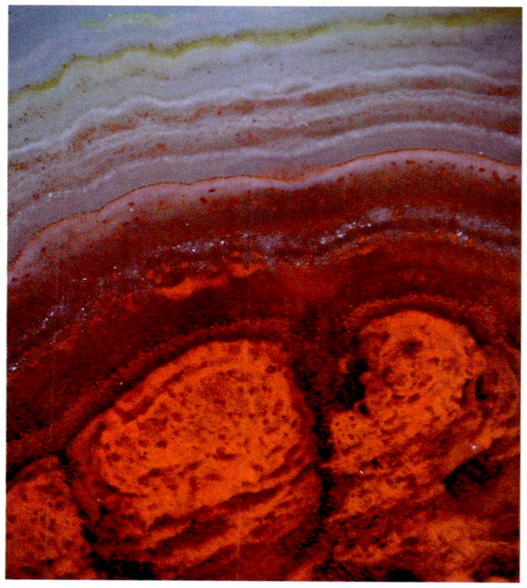

Moab, Utah, USA.

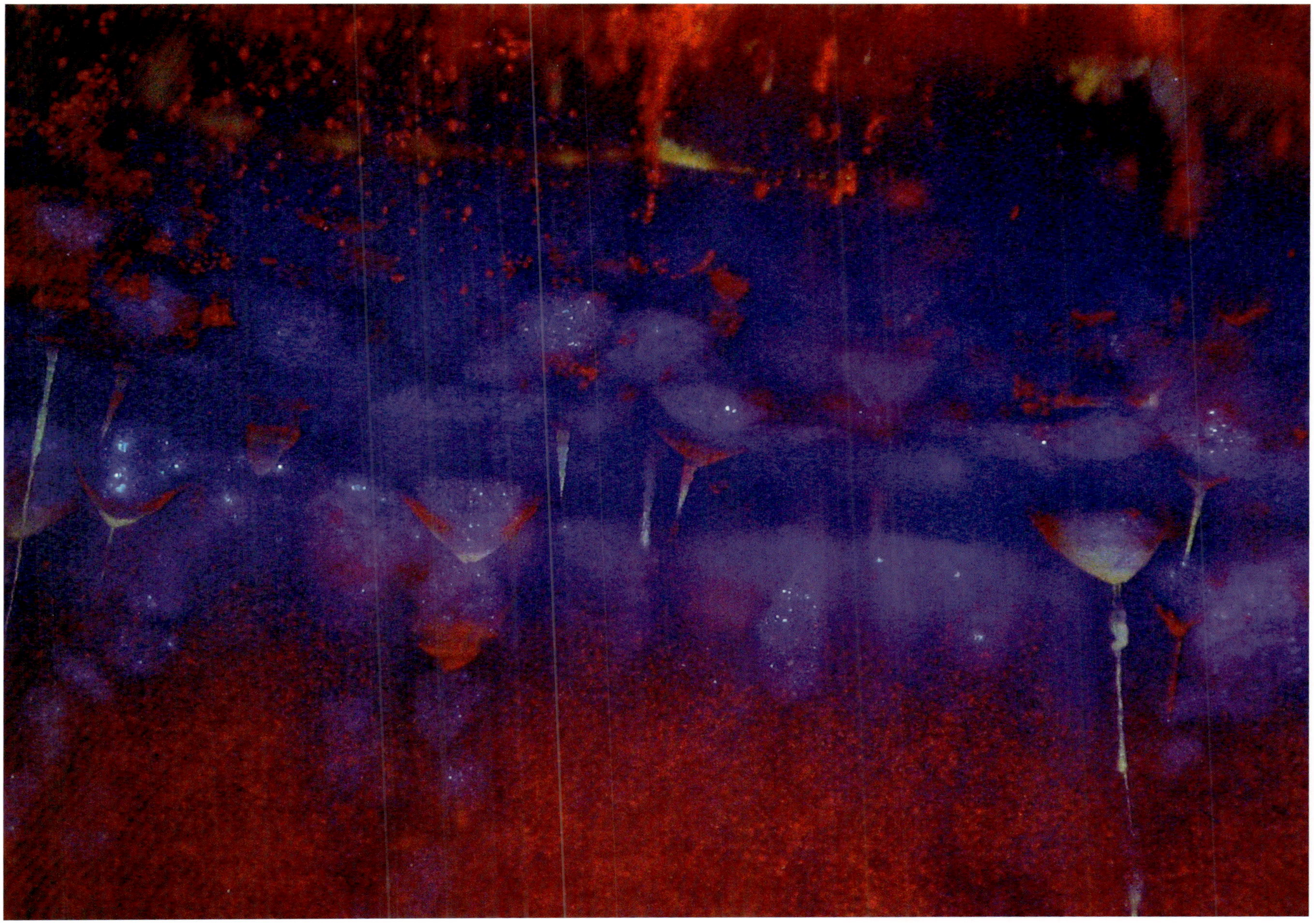

Moab, Utah, USA.

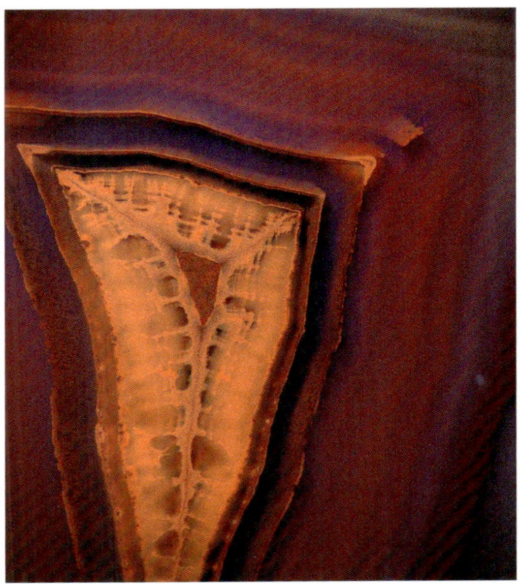

Moab, Utah, USA.

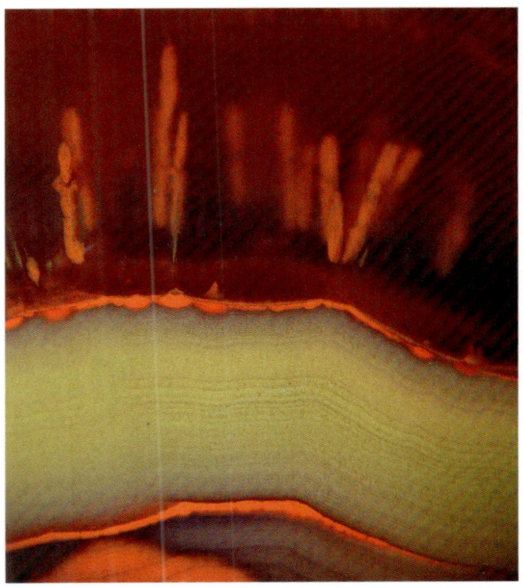

Moab, Utah, USA.

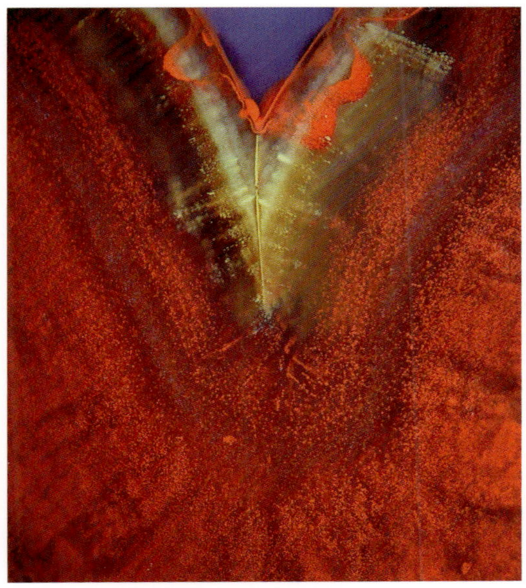

Yellow Cat, Utah, USA.

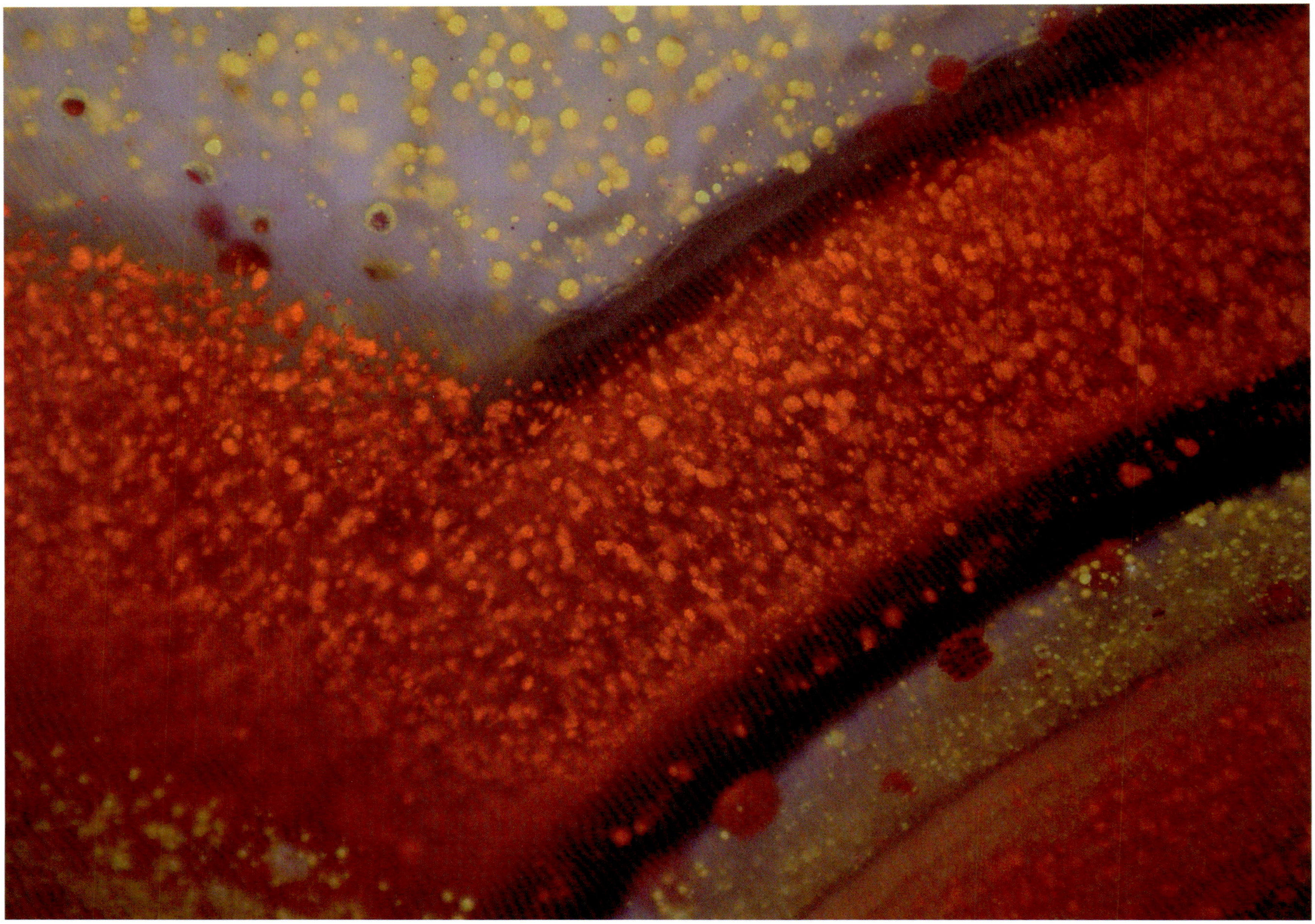

Moab, Utah, USA.

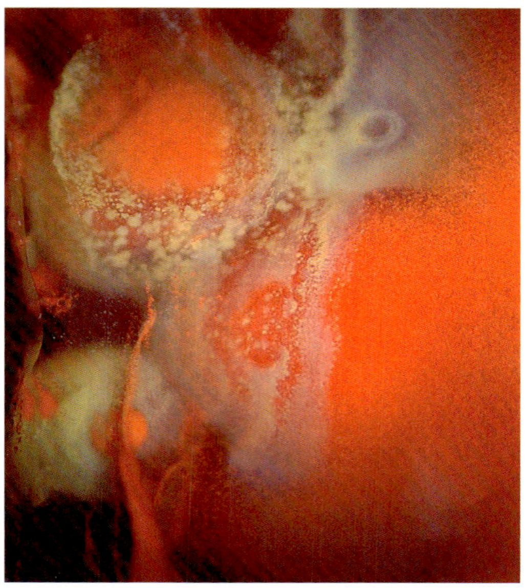

Moab, Utah, USA.

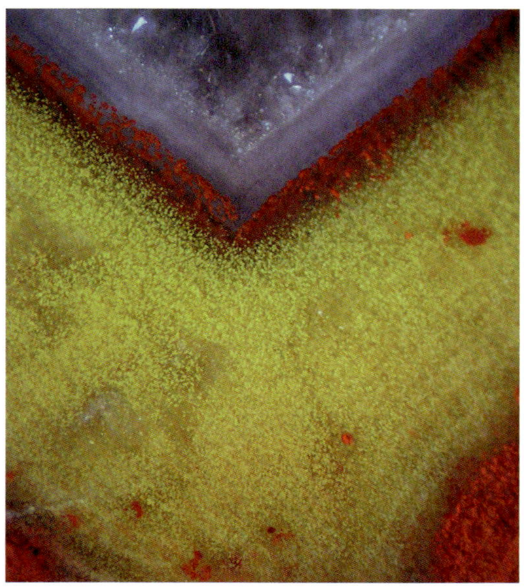

Moab, Utah, USA.

Moab, Utah, USA.

Moab, Utah, USA.

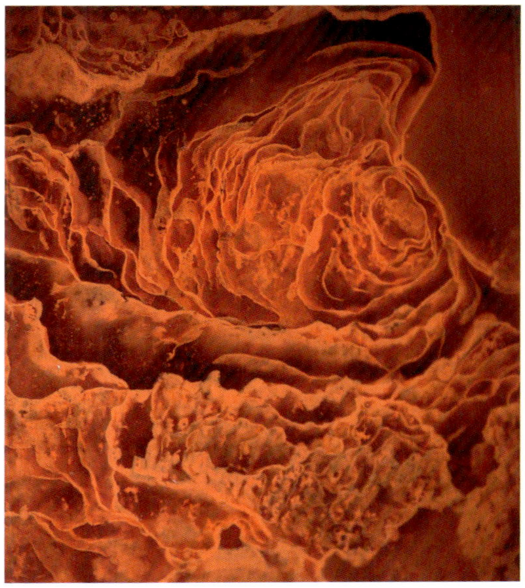

Castle Dale, Utah, USA.

Moab, Utah, USA.

Moab, Utah, USA.

Moab, Utah, USA.

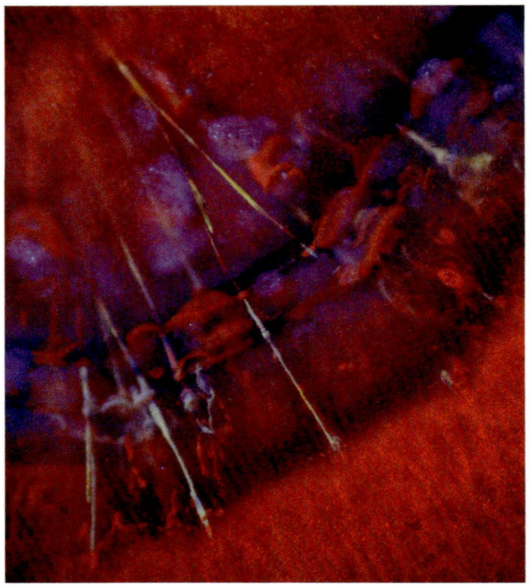
Moab, Utah, USA.

Moab, Utah, USA.

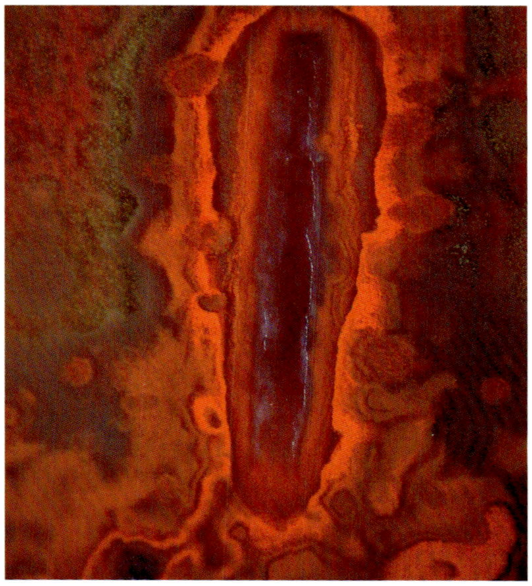
Moab, Utah, USA.

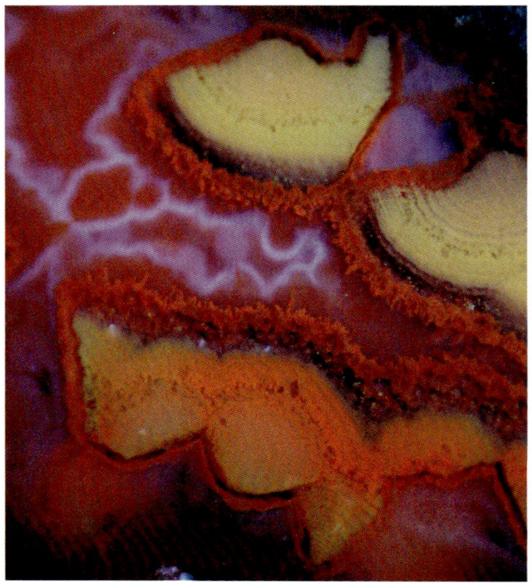
Moab, Utah, USA.

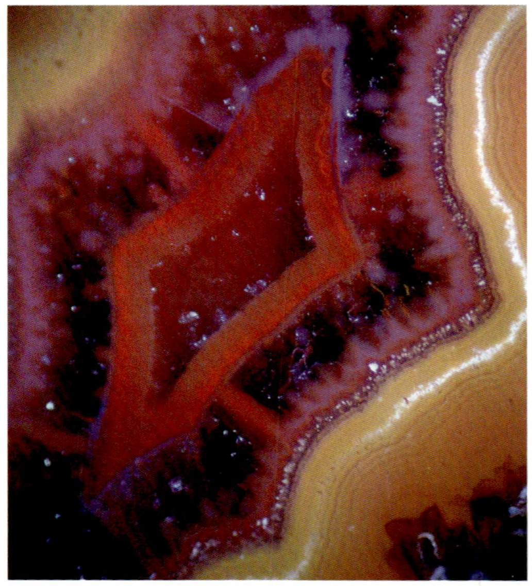
Moab, Utah, USA.

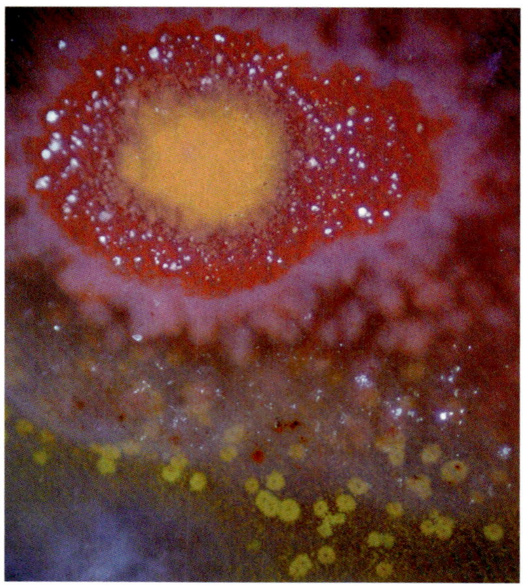
Moab, Utah, USA.

Moab, Utah, USA.

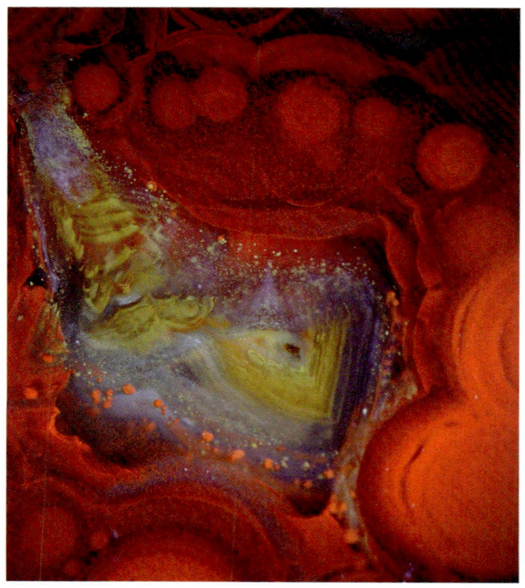
Moab, Utah, USA.

Moab, Utah, USA.

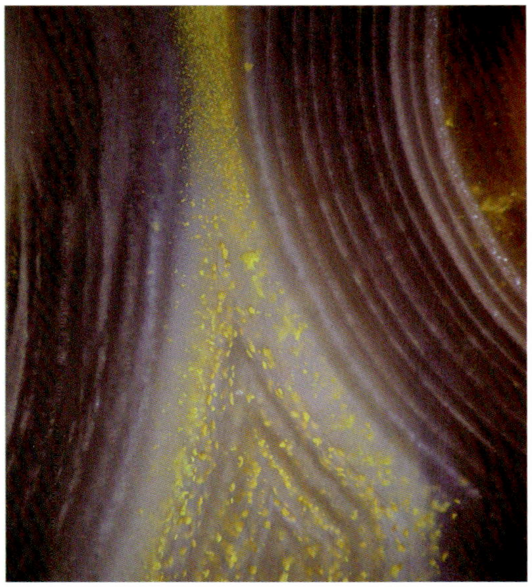

Moab, Utah, USA.

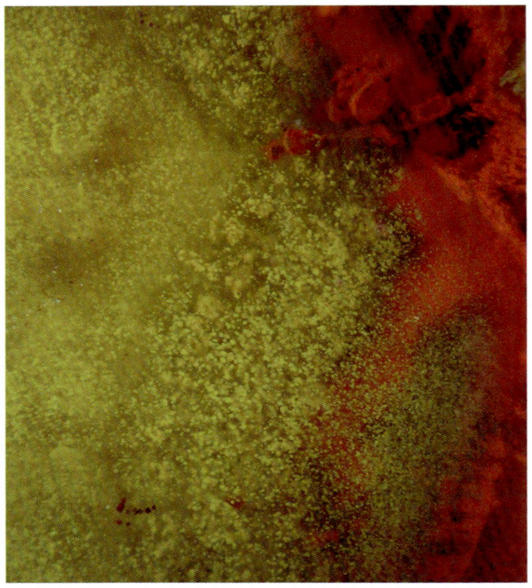

Moab, Utah, USA.

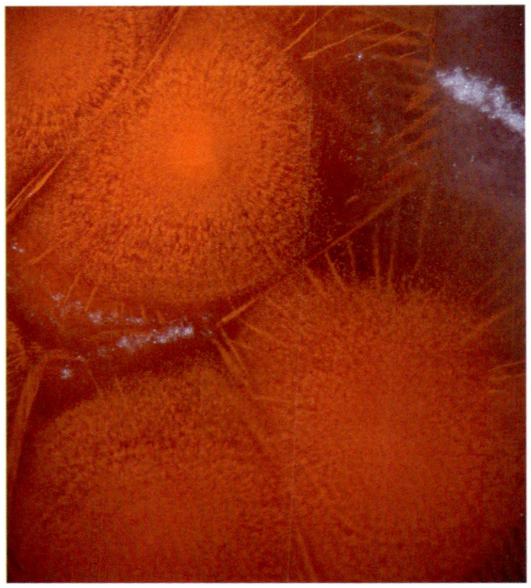

Yellow Cat, Utah, USA.

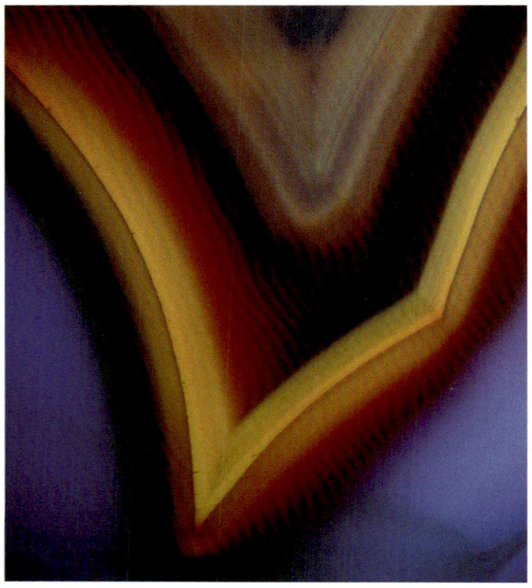

Condor agate. Patagonia, Argentina.

Condor agate. Patagonia, Argentina.

Condor agate. Patagonia, Argentina.

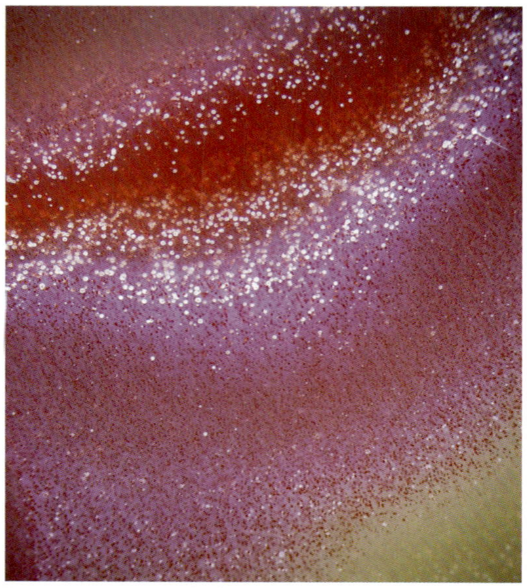

Laguna agate. Chihuahua, Mexico.

Laguna agate. Chihuahua, Mexico.

Dryhead agate. Montana, USA.

INTRODUCTION

CONTINUED

Our ancestors understandably misunderstood fossil woods. People living near the petrified forest of Thiruvakkarai, India, believed the petrified logs to be bones of a demon slain by the God Vishnu. Natives of the American Southwest held similar beliefs, attributing the huge logs of what is now Petrified Forest National Park to arrow and spear shafts dispatched by the Thunder God Shinauav and his enemies during a great battle. Some 4,000 years ago Etruscans placed a fossil cycad stem with other objects of reverence in a sepulchral chamber. Indigenous peoples also found utilitarian purposes for agatized wood, using it for building materials and tools. Arrow points knapped from fossil wood by Native Americans remain throughout the United States. As our ancestors became more civilized and educated, they appreciated fossil woods for their beauty and scientific interest. One of the first fossil wood deposits to be appreciated as such is the petrified forest at Chemnitz, Germany. This forest was discovered 300 years ago and is now considered by many to be the birthplace of petrified wood collecting. Specimens from Chemnitz were first collected as gemstones and made into jewelry for the Lords of Saxony. Scientists and laypersons alike eventually appreciated specimens from this vast forest for their importance to natural science. The forest now lies buried under the city and new specimens only see the light of day when haphazardly encountered in construction projects. Chemnitz retains and honors its natural heritage and is home to The Museum of Natural History of Chemnitz. This museum houses many fine representative specimens from this Permian forest, including *Cordaites*, conifers, *Medullosa*, *Grammatopteris*, *Psaronius*, *Calamites*, and various additional ferns and woods, plus a large reference collection relating to paleobotany. Because of the early discovery of this and other petrified forests in surrounding countries, Europeans have a rich tradition of, and appreciation for, fossil wood specimens.

Petrified forests around the world evoke a sense of wonder and excitement. Many national and state parks preserve large fossil wood deposits, allowing present and future generations to enjoy and appreciate them. The petrified forest in Arizona was located by the United States Army Corps of Engineers in 1849, immediately igniting the interest of many. To protect the forest from souvenir collectors and commercial operators who destroyed the logs looking for amethyst or crushed them into grit for sandpaper, the forest became a national monument in 1906. Even then the abundance of petrified wood seemed so endless that early park officials allowed each visitor to carry away forty pounds of wood. In 1962, the monument was elevated to National Park status. All petrified wood in the park is now ardently protected. We are fortunate to have such a preserve of fossil logs available for future generations to enjoy.

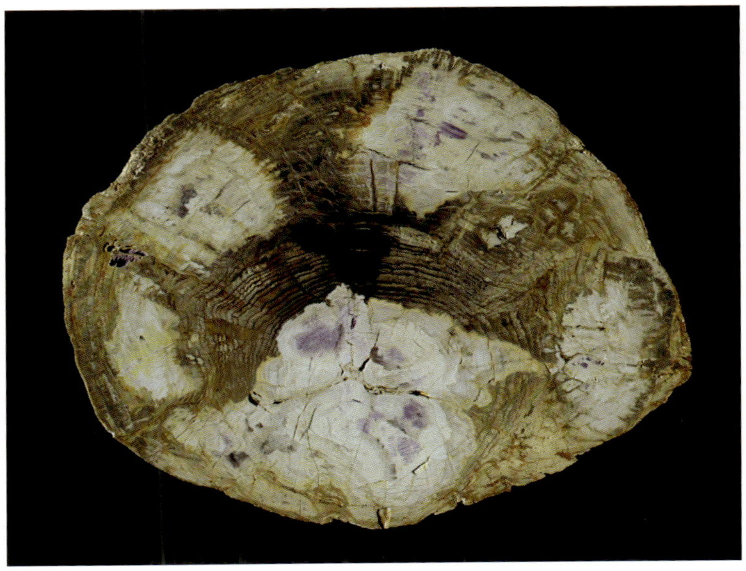

***Dadoxylon*. Permian. Chemnitz, Germany. Daniels specimen, 23 cm.**

Fossil wood arrowheads. Columbia River, Washington, USA. Daniels specimens, largest 4 cm.

Fossil wood arrowheads. Northern Arizona, USA. Daniels specimens, largest 7 cm.

Fossil wood arrowhead. Eastern Utah, USA. Daniels specimen, 3 cm.

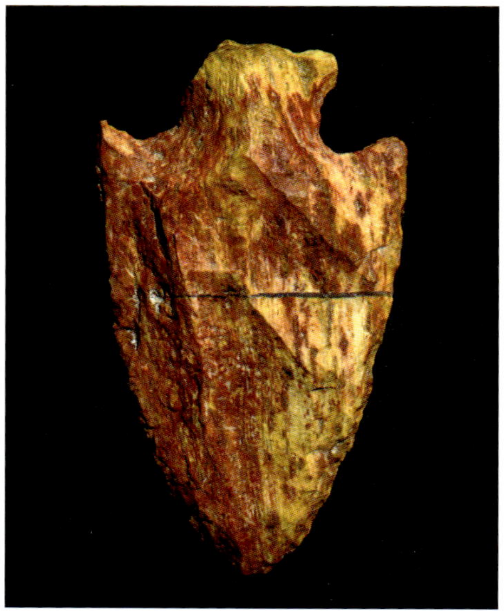

Fossil wood arrowhead. Angelina County, Texas, USA. Daniels specimen, 5 cm.

Fossil wood arrowhead. Lawrence County, Alabama, USA. Daniels specimen, 5 cm.

Fossil palm wood arrowhead. Louisiana, USA. Foshee specimen, 9 cm.

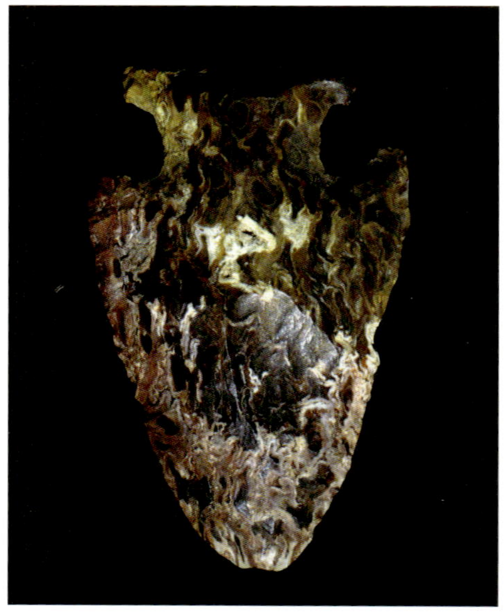

Fossil palm root arrowhead. Louisiana, USA. Foshee specimen, 5 cm.

Fossil wood arrowhead. Northern Arizona, USA. Daniels specimen, 4 cm.

Paperweight. Petrified Forest National Monument. Circa 1930-1944.

Decal. Petrified Forest National Monument.

Small dish. Petrified Forest National Park.

Official First Day Cover. United States Post Office. June 13, 1974.

Official First Day Cover. United States Post Office. June 13, 1974.

Postcard. Petrified Forest National Monument. Circa 1915-1917.

Postcard. Petrified Forest National Monument. Circa 1915-1917.

Postcard. Petrified Forest National Monument. Circa 1915-1917.

Postcard. Petrified Forest National Monument. Circa 1915-1917.

Postcard. Petrified Forest National Monument. Circa 1915-1917.

Postcard. Petrified Forest National Monument. Circa 1930-1944.

Postcard. Petrified Forest National Monument. Circa 1930-1944.

Postcard. Petrified Forest National Monument. Circa 1930-1944.

Postcard. Petrified Forest National Monument. Circa 1930-1944.

Postcard. Petrified Forest National Monument. Circa 1930-1944.

Postcard. Petrified Forest National Monument. Circa 1920-1955.

Postcard. Petrified Forest National Monument. Circa 1930-1944.

I

CHAPTER ONE

Fossil Wood Structure and Identification

Difficulties in identifying fossil wood abound; simple answers are few. Paleobotanists accomplish proper identification using a microscope to view transverse, tangential, and radial perspectives cut into thin sections. They then apply years of experience and specialized training to form conclusions. The costs and difficulty of preparing proper thin sections are substantial, and the knowledge base necessary for this level of identification is daunting. Most specimens of fossil wood in museums and private collections are prepared in thick transverse sections. This perspective presents a substantial amount of information, as well as the highest aesthetic appeal in many instances. For these reasons, we limit our fossil wood identification discussion to what you might be able to see and appreciate with the aid of a low-magnification hand lens or microscope while examining a cut and polished transverse section of fossil wood. In some instances, we include higher magnification micrographs to aid the discussion. This level of observation allows positive identification of many specimens. In other instances, we place a specimen in a category of increased definition. If you have a well-preserved specimen, you should be able to ascertain with some confidence if a wood is a gymnosperm [conifer or softwood] or an angiosperm [flowering tree or hardwood]. Differentiating one conifer from another at this level ranges from difficult to impossible, but you should be able to identify hardwoods to at least a grouping of genera and, in some cases, to a genus with some degree of confidence. Photographs of several types of fossil wood will assist in identifying extinct genera that have made their way into the limited field of described fossil woods. Some of these trees have been extinct for eons, some have living relatives, and some are of unknown affinity. The macro photography used in this book approximates a low-magnification view. Micro images take us deeper into a fascinating realm of cell and mineral structure. This closer view opens a big world, one with many additional avenues for further study if so desired.

The study of fossil wood structure, as opposed to the study of modern woods, adds considerable difficulty because of changes that transpired in the wood during the millions of years since it was a living plant. Because of poor preservation, even the most accomplished expert cannot identify all specimens of fossil wood. In some specimens, chalcedony or other minerals replace fossil wood structures to the point that no indicia of the original plants remain. Some will have structures so macerated, stressed, or compressed as to render identification out of the question. The wood may have undergone incredible tension, compaction, infiltration, heat, radiation, mineralization, and any number of environmental factors that can influence identification. Many available fossil woods were smothered as tree stems in volcanic ash or mudslides and then buried under thousands of feet of rock. Later, and often through violent mechanisms of plate tectonics and mountain building, these fossils were exposed by erosion, and in some cases carried by floodwaters, to lie in the open for thousands of years, subjected to the harsh elements of the Earth's surface. Is there any question why the integrity of cellular structure may not be in pristine condition?

Modern wood identification in its own right is an exacting science that requires substantial knowledge and attention to minute detail. When dealing with fossil woods, we start at the disadvantage of not working with actual wood. What we have to work with is rock, a fossil of wood that is millions of years old. We may not have perfect knowledge of how the tree grew and its original appearance and often not much information about the climate in which it grew. Many of the characteristics relied upon for identification by modern wood anatomists are not present, such as color, scent, specific gravity, hardness, and fluorescence. Whenever possible, we should attempt to determine the stratigraphic source of the fossil wood we wish to identify. If the fossil came from a geological formation that predates angiosperms, for example, then it is not an angiosperm.

Trees, our largest living organisms, evolved from tiny green algae that adapted to a terrestrial environment. Physical evidence of the first land plants comes from the Middle Silurian Period that dates to 425 million years before present. The earliest true trees were the progymnosperms that reproduced by spores.

Evolution of seed plants during the Paleozoic Era allowed trees to dominate the landscape. The two main types of seed plants today are the gymnosperms (seeds borne naked) and the angiosperms (seeds borne enclosed in an ovary). Gymnosperms, which date from the Late Devonian Period, include conifers, as well as *Cordaites*, cycads, *Ginkgo*, *Hermanophyton*, *Pentoxylon*, and *Rhexoxylon*. Conifers grow with one main stem with lateral side branches and include the largest individual living plants, the *Sequoia* (redwood), as well as the oldest, the bristlecone pine that can live to be 5,000 years old. Conifers conspicuously cover a broad range of habitat, from tropical forests to high mountaintops. Angiosperms (flowering plants) comprise two classes, each with plants with woody stems: dicotyledons include hardwoods and monocotyledons include palms and yuccas. The oldest unequivocal flowering plants date to the Lower Cretaceous Period. Presently, flowering plants are the dominant land plant. Hardwoods grow by branching and re-branching of the main stem. Conifers and hardwoods both have wood capable of adding to its circumference by secondary thickening, that is by adding new growth onto previous growth. In temperate climates, this secondary thickening usually results in annual growth rings. Tropical woods may have indistinct increments of growth or may have growth rings that evidence increments based on seasonal rainfall. Palms do not form secondary wood growth.

Continental Drift and the Study of Fossil Woods

Fossil wood studies shed light on paleoenvironments. The study of paleogeography assists our understanding of relationships between disparate fossil woods. The ancient world was not the same geographically as it is now. We find fossil palm wood in Wyoming, where now reign frigid winters unwelcoming to tropical foliage. Not only have climates changed, the size of, and relationships between, the continents are vastly different from the time fossil *Calamites* from Brazil were thriving plants. Where there were seas are now mountains; where tropical forests once flourished are now glaciers. The linkage of paleobotany and paleogeography is synergistic and indisputable.

In 1912, German meteorologist Alfred Wegener first proposed the concept of an ancient single massive continent, or protocontinent, he called Pangea. Paleontological studies suggesting similarities between plant (such as *Glossopteris*) and animal fossils in seemingly unrelated locations fueled Wegener's work. Fossil consistency had been noted in pairings of separate and distant locations, including South America and Africa, India and Madagascar, and Europe and North America. Wegener proposed the concept of continental drift, theorizing that continents over time slide around the Earth's surface, changing relative positions. Other scientists endeavored to buttress what appeared to be a logical theory lacking a supportable causal explanation.

In 1929, Arthur Holmes of the University of Edinburgh suggested thermal convection as the driving mechanism behind continental drift. Although correct, Holmes received little attention for this theory. As the Earth's mantle is heated by the molten core, it rises until cooled and then sinks. The surface of the Earth is composed of a system of plates that are in motion, but we seldom pay much attention to this normally imperceptible movement. Devastating tragedies, such as earthquakes and the tsunami that took the lives of more than 225,000 people living around the Indian Ocean in December of 2004, bring shocking reminders that our continental plates are on the move. Deep convection currents drive these plates in different directions. Expanding ocean floors sink at the edges, clashing with land masses. The heated, deep layer of rock is also the source of energy for lava in volcanoes, like those in the geologic history of much of the world. Volcanoes in many instances were responsible for the creation of marvelous specimens of fossil wood. Specimen Ridge in Yellowstone, for instance, comprises no less than 27 levels of petrified forests, each resultant from a series of volcanic eruptions. The perfectly preserved fossil wood and cones from the Cerro Quadrado of Argentina come to us courtesy of a tremendous volcanic eruption that buried the standing forest in hundreds of feet of ash, trapping trees with young cones attached to the branches and enabling the eons to transform wood anatomy into stone.

During the Permian Period, Pangea consisted of land masses that are now Africa, Antarctica, Australia, Madagascar, India, North America, South America, and much of Europe in one connected super continent. South America nestled into Africa; India bordered Madagascar, which bordered Africa; and North America connected to Europe. That is the way it was when the Earth experienced a cataclysmic event known as the P/T event (for Permian/Triassic), wiping out much plant and animal life. In the Late Jurassic,

INTERNATIONAL STRATIGRAPHIC CHART

ERA	PERIOD	EPOCH	MYBP
Cenozoic	Neogene	Holocene	0.0115
		Pleistocene	1.806
		Pliocene	5.332
		Miocene	23.03
	Paleogene	Oligocene	33.9 ±0.1
		Eocene	55.8 ±0.2
		Paleocene	65.5 ±0.3
Mezozoic	Cretaceous	Upper	99.6 ±0.9
		Lower	145.5 ±4.0
	Jurassic	Upper	161.2 ±4.0
		Middle	175.6 ±2.0
		Lower	199.6 ±0.6
	Triassic	Upper	228.0 ±2.0
		Middle	245.0 ±1.5
		Lower	251.0 ±0.4

ERA	PERIOD	EPOCH	MYBP
Paleozoic	Permian	Lopingian	260.4 ±0.7
		Guadalupian	270.6 ±0.7
		Cirsuralian	299.0 ±0.8
	Carboniferous	Upper Pennsylvanian	306.5 ±1.0
		Middle Pennsylvanian	311.7 ±1.1
		Lower Pennsylvanian	318.1 ±1.3
		Upper Mississippian	326.4 ±1.6
		Middle Mississippian	345.3 ±2.1
		Lower Mississippian	359.2 ±2.5
	Devonian		416.0 ±2.8
	Silurian		443.7 ±1.5
	Ordovician		488.3 ±1.7
	Cambrian		542.0 ±1.0
Precambrian			Lower limit not defined
Earth formed			Approx. 4,500

We use the term Tertiary Period frequently in this book. The time scale endorsed by the United States Geological Survey uses the term Quaternary Period to include the Holocene and Pleistocene epochs and the term Tertiary Period to include the Pliocene, Miocene, Oligocene, Eocene, and Paleocene epochs. MYBP is an abbreviation for millions of years before present.

Pangea divided into Laurasia and Gondwana. Gondwana consisted of current day Africa, Antarctica, Australia, India, Madagascar, and South America. Laurasia comprised disconnected masses that approximate modern Asia, Europe, and North America. During the Cretaceous Period, an inland sea stretched from the Arctic Ocean to the Caribbean, separating the ancestral Rocky Mountains from the eastern United States that was still connected to Europe. South America was now apart from Africa; Madagascar and India separated from Africa; India drifted toward Eurasia; and Antarctica and Australia were conjoined and off by themselves. By the Early Cenozoic Era, land masses approximated their current positions, yet with substantial differences. North America appeared similar to its modern configuration, although Florida and parts of Texas and the West Coast were under water. Africa, Europe, and India were apart from Asia. Asia was least consistent with its current form with an inland sea submerging much of the center of the continent. By the Miocene Epoch, the world assumed essentially its modern configuration, but, sorry Florida, you are still under water.

So what does this mean for the study of fossil woods? We should not be surprised by the similarity of specimen types found in Araguaina, Brazil, and Chemnitz, Germany. They grew on the same body of land, in similar climatic zones, during the same geologic period, on opposite sides of the equator. Nor should we be surprised that *Woodworthia* fossils discovered in Triassic Period formations in Arizona and in Zimbabwe are similar. When attempting to identify unusual fossils it is helpful to consider similarity to identified specimens from other locations in the context of the physical relationships of the respective land masses at the time.

Macro movements of huge land masses had micro impacts that affected the formation and exposure of fossil woods. In the southwestern United States, Early Triassic mountains arose in southern and central Arizona. Debris-laden waters flowing to the north carried sand and gravel across a broad coastal plain. Later, volcanoes imparted enormous quantities of volcanic ash that became the raw material for the Chinle Formation where the trees of Petrified Forest National Park and many others fossilized. In Jurassic times, the southwestern United States was home to many shallow lakes and ponds in a moist climate rich in vegetation. Introduction of volcanic ash resulted in a sequence of floodplain deposits that formed the Upper Morrison Formation. These events created the amazingly well-preserved and beautiful fossil woods found throughout much of Utah and Western Colorado. The fossil-bearing deposits at one time reposed under thousands of feet of soil and rock. Mechanisms related to continental drift exposed large sectors of this formation, allowing collection of these fossils. In some instances, massive blocks of land uplifted. Laccoliths created the Henry Mountains, where igneous intrusions pushed overlying sedimentary layers into domes, exposing large sequences of strata, including Morrison Formation exposures replete with fossil wood. As incredible as it seems, the Chinle Formation that once reposed half a mile beneath the sea is now a mile above sea level with many exposures at the surface of the planet, allowing easy access to ancient forests.

Land mass of Pangea as it existed during the Late Permian Period, some 255 million years before present. The white outlines approximate the boundaries of what became North America, Europe, South America, Africa, Antarctica, India, Madagascar, and Australia.

Land mass of Gondwana and a portion of Laurasia (upper land masses) as they existed during the Late Jurassic Period, some 150 million years before present, after Pangea began to break apart. The white outlines approximate the boundaries of what became North America, Europe, South America, Africa, Antarctica, India, Madagascar, Australia, and New Zealand.

Delicate Arch in Arches National Park at Moab, Utah, is composed of Entrada Sandstone. It sits on the northern flank of the Salt Valley Anticline, produced by the incredible force of salt when it tilted and bowed up overlying Mesozoic rocks. The view through and behind the arch shows the La Sal Mountains. The highest elevation is Mt. Peale at 12,721 feet. The La Sal Mountains were formed during the Tertiary Period by a series of laccolithic intrusions whereby magma rose along fractures and spread out laterally along bedding planes of enclosing sedimentary rocks. This resulted in a bulging central plug of diorite porphyry that tapers farther from the core. Such laccolithic intrusions and salt anticlines bow up and otherwise incline large areas of sedimentary rocks, thereby increasing erosion and exposing wood-bearing sequences.

FOSSIL WOOD IDENTIFICATION DEFINITIONS

Certain Tertiary Period fossil wood locations provide well-preserved fossil woods of many genera that lend themselves to comparison with modern woods. Consequently, amateur wood anatomists in those areas, such as the northwestern United States, enjoy a rich heritage of fossil wood identification. It is important to bear in mind, however, that these species of wood evolved over time. When a section of fossil wood from a bygone epoch is studied and similarities with a modern genus or species are apparent, it is probably not accurate to identify the fossil by the name of the modern tree. In this instance, we will call a certain fossil wood from the Miocene Epoch "sycamore-like" rather than simply calling it sycamore. Our identification section begins with term definitions, followed by discussions of contemporary temperate zone conifers and hardwoods. Modern wood identification knowledge is helpful when examining Tertiary Period fossil woods because of significant similarities. This information will aid in the discussion of older fossil woods that we principally identify by reference to macro photographs and micro images. After Tertiary woods, we will proceed back through time as though digging into the Earth's stratigraphic layers and discuss woods of each geologic era as we encounter them. Please bring a good shovel, as we will be digging through thousands of feet of strata.

TOP: **unidentified wood. Patagonia, Argentina. Daniels specimen, 19 cm.**
MIDDLE: ***Woodworthia*. Zimbabwe. Daniels specimen, 18 cm.**
BOTTOM: ***Woodworthia*. Utah, USA. Daniels specimen, 30 cm.**

Term Definitions

bole: the main stem or trunk of a tree.

deciduous tree: a tree that seasonally loses its leaves, typically in the fall.

earlywood: Temperate-zone trees usually grow rapidly in the early season and more slowly in the hot summer months. The wood formed in the spring is lower in density, with larger cells, and is called earlywood or springwood. The more dense wood formed later in the season is called latewood or summerwood. The transition may be gradual or abrupt.

fiber: an elongated cell with pointed ends and thick walls.

fusiform ray: a spindle-shaped ray that often includes a horizontal resin canal.

hardwood: wood produced by angiosperms, the botanical group of trees that bear broad leaves. The term has no relationship to the actual hardness of the wood. In many instances, the wood of a hardwood tree is softer than the wood from a softwood tree.

heartwood: The older, nonfunctional central wood in a tree. Sapwood is the new wood added each year that is functional. Sapwood of previous years becomes nonfunctional and forms outer layers of heartwood. In living trees, the color differential between sapwood and heartwood can be quite distinct. The color of the heartwood comes from deposition of extractives and exists in shades of yellow, orange, red, and brown in modern woods. These extractives also reduce the permeability of some woods, making some, such as white oak and western red cedar, more durable and less absorbent. Heartwood may also be more resistant to fungal and insect infestations. The heartwood color boundary approximates the growth ring contour but does not precisely coincide with it. Sapwood is widest toward the top of a tree and is wider in healthy, vigorous trees than in trees that may be at a disadvantage in obtaining nutrients and the radiant energy of the sun. With the exceptions of balsam fir, hemlock, and spruce, conifers now growing in North America generally have distinct heartwood. Certain species of trees have substantially more sapwood rings, which can be a useful tool in modern wood identification. Black locust, for example, typically has just a few rings of sapwood while sugar maple may have 30 to 40, and sweet birch may have 60 to 80. How the characteristics of sapwood and heartwood translate to the study of fossil wood is undetermined. The lack of permeability of heartwood may result in mineralization different from sapwood in some limited situations, depending on the fossilizing mechanism. This difference might infrequently aid in identification of woods with affinity to living woods and help explain some fossil wood patterns. Another unanswered question is whether the presence of extractives affects the color or texture of the mineralized fossil.

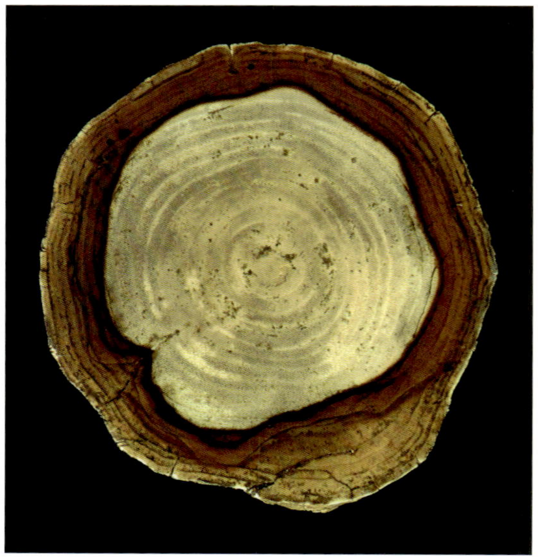

Fossil hardwood. Oligocene/Miocene. Sweet Home, Oregon, USA. Daniels specimen, 7 cm.

latewood: see earlywood.

longitudinal: lengthwise through the stem.

longitudinal epithelial cells: thin-walled parenchyma cells that surround longitudinal resin canals.

parenchyma: tissue consisting of relatively thin-walled cells that primarily store and distribute carbohydrates, although they can serve other purposes. They are the living cells of wood. Axial (longitudinal) parenchyma occurs in strands along the grain and may be visible with a hand lens on the transverse surface of wood as tiny dots, as sheaths around pores, or as broken or continuous lines or bands.

petiole: the base of the stalk of a leaf.

pith: the central cylinder of a stem.

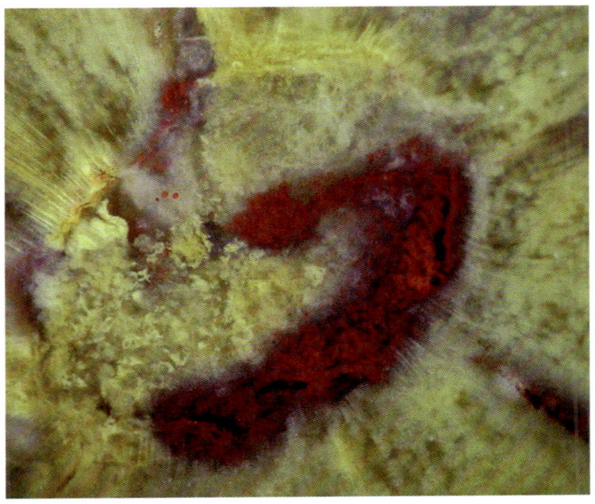

Northern Arizona, USA; pith in fossil wood.

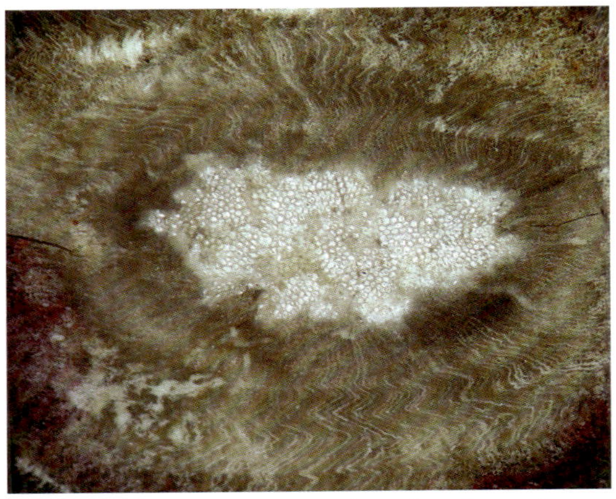

Queensland, Australia; pith in fossil wood.

Queensland, Australia; pith in fossil wood.

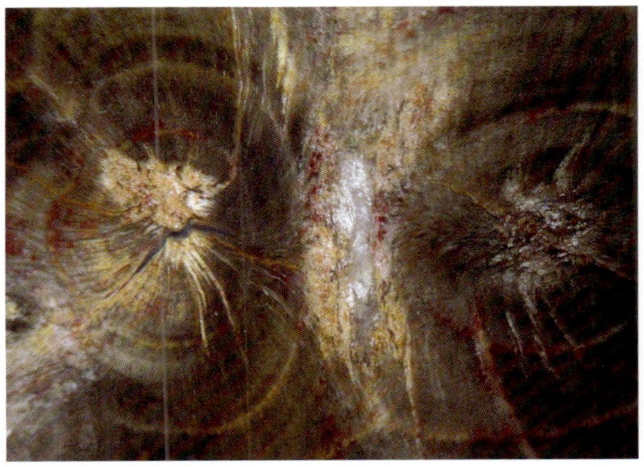

Queensland, Australia; double pith in fossil wood.

pore: cross section of a vessel. The term is often used interchangeably with vessel.

radial section: cut along the grain parallel to the rays.

ray: All species of wood have rays. These are thin aggregations of cells extending radially. They appear as thin to broad lines emanating inward from the exterior of the tree in transverse section, essentially at right angles to the growth rings. When consisting of one row of cells, a ray is uniseriate. When consisting of two rows of cells, a ray is biseriate. When consisting of more than two rows of cells, a ray is multiseriate. An aggregate ray is a broad ray transected by longitudinal cells, such as fibers, tracheids, or vessels. It appears under magnification as a composite of many smaller rays.

resin canal: A resin canal is not actually a wood element but rather a cavity or intercellular space surrounded by secreting epithelial cells. In heartwood, tylosoids usually plug the resin canals.

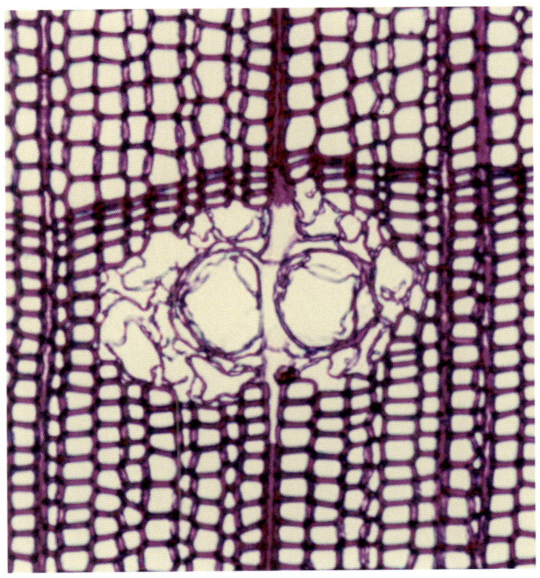

Resin canal in *Pinus edulis* (piñon pine).

sapwood: the woody outer portion of the tree in which xylem cells are functional and, hence, physiologically active. Sapwood is generally lighter in color than heartwood in living woods.

seriation: The width of the ray as measured in number of cells.

short shoot: short branch bearing leaves or fruiting structures.

softwood: wood produced by the botanical group of trees (conifers) that, in most species, have needle-like or scale-like leaves. The term has no relationship to the actual softness of the wood.

stele: the primary core of a tree immediately surrounding the pith.

tangential section: cut along the grain at right angles to the rays.

tracheid: fibrous lignified cells; the principal cell type in softwoods and a minor cell type in hardwoods.

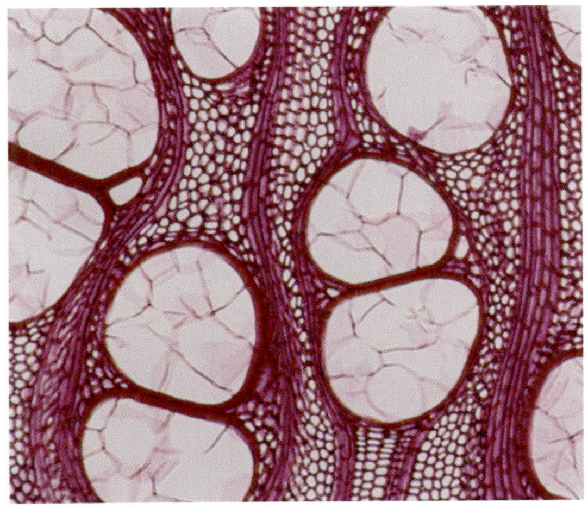

Tyloses in *Robinia pseudoacacia* (black locust).

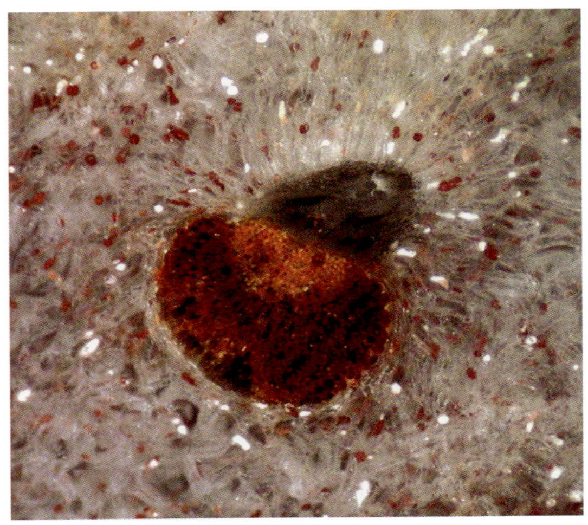

Vascular bundle in Louisiana *Palmoxylon*.

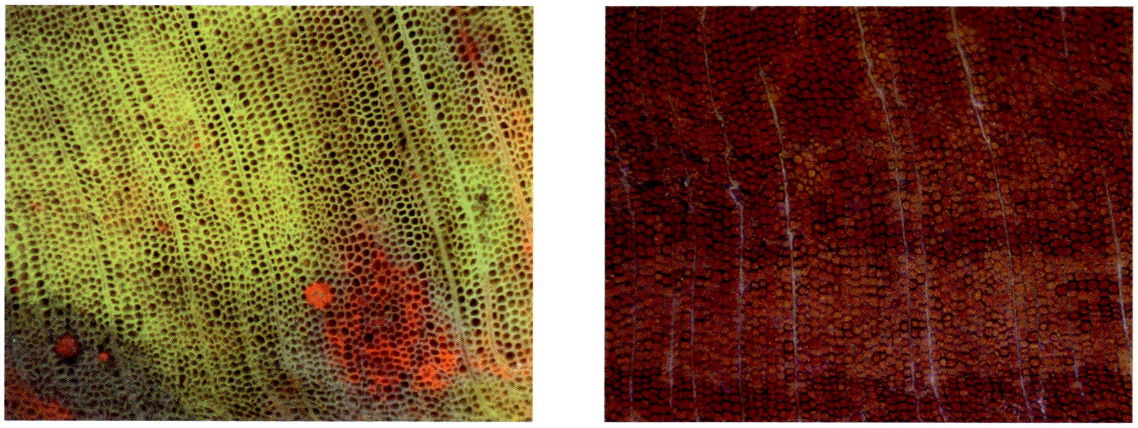

Transverse sections of Triassic Period fossil wood from northern Arizona. Note how the appearance can differ from looking as though the cells are hollow to having a full, bloated appearance. This difference can appear even within a single specimen and results from differences in mineralization.

transverse section (cross-section): cut at right angles to the grain, revealing the full annual rings.

tyloses: sac-like or cyst-like structures in hardwoods that sometimes develop in a vessel, typically in conjunction with heartwood formation or because of injury. Evidence of tyloses may not survive the fossilization process even in otherwise well-preserved specimens.

tylosoids: structures in resin canals that resemble tyloses in hardwoods.

vascular bundle or **bundle**: A grouping of cells comprising both xylem and phloem that conduct water and nutrients in some plants, such as palm.

vessel: a tube-like conductor of water and nutrients within the xylem of most angiosperms.

IDENTIFICATION OF TERTIARY PERIOD FOSSIL SOFTWOODS

Identification of fossil conifers is an area most suited for professionals equipped with considerable knowledge, training, experience, and a powerful microscope. Inspection of a polished transverse section with a hand lens or a low-power microscope provides little means of differentiation, yet an amateur can discover some clues to identification in a well-preserved specimen. The type of detail necessary for positive identification of conifers generally mandates study of minute connective passages, termed pits, located between ray tracheids and longitudinal tracheids, between longitudinal tracheids and ray parenchyma cells, and connecting one longitudinal tracheid to another tracheid. The shape and characteristics of these microscopic pits are unique for specific species of trees. Coniferous xylem consists of rigid cells called tracheids, which provide the main support for the vascular cylinder or stele. This structure allows a tree to stand upright. The amount of tracheids composing the wood structure is enormous. Coniferous wood cells are as tiny as 25 micrometers (1/1000th of an inch). One cubic foot of spruce wood contains 10 to 14 billion tracheids. Within living conifers exist no less than 46 genera and 600 species of trees with almost worldwide distribution. There are more than 110 species of pine alone. Adding to this complexity is the number of extinct coniferous trees. Few experts have the knowledge and experience necessary to make accurate identifications of large numbers of conifers.

Resin Canals

There are two types of resin canals in conifers, normal and traumatic. Traumatic resin canals result from injury or stress to the wood, are uncommon, and generally appear in transverse section as a line of several to many along a growth ring. They are absent elsewhere and, consequently, are distinguishable from normal resin canals.

Douglas fir, larches, pines, and spruce always have normal resin canals. They are larger, more abundant, and evenly spaced in pines and are usually present in all growth rings. Douglas fir and spruces sometimes have solitary resin canals evenly distributed in the latewood, yet in some trees they group in tangential rows of 5 to 30 or more. Resin canals in larches are single or in groups of several. In Douglas fir, larches, and spruce, resin canals may be absent in some growth rings. The sizes of the resin canals in conifers depend on the age of the tree and the growth rate, rendering their usefulness as a diagnostic tool less helpful in fossil woods where the sample is normally small, from an unknown growth position, and from an ancient forest of unknown habit. Epithelial cells in pine are thin walled, while those of Douglas fir, larch, and spruce are thick walled. In fossil woods, resin canals appear as small dots in a transverse section, predominately in the middlewood to latewood of an annual ring. In pines, they are conspicuous and observable even without magnification. Common temperate zone conifers with no resin canals can include bald cypress (*Taxodium*), cedar (*Juniperus* or *Thuja*), hemlock (*Tsuga*), redwood (*Sequoia*), true fir (*Abies*), or yew (*Taxus*). Use magnification to look for resin canals, as those in woods other than pine are sparse and difficult to observe.

NORMAL RESIN CANALS				
NORMAL	NORMAL BUT ABSENT IN SOME GROWTH RINGS	THIN-WALLED	THICK-WALLED	NO NORMAL RESIN CANALS
Douglas fir	Douglas fir	pine	Douglas fir	bald cypress
larch	larch		larch	cedar
pine	spruce		spruce	hemlock
spruce				redwood
				true fir
				yew

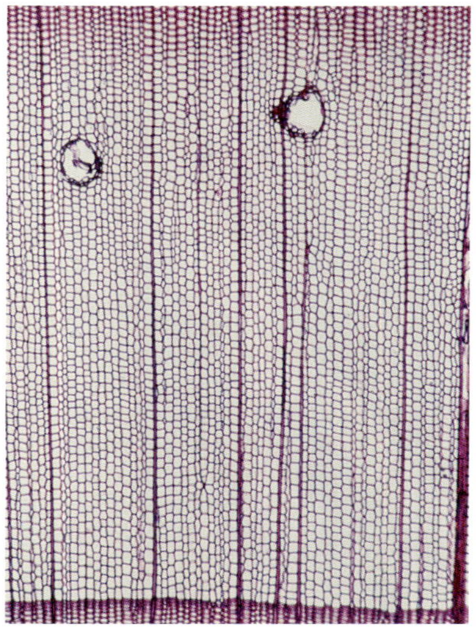

Picea glauca (white spruce), magnification × 40.

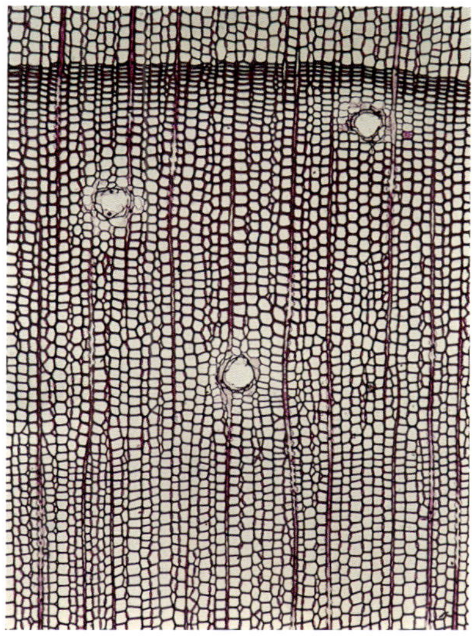

Pinus strobus (eastern white pine), magnification × 40.

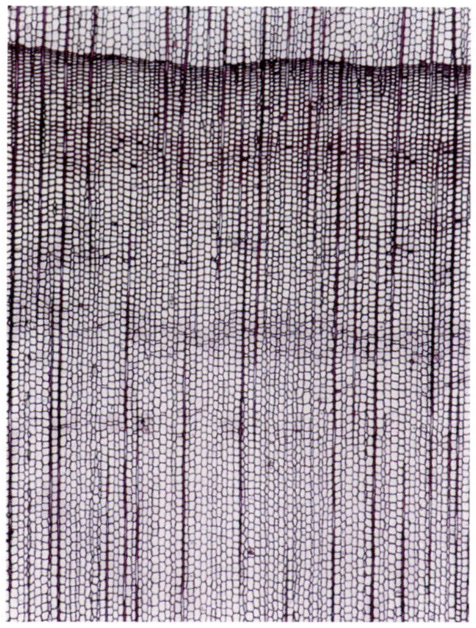

Chamaecyparis thyoides (Atlantic white cedar), magnification × 40.

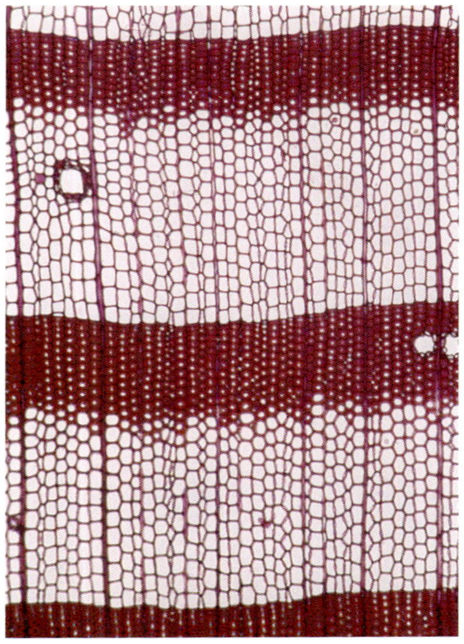

Larix laricina (tamarack or larch), magnification × 40.

Pine-like fossil wood. Eastern Oregon, USA. Tertiary. Daniels specimen.

Juniperus occidentalis (western juniper), magnification × 40.

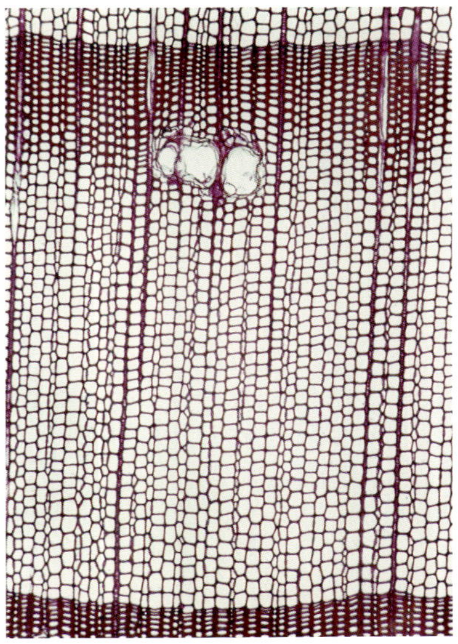

Pinus edulis (piñon pine), magnification × 40.

Tsuga heterophylla (western hemlock), magnification × 40.

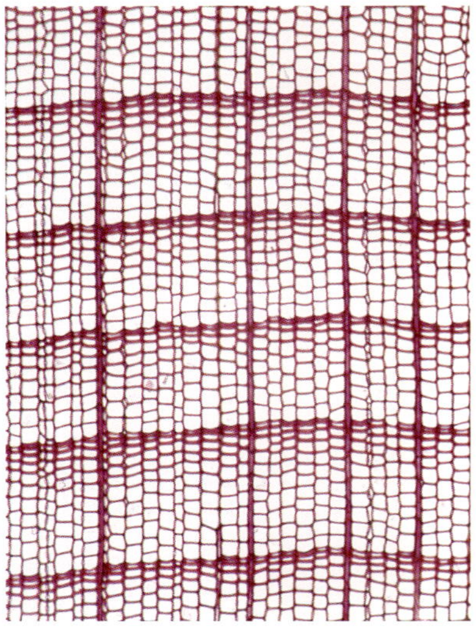

Abies procera (noble fir), magnification × 40.

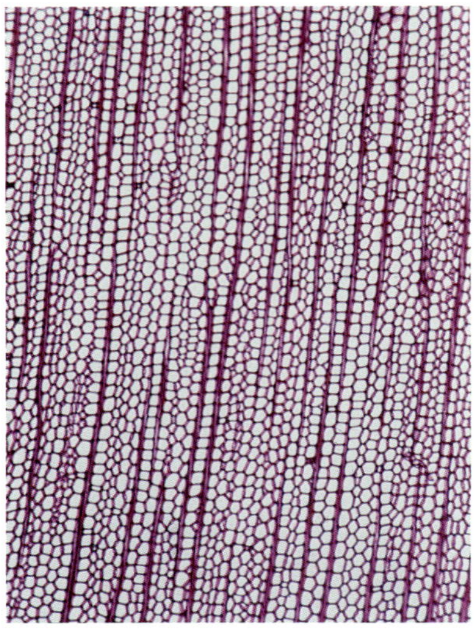

Metasequoia (dawn redwood), magnification × 40.

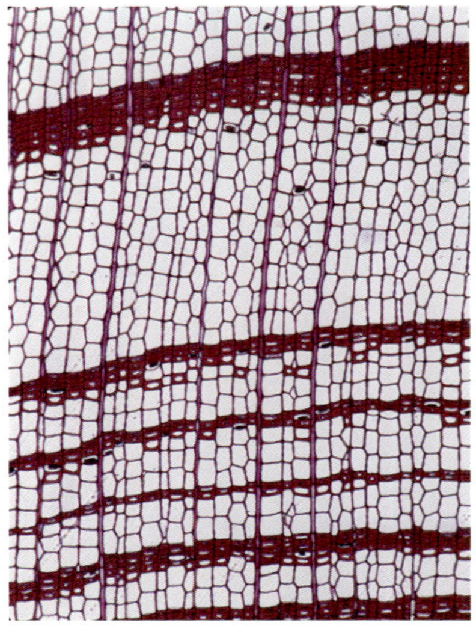

Taxodium distichum (bald cypress), magnification × 40.

Abies balsamea (balsam fir), magnification × 40.

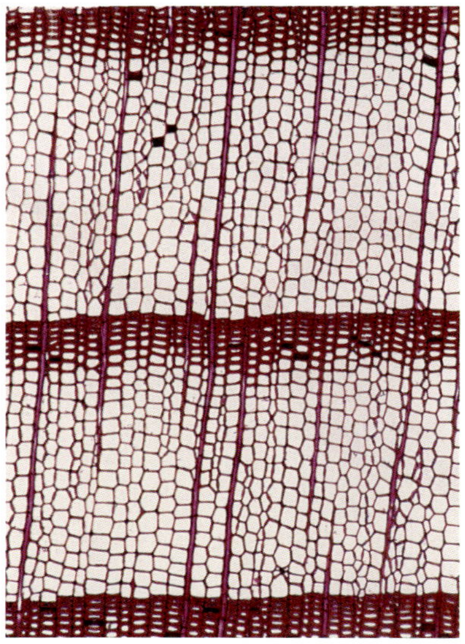

Sequoiadendron giganteum (giant sequoia), magnification × 40.

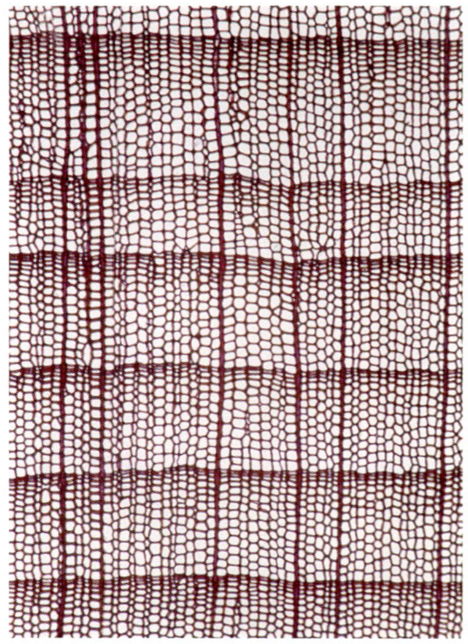

Taxus brevifolia (Pacific yew), magnification × 40.

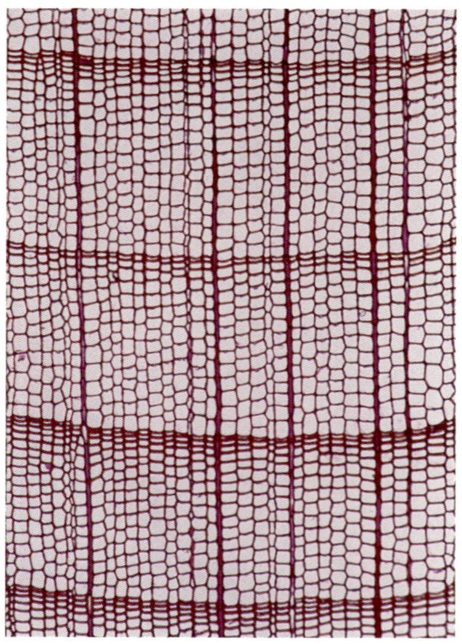

Libocedrus decurrens (incense cedar), magnification × 40.

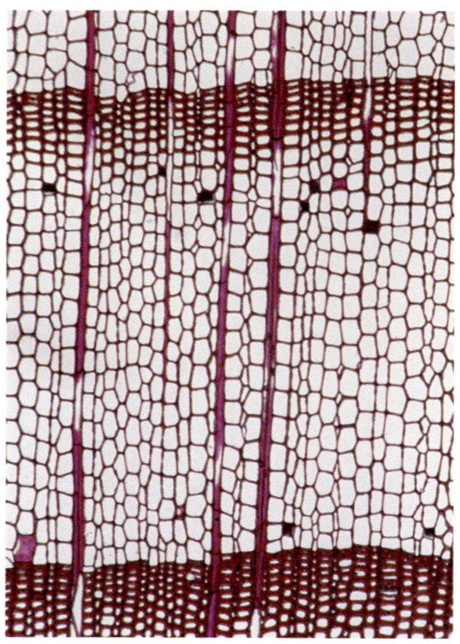

Sequoia sempervirens (redwood), magnification × 40.

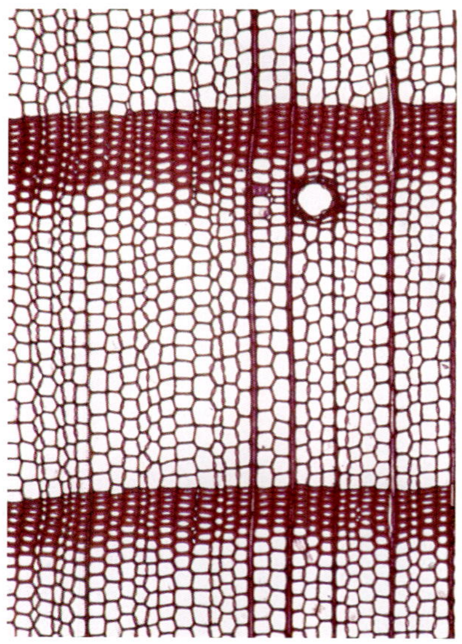

Pseudosuga menziesii (Douglas fir), magnification × 40.

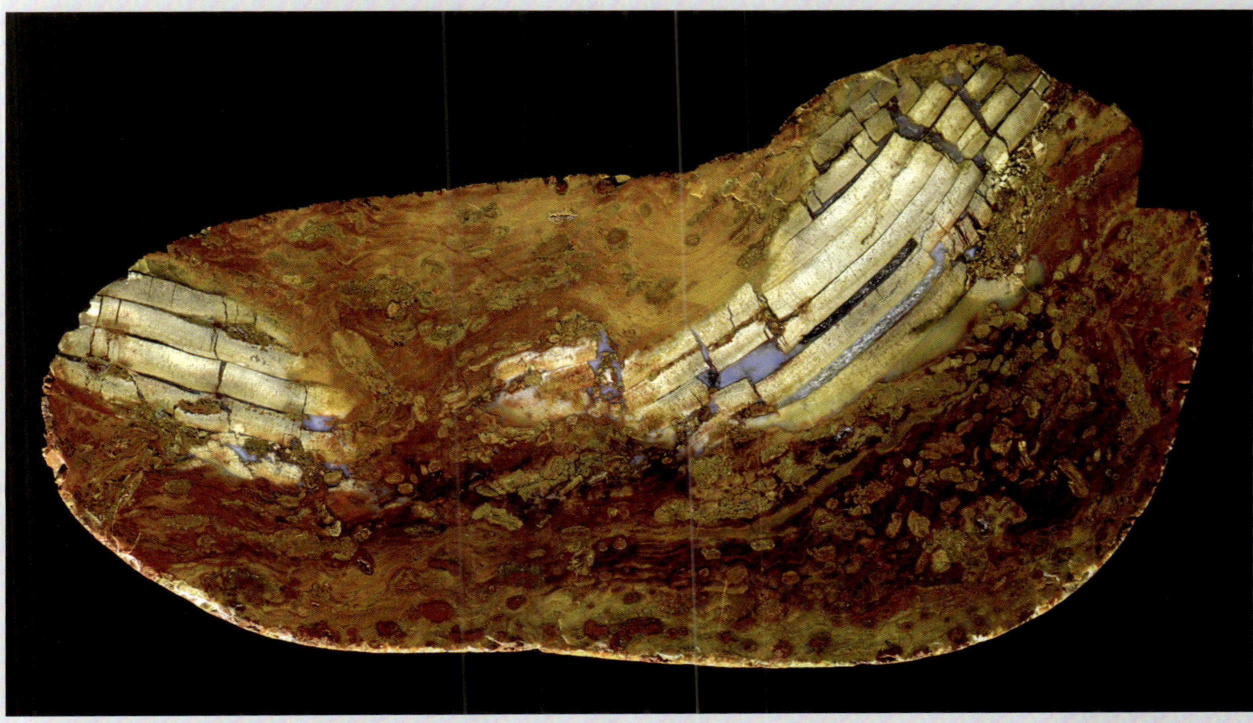

Fossil conifer. Tertiary. Oregon, USA. Daniels specimen, 17 cm.

Micro image showing coniferous cell structure. Fossil conifer. Tertiary. Oregon, USA. Daniels specimen. Growth ring width 3.7 mm.

Higher resolution micro image showing coniferous cell structure. Fossil conifer. Tertiary. Oregon, USA. Daniels specimen. Largest tracheids approximately 0.05 mm.

Micro image showing *Sequoia*-like cell structure. Columbia River Basalt, Miocene. Yakima, Washington, USA. Daniels specimen. Growth ring width 1.1 mm.

Sequoia-like fossil wood. Columbia River Basalt, Miocene. Yakima, Washington, USA. Daniels specimen, 20 cm.

Micro image showing pine-like cell structure. Miocene. Eastern Oregon, USA. Daniels specimen. Resin canals 0.3 mm.

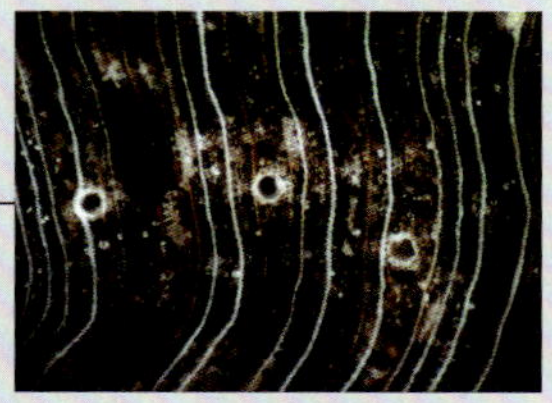

Higher resolution micro image showing pine-like cell structure. Miocene. Eastern Oregon, USA. Daniels specimen. Resin canals 0.3 mm.

Fossil pine-like wood. Miocene. Eastern Oregon, USA. Daniels specimen, 10.5 cm.

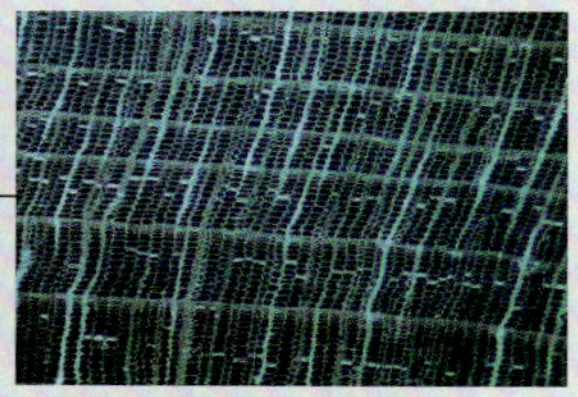

Micro image showing fir-like cell structure. Miocene. Ratka, Hungary. Daniels specimen. Average width of growth rings 0.5 mm.

Fossil fir-like wood. Miocene. Ratka, Hungary. Daniels specimen, 7.5 cm.

Fossil *Sequoia*-like wood. Tertiary. Ely, Nevada, USA. Daniels specimen. Width of growth ring approximately 0.4 cm. Average tracheid approximately 0.04 mm.

Earlywood to Latewood Transition

In conifers, transition from earlywood to latewood within an annual ring can be an indication of genus. It is gradual in balsam fir, cedar, hemlock, juniper, soft pines (white pine and sugar pines), spruce, and yew. It is abrupt in Douglas fir, hard pines (red pine, yellow pine, ponderosa pine, and lodgepole pine), and larch. Redwood and bald cypress have similar narrow, dense latewood with abrupt transition from earlywood to latewood and, consequently, distinct growth ring boundaries.

EARLYWOOD TO LATEWOOD TRANSITION	
GRADUAL	ABRUPT
balsam fir	bald cypress
cedar	Douglas fir
hemlock	hard pines
juniper	larch
soft pines	redwood
spruce	
yew	

Tracheids

While both softwoods and hardwoods have tracheids, the tracheids in conifers are by far the dominant element in a transverse view. Conifer tracheids line up in uniform rows or chains.

Tracheid size can also be indicative of the conifer species. For coniferous woods currently growing in the United States, the largest tracheid diameters appear in redwood and the smallest in yew, at one-third the size of those in the redwood.

Rays

All species of wood have rays. Rays in coniferous woods are usually either uniseriate or fusiform. Trees with fusiform rays will always have longitudinal resin canals. The rays of wood without resin canals are generally uniseriate, although biseriate rays infrequently occur in many conifers, including those without resin canals, and are common in redwood. The spacing of rays as viewed in a transverse section remains fairly constant and is not a valuable tool for identification as the differences are slight.

Parenchyma

The position and distribution of axial parenchyma cells in conifers are helpful to wood identification but difficult to discern in many fossil wood specimens. When parenchyma cells are scattered widely among tracheids, the parenchyma are "diffuse." When grouped, even loosely, in a band or row essentially following the curve of the growth ring, the parenchyma cells are "banded." When restricted to the outer aspect of the growth ring, they are "terminal" or "marginal." Pine, spruce, and yew do not have axial parenchyma. Parenchyma cells can be found sporadically—terminally, diffuse, or in combination—in Douglas fir, hemlock, larch, and true fir, but typically only in latewood. In cedar, cypress, and juniper, parenchyma cells are abundant as banded or diffuse but unevenly distributed. Axial parenchyma cells are most frequent in redwood and bald cypress, ranging from diffuse to banded.

IDENTIFICATION OF TERTIARY PERIOD FOSSIL HARDWOODS

Hardwood structure is more complex and varied than conifer structure, providing a broad field of interest to the fossil wood enthusiast. During the Eocene through Miocene epochs, large forests of deciduous trees thrived in temperate zones, including what is now the northwestern United States. Hardwood structure is common in Tertiary Period fossil woods, of which there are fortunately many fine, well-preserved specimens. Tertiary Period fossil woods from the Front Range of Colorado, Nevada, Oregon, Washington, and Wyoming are among the world's best. Nice specimens also hail from Belgium, Hungary, Turkey, Indonesia, and other locations around the globe. As with conifers, there are many extant and extinct species of hardwoods. Numerous genera and some 250,000 extant species (as many as 300,000 by some estimates) reside in the world of flowering plants (angiosperms). Many of these are hardwood trees. Eight hundred species of hardwood are now indigenous to the United States, and thousands more thrive in the tropics. There are 400 species of oak alone.

The focus of this section is to discuss the extremely complex features of fossil hardwood identification in general terms. We describe several modern hardwood species with distinguishing characteristics. The purpose is not to claim that Tertiary Period fossils are identical to their modern counterparts but to help an amateur fossil wood anatomist or beginning student of paleobotany discern categories of modern woods as an aid to classifying prehistoric woods. We should limit identification of fossil woods based on these comparisons to gross categories, such as a "ring porous hardwood" or "chestnut-like hardwood."

Transverse section features that distinguish hardwoods from conifers are the presence of pores (vessel elements), obscure radial alignment of longitudinal cells, complex overall structure, and a great variety of ray types.

Pores

When a longitudinal vessel is cut transversely, the resulting opening is a pore. In transverse sections, pores are the largest wood elements visible. The largest pores are clearly visible without magnification, and smaller pores are visible with aid of a hand lens. Pores are either solitary or grouped. Solitary pores are alone, touching no other pore, and are usually rounded to oval. When oval, they are longer in a radial direction. In woods with thin-walled vessels and thick-walled fibers, the pores are angular. Several pores contiguous in a radial direction are called a pore multiple. When these pores are compressed together, they may appear to be one divided vessel. A radial series of single or multiple pores is a "pore chain." The latewood pores of certain woods (including all elms and hackberry) group in concentric wavy

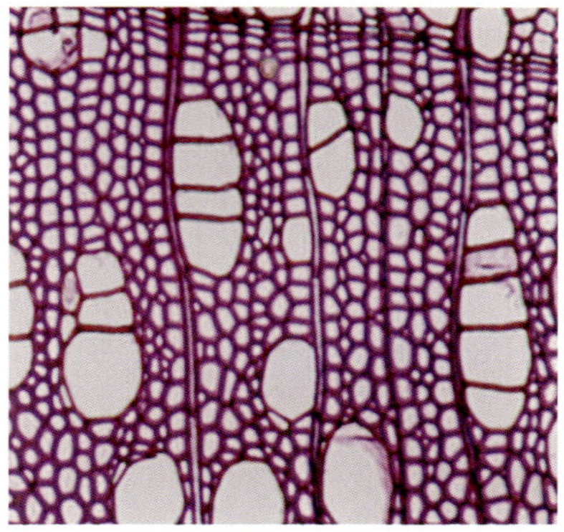

Populus tremuloides (quaking aspen): pore chain.

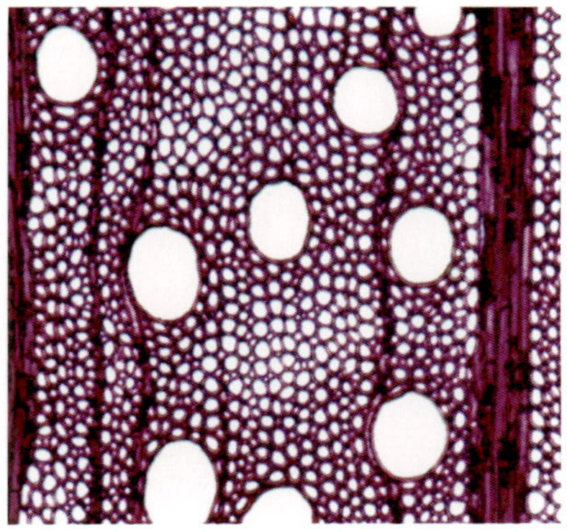

Acer saccharum (sugar maple): solitary pores.

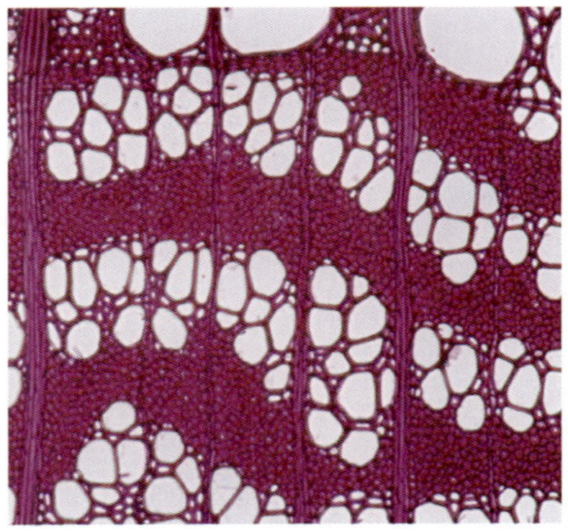

Ulmus americana (American elm): ulmiform pores.

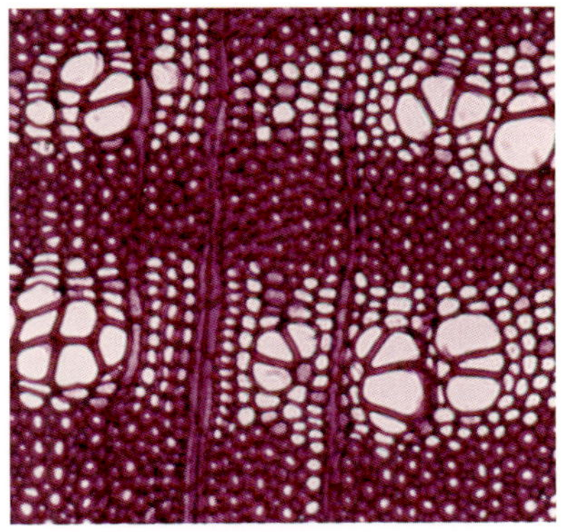

Gleditsia triacanthos (honey locust): nested pores.

bands called ulmiform or, simply, wavy bands. The term "nested pores" or "pore cluster" is used where many pores touch each other, both radially and tangentially. Tyloses form in the vessels of some hardwoods and can be a diagnostic tool. See Appendix A for a list of specific woods by pore arrangement.

Tracheids

The two types of hardwood tracheids are vascular tracheids and vasicentric tracheids. Vascular tracheids are the approximate size of small latewood vessel elements. The concentric wavy bands of latewood elm and hackberry consist of small vessel elements and vascular tracheids. In transverse section, vascular tracheids and pores appear identical. Vasicentric tracheids look similar to, and are associated with, axial parenchyma. They are frequently located close to large earlywood vessels of ring porous woods like oak and chestnut.

Fibers

Another major component of hardwoods is fibers. Fibers are thick-walled, fibrous, small-diameter cells that constitute 50 percent or more of some hardwoods by volume. Microscopic characteristics of fibers, such as the presence or lack of spiral thickenings in their inner walls, are diagnostic in microscopic analyses of wood. They provide no assistance in identifying fossil wood features in transverse sections with low-power magnification.

Parenchyma

Most hardwoods possess greater amounts of longitudinal or axial parenchyma cells than do conifers. Exceptions are poplars and willow, which have only sparse longitudinal parenchyma, and cherry, which rarely contains any parenchyma. Observation of longitudinal parenchyma cell arrangements and their relationship with pores, as visible in a transverse section, is helpful in hardwood identification. Appendix A presents a list of specific woods by parenchyma groupings. Because of the small size of parenchyma cells, use of a handheld microscope is advisable. Parenchyma cells are classified in the following categories:

Apotracheal parenchyma: longitudinal parenchyma arranged independent of vessels and separated from pores by rays or fibers.

- **Diffuse apotracheal parenchyma** (*not visible with a hand lens*): isolated single parenchyma cells irregularly distributed among fibers.

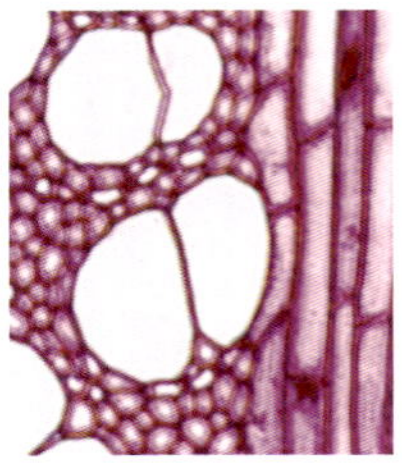

Platanus occidentalis **(American sycamore): diffuse apotracheal parenchyma.**

- **Diffuse-in-aggregates apotracheal parenchyma** (*not visible with a hand lens in some species*): grouped in short tangential lines, sometimes from ray to ray.

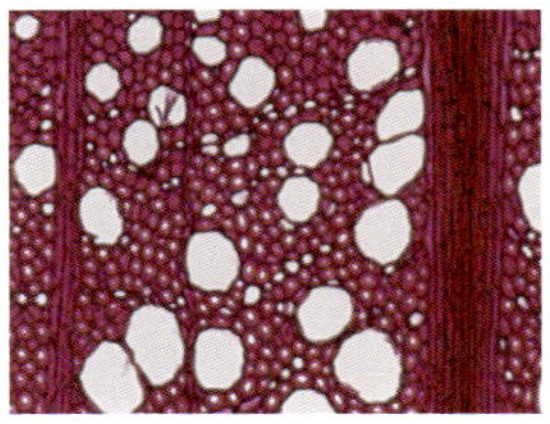

Fagus grandifolia **(American beech): diffuse-in-aggregates apotracheal parenchyma.**

Paratracheal parenchyma: longitudinal parenchyma associated with the vessels or vascular tracheids.

- **Scanty paratracheal parenchyma** (*not visible with a hand lens*): confined to a few cells around the vessels.

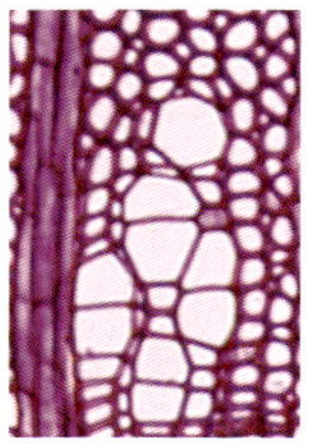

Ilex opaca **(American holly): scanty paratracheal parenchyma.**

- **Vasicentric paratracheal parenchyma** (*usually visible with a hand lens*): form a complete or partial sheath around a vessel.

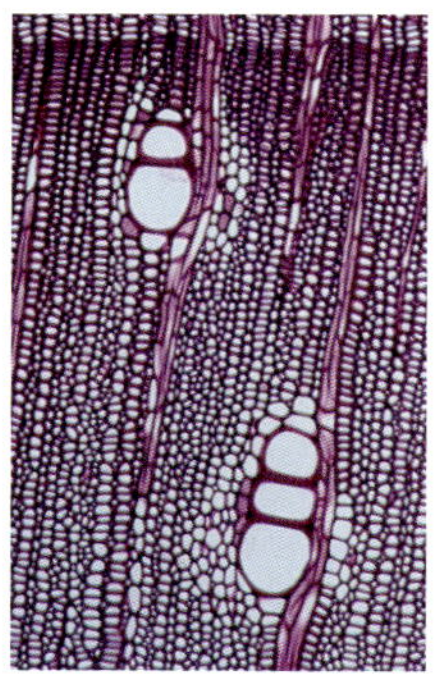

Umbellularia californica **(California laurel): vasicentric paratracheal parenchyma.**

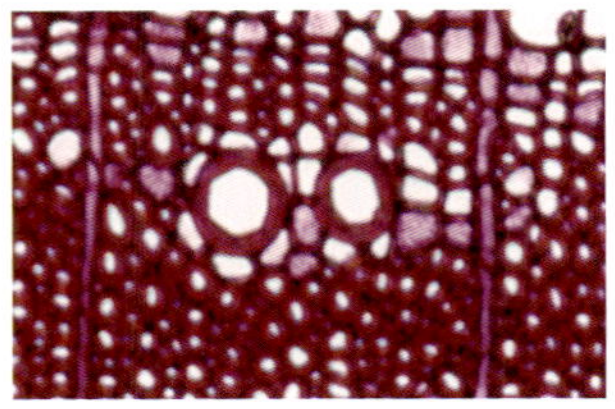

Fraxinus latifolia **(Oregon ash): vasicentric paratracheal parenchyma.**

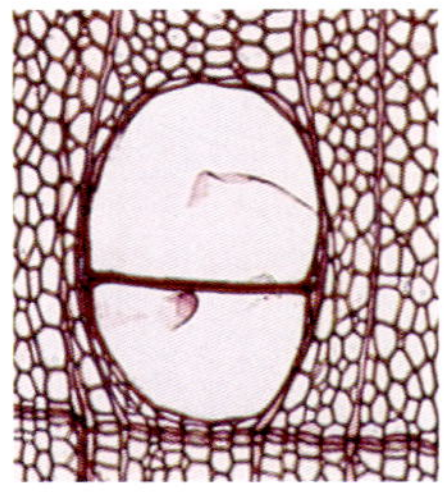

Juglans cinerea **(butternut): vasicentric paratracheal parenchyma.**

- **Aliform paratracheal parenchyma**: extend from the edges of a pore forming wing-like lateral extensions; common in tropical hardwoods.

- **Confluent paratracheal parenchyma**: form irregular tangential or diagonal bands that connect two or more pores.

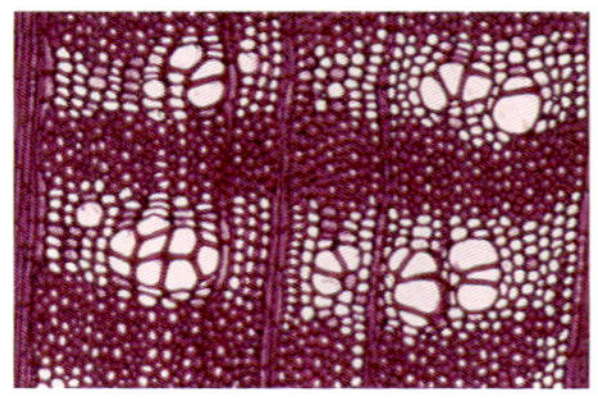

Gleditsia triacanthos **(honey locust): confluent paratracheal parenchyma.**

Banded parenchyma: form concentric (tangential) lines or bands. Because some bands make contact with pores and some do not, banded parenchyma are neither paratracheal nor apotracheal. Reticulate parenchyma cells occur in woods where bands of parenchyma contact rays of the same approximate width to form a mesh or netlike pattern.

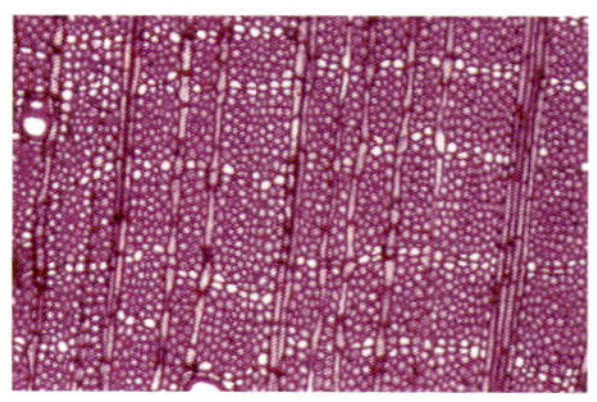

Carya cordiformis **(bitternut hickory): banded parenchyma.**

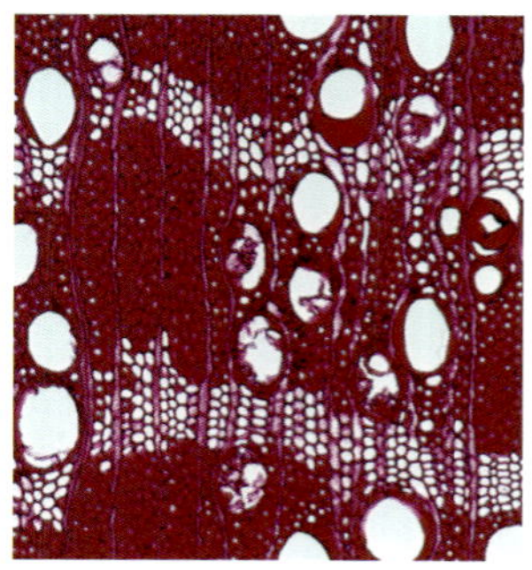

Casuarina quadrivalvis **(she oak): banded parenchyma.**

Vantage, Washington, USA; Tertiary fossil wood: banded parenchyma.

Marginal parenchyma: longitudinal parenchyma occurring as occasional cells or forming an essentially continuous band of one or more cells in width at the margin of a growth ring.

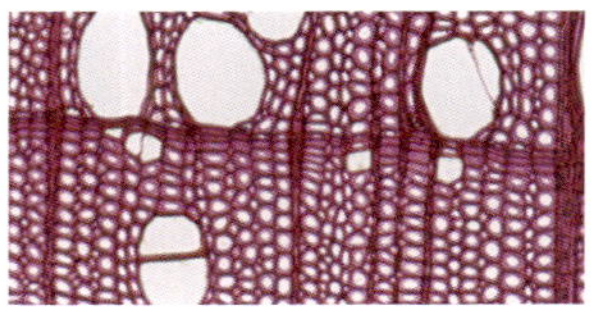

Betula alleghaniensis **(yellow birch): marginal parenchyma.**

Hardwood Rays

Hardwood rays have a greater variety in width than softwood rays. Consequently, rays are an important aid in fossil hardwood identification. Whereas the rays of nearly all conifers are uniseriate, most hardwood rays are biseriate to multiseriate. Exceptions are chestnut, poplar, and willow. Rays in oak are often 30 or more cells wide. Some hardwoods, including American beech, oaks, and sugar maple, have uniseriate rays in combination with multiseriate rays. To determining the precise width of a ray in cells requires microscopic examination in most instances. Noded rays bulge or swell at the growth ring. See Appendix A for groupings of specific woods by rays.

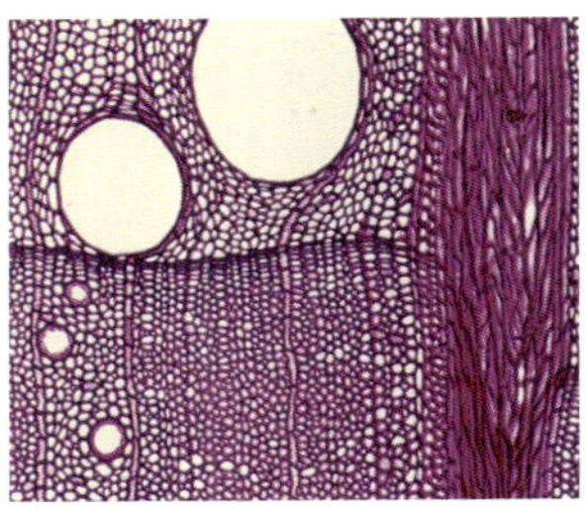

Quercus velutina **(black oak): rays are uniseriate plus 14 seriate.**

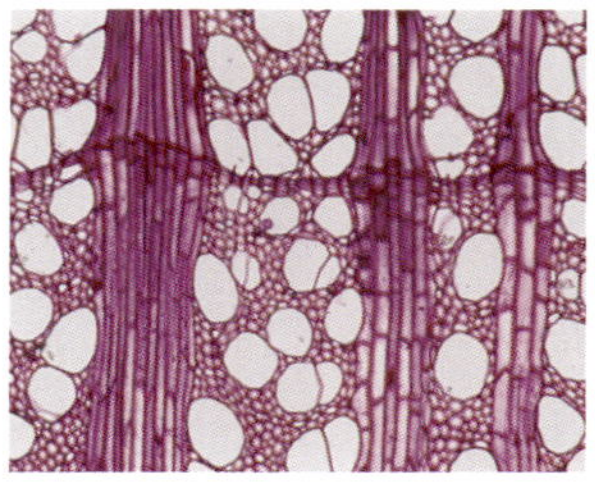

Platanus occidentalis **(American sycamore): noded multi-seriate rays.**

For our purposes, the aspect that distinguishes angiosperms from other fossil woods is the presence of vessels (pores in transverse section) in the wood, although a few vascular angiosperms do not have vessels and a few gymnosperms do. Once we recognize a specimen as a hardwood by the presence of pores, we categorize it as ring porous, semi ring porous, or diffuse porous. Wood of any particular species normally falls within one of these groupings, but the wood of some species may vary from one category to another. A factor to bear in mind in the study of fossil woods is that woods of a single species may be ring porous when grown in a temperate climate and diffuse porous when grown in a warmer environment. Ring porous woods may not demonstrate ring porous habit in the first few years of growth.

Ring Porous Hardwoods

If a hardwood is ring porous, the largest, most prominent pores reside within the earlywood. Early season pores are considerably larger than those of late season. They may form a single row or two or more rows. The transition from earlywood to latewood in a ring porous wood may be sharply defined by a clear diminution in the size of the pores or may be gradual. A wood with earlywood pores concentrated at the growth ring that are no larger than pores in latewood is diffuse porous.

Oak-like and chestnut-like: One group of ring porous hardwoods is the oak-chestnut group. Both oak (exclusive of live oak and tanoak) and chestnut are ring porous woods with large earlywood pores and radially arranged latewood pores. Oaks have very large multi-seriate rays, usually 20 seriate and larger, as well as fine, mostly uniseriate rays. Chestnuts are also ring porous with large earlywood pores and radially arranged latewood pores, but they predominately have uniseriate rays and no huge oak-like rays. In red oaks, the latewood pores are larger, more rounded, thick walled, and individually distinct; heartwood tyloses are absent or sparse. In white oaks, the latewood pores are very small, numerous, and indistinct; heartwood tyloses are plentiful. The pith in oaks frequently forms a five-pointed star.

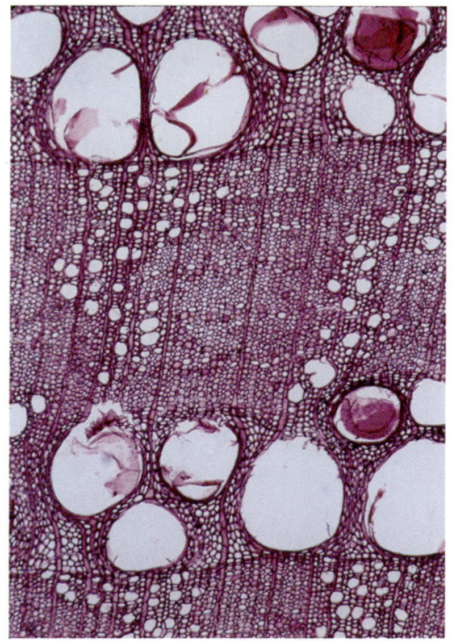

Quercus chapmanii **(Chapman oak), magnification x 40.**

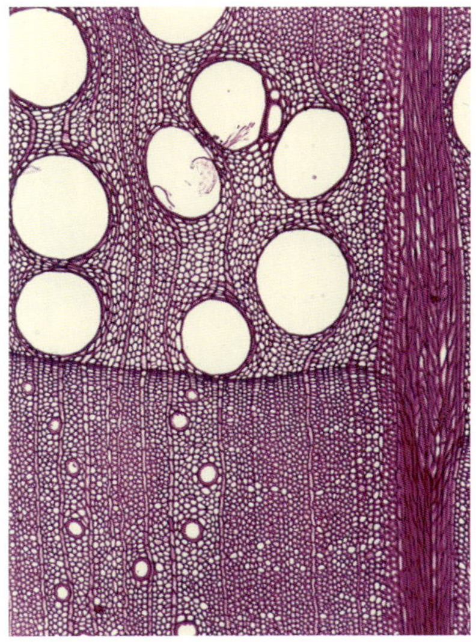

Quercus velutina **(black oak), magnification x 40.**

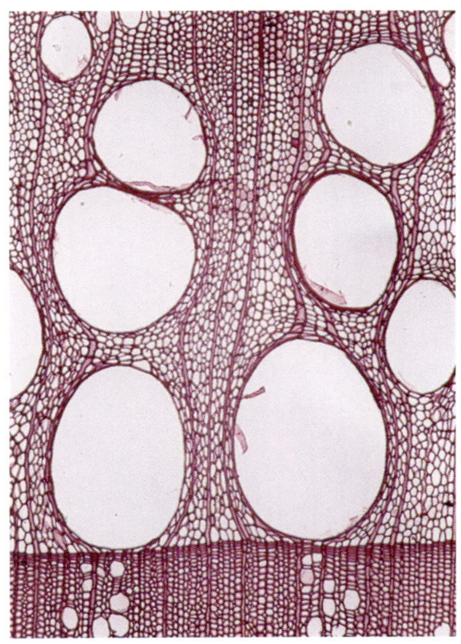

Castanea dentata **(chestnut), magnification x 40.**

Macro photograph showing oak-like cell structure in fossil wood. Miocene. Oregon, USA

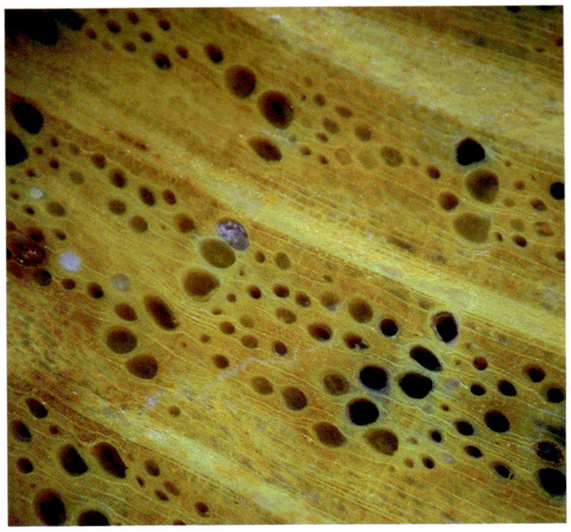

Micro image showing oak-like cell structure in fossil wood. Pliocene. Deschutes, Oregon, USA. Daniels specimen.

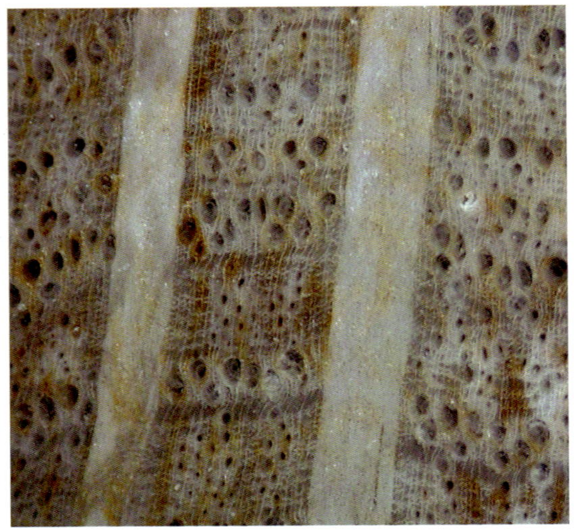

Micro image showing oak-like cell structure in fossil wood. Tertiary. Swartz Canyon, Oregon, USA. Daniels specimen.

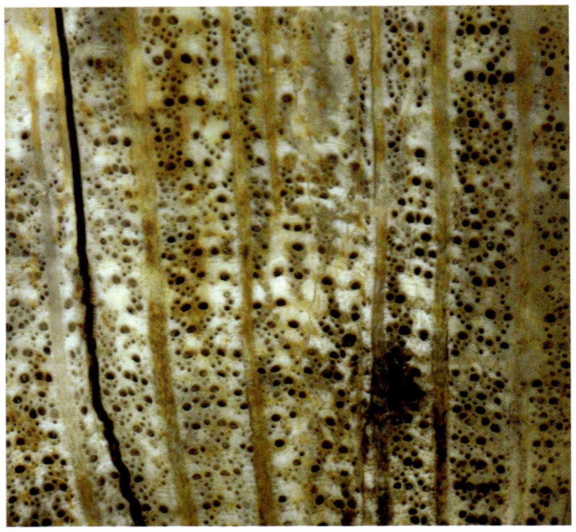

Macro photograph showing oak-like cell structure in fossil wood. Pliocene. Deschutes, Oregon, USA. Daniels specimen.

Micro image showing oak-like cell structure in fossil wood. Pliocene. Deschutes, Oregon, USA. Daniels specimen.

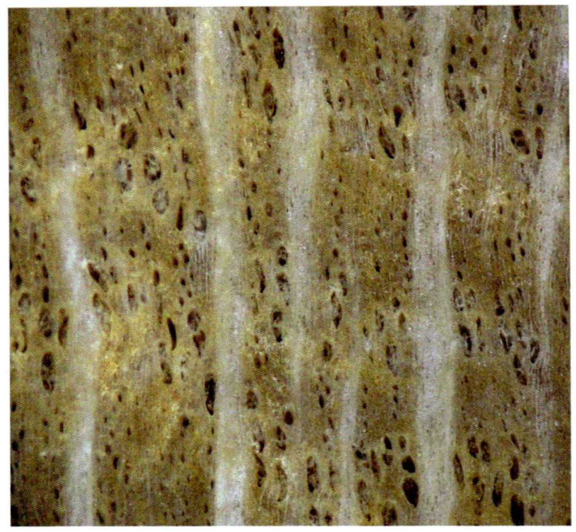

Micro image showing oak-like cell structure in fossil wood. Miocene. Stinking Water, Oregon, USA. Daniels specimen.

Macro photograph showing oak-like cell structure in fossil wood. Pliocene. Deschutes, Oregon, USA. Daniels specimen.

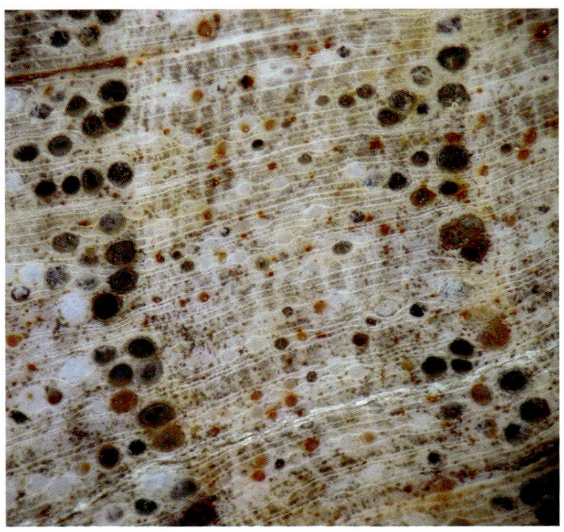

Micro image showing oak-like cell structure in fossil wood. Pliocene. Deschutes, Oregon, USA. Daniels specimen.

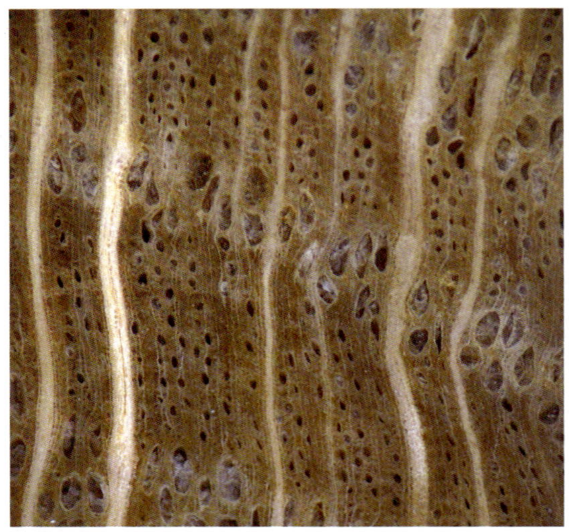

Micro image showing oak-like cell structure in fossil wood. Miocene. Stinking Water, Oregon, USA. Daniels specimen.

Double-hearted oak-like fossil wood. Pliocene. Deschutes, Oregon, USA. Daniels specimen, 20 cm.

Micro images showing oak-like cell structure. Pliocene. Deschutes, Oregon, USA. Daniels specimen. Width of growth rings approximately 0.8 to 0.9 mm.

Oak-like fossil wood. Tertiary. Swartz Canyon, Oregon, USA. Daniels specimen, 23 cm.

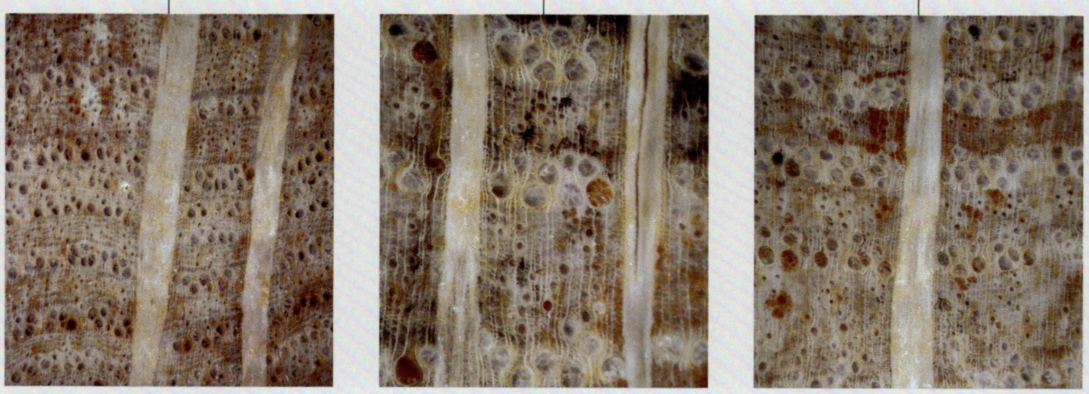

Micro images showing oak-like cell structure. Tertiary. Swartz Canyon, Oregon, USA. Daniels specimens. Average width of rays approximately 0.5 mm.

Elm-like: Elm and hackberry are ring porous woods with distinguishing tangential wavy bands of pores throughout the latewood. In American elm and winged elm, the earlywood band of pores is one pore wide. This earlywood band is visible to the unaided eye in American elm but is not visible without a hand lens for other woods in this category. Transition to latewood is abrupt. In slippery elm, the earlywood band of pores is more than one and as many as five pores wide. Slippery elm and hackberry are difficult to distinguish. The largest rays in hackberry are always at least 7 seriate and can be 10 or 12 seriate. In slippery elm, the largest rays are generally 4 or 5 seriate and are never more than 7 seriate.

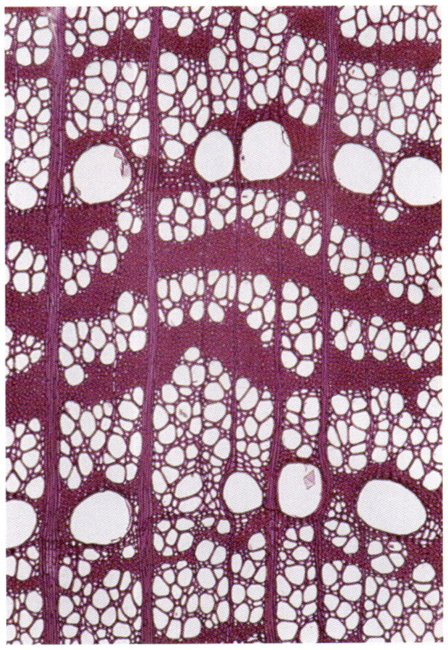

Ulmus americana (American elm), magnification x 40.

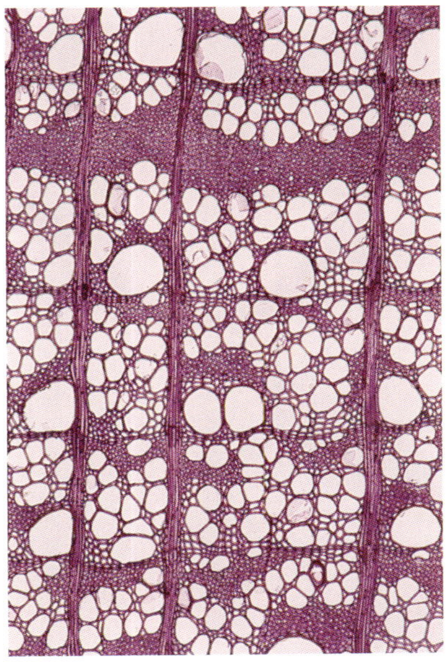

Ulmus thomasii (rock elm), magnification x 40.

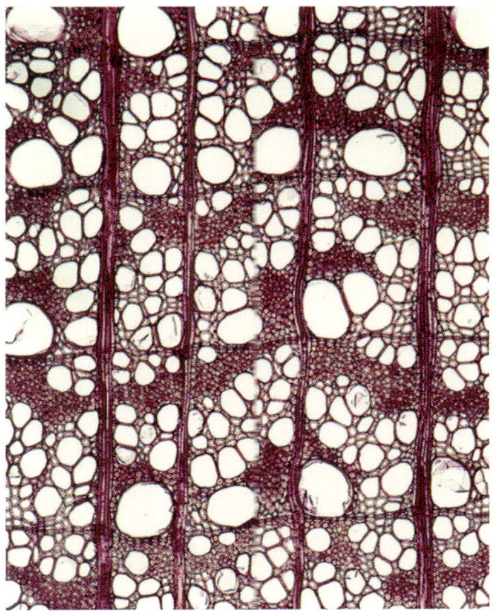

Ulmus rubra (slippery elm), magnification x 40.

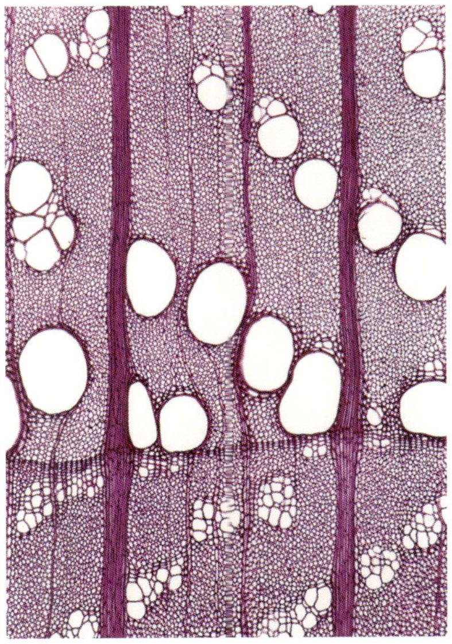

Celtis occidentalis (hackberry), magnification x 40.

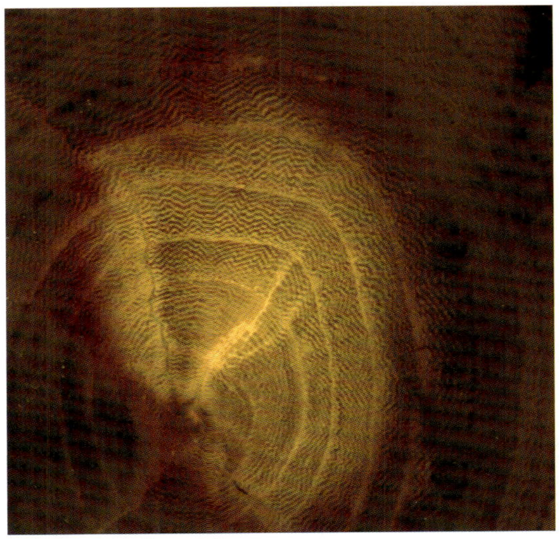

Macro photograph showing elm-like cell structure in fossil wood. Miocene. Hoodoo Basin, Oregon, USA. Daniels specimen.

Macro photograph showing elm-like cell structure in fossil wood. Miocene. Grassy Mountain, Oregon, USA. Daniels specimen.

Macro photograph showing elm-like cell structure in fossil wood. Miocene. Grassy Mountain, Oregon, USA. Daniels specimen.

Macro photograph showing ulmiform pore arrangement. Columbia River Basalt, Miocene. Maryhill, Washington, USA. Daniels specimen.

Elm-like fossil wood. Columbia River Basalt, Miocene. Maryhill, Washington, USA. Daniels specimen, 40.5 cm.

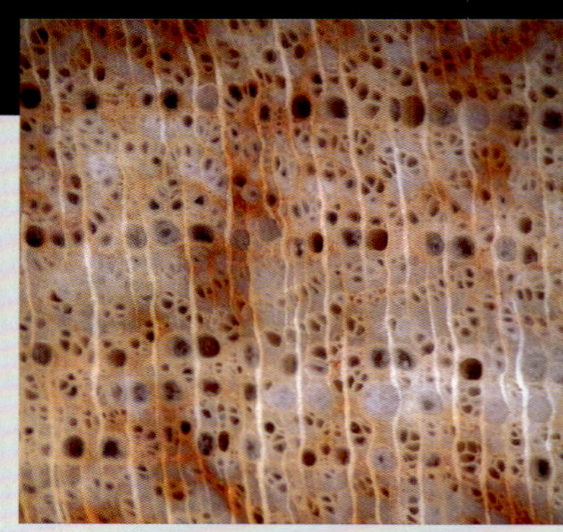

Micro images showing ulmiform pore arrangement. Columbia River Basalt. Miocene. Maryhill, Washington, USA. Daniels specimens. Largest pores approximately 0.2 mm.

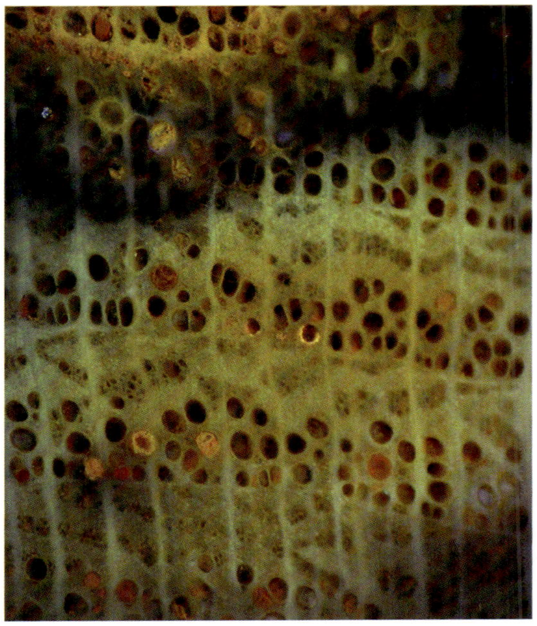

Micro image showing elm-like cell structure in fossil wood. Miocene. Grassy Mountain, Oregon, USA. Daniels specimen.

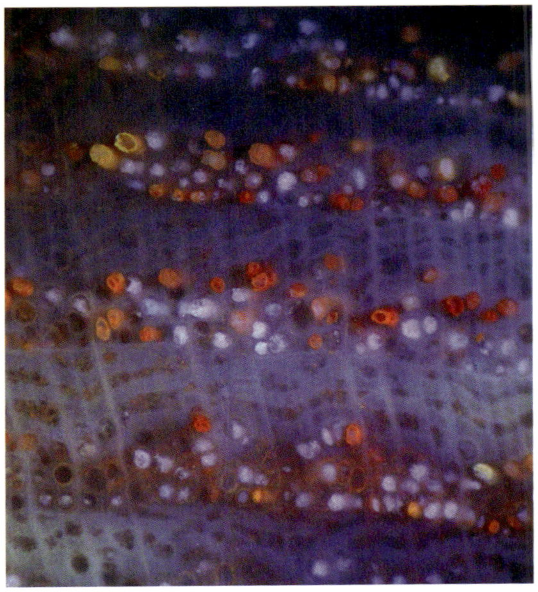

Micro image showing elm-like cell structure in fossil wood. Miocene. Grassy Mountain, Oregon, USA. Daniels specimen.

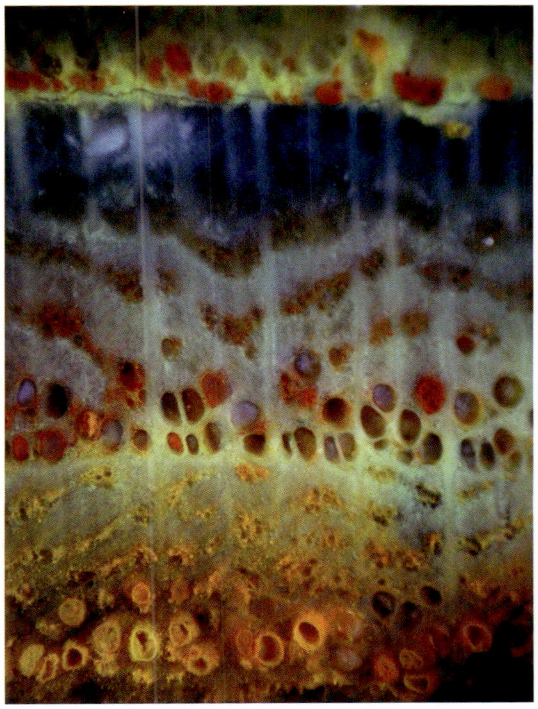

Micro image showing elm-like cell structure in fossil wood. Miocene. Grassy Mountain, Oregon, USA. Daniels specimen.

Hickory-like: Hickory and pecan (genus *Carya*) are ring porous woods with banded parenchyma cells in long, straight, somewhat continuous tangential lines. In combination with the thin rays, these bands of parenchyma form a net-like or web-like pattern. Earlywood pores are large in an intermittent row along the growth boundary. Latewood pores are few and solitary or in radial multiples of two to several. No parenchyma cells exist in the earlywood of hickory, whereas

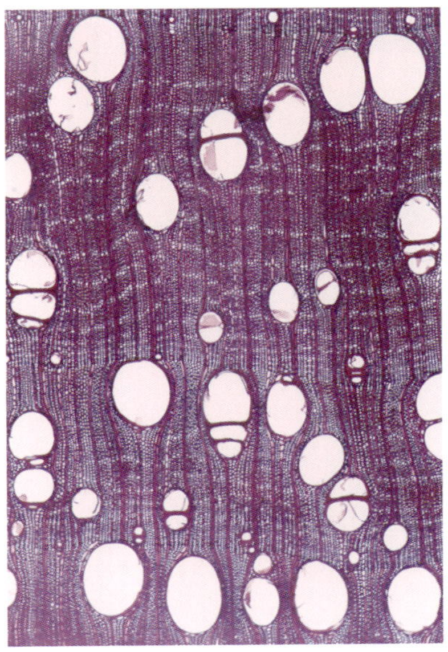

Carya illinoinensis (pecan), magnification x 40.

parenchyma cells are present in the earlywood of pecan. Hickory earlywood is one pore wide with an abrupt shift to latewood. Pecan displays a gradual decrease in pore size from large earlywood pores and could, therefore, be grouped with semi ring porous woods.

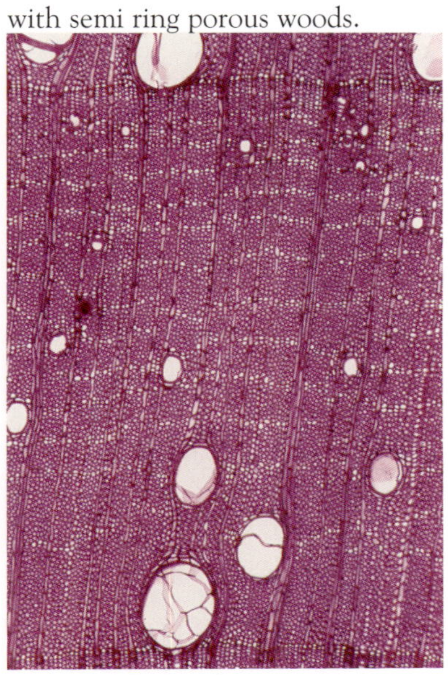

Carya cordiformis (bitternut hickory), magnification x 40.

Macro photograph showing hickory-like cell structure. Pliocene. Deschutes, Oregon, USA. Daniels specimen.

Hickory-like fossil wood. Columbia River Basalt, Miocene. Saddle Mountain, Washington, USA. Daniels specimen, 33.5 cm.

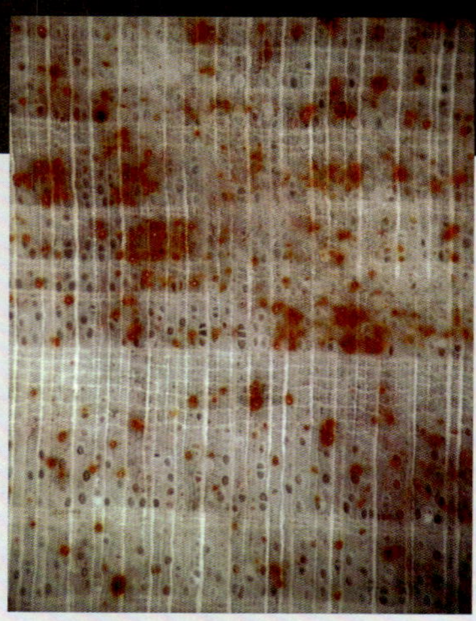

Micro images showing hickory-like cell structure. Columbia River Basalt, Miocene. Saddle Mountain, Washington, USA. Daniels specimen. Growth rings range in size from 0.4 to 1.6 mm.

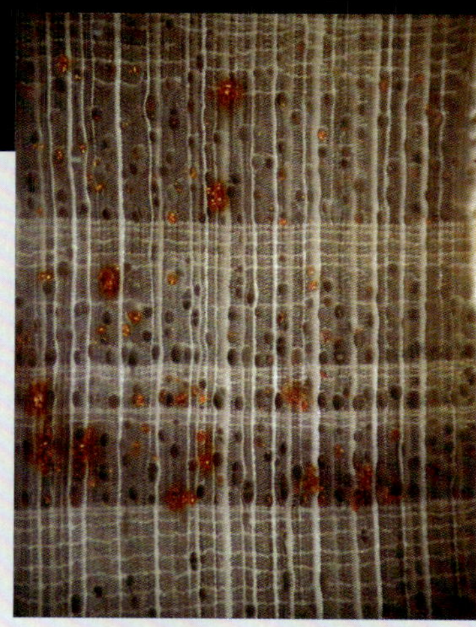

The numerous remaining ring porous woods are more difficult to identify individually and will require further study and the use of microscopic thin sections. Other ring porous woods include ash, black locust, honey locust, Osage orange, red mulberry, and sassafras.

Hickory-like fossil wood. Pliocene. Deschutes, Oregon, USA. Daniels specimen, 18.5 cm.

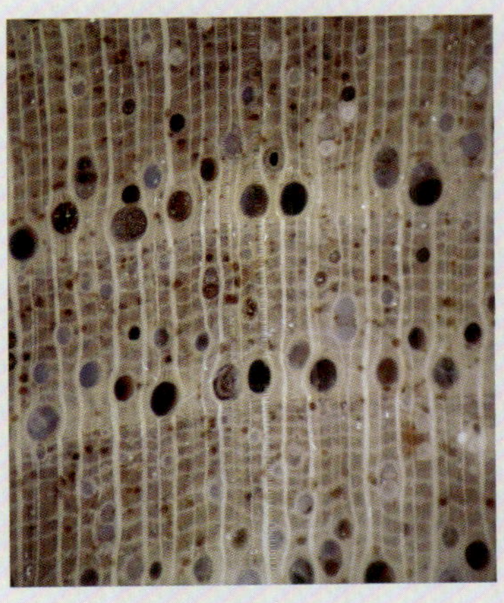

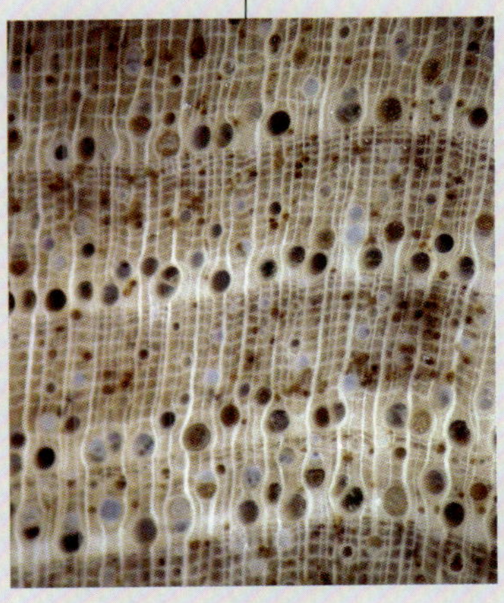

Micro images showing hickory-like cell structure. Pliocene. Deschutes, Oregon, USA. Daniels specimen. Growth rings range from 0.6 to 1.2 mm.

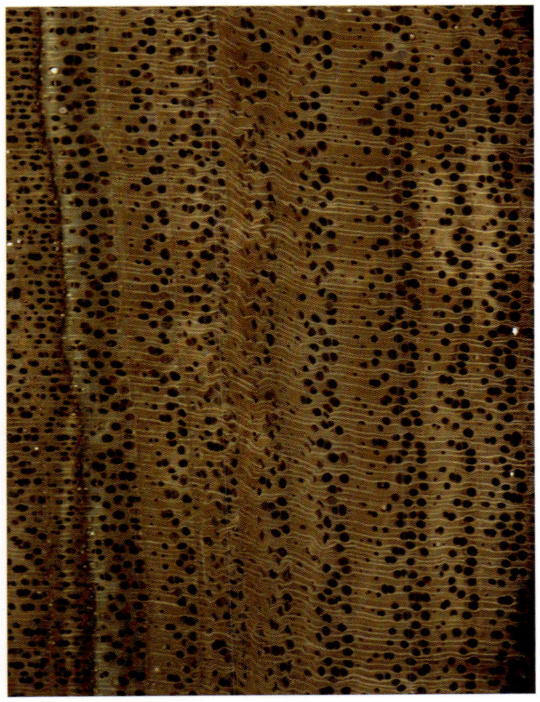

Fossil wood showing ring porous cell structure. Miocene. McDermitt, Oregon, USA. Daniels specimen.

Fossil wood showing ring porous cell structure. Oligocene/Miocene. Sweet Home, Oregon, USA. Daniels specimen.

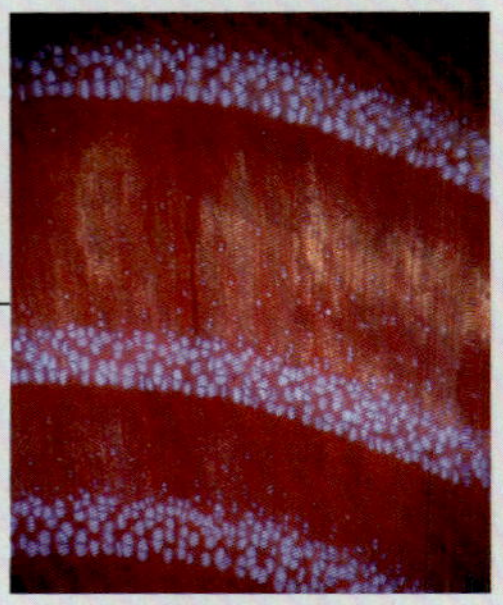

Micro image showing ring porous cell structure. Miocene. Succor Creek, Oregon/ Idaho, USA. Daniels specimen. Largest growth ring approximately 3 mm.

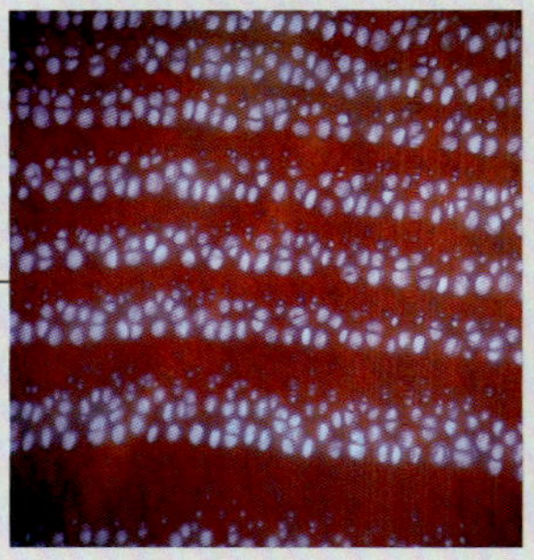

Micro image showing ring porous cell structure. Miocene. Succor Creek, Oregon/Idaho, USA. Daniels specimen. Largest pores approximately 0.2 mm.

Ring porous fossil hardwood. Miocene. Succor Creek, Oregon/Idaho, USA. Daniels specimen, 7.5 cm.

Fossil wood showing ring porous cell structure. Miocene. Grassy Mountain, Oregon, USA. Daniels specimen.

Fossil wood showing ring porous cell structure. Miocene. Grassy Mountain, Oregon, USA. Daniels specimen.

Fossil wood showing ring porous cell structure. Miocene. Vantage, Washington, USA. Daniels specimen.

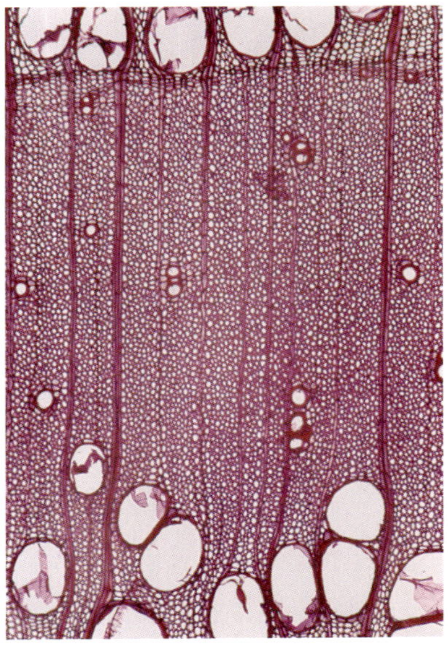

Fraxinus americana (white ash), magnification x 40.

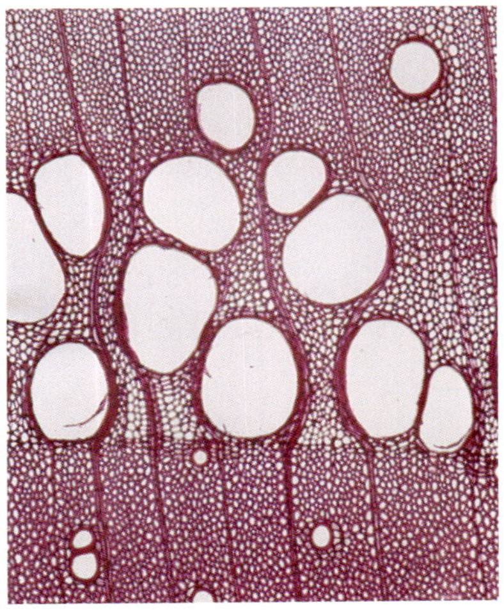

Fraxinus nigra (black ash), magnification x 40.

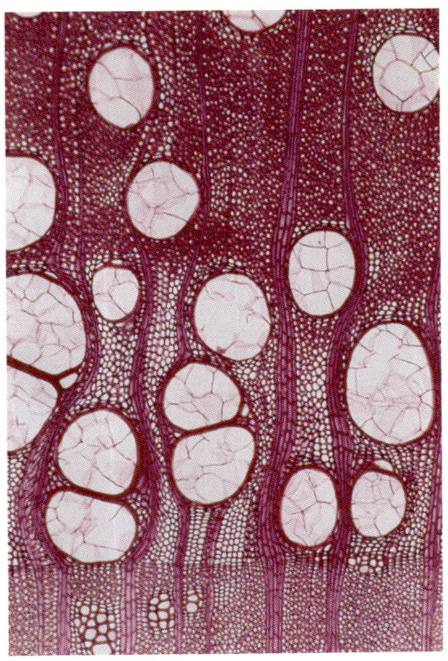

Robinia pseudoacacia (black locust), magnification x 40.

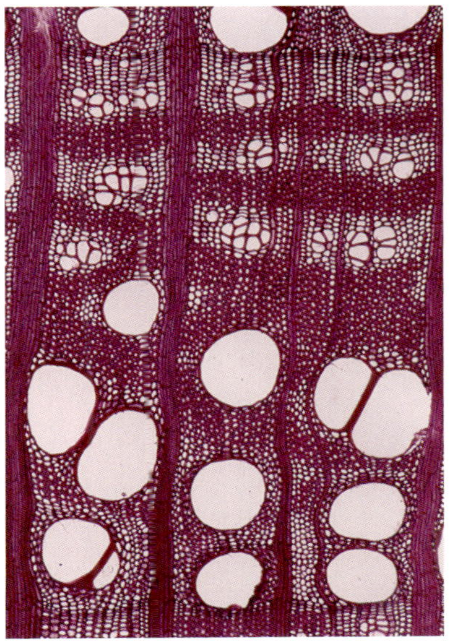

Gleditsia triacanthos (honey locust), magnification x 40.

Semi Ring Porous Hardwoods

The pores of semi ring porous wood decrease gradually in size from large to significantly smaller with no clear differentiation between earlywood and latewood and no abrupt change in size. Examples of semi ring porous woods are black walnut, butternut, hickory, live oak, persimmon, tanoak, and inconsistently aspen, black cherry, and black willow.

Live oak-like and tanoak-like: Live oak (a true oak) and tanoak (not a true oak) are semi ring porous woods with relatively large rays, separated by many smaller, mostly uniseriate rays. In both tanoak and live oak, the growth rings are indistinct with little change in pore size. In live oak, the pores appear in an irregular radial arrangement. Tanoak pores are visible in a wavering radial arrangement. You can differentiate tanoak from live oak by the irregular distinct tangential lines of banded parenchyma cells arranged between radially distributed tanoak pores.

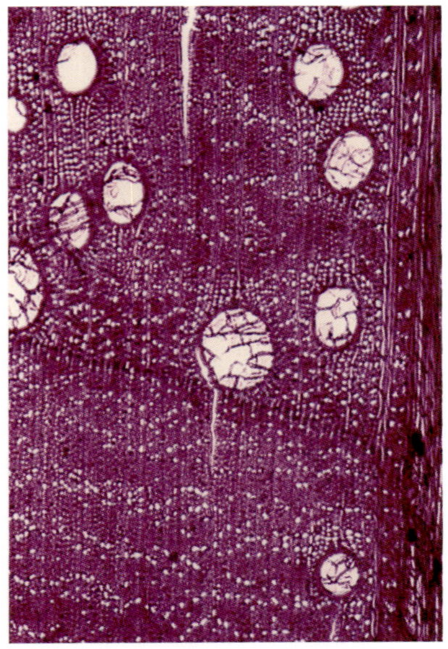

Quercus virginiana (live oak), magnification x 40.

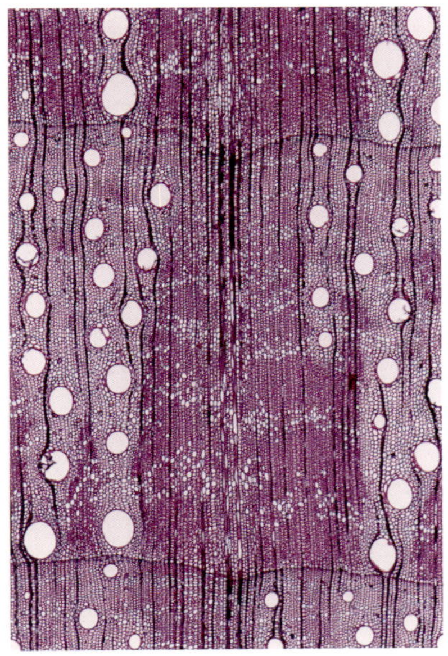

Lithocarpus densiflorus (tanoak), magnification x 21.

Live oak-like fossil wood. Catahoula Formation, Oligocene. Louisiana, USA. Daniels specimen, 19.5 cm.

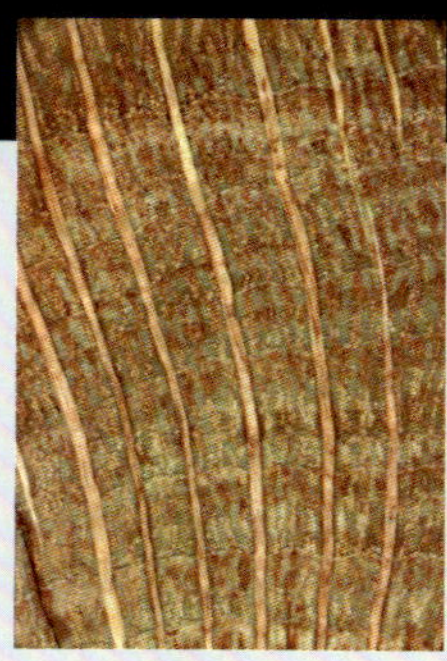

Macro photograph showing live oak-like cell structure. Catahoula Formation, Oligocene. Louisiana, USA. Daniels specimen. Widest ray approximately 0.45 mm.

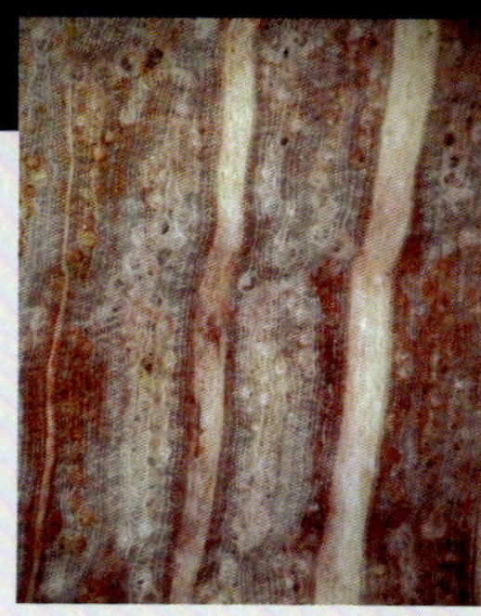

Micro image showing live oak-like cell structure. Catahoula Formation, Oligocene. Louisiana, USA. Daniels specimen. Widest ray approximately 0.45 mm.

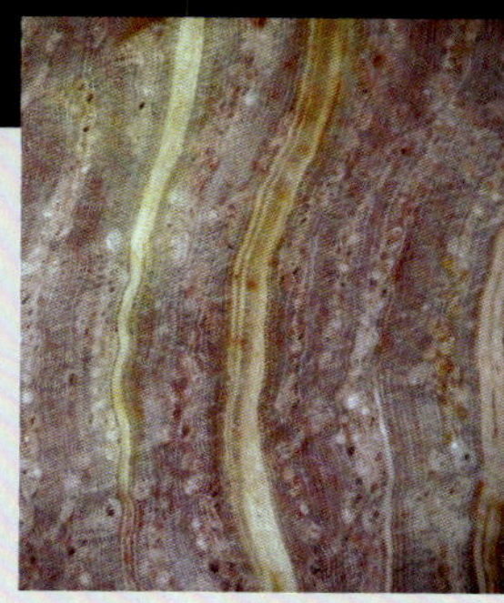

Micro image showing live oak-like cell structure. Catahoula Formation, Oligocene. Louisiana, USA. Daniels specimen. Widest ray approximately 0.45 mm.

Walnut-like: Two forms of the genus *Juglans*, walnut and butternut, are semi ring porous woods with fairly large first pores in earlywood. Pore size decreases gradually into latewood. Pores exhibit a marked decrease in size from the outer latewood of one ring to the earlywood of the next. Pores are solitary and in radial multiples of two to several. Tyloses are common, and growth rings are distinct. Rays are fine and indistinct to the unaided eye. Parenchyma cells are banded or arranged in short, thin tangential, often wavy, lines of diffuse-in-aggregate.

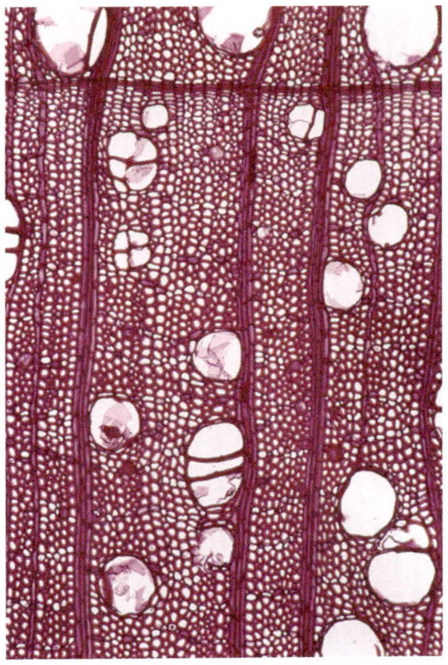

Juglans nigra (black walnut), magnification x 40.

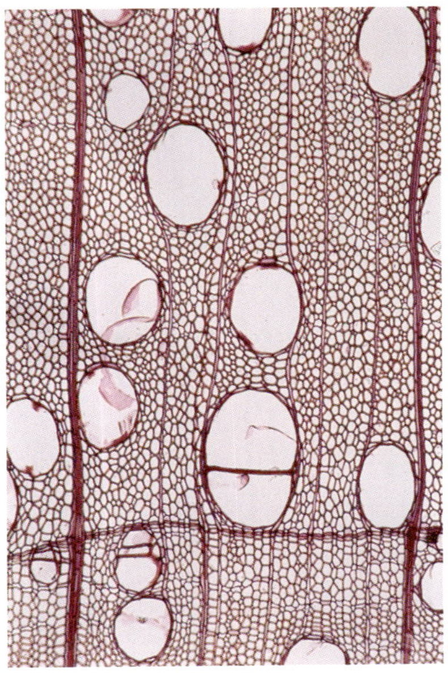

Juglans cinerea (butternut), magnification x 40.

Walnut-like fossil wood. Columbia River Basalt, Miocene. Vantage, Washington, USA. Daniels specimen, 23.5 cm.

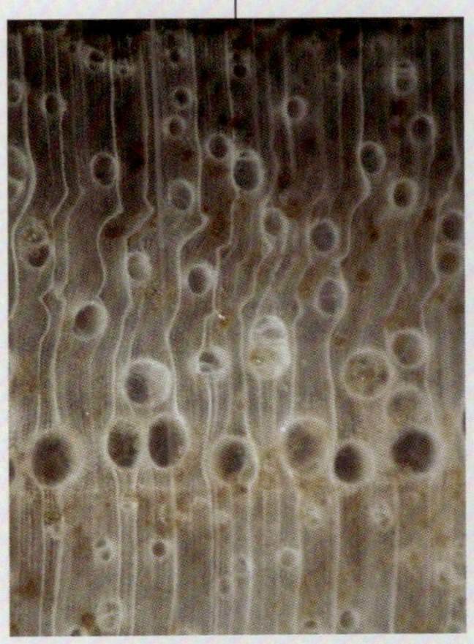

Micro image showing walnut-like cell structure. Columbia River Basalt, Miocene. Vantage, Washington, USA. Daniels specimen. Large pores at earlywood growth ring boundary 0.2 mm.

Micro image showing walnut-like cell structure. Columbia River Basalt, Miocene. Vantage, Washington, USA. Daniels specimen. Large pores at earlywood growth ring boundary 0.2 mm.

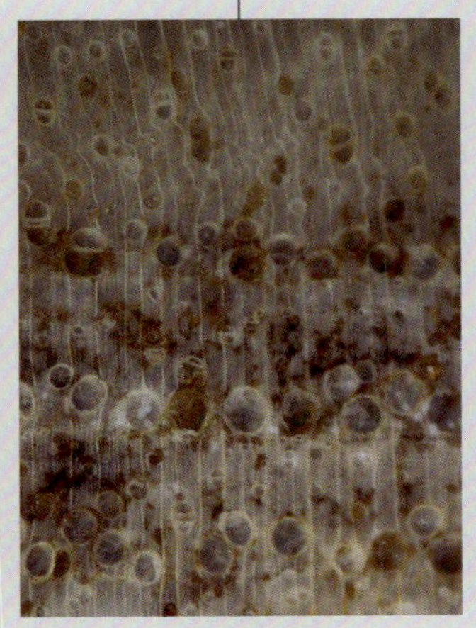

Micro image showing walnut-like cell structure. Columbia River Basalt, Miocene. Vantage, Washington, USA. Daniels specimen. Large pores at earlywood growth ring boundary 0.2 mm.

HARDWOOD IDENTIFICATION CHART

RING POROUS HARDWOODS If a hardwood is ring porous, the largest, most prominent pores reside within the earlywood. Early season pores are considerably larger than those of late season. They may form a single row or two or more rows. The transition from earlywood to latewood in a ring porous wood may be sharply defined by a clear diminution in the size of the pores or may be gradual. A wood with earlywood pores concentrated at the growth ring that are no larger than pores in latewood is diffuse porous.

	PORES	RAYS	PARENCHYMA
Oak-like	Large earlywood pores — radially arranged latewood pores.	Large multi-seriate rays, 20 seriate and larger, plus fine, mostly uniseriate rays.	
Chestnut-like	Large earlywood pores — radially arranged latewood pores.	No large rays — predominately uniseriate rays.	
Elm-like	Distinguishing tangential wavy bands of pores throughout latewood; abrupt transition to latewood. American elm and winged elm: earlywood band of pores is 1 pore wide. Sippery elm and hackberry: earlywood bands of pores is more than 1 and as many as 5 pores wide.	American elm and winged elm: 1-7 seriate; hackberry: at least 7 seriate, up to 10-12 seriate; slippery elm: largest 4-5 seriate, never more than 7 seriate.	
Hickory-like	Earlywood pores large in an intermittent row along the growth ring boundary; latewood pores few, solitary or in radial multiples of 2 to several — abrupt shift to latewood.	1-4 seriate.	Banded in long, straight, somewhat continuous tangential lines. In combination with the thin rays, these bands of parenchyma form a net-like or web-like pattern — no parenchyma in earlywood.
Pecan-like	Earlywood pores large in an intermittent row along the growth ring boundary; latewood pores few, solitary or in radial multiples of 2 to several; gradual decrease in pore size from large earlywood pores.	1-4 seriate.	Banded in long, straight, somewhat continuous tangential lines. In combination with the thin rays, these bands of parenchyma form a net-like or web-like pattern — parenchyma present in earlywood.

SEMI RING POROUS HARDWOODS The pores of semi ring porous wood decrease gradually in size from large to significantly smaller with no clear differentiation between earlywood and latewood and no abrupt change in size. Examples of semi ring porous woods are black walnut, butternut, hickory, live oak, persimmon, tanoak, and inconsistently aspen, black cherry, and black willow.

Live oak-like	Growth rings indistinct with little change in pore size — pores in an irregular radial arrangement.	Large rays separated by many smaller, mostly uniseriate rays.	Broken, irregular lines of banded parenchyma in latewood, barely visible with hand lens.
Tanoak-like	Growth rings indistinct with little change in pore size — pores in wavering radial arrangement.	Large rays separated by many smaller, mostly uniseriate rays.	Ragged, distinct tangential lines of banded parenchyma arranged between radially distributed pores — visible to unaided eye.
Walnut-like	Fairly large first pores in earlywood, decreasing gradually into latewood — pores solitary and in radial multiples of 2 to several — distinct growth rings.	Fine and indistinct to the unaided eye; 1-5 seriate in walnut; 1-4 seriate in butternut.	Banded or arranged in short, thin, often wavy lines of diffuse-in-aggregate.

DIFFUSE POROUS HARDWOODS In diffuse porous woods, pores are uniform in size and evenly distributed from earlywood to latewood. Diffuse porous woods include American holly, aspen, basswood (linden), beech, birch, black cherry, blackgum, black willow, California laurel (Pacific myrtle), cottonwood, dogwood, live oak, magnolia, maple, red alder, sweetgum, sycamore, tanoak, and yellow poplar (tulip tree).

Sycamore-like	Small and numerous; solitary or in multiples that are both radial and tangential, merging into clusters on some specimens. Pore size and distribution uniform from earlywood up to the growth ring where they decrease significantly in both size and number, resulting in a well-defined latewood boundary.	Numerous, large, clearly noded rays, nearly uniform in size, up to about 14 seriate.	
Beech-like	Small and numerous; solitary or in multiples that are both radial and tangential, merging into clusters on some specimens. Pore size and distribution uniform from earlywood up to the growth ring where they decrease significantly in both size and number, resulting in a well-defined latewood boundary.	Numerous, large, clearly noded rays — rays of two distinct sizes with narrow rays 1-5 seriate separating wide rays 15-25 seriate.	

Diffuse Porous Hardwoods

In diffuse porous woods, pores are uniform in size and evenly distributed from earlywood to latewood. Diffuse porous woods include American holly, aspen, basswood (linden), beech, birch, black cherry, blackgum, black willow, California laurel (Pacific myrtle), cottonwood, dogwood, live oak, magnolia, maple, red alder, sweetgum, sycamore, tanoak, and yellow poplar (tulip tree).

Sycamore-like and beech-like: Both sycamore and beech have numerous, large, clearly noded rays. Pores are numerous and solitary or in
multiples that are both radial and tangential, merging into clusters in some specimens. Pore size and distribution is uniform from earlywood into latewood and continues to the growth ring margin where they decrease significantly in both size and number, resulting in a well-defined latewood boundary. Sycamore pores are small, typically crowded, and solitary in multiples of 2 to 5. Pores in beech are scattered. Sycamore rays are nearly uniform in size. In beech, the largest rays are larger than the largest sycamore rays and are more widely and randomly distributed. The uneven ray size in beech is apparent in transverse section. Rays in sycamore are up to about 14 seriate. In beech, rays are of two distinct sizes with the narrow rays 1 to 5 seriate separating wide rays 15 to 25 seriate.

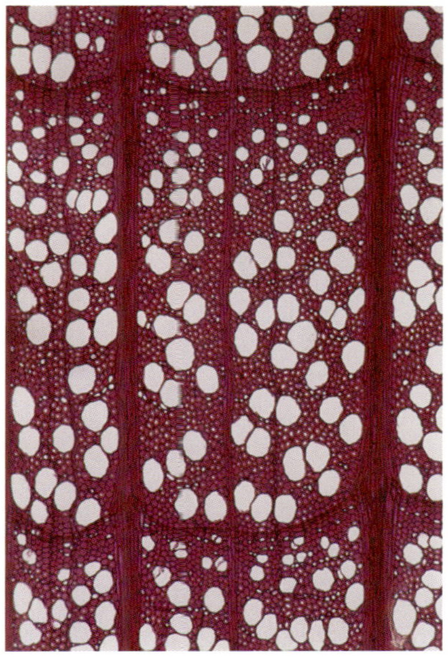

Fagus grandifolia (American beech), magnification x 40.

Macro photograph showing sycamore-like cell structure in fossil wood. Eocene. Post, Oregon, USA. Dietz specimen.

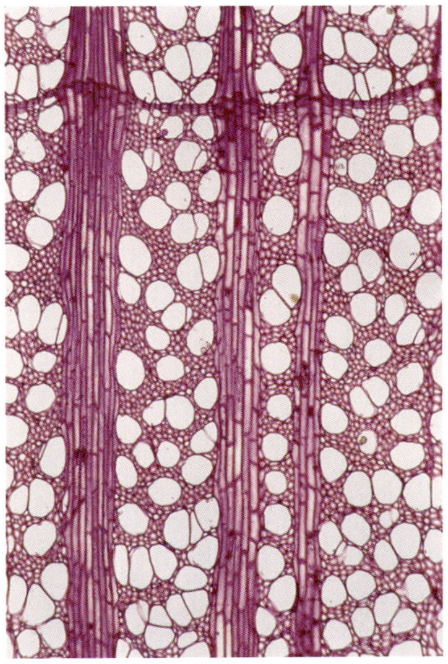

Platanus occidentalis (American sycamore), magnification x 40.

Macro photograph showing sycamore-like cell structure in fossil wood. Eocene. Douglas County, Colorado, USA. Daniels specimen.

Macro photograph showing beech-like cell structure in fossil wood. Eocene. Grassy Mountain, Oregon, USA. Daniels specimen.

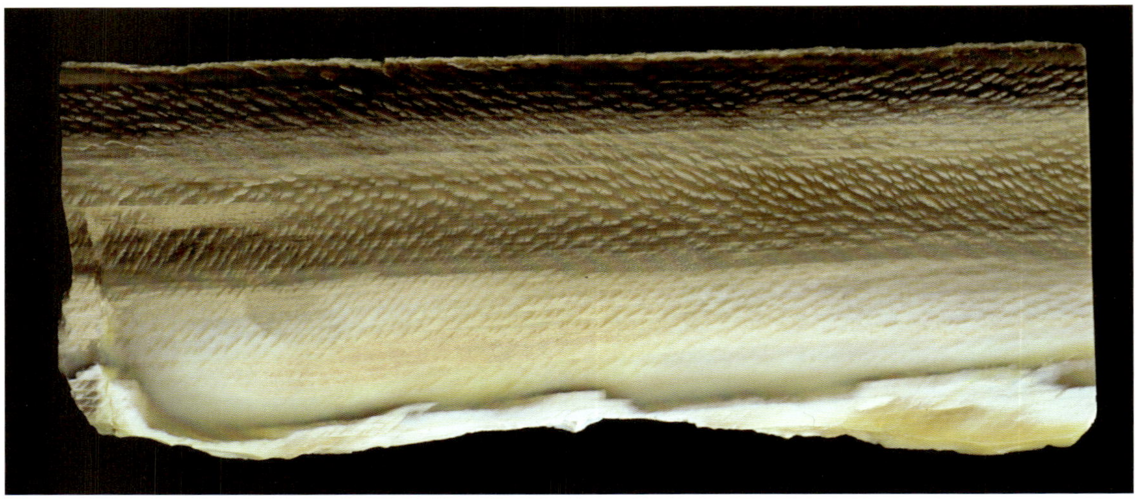

Sycamore-like ray fleck in fossil wood radial section. Columbia River Basalt, Miocene. Saddle Mountain, Washington, USA. Daniels specimen, 19.5 cm.

Micro image showing sycamore-like cell structure in fossil wood. Eocene. Post, Oregon, USA. Dietz specimen.

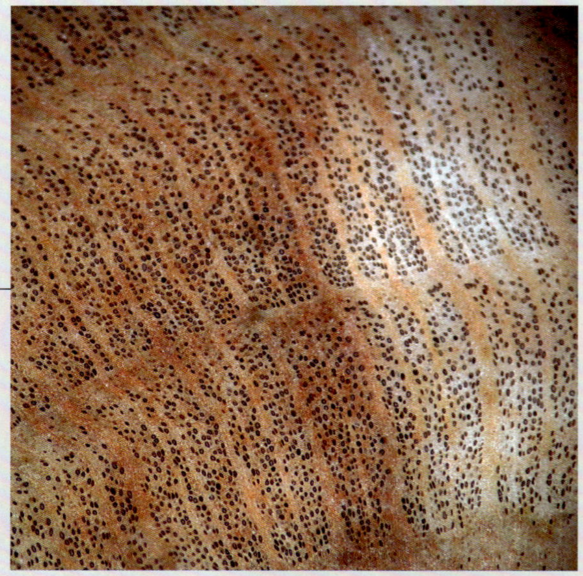

Micro image showing sycamore-like cell or beech-like structure. Little Butte Volcanic Series, Oligocene/Miocene. Sweet Home, Oregon, USA. Daniels specimen. Growth rings approximately 2.2 mm.

Sycamore-like or beech-like fossil wood. Little Butte Volcanic Series, Oligocene/Miocene. Sweet Home, Oregon, USA. Daniels specimen, 7.5 cm.

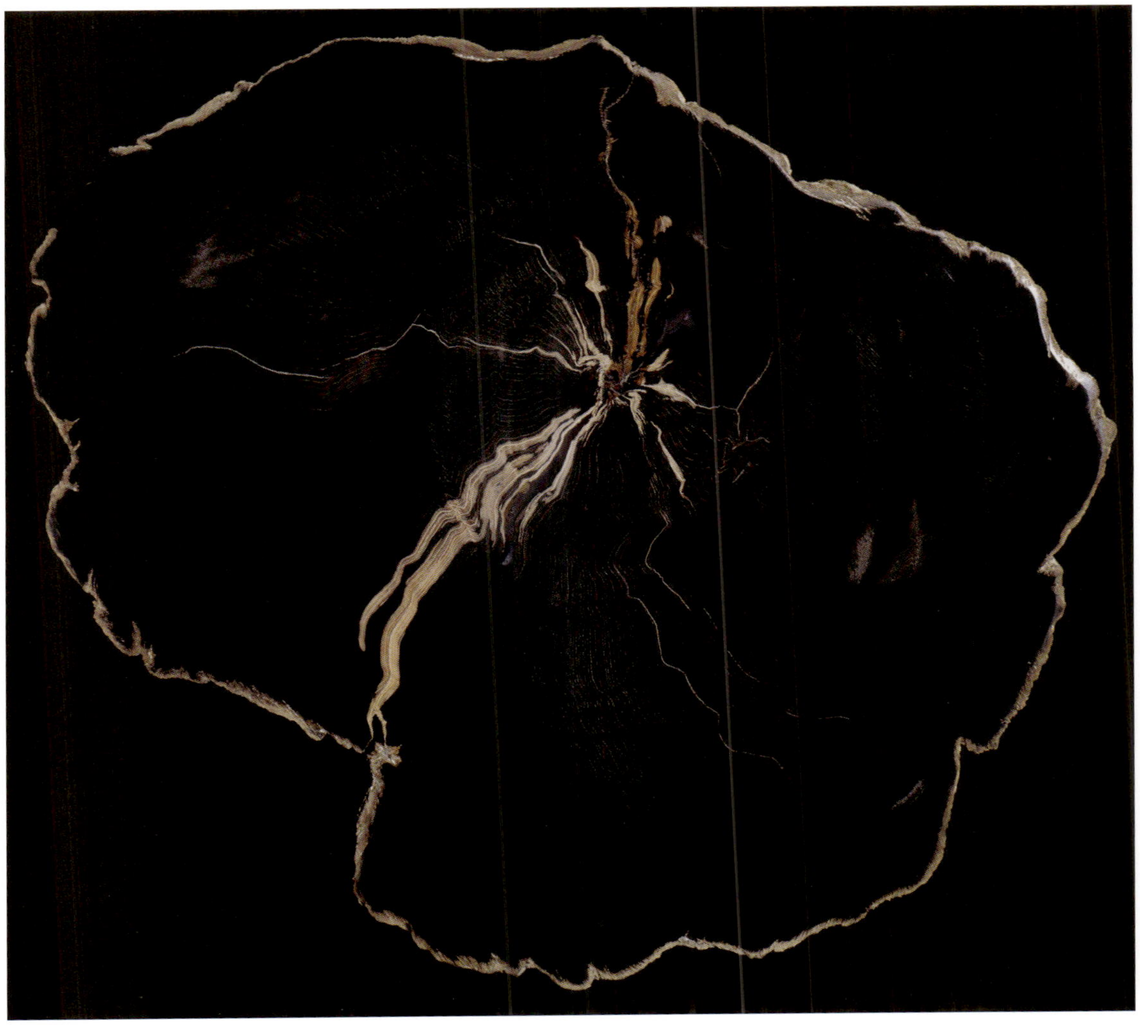

Sycamore-like fossil wood. Green River Formation, Eocene. Flaming Gorge, Utah, USA. Albers specimen, 22 cm.

Most of the other diffuse porous hardwoods have too few distinguishing characteristics to allow definitive identification of a polished transverse section with a hand lens. This group includes alder, aspen, basswood, birch, black cherry, California laurel, cottonwood, dogwood, holly, magnolia, maple, sweetgum, willow, and yellow poplar. Among these woods are differences and similarities. Some of the differences may assist individual identification. See Appendix A for a discussion of factors that may aid further discrimination between diffuse porous hardwoods.

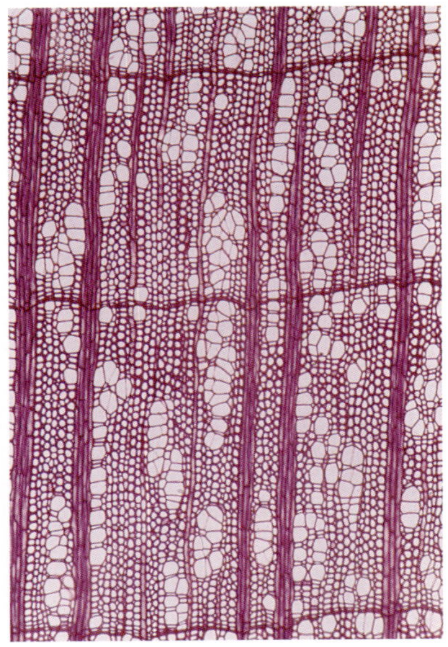

Ilex opaca (American holly), magnification × 40.

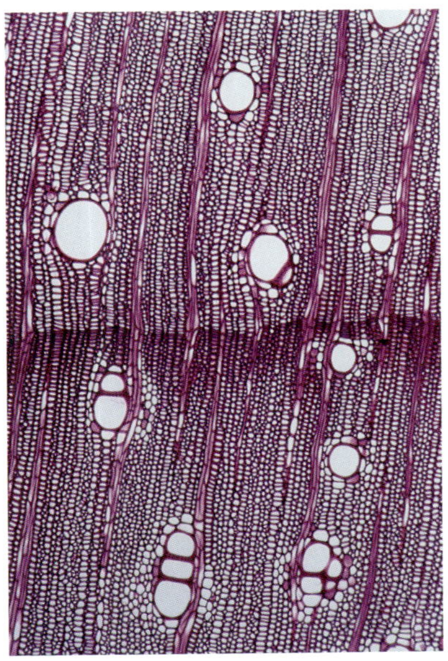

Umbellularia californica (California laurel), magnification × 40.

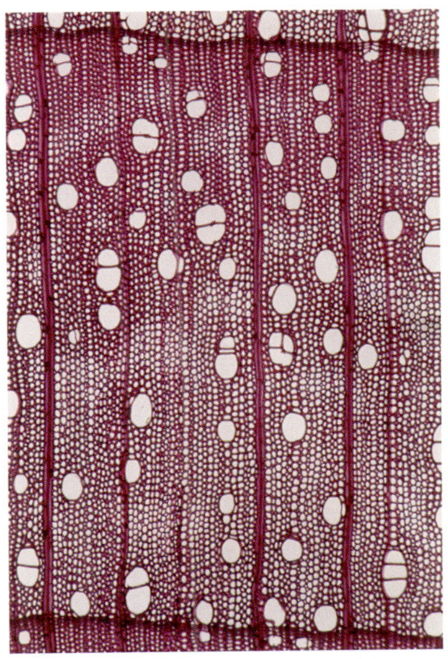

Acer saccharum (sugar maple), magnification × 40.

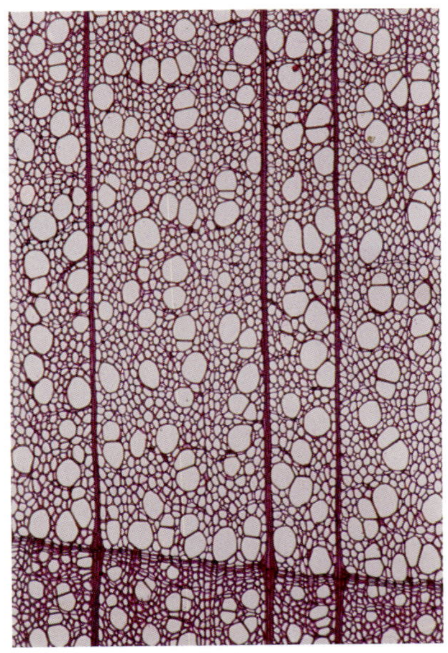

Tilia americana (American basswood), magnification × 40.

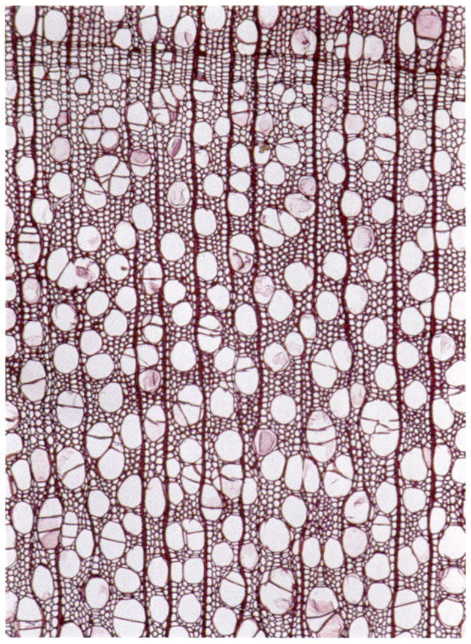

Salix nigra (black willow), magnification × 40.

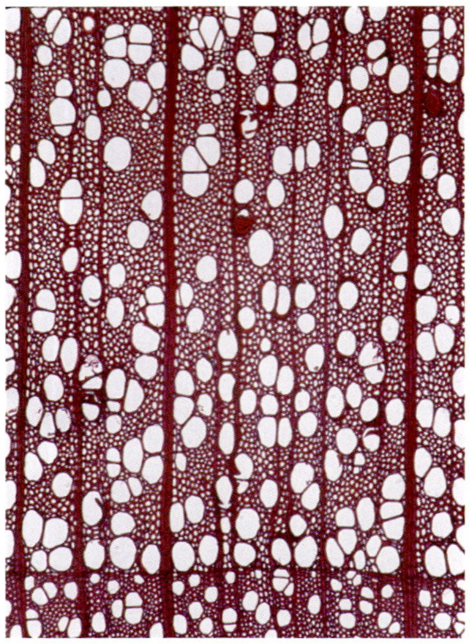

Prunus serotina (black cherry), magnification × 40.

Macro photograph showing diffuse porous cell structure in fossil wood. Miocene. Grassy Mountain, Oregon, USA. Daniels specimen.

Liquidambar styraciflua (sweet-gum), magnification × 40.

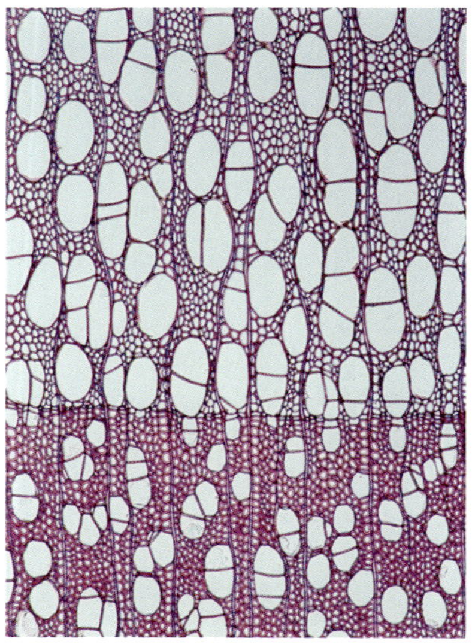

Populus deltoides (eastern cotton-wood), magnification × 40.

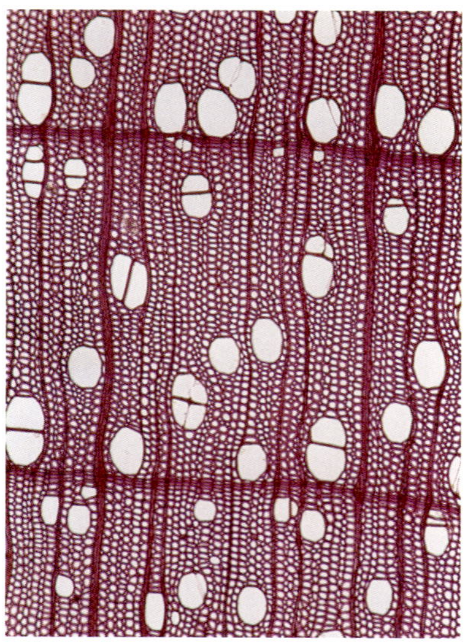

Betula alleghaniensis (yellow birch), magnification × 40.

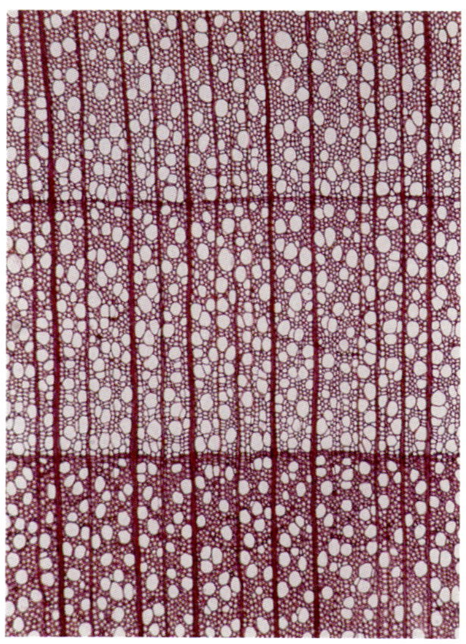

Sorbus americana (American mountain ash), magnification × 40.

Macro photograph showing diffuse porous cell structure in fossil wood. Oligocene/Miocene. Sweet Home, Oregon, USA. Daniels specimen.

Maple-like fossil wood. Columbia River Basalt, Miocene. Yakima, Washington, USA.
Daniels specimen, 24 cm.

Micro image showing maple-like cell structure. Columbia River Basalt, Miocene. Yakima, Washington, USA. Daniels specimen. Largest solitary pores approximately 0.1 mm.

Form Genus Descriptions

Form genera practice applies different names to dissociated plant structures. We use the suffix *oxylon* as a form genus descriptor when relating a fossil wood to an extant genus based on perceived similarities. This practice continues even though there may be distinct differences between the fossil genus and the living wood. Some categories for fossil conifers follow:

- *Dadoxylon*: Used for woods of Paleozoic Era conifers with similarities to extant *Araucaria*. Woods of *Dadoxylon* and *Araucarioxylon* are indistinguishable. The term *Dadoxylon* refers only to Paleozioc Era conifers. Both *Dadoxylon* and *Araucarioxylon* lack resin canals.

- *Araucarioxylon*: Mesozoic Era woods with similarities to extant *Araucaria* (and with *Dadoxylon* characteristics). No resin canals. Rays typically uniseriate but range to quadraseriate. Little or no parenchyma. Small pith.

Micro image showing *Araucarioxylon* cell structure in fossil wood. Jurassic. Santa Cruz, Argentina. Daniels specimen. Tracheids average approximately 0.0625 mm.

- *Podocarpoxylon*: Wood more similar to *Podocarpus* than to other conifers. Lacks resin canals; rays are normally uniseriate, occasionally biseriate.

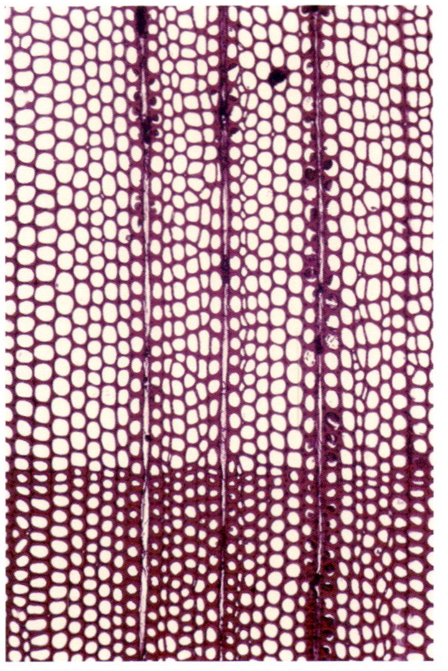

***Araucaria angustifolia* (parana pine), magnification × 40.**

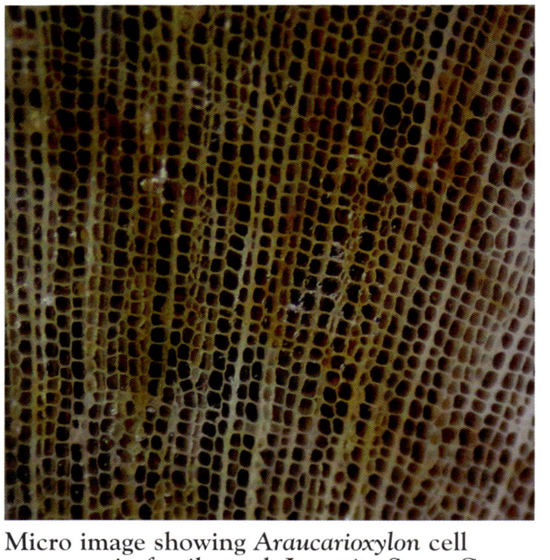

Micro image showing *Araucarioxylon* cell structure in fossil wood. Jurassic. Santa Cruz, Argentina. Daniels specimen. Tracheids average approximately 0.0625 mm.

- *Pityoxylon*: Most similar to modern Douglas fir, larch, pine, and spruce. Resin canals are present.

- *Taxodioxylon*: Fossil woods with characteristics of extant Taxodiaceae [including *Metasequoia*, *Sequoia*, *Sequoiadendron* (giant sequoia), and *Taxodium* (bald cypress)] and Cupressaceae [including cedar, cypress, and juniper]. Annual rings normally well defined with abundant vertical rows of parenchyma. Some authorities split this form genus into *Taxodioxylon* and *Cupressinoxylon*, defining *Cupressinoxylon* as wood with characteristics of extant cedar, cypress, and juniper; noting narrow, well-defined growth rings; wood parenchyma frequently with resin; no normal resin canals but possibly some traumatic resin canals.

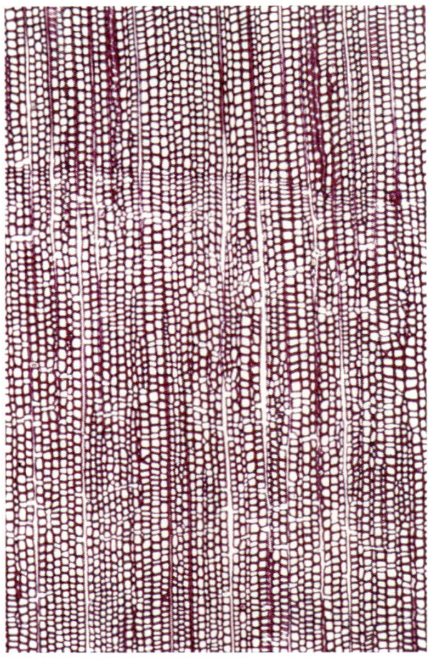

***Podocarpus salignus* (caobilla), magnification × 40.**

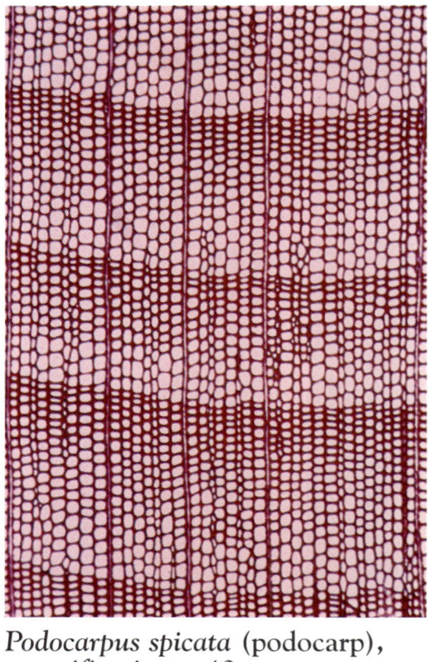

***Podocarpus spicata* (podocarp), magnification × 40.**

Identification of Remarkable and Unusual Tertiary Period Fossil Woods

Angiosperms are divided into two classes — monocotyledons (monocots) and dicotyledons (dicots). The previously discussed hardwoods are dicots. Many monocots, including palms and yuccas, are also arborescent plants. *Palmoxylon* is common in some North American Tertiary Period wood-collecting areas and rarely or never encountered in others. Oligocene *Palmoxylon* from Louisiana is among the most beautiful and perfectly preserved fossil woods that exist. Fossil *Palmoxylon* are present in Arizona, California, Colorado, New Mexico, Oregon, South Dakota, Texas, Utah, Wyoming, and such divergent locations as Argentina and Indonesia. Botanists and paleobotanists are aware of more than 2,500 species of modern palm and many extinct species of *Palmoxylon*. The appearance within the stem of separate species is frequently similar. Palm structure is quite uniform, consisting of stems without secondary growth comprising fibrovascular bundles embedded in a mass of parenchymatous ground tissue. The scattered bundles contain a small vascular segment with two to four large vessels and a typically larger fibrous bundle cap. Some species also include small bundles containing only fibers. A small *Palmoxylon* from Wyoming grew in clusters of small, jointed stems and appears similar in structure to the modern cluster palm. Local collectors frequently, yet incorrectly, refer to it as "cane." The accompanying photographs and micrographs reveal some distinct anatomical differences between palm varieties.

Palmoxylon. Green River Formation, Eocene. Tule Butte, Wyoming, USA. Albers specimen, 21 cm.

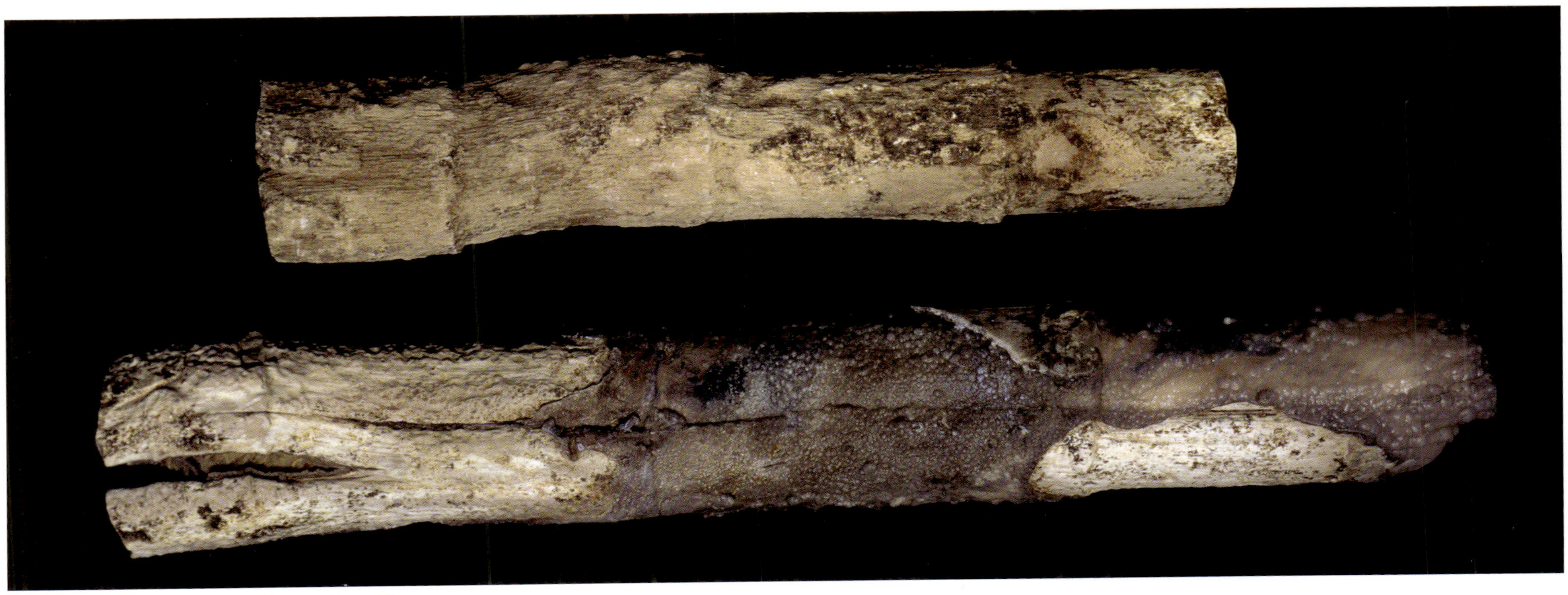

Palmoxylon.Green River Formation, Eocene. Farson, Wyoming, USA. Albers specimens, 17 and 14 cm.

Palmoxylon. Cretaceous. Rio Negro, Argentina. Daniels specimen, 10.5 cm.

Palmoxylon. Catahoula Formation, Oligocene. Louisiana, USA. Daniels specimen, 6 cm.

Palmoxylon. Green River Formation, Eocene. Farson, Wyoming, USA. Daniels specimen, 3 cm.

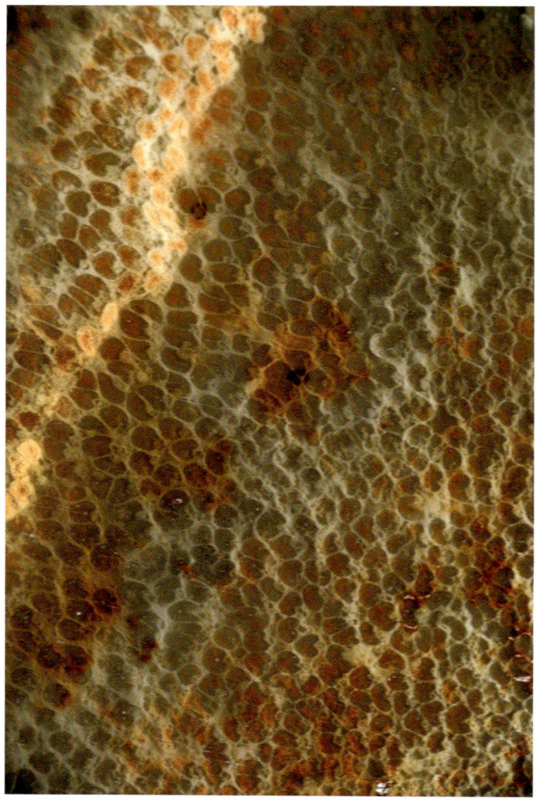

Macro photograph showing *Palmoxylon* structure in fossil wood. Cretaceous. Rio Negro, Argentina. Daniels specimen.

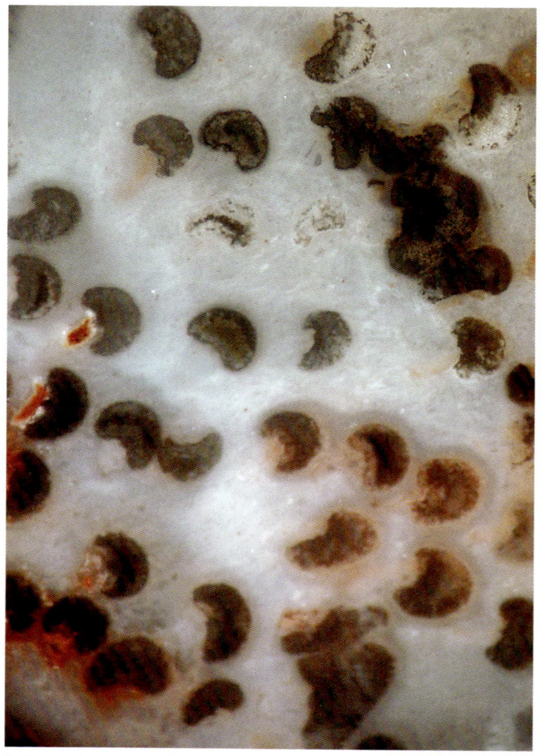

Micro image showing *Palmoxylon* cell structure in fossil wood. Cretaceous. Rio Negro, Argentina. Daniels specimen. The vascular bundles in this *Palmoxylon* are particularly small, ranging from approximately 0.3 to 0.8 mm.

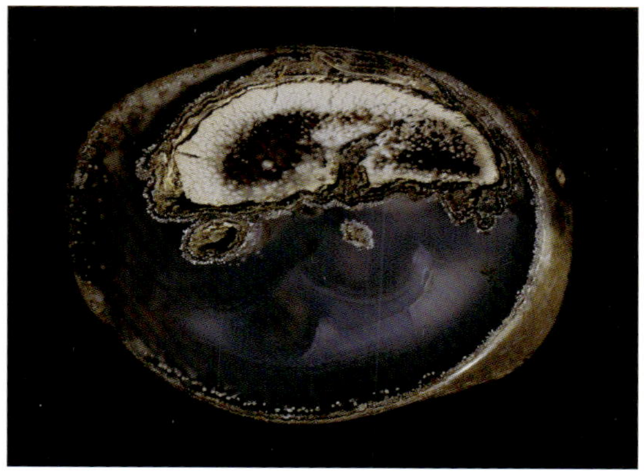

Palmoxylon. Green River Formation, Eocene. Farson, Wyoming, USA. Daniels specimen, 4 cm.

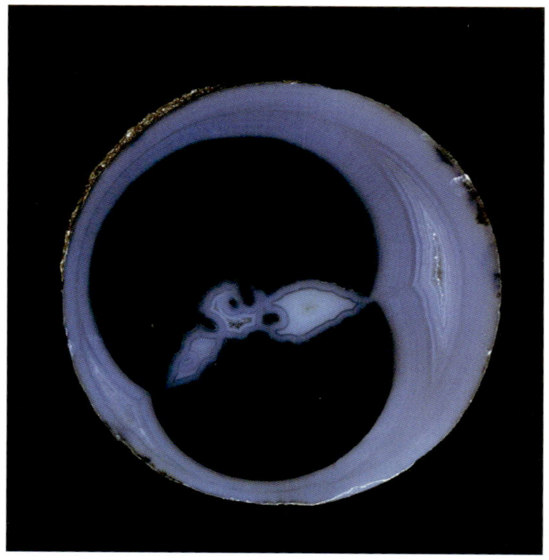

Palmoxylon. Green River Formation, Eocene. Farson, Wyoming, USA. Daniels specimen, 3.5 cm.

Palmoxylon. Green River Formation, Eocene. Farson, Wyoming, USA. Albers specimen, 4 cm.

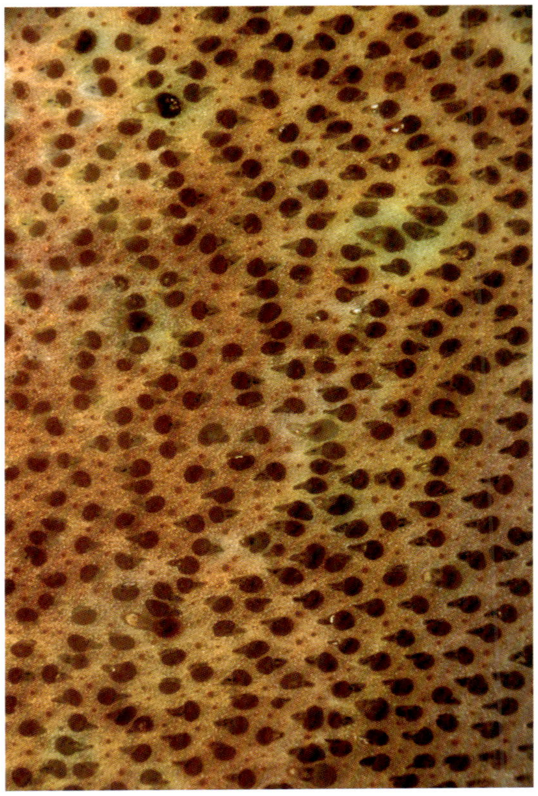

Macro photograph showing *Palmoxylon* structure in fossil wood. Catahoula Formation, Oligocene. Louisiana, USA. Daniels specimen.

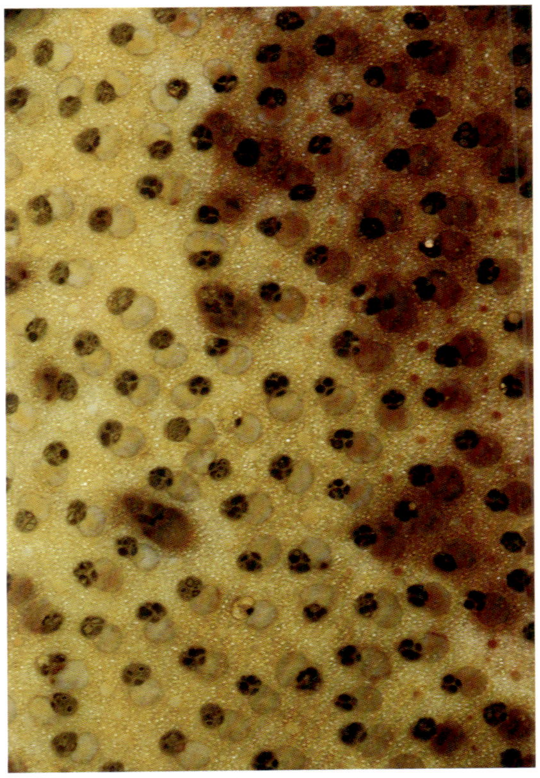

Macro photograph showing *Palmoxylon* structure in fossil wood. Catahoula Formation, Oligocene. Louisiana, USA. Daniels specimen.

Palmoxylon. Green River Formation, Eocene. Farson, Wyoming, USA. Daniels specimen, 4 cm.

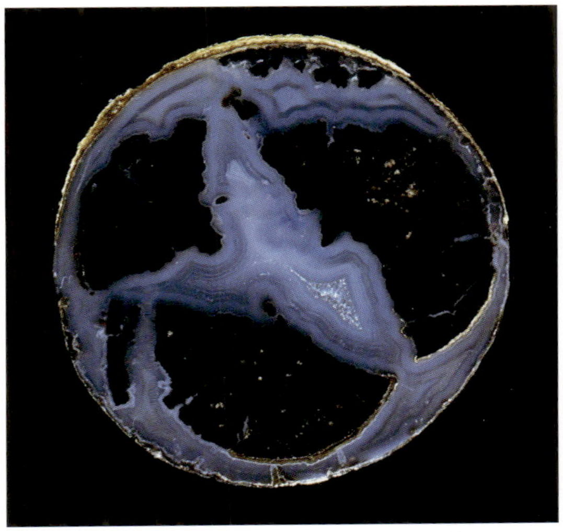

Palmoxylon. Green River Formation, Eocene. Farson, Wyoming, USA. Daniels specimen, 3.5 cm.

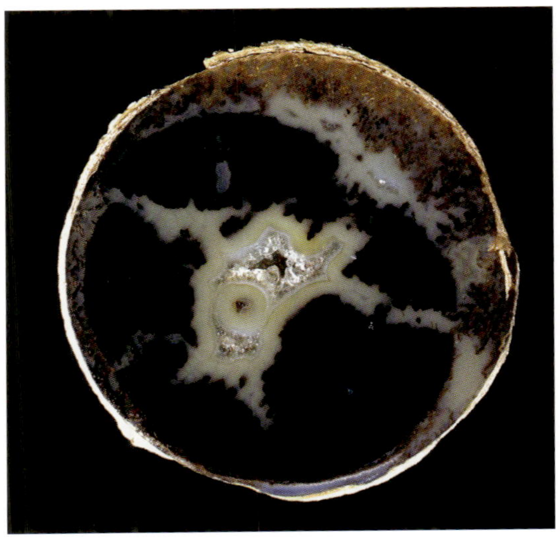

Palmoxylon. Green River Formation, Eocene. Farson, Wyoming, USA. Daniels specimen, 4 cm.

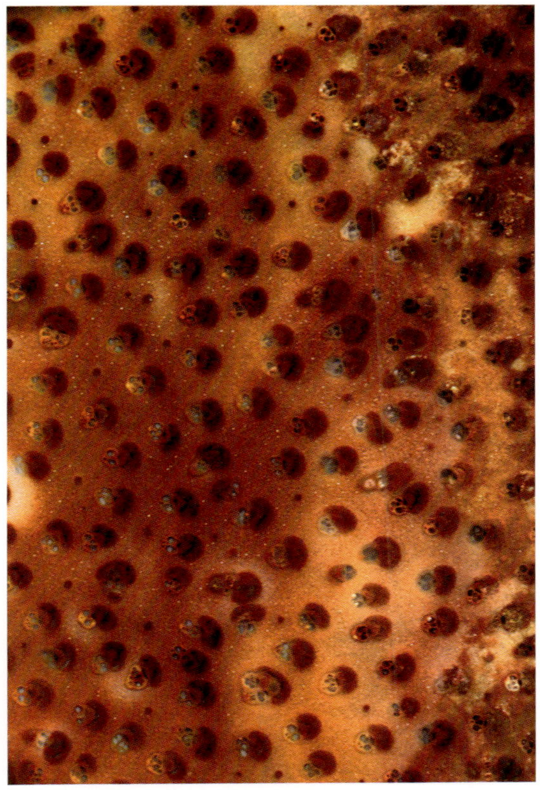

Macro photograph showing *Palmoxylon* structure in fossil wood. Catahoula Formation, Oligocene. Louisiana, USA. Daniels specimen.

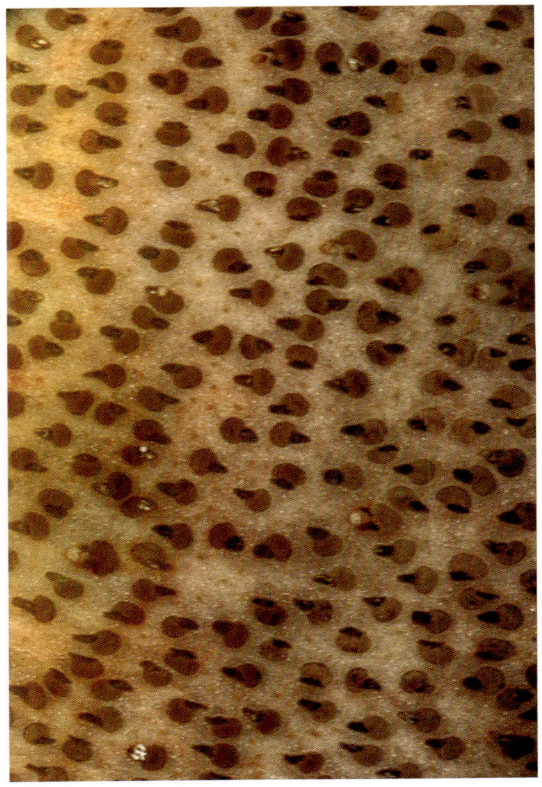

Macro photograph showing *Palmoxylon* structure in fossil wood. Catahoula Formation, Oligocene. Louisiana, USA. Daniels specimen.

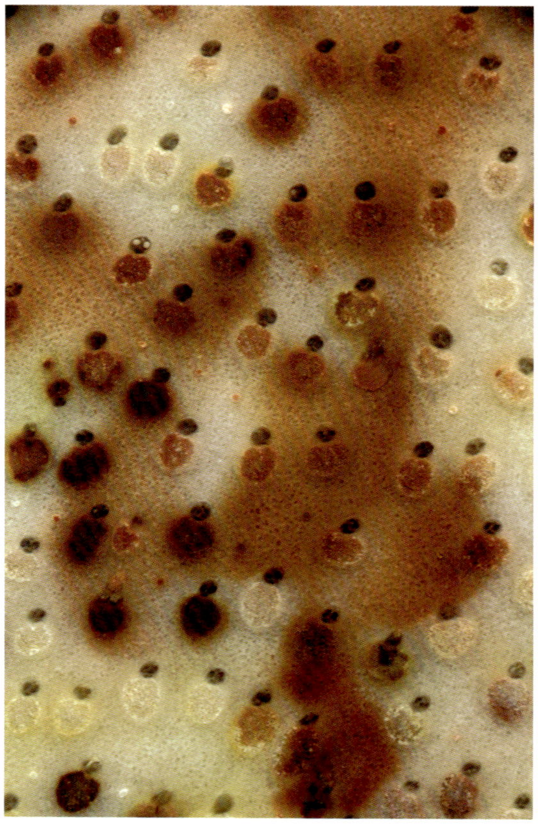

Macro photograph showing *Palmoxylon* structure in fossil wood. Catahoula Formation, Oligocene. Louisiana, USA. Daniels specimen.

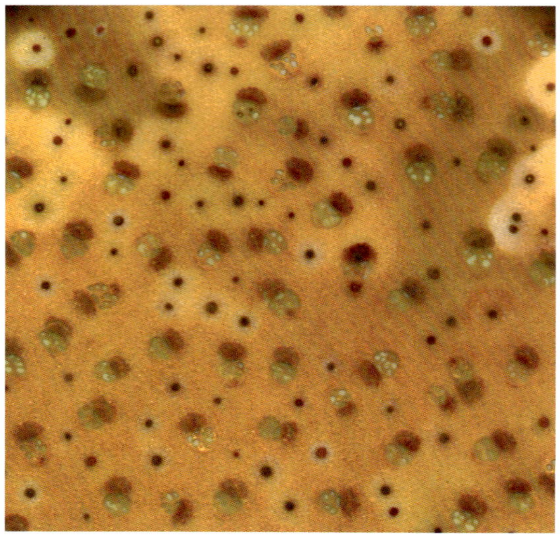

Macro photograph showing *Palmoxylon* structure in fossil wood. Catahoula Formation, Oligocene. Louisiana, USA. Daniels specimen.

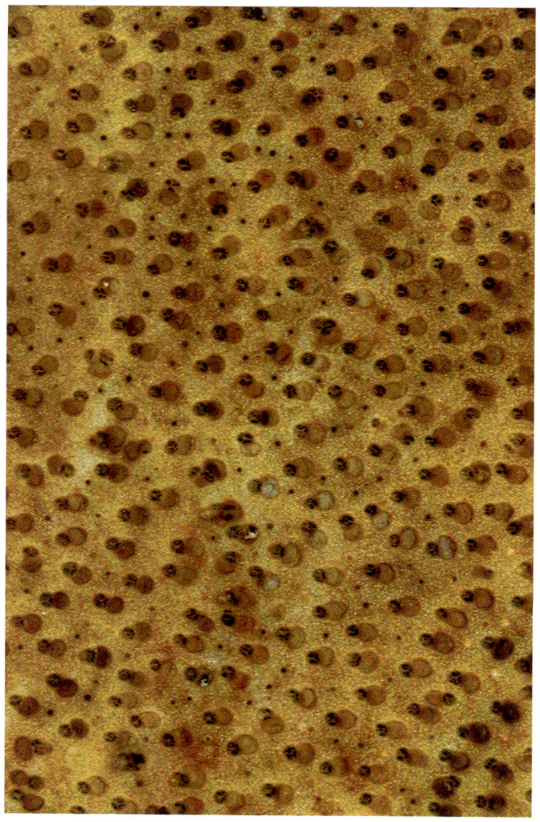

Macro photograph showing *Palmoxylon* structure in fossil wood. Catahoula Formation, Oligocene. Louisiana, USA. Daniels specimen.

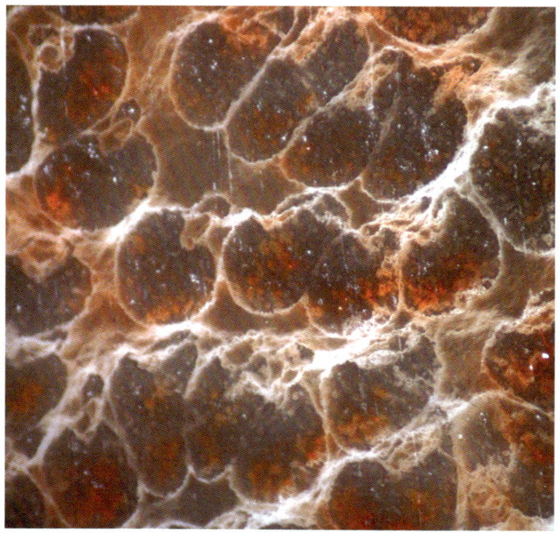

Micro image showing *Palmoxylon* cell structure in fossil wood. Cretaceous. Rio Negro, Argentina. Daniels specimen.

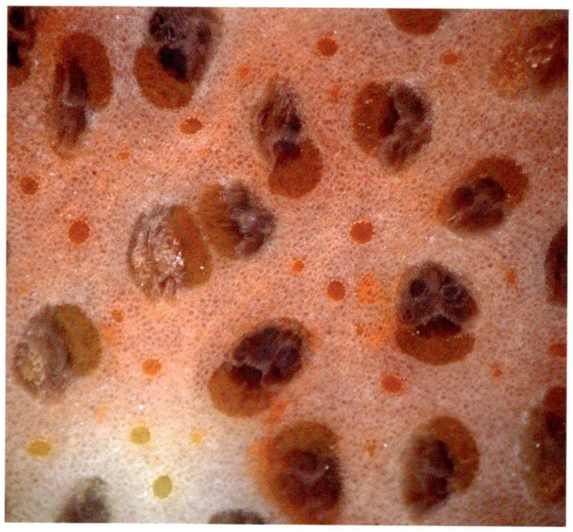

Micro image showing *Palmoxylon* cell structure in fossil wood. Catahoula Formation, Oligocene. Louisiana, USA. Daniels specimen.

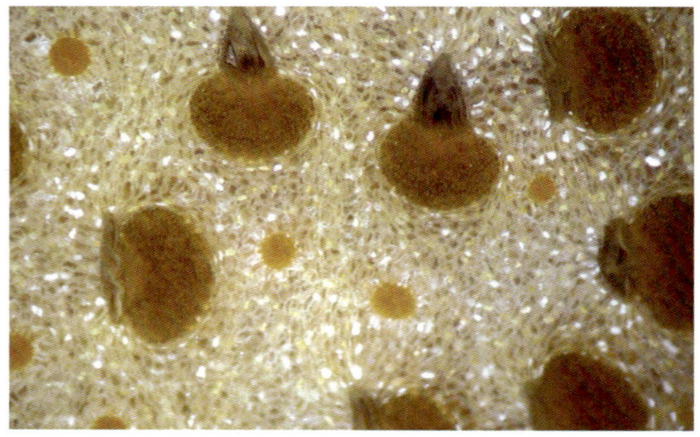

Micro image showing *Palmoxylon* cell structure in fossil wood. Catahoula Formation, Oligocene. Louisiana, USA. Daniels specimen.

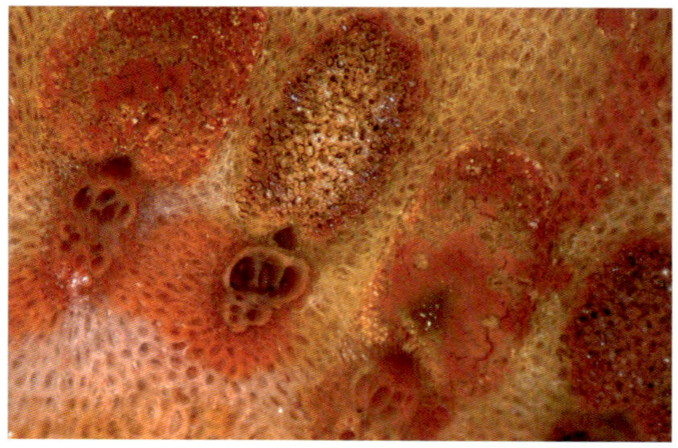

Micro image showing *Palmoxylon* cell structure in fossil wood. Catahoula Formation, Oligocene. Louisiana, USA. Daniels specimen.

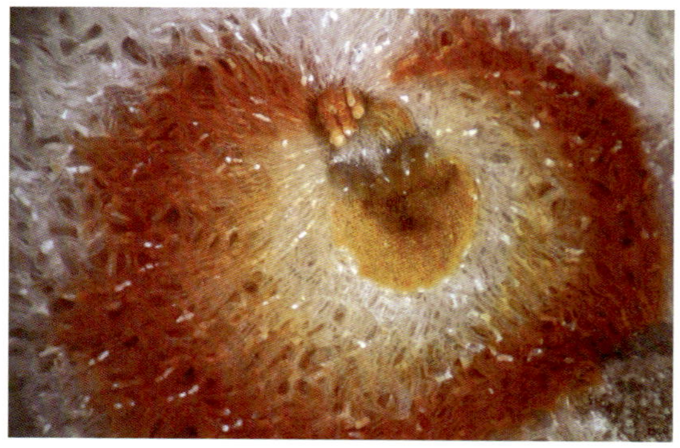

Micro image showing *Palmoxylon* cell structure in fossil wood. Catahoula Formation, Oligocene. Louisiana, USA. Daniels specimen.

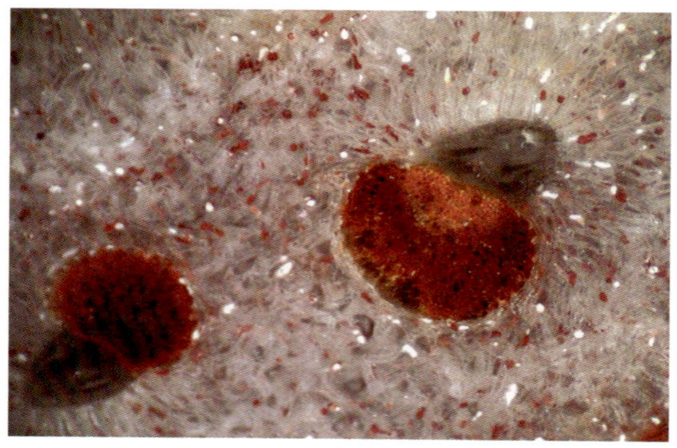

Micro image showing *Palmoxylon* cell structure in fossil wood. Catahoula Formation, Oligocene. Louisiana, USA. Daniels specimen.

Fossil yucca wood is extraordinarily rare, discovered only in a small area of northwest Nevada in a Miocene or Lower Pliocene formation. It has similarities to the largest tree in the modern yucca family, the Joshua tree (*Yucca brevifolia*), and is likely an ancestor. These fossil stems, roots, and leaves were studied by Professors William D. Tidwell of Brigham Young University and Lee R. Parker of California Polytechnic State University, who then authored an article in the *Review of Palaeobotany and Palynology* published in 1990. In describing *Protoyucca shadishii*, Professors Tidwell and Parker reported this find as fossils of an arborescent monocotyledon with secondary growth, a rarity in the fossil record. "The stems consist of central cylinders containing both primary and secondary tissues within their own respective cylinders. The primary cylinder in the middle of these stems contains bundles situated in what was once parenchyma and, in turn, is surrounded by a cylinder of secondary tissues composed of densely packed bundles within poorly preserved parenchymatous tissue. A secondary thickening meristem, secondary cortex and in some specimens, periderm occurs outside the cylinder containing secondary bundles." (Tidwell and Parker) The primary cylinder shows gaps with the appearance of a wheel with thick spokes of radiating tissue. The secondary cylinder is solid and concentrically arranged.

Travel south to the Sonoran Desert of southern Nevada, southern California, and western Arizona and you will find the current range of the Joshua tree, lending support to the theory that the *Protoyucca* discovery site was the ancestral home of this marvelous tree that migrated south with climate changes. In its current Sonoran Desert environment, nighttime temperatures drop well below freezing and summertime temperatures rarely reach 100° Fahrenheit. Some Joshuas grow to heights of more than 40 feet, with trunk circumferences of 12 or more feet, and live several hundred years. Sadly, environmental degradation, including pollution drifting eastward from cities in Southern California, threatens these marvelous plants.

TOP
Protoyucca shadishii.
Miocene/Pliocene. Nevada, USA.
Daniels specimen, 13 cm.

BOTTOM
Protoyucca shadishii.
Miocene/Pliocene. Nevada, USA.
Daniels specimen, 15 cm.

The world's tallest Joshua tree and Frank Daniels. Joshua Tree National Park, California, USA.

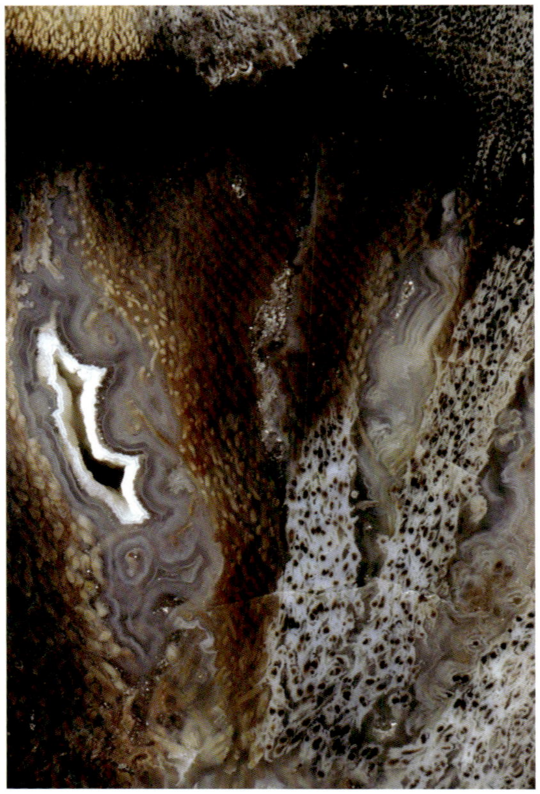

Macro photograph showing *Protoyucca shadishii* cell structure. Miocene/Pliocene. Nevada, USA. Daniels specimen.

Protoyucca shadishii flower. Miocene/Pliocene. Nevada, USA. Shadish specimen, 9.5 cm.

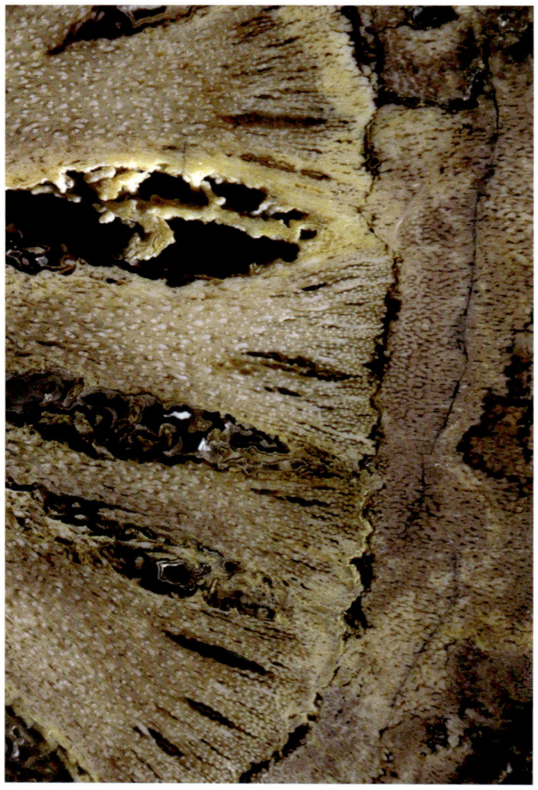

Macro photograph showing *Protoyucca shadishii* cell structure. Miocene/Pliocene. Nevada, USA. Daniels specimen.

Micro image showing *Protoyucca shadishii* cell structure. Miocene/Pliocene. Nevada, USA. Daniels specimen.

Micro image showing *Protoyucca shadishii* cell structure. Miocene/Pliocene. Nevada, USA. Daniels specimen. Width of view approximately 1 cm.

Joshua tree blossom. Joshua Tree National Park, California, USA.

Flowering Joshua tree. Joshua Tree National Park, California, USA.

Fossil ginkgo wood is a rarity much sought by fossil wood collectors. Although the North American fossil leaf record for ginkgo proves they were widespread during the Mesozoic and Cenozoic eras, it was unknown as a living tree in the West until the 1690 discovery by western botanists of living trees in Japan, where Buddhist monks cultivated the trees in temple gardens. A more recent discovery in a remote region of southeastern China at the borders of Chekiang and Anhwei provinces may establish proof that a small wild stand of ginkgos survived; however, there is debate about whether this stand survived unassisted by human intervention. Regardless, the term "living fossil" appropriately applies to extant ginkgo. Some ginkgo trees in the Far East are estimated to be 1,200 years old and grow to nearly 30 meters (100 feet) in height with a circumference of 7.5 meters (25 feet). Just one species remains out of no less than 16 genera. Reproductive structures and leaves grow from branches and short shoots. Ginkgo is a deciduous gymnosperm. It is also dioecious, meaning there are both male and female trees. The wood structure resembles a conifer without resin canals, yet can be distinguished from conifers through microscopic examination of tracheid chains. Ginkgo tracheids do not align in neat, even rows as in conifers. Individual ginkgo tracheids within a single row vary considerably in size from one to the next in a seemingly random manner. Each radial tracheid row includes small and large diameter cells. As described by Professor George Beck of Central Washington College in the 1940s, "the rows take on the appearance of irregularly rounded, odd-sized beads." Ginkgo rays are uniseriate.

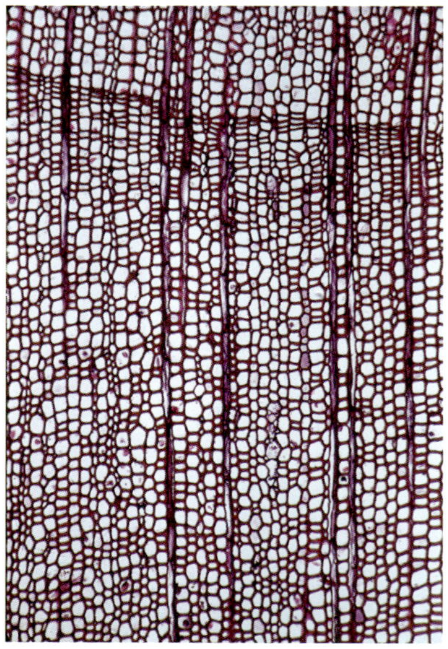

Ginkgo biloba (maidenhair tree), magnification x 40.

Micro image showing ginkgo-like cell structure in fossil wood. Columbia River Basalt, Miocene. Vantage, Washington, USA. Daniels specimen. Average earlywood tracheid size 0.045 mm.

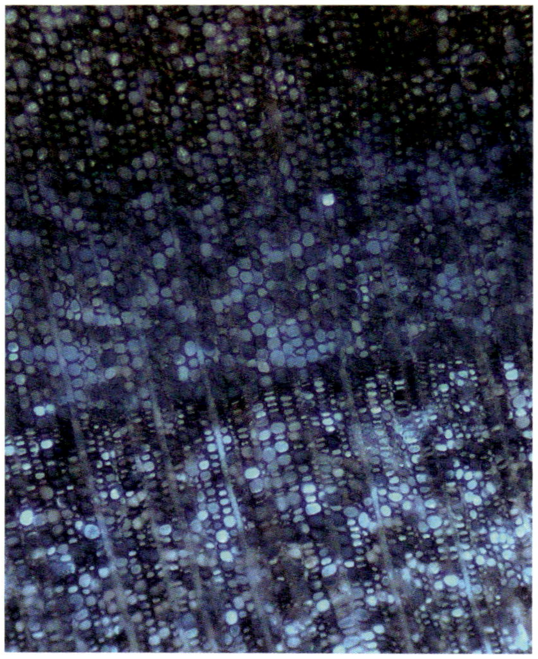

Micro image showing ginkgo-like cell structure in fossil wood. Columbia River Basalt, Miocene. Vantage, Washington, USA. Daniels specimen.

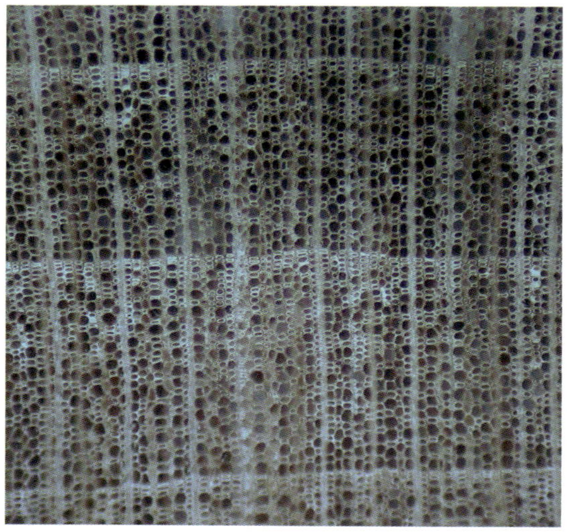

Micro image showing ginkgo-like cell structure in fossil wood. Columbia River Basalt, Miocene. Vantage, Washington, USA. Daniels specimen. Growth rings 0.8 mm.

Although ginkgo is an ancient tree that has been around since at least the Late Triassic, confirmed fossil ginkgo wood is rare. In transverse view, ginkgo is easily confused with *Podocarpoxylon*. Professor Beck differentiated the fossil ginkgo wood (*Ginkgo bonesii*) from the Clarno Formation in Oregon from the fossil ginkgo wood (*Ginkgo beckii*) found near Vantage, Washington, in part by examination of the tracheids. In *Ginkgo bonesii*, he found the smaller-sized tracheids have a tendency to align in "short tangential or oblique bands."

Unidentified fruit/cone. Wagon Bed Formation, Eocene. Beaver Rim, Wyoming, USA. Albers specimen, 8 cm.

Unidentified fruit/cone. Wagon Bed Formation, Eocene. Beaver Rim, Wyoming, USA. Albers specimen, 7.5 cm.

Two rare Tertiary Period fossil plants from Wyoming are an unidentified monocot called "pre-palm" by local collectors and *Acrostichum* (giant fern). "Pre-palm" fossil wood is somewhat of a mystery. Possibly related fossil fruit found in proximity to the "pre-palm" stems add to the interest and mystery of this unusual fossil. Leaves of *Acrostichum* grew to 1.5 meters (5 feet) long. When rarely discovered, the typical find is fossil *Acrostichum* as small stem portions of the large fronds, although trunk fragments infrequently occur. An uncommon fossil wood common in deposits of the Eocene Green River Formation in Wyoming is *Shinoxylon*, a member of the pepper tree family. *Shinoxylon* is a diffuse porous hardwood with distinct growth rings and vessels that occur as large tyloses-filled vessels singly or in multiples of two to four, along with small vessels in multiples into the twenties.

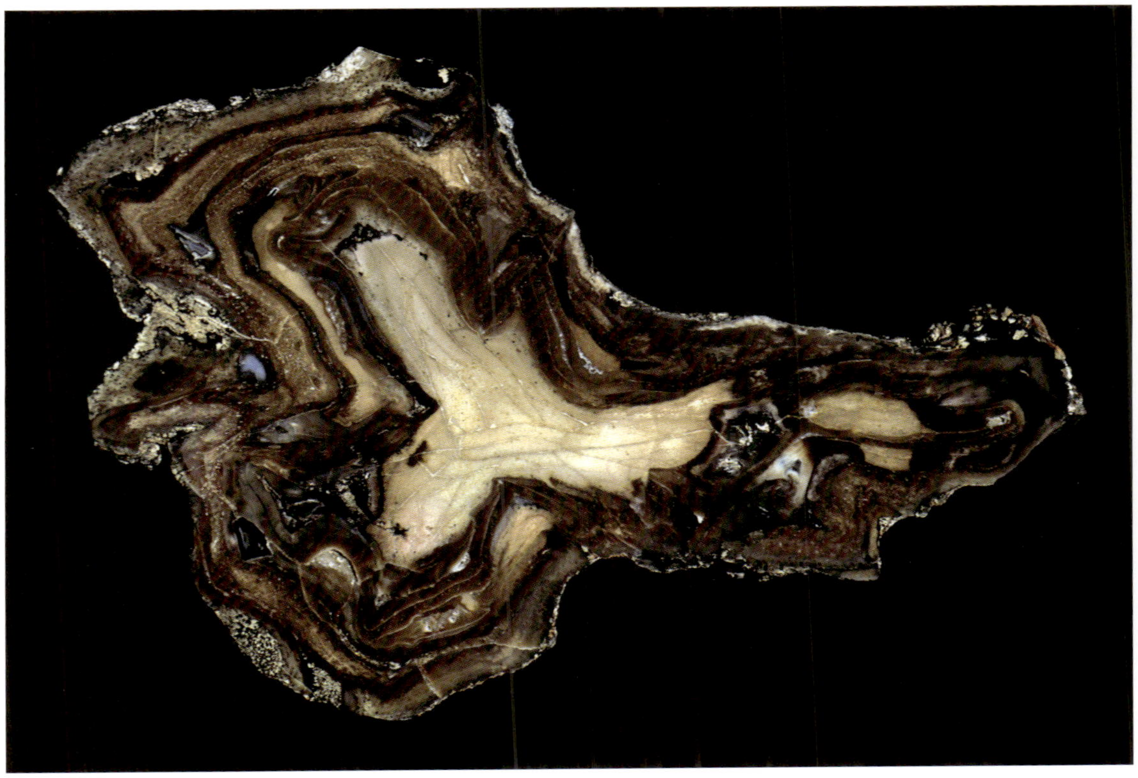

Unidentified "pre-palm." Wagon Bed Formation, Eocene. Beaver Rim, Wyoming, USA, Daniels specimen, 4 cm.

Unidentified fruit/cone. Wagon Bed Formation, Eocene. Beaver Rim, Wyoming, USA. Albers specimen, 7.5 cm.

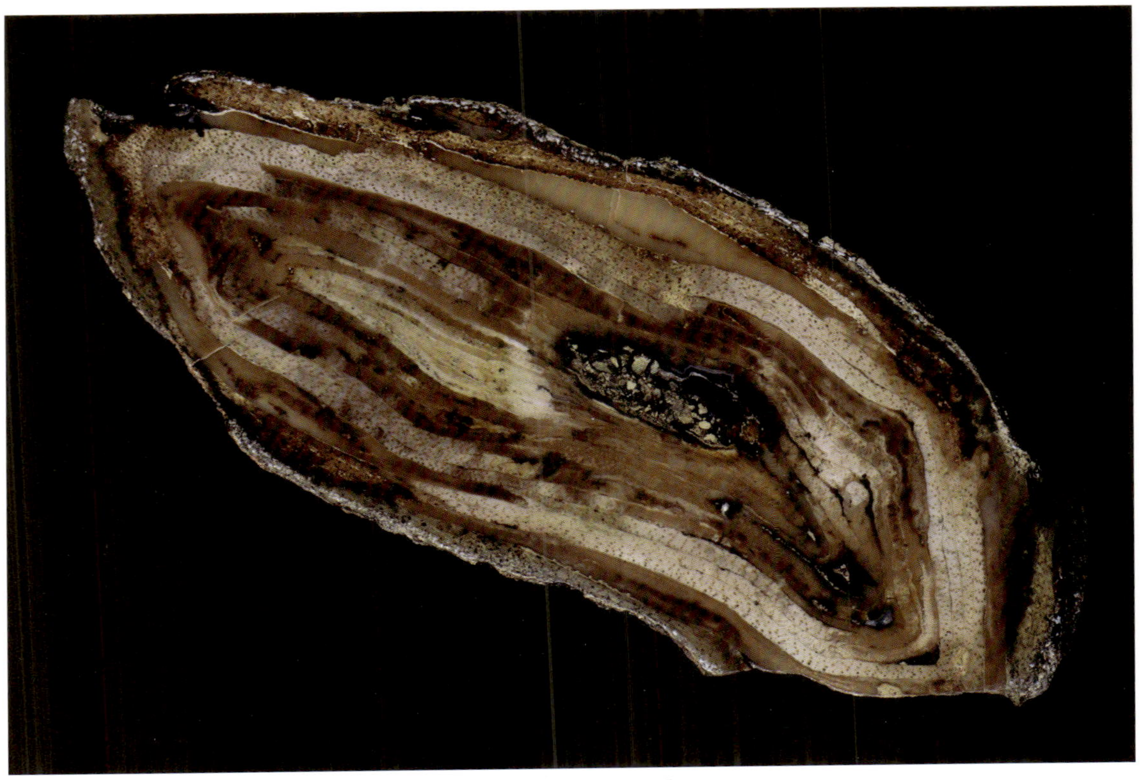

Unidentified "pre-palm." Wagon Bed Formation, Eocene. Beaver Rim, Wyoming, USA. Albers specimen, 22 cm.

Unidentified fruit/cone. Wagon Bed Formation, Eocene. Beaver Rim, Wyoming, USA. Albers specimen, 8 cm.

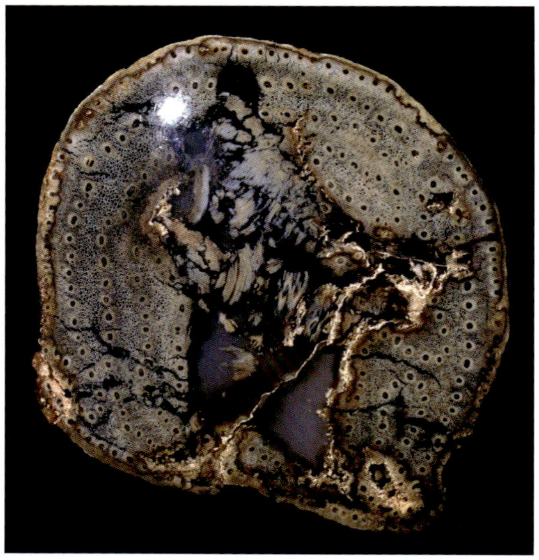

Acrostichum stalk. Green River Formation, Eocene. Eden Valley, Wyoming, USA. Albers specimen, 3.5 cm.

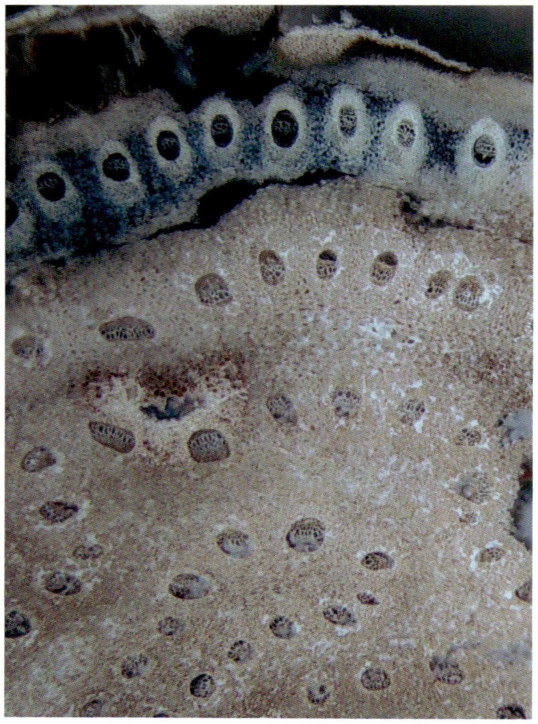

Micro image showing cell structure in *Acrostichum* stalk. Green River Formation, Eocene. Eden Valley, Wyoming, USA. Daniels specimen. Largest solitary vascular bundles approximately 1.2 mm.

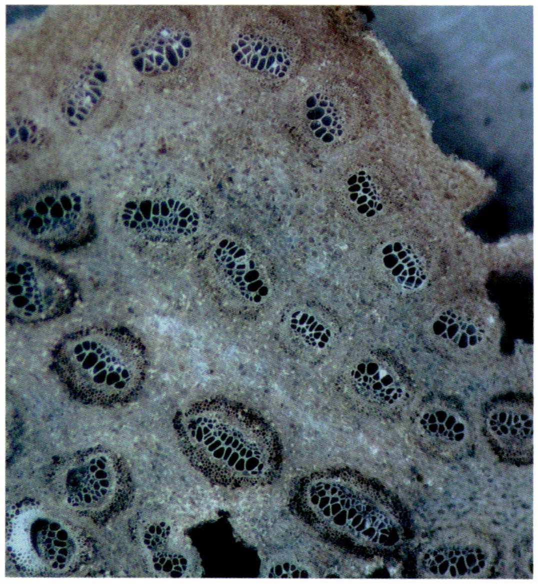

Micro image showing cell structure in *Acrostichum* stalk. Green River Formation, Eocene. Eden Valley, Wyoming, USA. Daniels specimen. Largest solitary vascular bundles approximately 1.2 mm.

An unusual fossil fern found in Texas in Eocene Epoch formations is *Cyathodendron texanum*. This fern grew upright with an exterior of spirally arranged leaf bases.

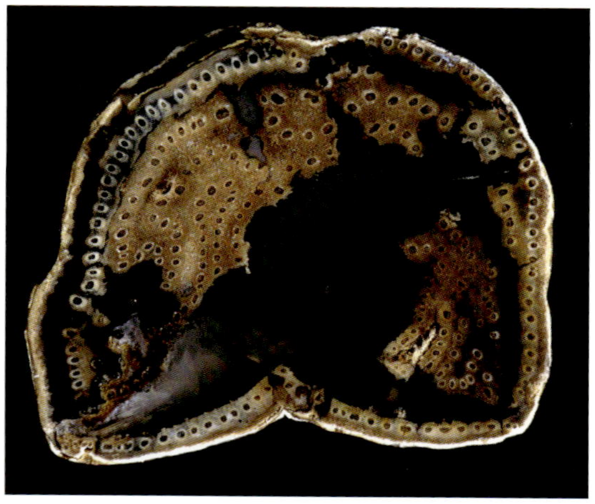

Acrostichum stalk. Green River Formation, Eocene. Eden Valley, Wyoming, USA. Daniels specimen, 3 cm.

Pair of *Cyathodendron* stems. Eocene. Texas, USA. Daniels specimens, 12 and 15 cm.

Acrostichum stalk. Green River Formation, Eocene. Eden Valley, Wyoming, USA. Albers specimen, 3 cm.

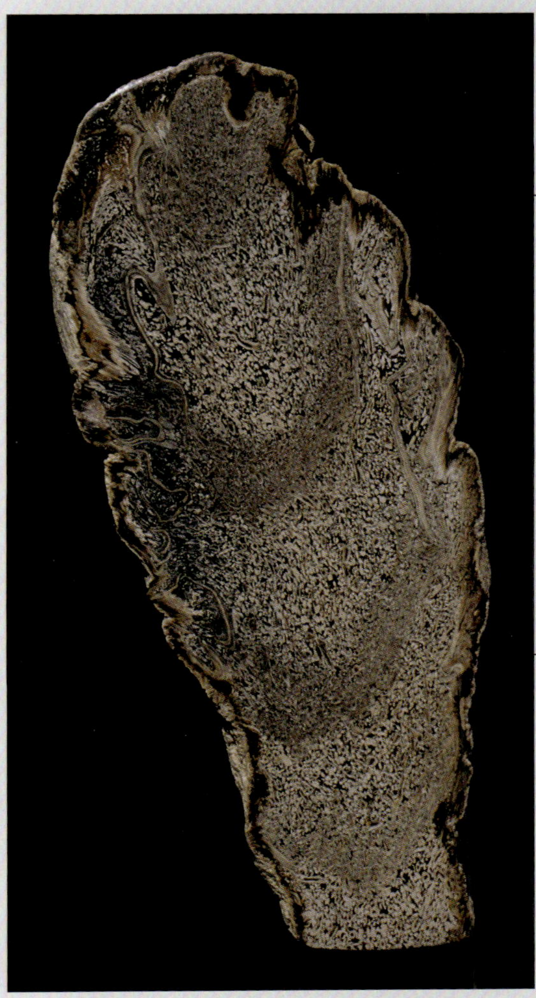

Macro photograph showing detail in *Cyathodendron* stem. Eocene. Texas, USA. Daniels specimen.

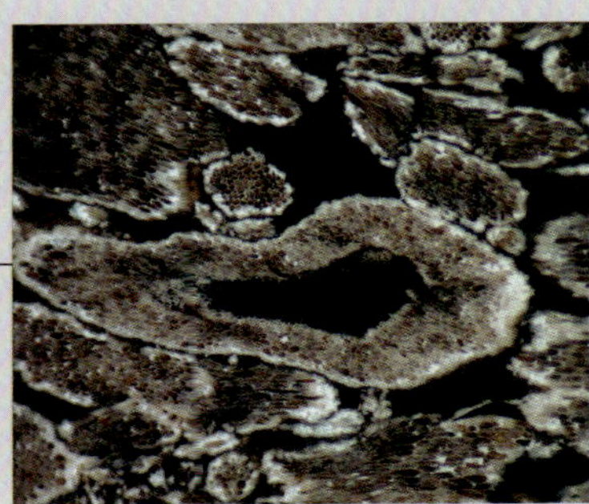

Micro image showing detail in *Cyathodendron* stem. Eocene. Texas, USA. Daniels specimen.

Cyathodendron stem. Eocene. Texas, USA. Daniels specimen, 18 cm.

Macro photograph showing detail of fossil *Trochodendron* structure. Trout Creek Formation, Miocene. McDermitt, Oregon, USA. Daniels specimen.

Macro photograph showing detail of fossil *Trochodendron* structure. Trout Creek Formation, Miocene. McDermitt, Oregon, USA. Daniels specimen.

Cyathodendron stem. Eocene. Texas, USA. Daniels specimen, 13 cm.

Trochodendron is unusual for a hardwood because it lacks vessels. The structure is rows of tracheids with distinct growth rings and both uniseriate and very broad rays. Fossil *Trochodendron* specimens are found in a few Tertiary Period collecting locales in Oregon, most notably Sweet Home and McDermitt.

Micro image showing detail of fossil *Trochodendron* structure. Trout Creek Formation, Miocene. McDermitt, Oregon, USA. Daniels specimen.

Micro image showing detail of fossil *Trochodendron* structure. Trout Creek Formation, Miocene. McDermitt, Oregon, USA. Daniels specimen.

A fossil plant encountered in Permian to modern formations is a member of the Osmundaceae family, an ancient fern family with living representatives, including those of the genus *Osmunda*. Osmundaceae has the most extensive fossil record of any fern with more than 150 species of fossils, 50 known as fossil stems. Osmundaphytes typically display a complex central xylem cylinder surrounded by C-shaped or omega-shaped leaf traces and intertwined roots. Osmundaphytes are readily recognizable in reasonably well-preserved fossil remains. Lovely specimens come from Jurassic formations in Queensland and Tasmania, Australia. Because of the wide geographic distribution of these ferns in past geologic ages, fossil stems reside in many locations worldwide, including Jurassic to Tertiary formations in the western United States.

Macro photograph showing detail in fossil *Osmunda*. Jurassic. Queensland, Australia. Daniels specimen.

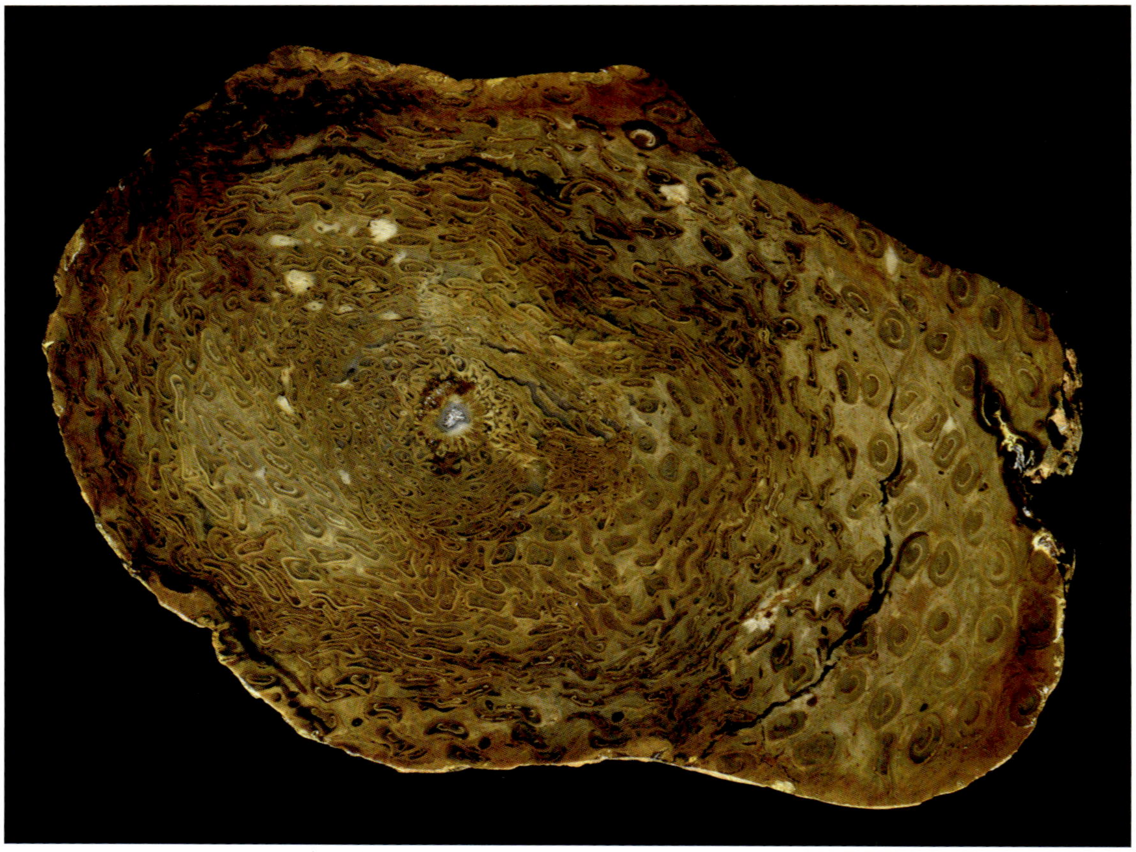

***Osmunda*. Jurassic. Queensland, Australia. Daniels specimen, 14.5 cm.**

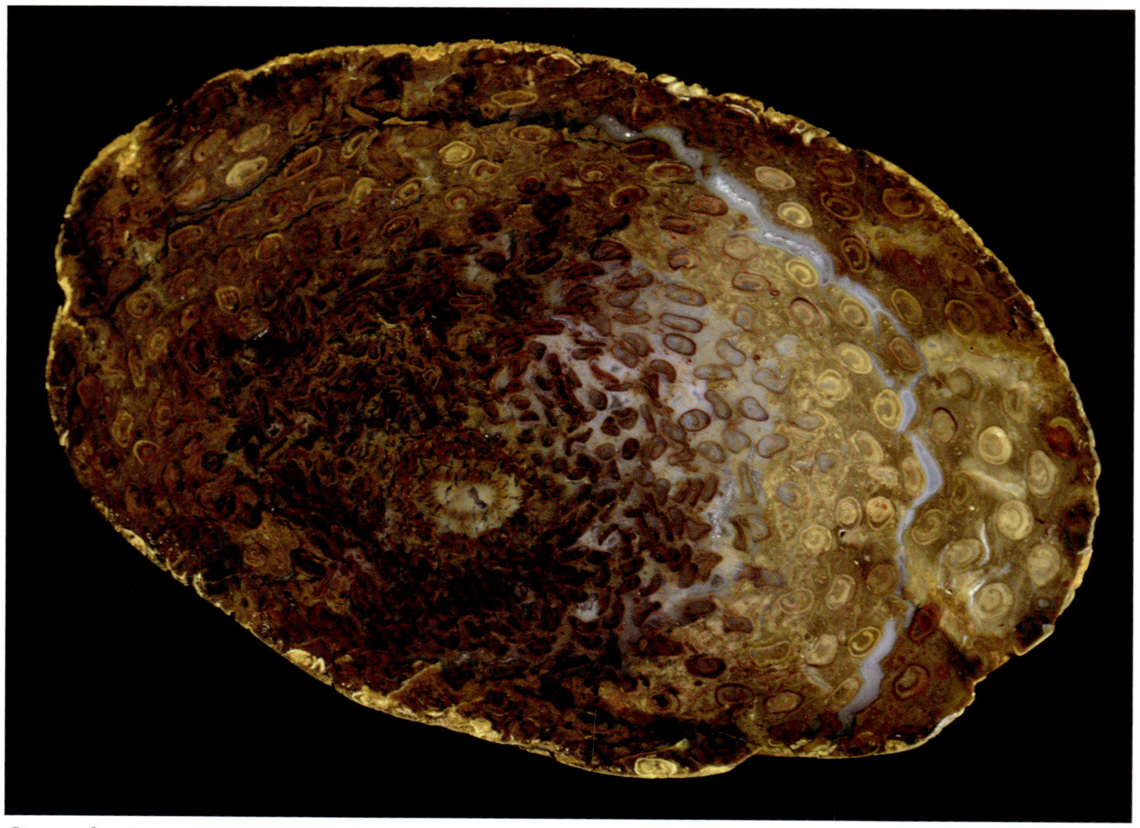

***Osmunda*. Jurassic. Queensland, Australia. Daniels specimen, 18 cm.**

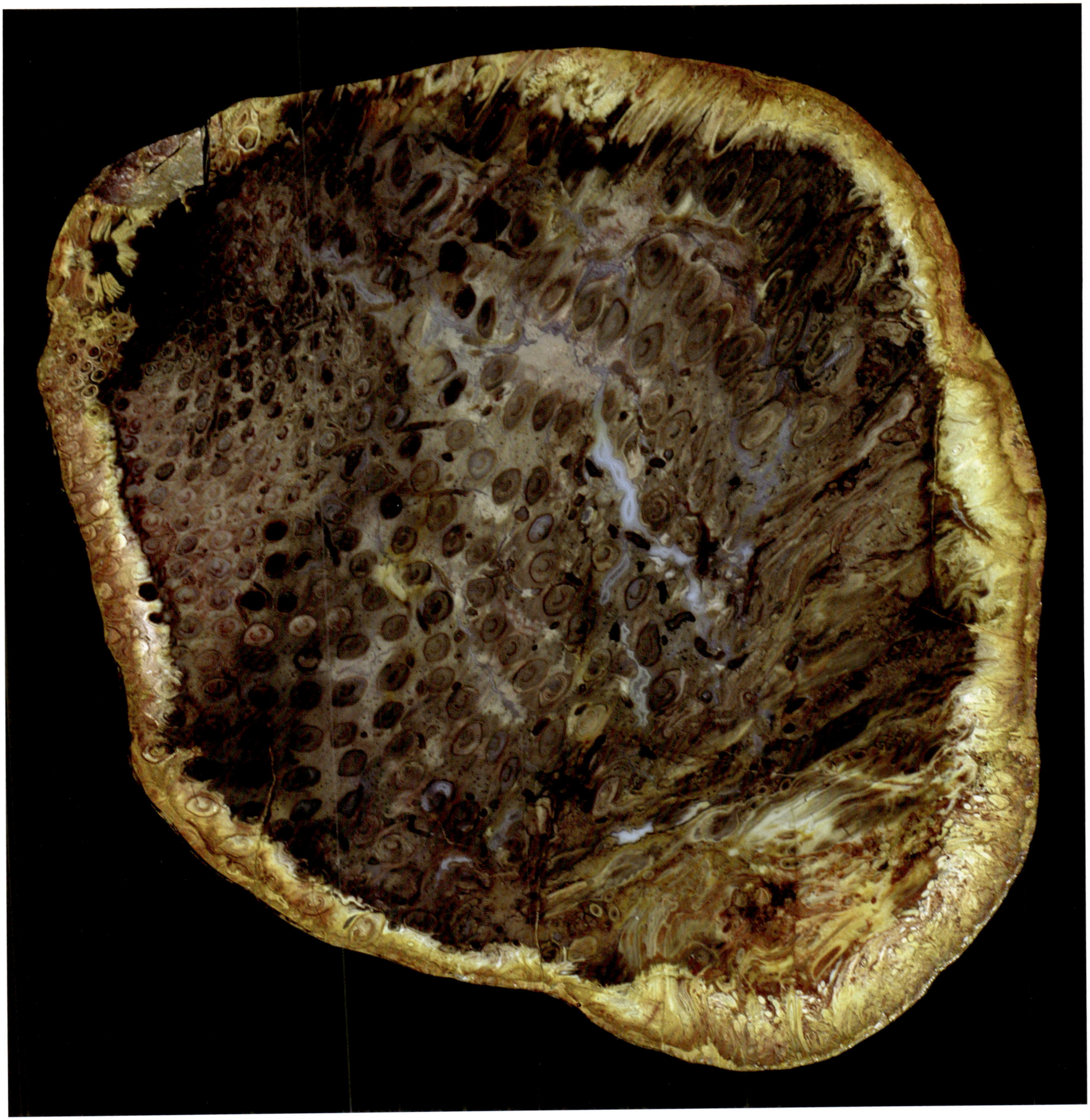

Osmunda. Jurassic. Queensland, Australia. Viney specimen, 20 cm.

Aurealcaulis moorei. Lower Eocene. New Mexico, USA. Dayvault specimen, 12.5 cm.

Osmunda. Clarno Formation, Eocene. Eastern Oregon, USA. Murphy specimen, 9 cm.

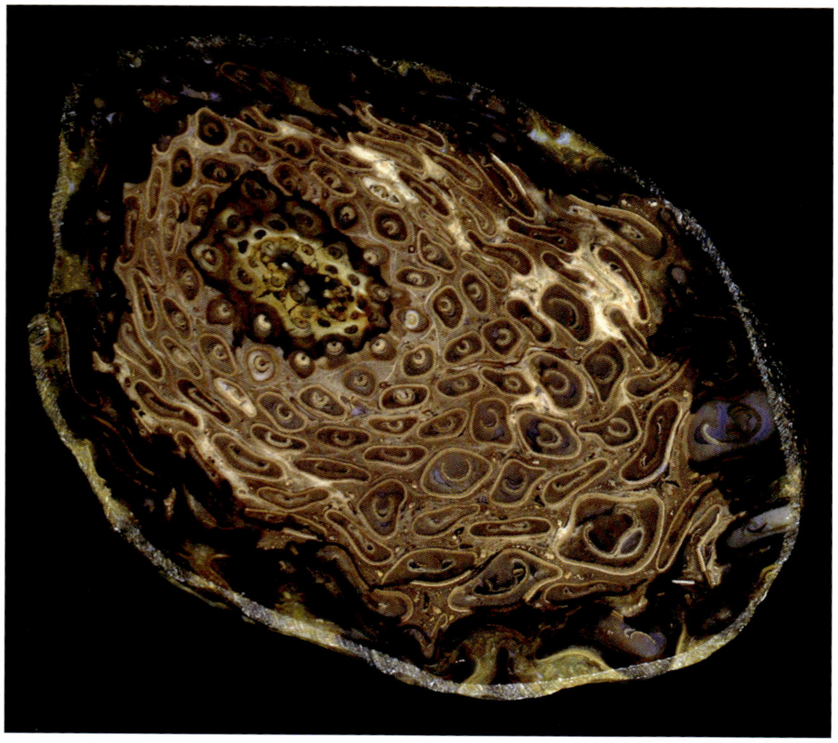

Osmunda. Jurassic. Queensland, Australia. Daniels specimen, 7.5 cm.

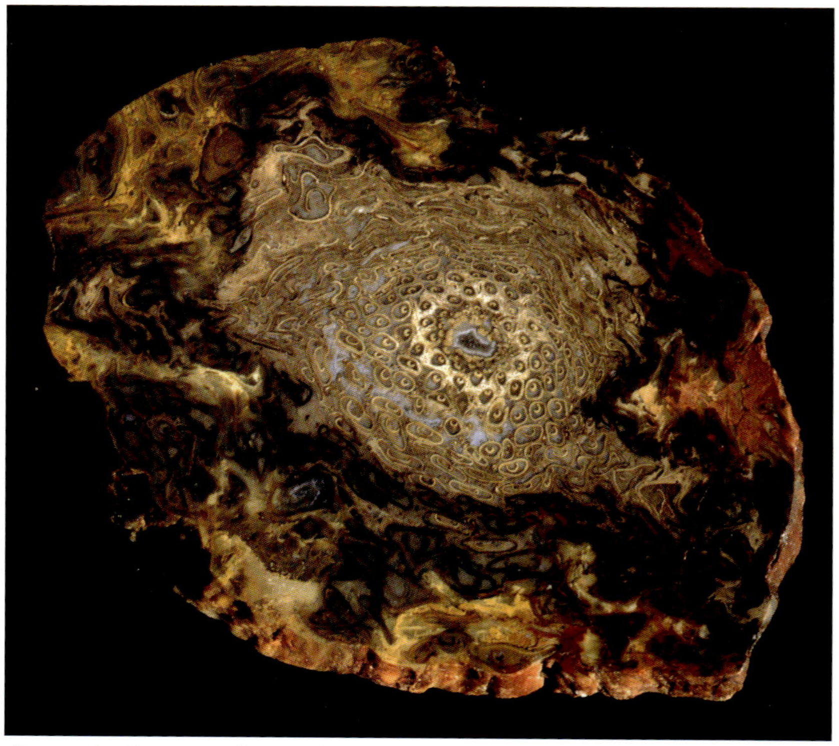

Osmunda. Jurassic. Queensland, Australia. Daniels specimen, 10.5 cm.

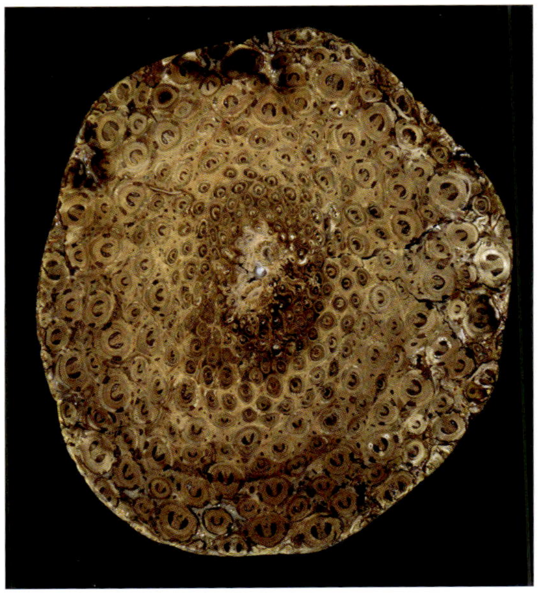

Osmunda. Jurassic. Queensland, Australia. Daniels specimen, 6.5 cm.

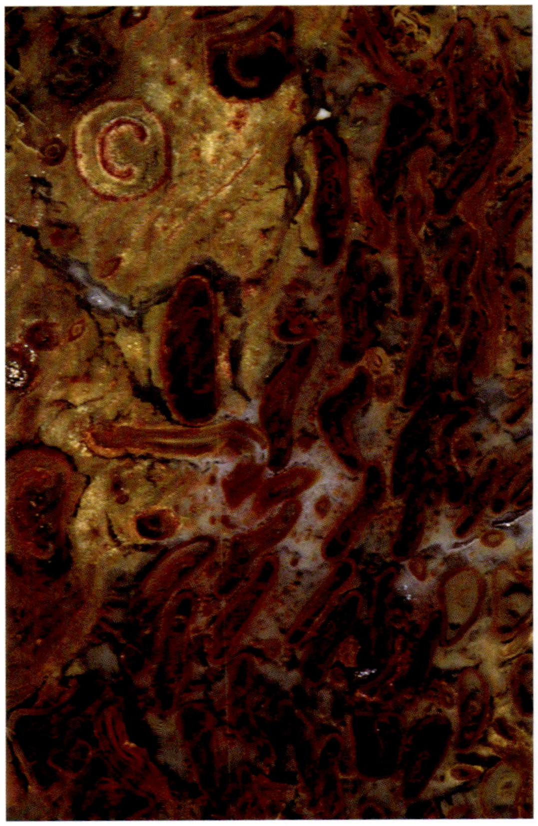

Macro photograph showing detail in fossil *Osmunda*. Jurassic. Queensland, Australia. Daniels specimen.

Macro photograph showing stele detail in fossil *Osmunda*. Jurassic. Queensland, Australia. Daniels specimen.

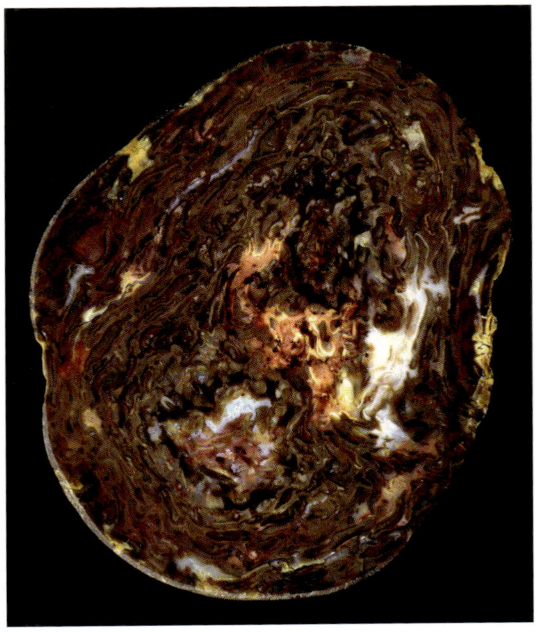

Osmundaa. Jurassic. Queensland, Australia. Daniels specimen, 6 cm.

Macro photograph showing detail in fossil *Osmunda*. Jurassic. Queensland, Australia. Daniels specimen.

Macro photograph showing detail in fossil *Osmunda*. Jurassic. Queensland, Australia. Daniels specimen.

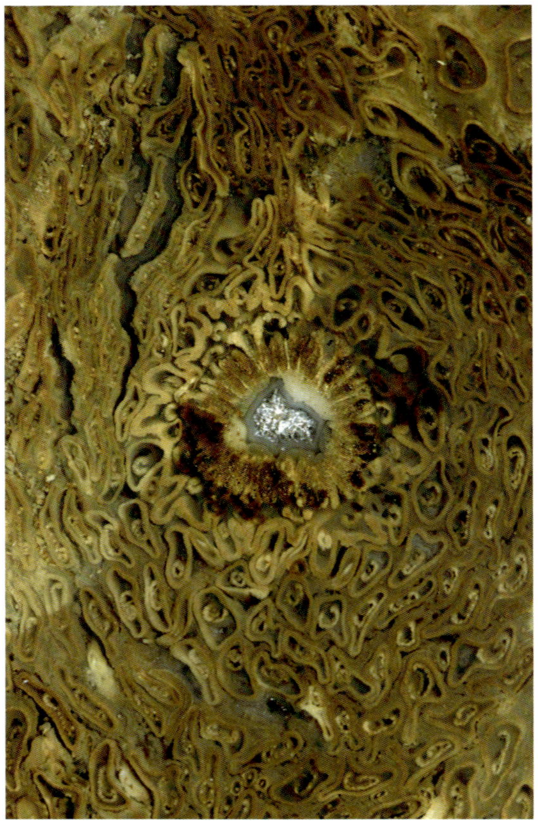

Macro photograph showing detail in fossil *Osmunda*. Jurassic. Queensland, Australia. Daniels specimen.

Macro photograph showing stele detail in fossil *Osmunda*. Jurassic. Queensland, Australia. Daniels specimen.

Micro image showing stele detail in fossil *Osmunda*. Jurassic. Queensland, Australia. Daniels specimen.

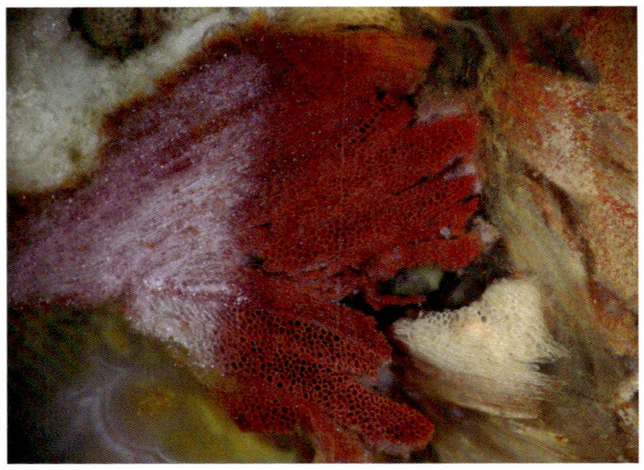

Micro image showing stele detail in fossil *Osmunda*. Jurassic. Queensland, Australia. Daniels specimen.

Micro image showing petiole detail in fossil *Osmunda*. Jurassic. Queensland, Australia. Daniels specimen.

Micro image showing rootlet detail in fossil *Osmunda*. Jurassic. Queensland, Australia. Daniels specimen.

Micro image showing stele detail in fossil *Osmunda*. Jurassic. Queensland, Australia. Daniels specimen.

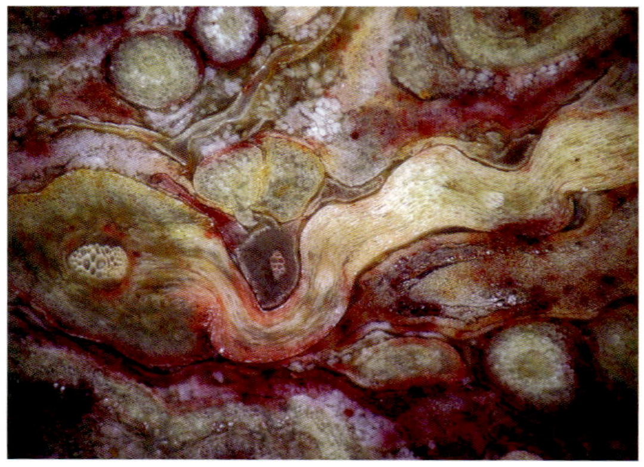

Micro image showing rootlet detail in fossil *Osmunda*. Jurassic. Queensland, Australia. Daniels specimen.

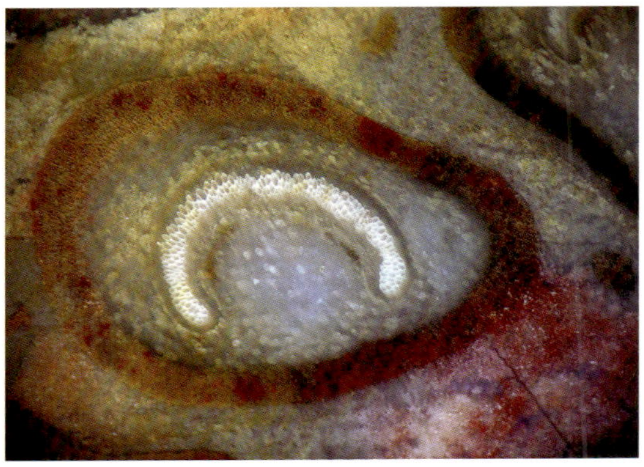

Micro image showing petiole detail in fossil *Osmunda*. Jurassic. Queensland, Australia. Daniels specimen.

Identification of Mesozoic Era Fossil Woods

Among the most beautiful of the world's fossil woods are the Jurassic and Triassic Period woods of the southwestern United States. These formations precede any significant evolution of angiosperms, and the woods are simpler in structure with fewer identified genera. The identified Jurassic woods of Colorado and Utah are predominately coniferous. Some Jurassic locations in Colorado and Utah contain more exotic fossil plants, including cycadeoids, *Osmunda*, *Hermanophyton*, and an unidentified polyvascular fossil wood similar to *Pentoxylon*. Fossil wood hunters frequently encounter fossil dinosaur bone in the wood-rich Jurassic Morrison Formation.

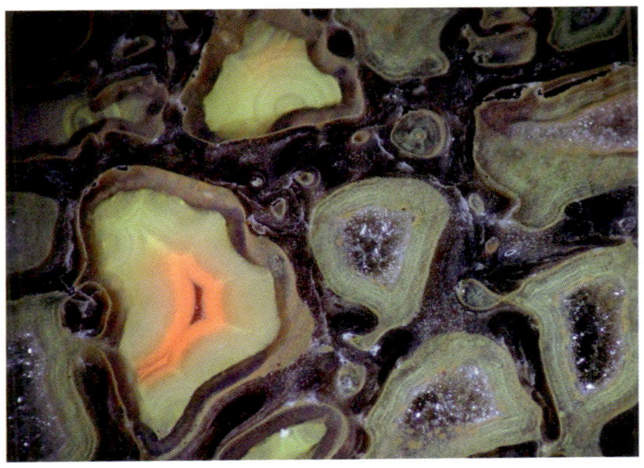

Micro image showing cell detail in dinosaur bone. Morrison Formation, Jurassic. Colorado/Utah, USA. Daniels specimen.

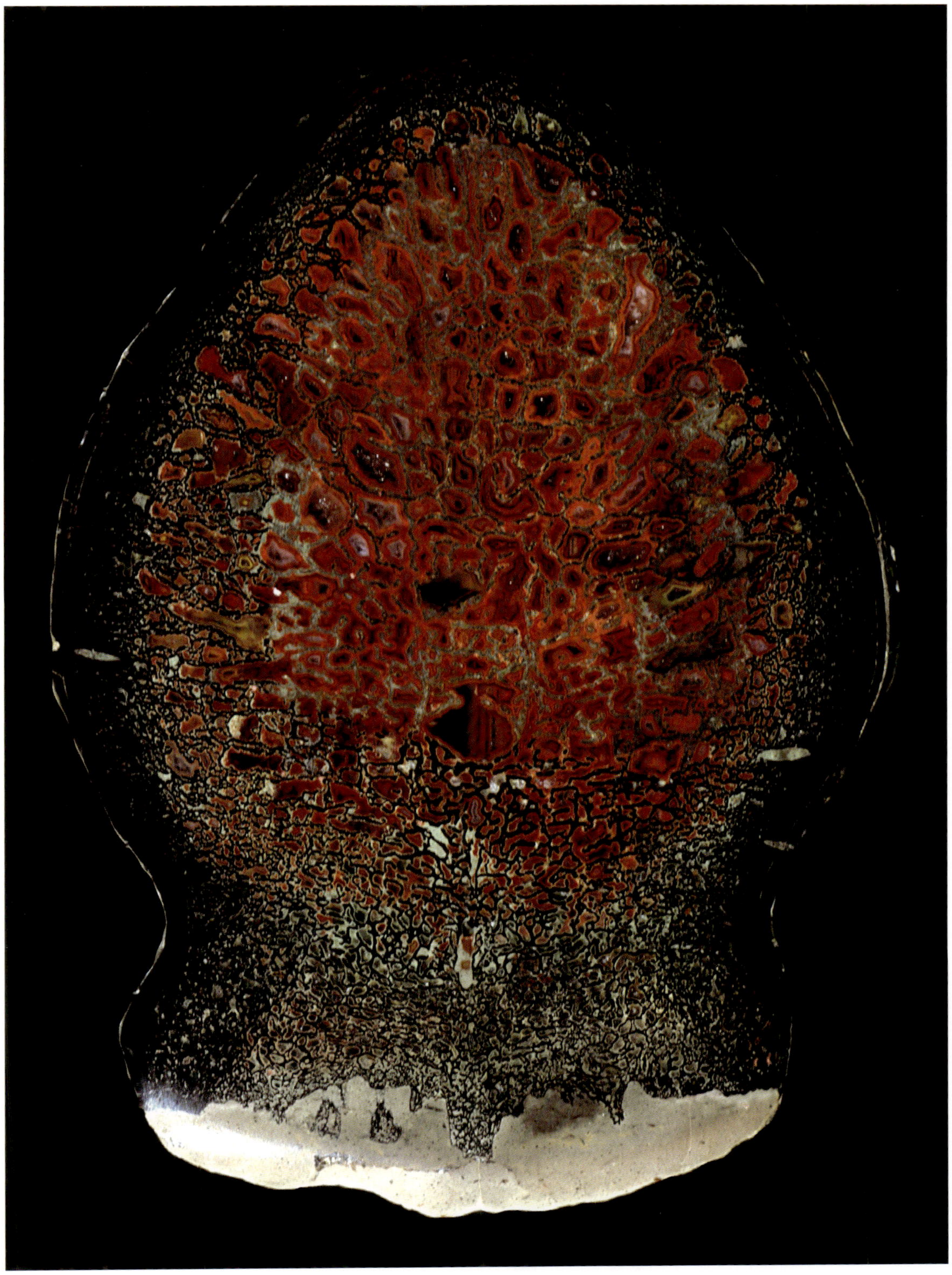

Dinosaur bone. Morrison Formation, Jurassic. Colorado/Utah, USA. Daniels specimen, 11.5 cm.

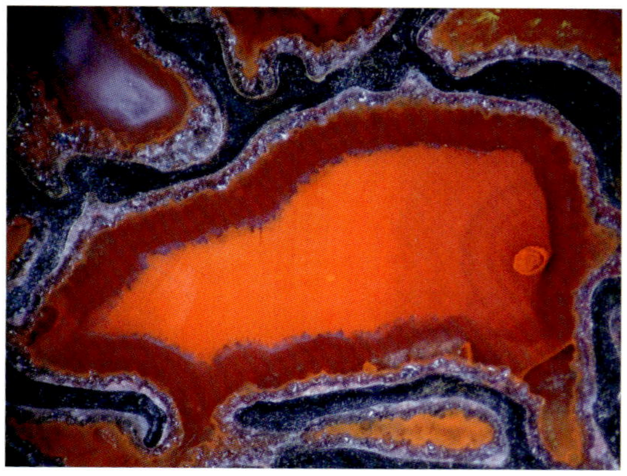

Micro image showing cell detail in dinosaur bone. Morrison Formation, Jurassic. Colorado/Utah, USA. Daniels specimen.

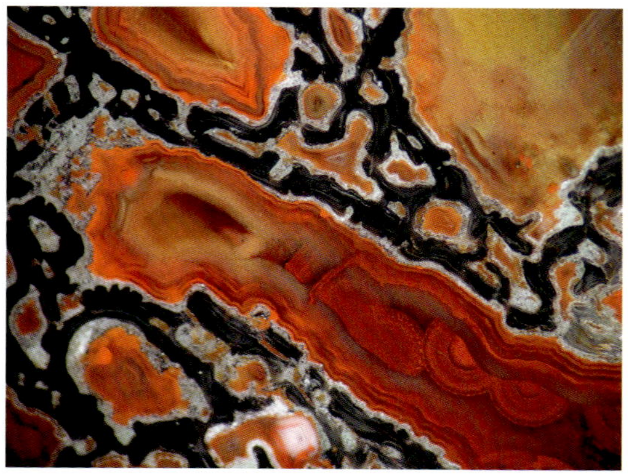

Micro image showing cell detail in dinosaur bone. Morrison Formation, Jurassic. Colorado/Utah, USA. Daniels specimen.

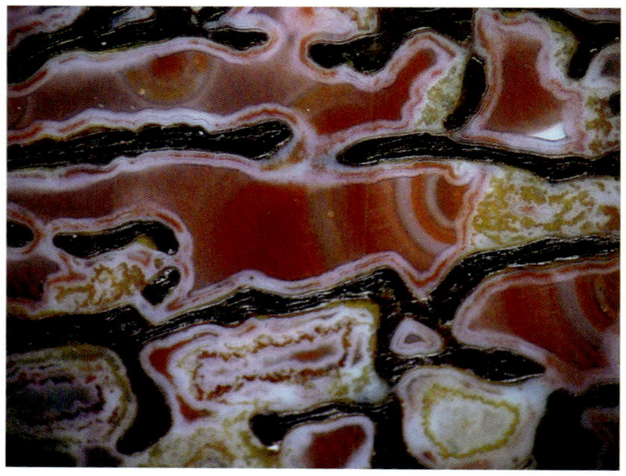

Micro image showing cell detail in dinosaur bone. Morrison Formation, Jurassic. Colorado/Utah, USA. Daniels specimen.

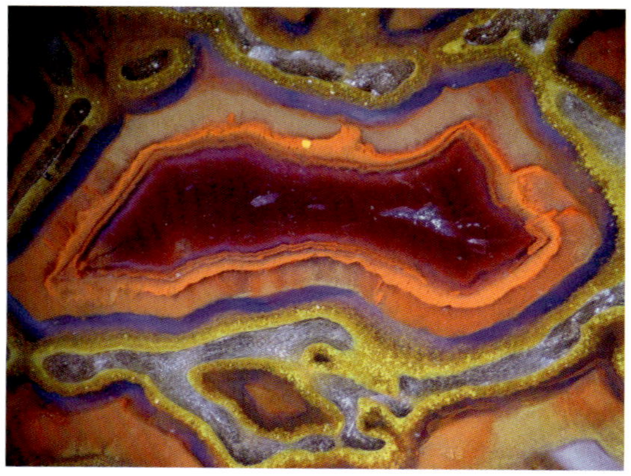

Micro image showing cell detail in dinosaur bone. Morrison Formation, Jurassic. Colorado/Utah, USA. Daniels specimen.

Micro image showing cell detail in dinosaur bone. Morrison Formation, Jurassic. Colorado/Utah, USA. Daniels specimen.

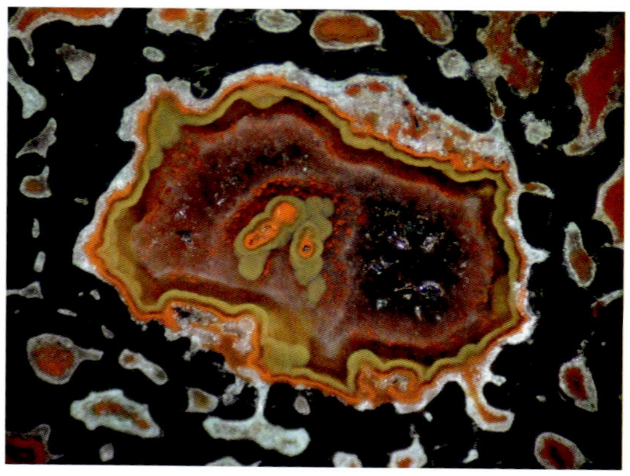

Micro image showing cell detail in dinosaur bone. Morrison Formation, Jurassic. Colorado/Utah, USA. Daniels specimen.

Micro image showing cell detail in dinosaur bone. Morrison Formation, Jurassic. Colorado/Utah, USA. Daniels specimen.

The Jurassic Period is sometimes referred to as the Age of Cycads. Although we group cycadophytes (the term used for cycad and cycad-like plants) with gymnosperms, they are not closely related to any other group of living plants. The plants have thick, stout trunks with an external armor of hardened, spirally arranged, persistent leaf bases and massive piths. The major portion of the stem consists of cortex and pith. Leaf bases remain as fronds die off, forming the persistent armor of the plant. These tough leaf bases give the plant its main support and are the most likely part of the plant to fossilize. Fossil cycads with the pith intact are rarely discovered. The crown consists of large, palm-like leaves. Some cycadophytes have short, squat stems while others have tall, columnar stems. The 200 living species of cycads are merely a relic of the large populations that once existed.

The two cycadophyte orders are Cycadales and Cycadeoidales, the cycads and cycadeoids, respectively. The first fossil cycad evidence dates to the Early Permian Period. Most petrified cycadophytes likely to be encountered are cycadeoids, which were widespread from the Early Triassic to the Late Cretaceous periods, when they disappeared from the fossil record. Evidence of the wide Mesozoic geographic distribution of cycadophytes includes specimens discovered in Antarctica, Argentina, Australia, Austria, Belgium, France, Germany, Great Britain, Greenland, India, Italy, Mexico, Poland, South Africa, and United States. United States fossil cycadophyte discovery sites include locations in California, Colorado, Maryland, New Mexico, South Dakota, Texas, Utah, and Wyoming. Cycadophytes are of particular interest to collectors of fossil woods because of their rarity and the beauty of polished specimens of tangential sections cut through the armor.

Identification of fossil cycadophytes can be simple or difficult, depending upon the portion preserved. The trunks grew in a variety of shapes and forms. Mesozoic Era cycads sometimes branched. Cycadeoid stems are unbranched or sparsely branched. The most salient feature for many cycadophytes is the diamond pattern formed by the leaf bases. This pattern is frequently discernable on an uncut specimen, but not all cycadophytes have well-developed diamonds. Cycadales bear seeds at the top of a stem or on a large, seed-bearing frond, some in cones and others without distinct cone structure. Cones in cycadeoids are produced in the trunk between leaf bases.

Cycadophyte seeds. Cretaceous. Rio Negro, Argentina. Daniels specimens, largest 3 cm.

Cycadeoid. Cretaceous. Rio Negro, Argentina. Daniels specimen, 35 cm; 40 pounds.

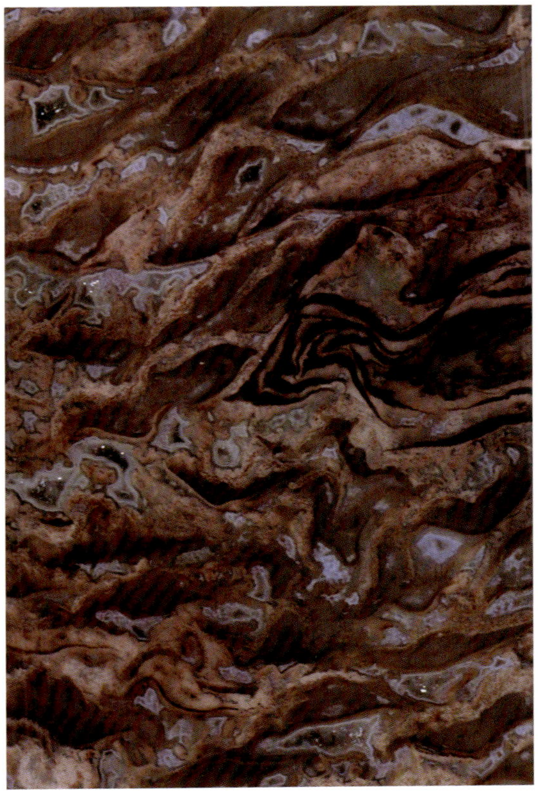

Macro photograph showing cycadeoid cell structure including partial cone. Cretaceous. Rio Negro, Argentina. Daniels specimen.

Cycadeoid with numerous cones. Cretaceous. Rio Negro, Argentina. Daniels specimen, 35.5 cm.

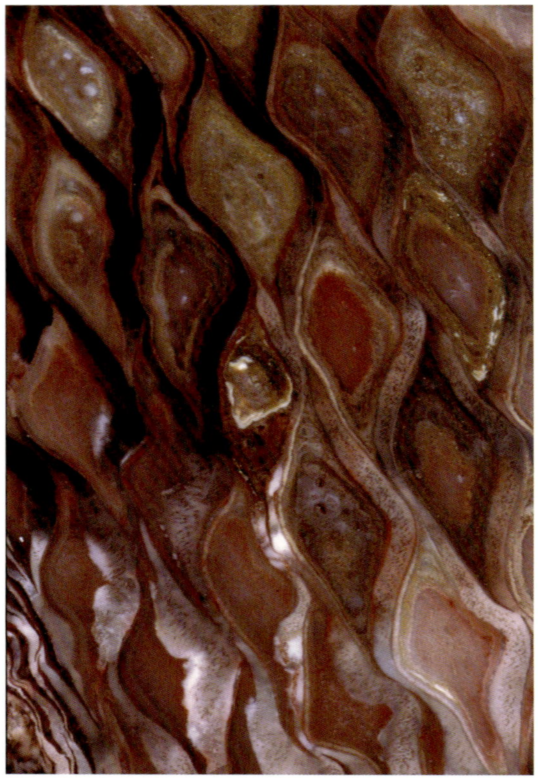

Macro photograph showing cycadeoid cell structure including partial cone. Cretaceous. Rio Negro, Argentina. Daniels specimen. Length of view approximately 2.5 cm. The diamonds in this specimen are quite small for a cycadeoid, averaging less than 1.0 cm in length.

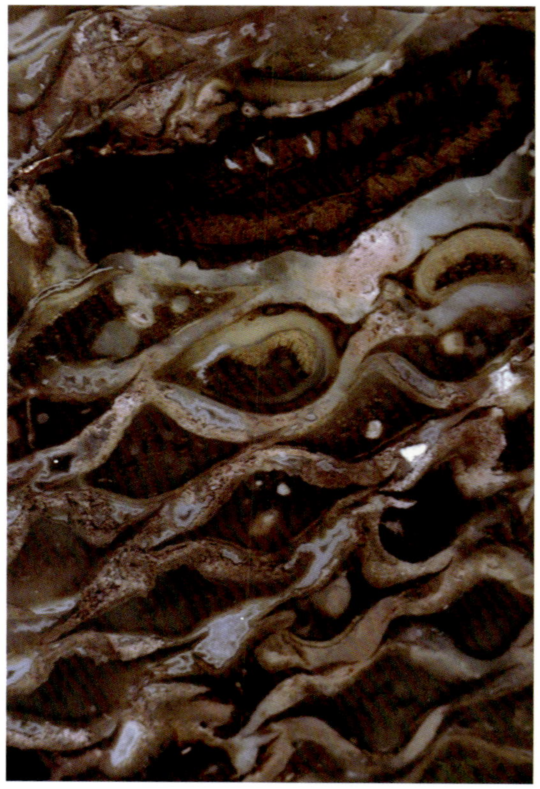

Macro photograph showing cycadeoid cell structure including cone. Cretaceous. Rio Negro, Argentina. Daniels specimen.

Cycadeoid with cones. Cretaceous. Rio Negro, Argentina. Daniels specimen, 13.5 cm.

Cycadophyte. Cretaceous. Rio Negro, Argentina. Daniels specimen, 10 cm.

Cycadeoid with cones. Cretaceous. Rio Negro, Argentina. Daniels specimen, 11 cm.

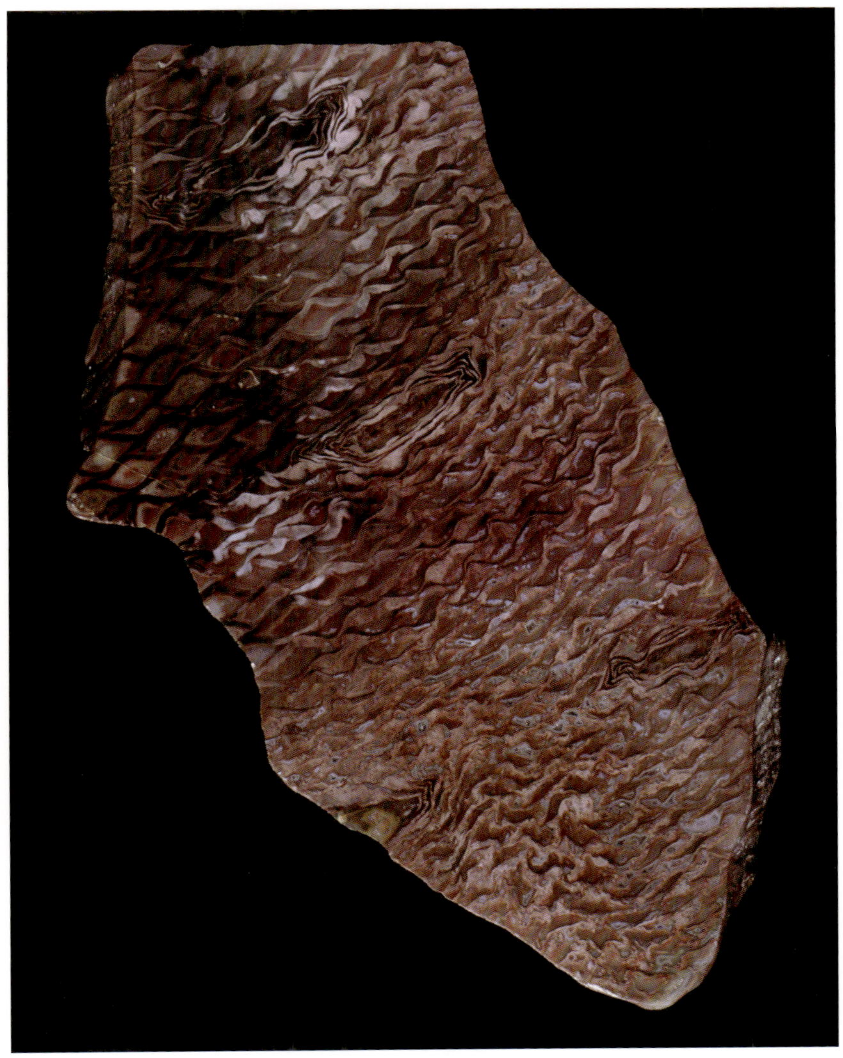

Cycadeoid with cones. Cretaceous. Rio Negro, Argentina. Daniels specimen, 14 cm.

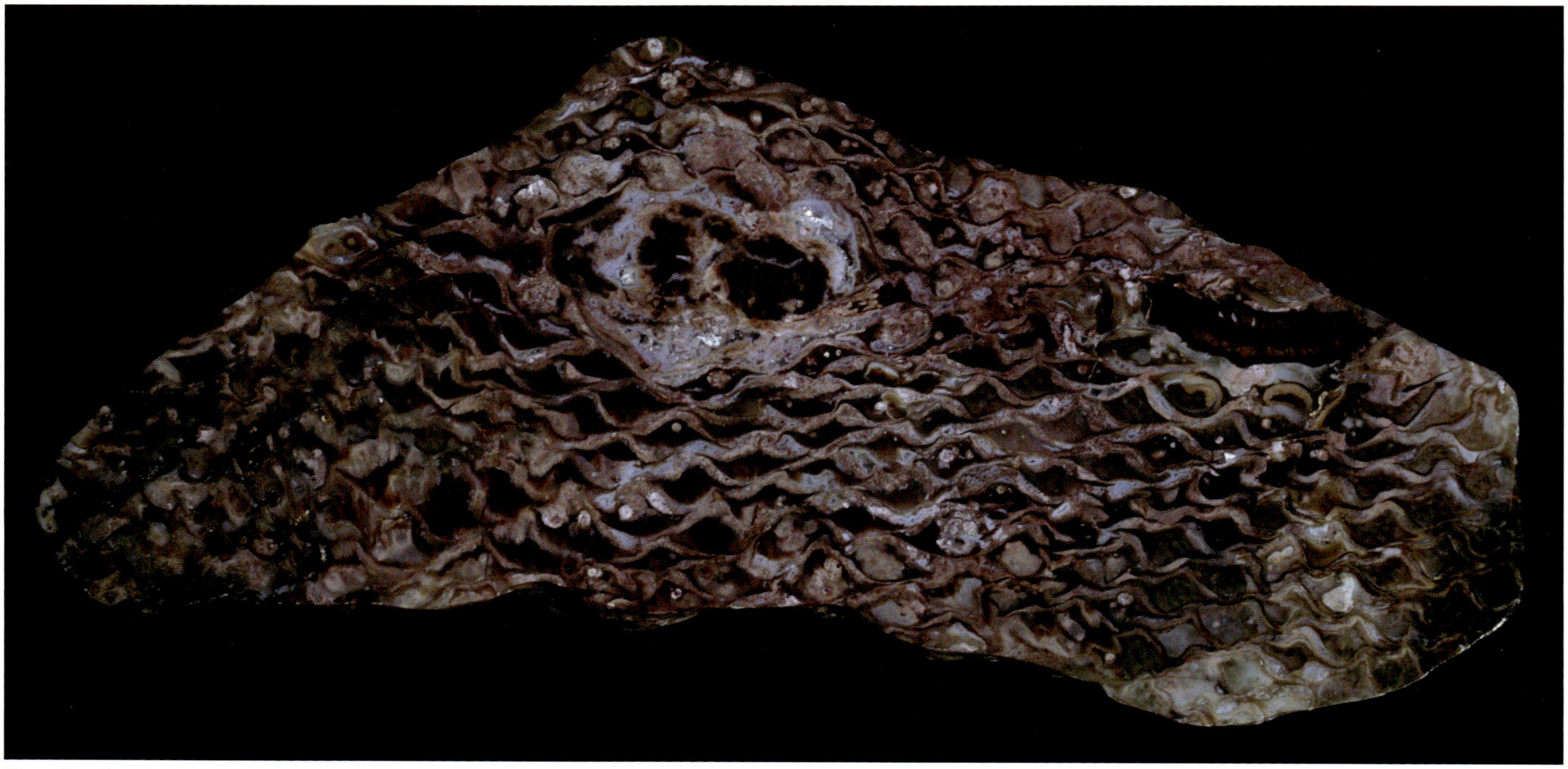

Cycadeoid with cones. Cretaceous. Rio Negro, Argentina. Daniels specimen, 13.5 cm.

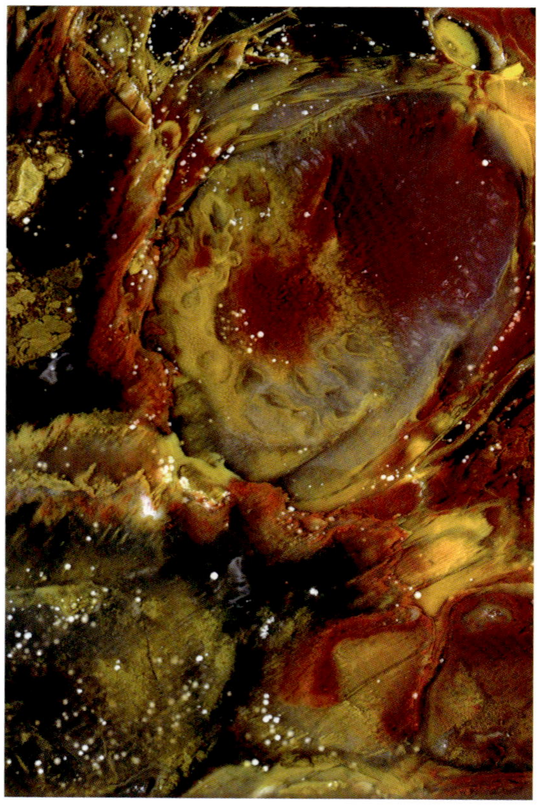

Macro photograph showing cycadeoid detail including cone. Morrison Formation, Jurassic. Henry Mountains, Utah, USA. Dayvault specimen.

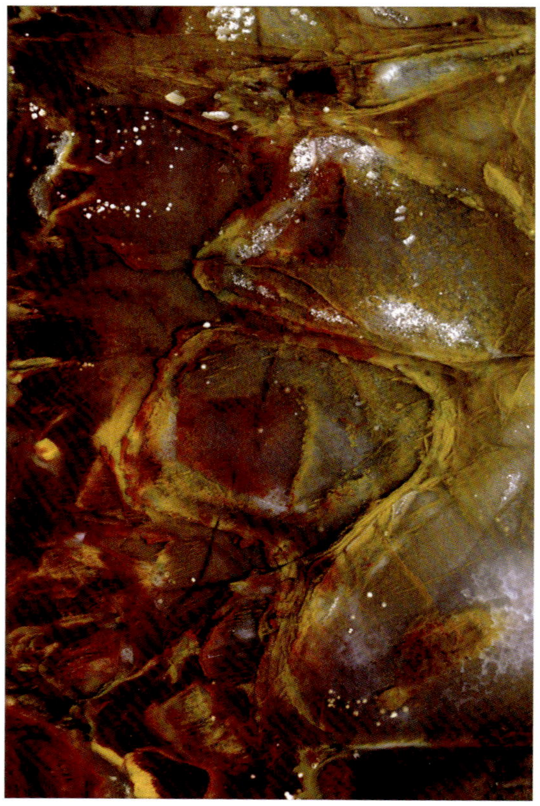

Macro photograph showing cycadeoid detail. Morrison Formation, Jurassic. Henry Mountains, Utah, USA. Dayvault specimen.

Macro photograph showing cycadeoid cell structure including cone. Cretaceous. Rio Negro, Argentina. Daniels specimen.

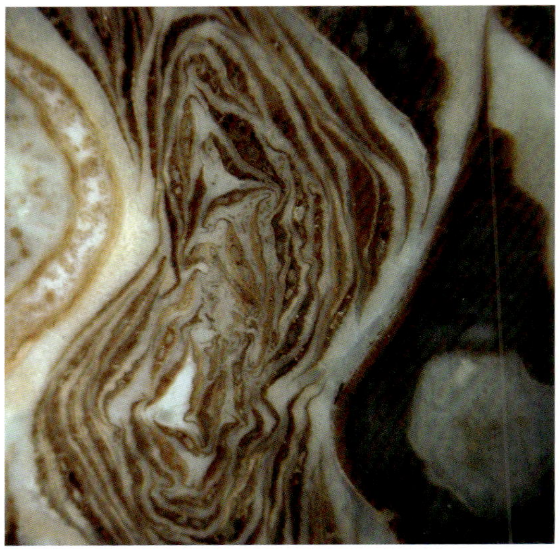

Micro image showing cycadeoid cone detail. Cretaceous. Rio Negro, Argentina. Daniels specimen.

Micro image showing cycadeoid detail. Cretaceous. Rio Negro, Argentina. Daniels specimen.

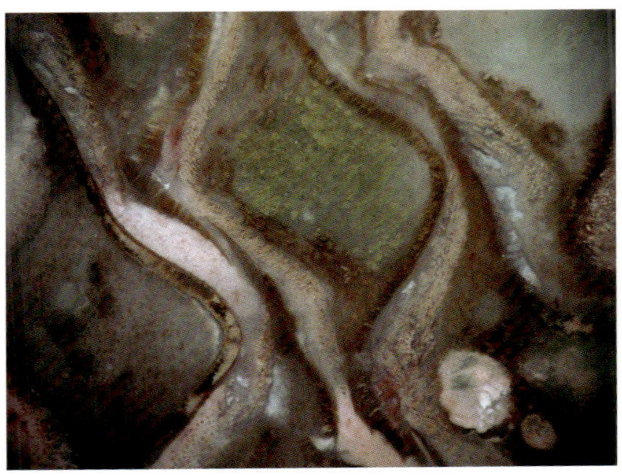

Micro image showing cycadeoid detail. Cretaceous. Rio Negro, Argentina. Daniels specimen.

Hermanophyton specimens from southwestern Colorado compete in silicification and beauty with the finest fossil wood specimens in existence. When viewed in transverse section, the image is like a lovely flower with symmetrical petals, infrequently blue with traces of red. Fortunately, early collectors of these fossils reported their finds to academic paleobotanists, resulting in an excellent research paper written by Professors William D. Tidwell and Sidney R. Ash and published in 1990 in the journal *Palaeontographica*. As they noted, "...the fossil has attracted considerable interest from amateur collectors in the western United States because of its unusual anatomy and apparent rarity." *Hermanophyton* specimens reside in just a few locations in exposures of the Jurassic Morrison Formation in southwestern Colorado and southeastern Utah. The McElmo Canyon area in Colorado produced the finest quality specimens. *Hermanophyton* fossils have been found in proximity to fossil conifers, cycadeoids, yet unidentified polyvascular wood fossils, and silicified dinosaur bone. Specimens range in size from 30 to more than 225 millimeters in diameter and can be more than 10 meters in length. "They consist of a relatively narrow pith, a xylem cylinder which is composed of 9-13 wedge-shaped segments of xylem separated by broad primary rays, a two-layer cortex, and a wide ramentum composed of hairs and club-shaped appendages. Small, round, persistent leaf bases are imbedded in the ramentum." (Tidwell and Ash) The authors surmised that this gymnosperm "of obscure affinity" grew in a "reasonably moist, forested environment" and "was probably a short to medium tall, narrow-stemmed tree crowned with small leaves which dropped off as the plant grew leaving numerous, small, persistent leaf bases on the lower part of the stem."

***Hermanophyton glismannii*. Morrison Formation, Upper Jurassic. Cortez, Colorado, USA. Dayvault specimen, 15.5 cm.**

Hermanophyton. Morrison Formation, Upper Jurassic. Cortez, Colorado, USA. Daniels specimen, 7 cm.

Unidentified polyvascular wood. Morrison Formation, Upper Jurassic. Cortez, Colorado, USA. Dayvault specimen, 8.5 cm.

Hermanophyton taylorii. Morrison Formation, Upper Jurassic. Cortez, Colorado, USA. Daniels specimen, 9 cm.

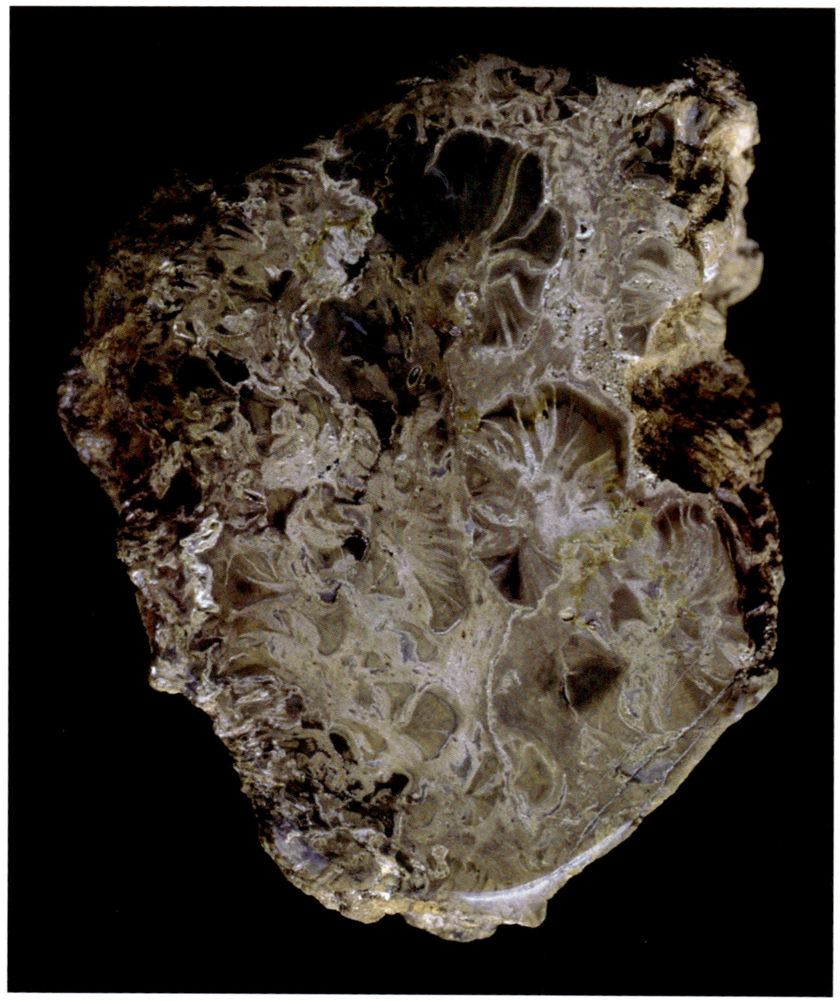

Unidentified polyvascular wood. Morrison Formation, Upper Jurassic. Cortez, Colorado, USA. Dayvault specimen, 10.5 cm.

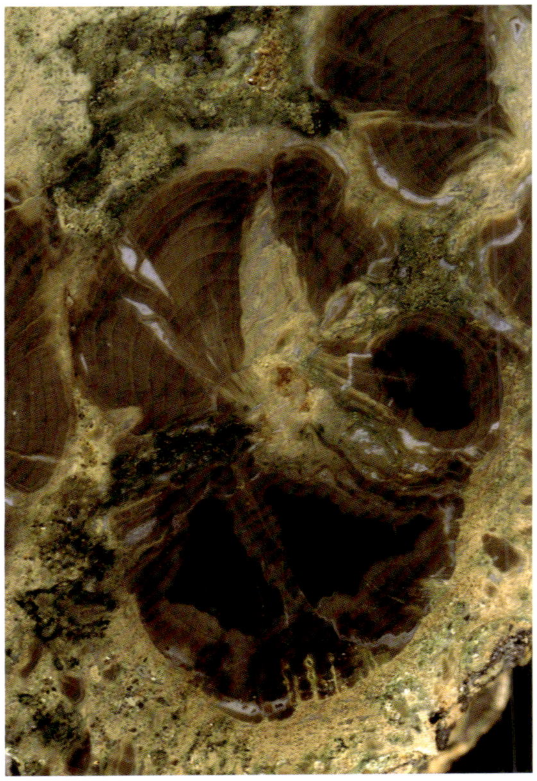

Macro photograph showing detail of unidentified polyvascular wood. Morrison Formation, Upper Jurassic. Cortez, Colorado, USA. Dayvault specimen.

Macro photograph showing *Hermanophyton* detail. Morrison Formation, Upper Jurassic. Cortez, Colorado, USA. Daniels specimen.

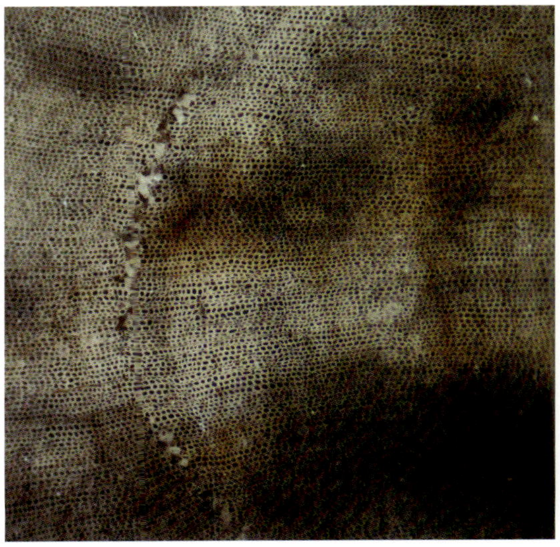

Micro image showing *Hermanophyton* detail. Morrison Formation, Upper Jurassic. Cortez, Colorado, USA. Daniels specimen.

Micro image showing *Hermanophyton* detail. Morrison Formation, Upper Jurassic. Cortez, Colorado, USA. Daniels specimen.

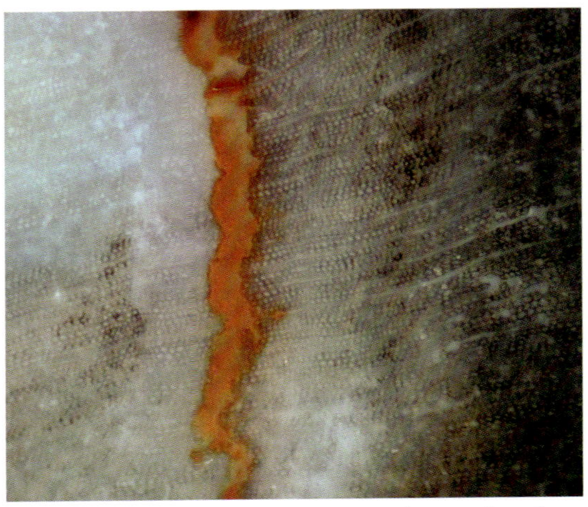

Micro image showing *Hermanophyton* detail. Morrison Formation, Upper Jurassic. Cortez, Colorado, USA. Daniels specimen.

Micro image showing unidentified conifer detail. Morrison Formation, Upper Jurassic. Cortez, Colorado, USA. Dayvault specimen.

Micro image showing detail of unidentified polyvascular wood. Morrison Formation, Upper Jurassic. Cortez, Colorado, USA. Dayvault specimen.

On the basis of interest and beauty, Australia's Mesozoic Era fossils rival those of any location. Tasmanian Jurassic horizons produce coniferous wood, osmundaphytes, and small apparent tree ferns. "Shipworm wood" from Western Australia's Lower Cretaceous Windalia Radiolarite Formation, found along the western fringes of the Kennedy Ranges in the Gascoyne region, is coniferous wood with borings of a bivalve mollusk such as *Teredo* or *Bankia*, commonly known as shipworms. These borings are often colorfully infilled with silica. Jurassic formations in Queensland, in particular, have produced many fine specimens of coniferous woods, finely preserved osmundaphytes, and the unusual extinct unnamed gymnosperm informally called *Pentoxylon* by collectors because of perceived similarities with the Jurassic fossil *Pentoxylon* from India.

Shipworm wood. Windalia Radiolarite Formation. Lower Cretaceous. Kennedy Ranges, Western Australia, Australia. Daniels specimen, 14 cm.

Shipworm wood. Windalia Radiolarite Formation. Lower Cretaceous. Kennedy Ranges, Western Australia, Australia. Daniels specimen, 22 cm.

Shipworm wood. Windalia Radiolarite Formation. Lower Cretaceous. Kennedy Ranges, Western Australia, Australia. Daniels specimen, 17.5 cm.

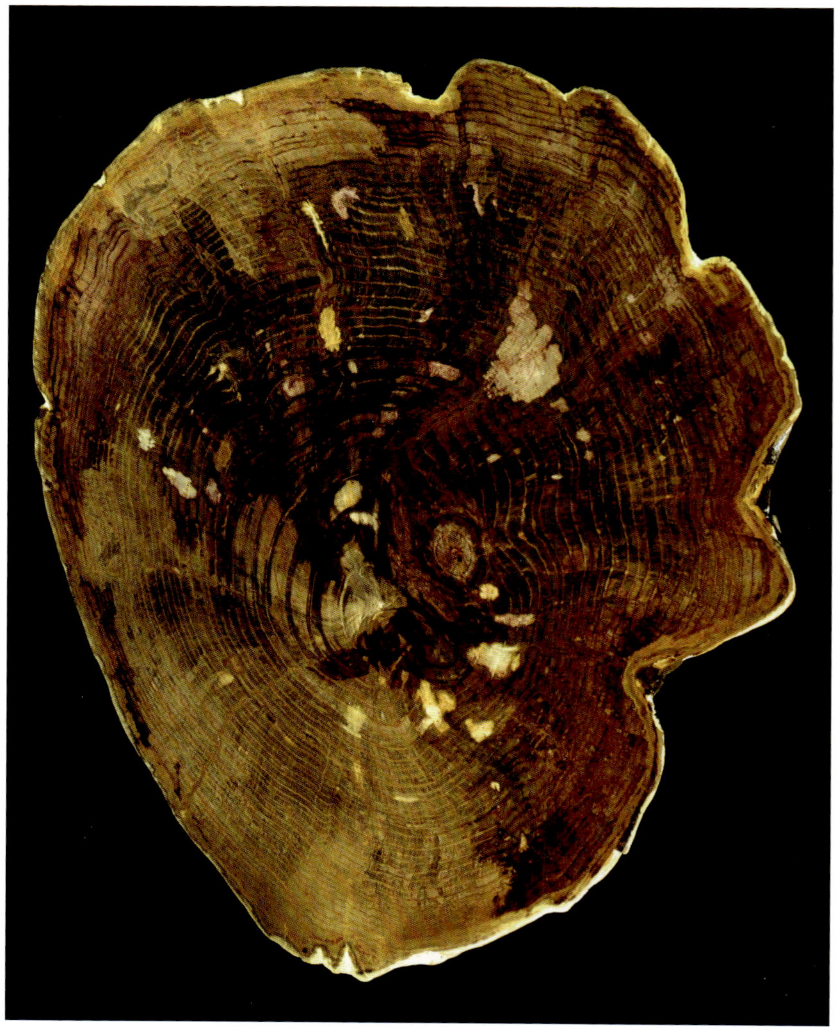

Unidentified conifer. Jurassic. Queensland, Australia. Daniels specimen, 20 cm.

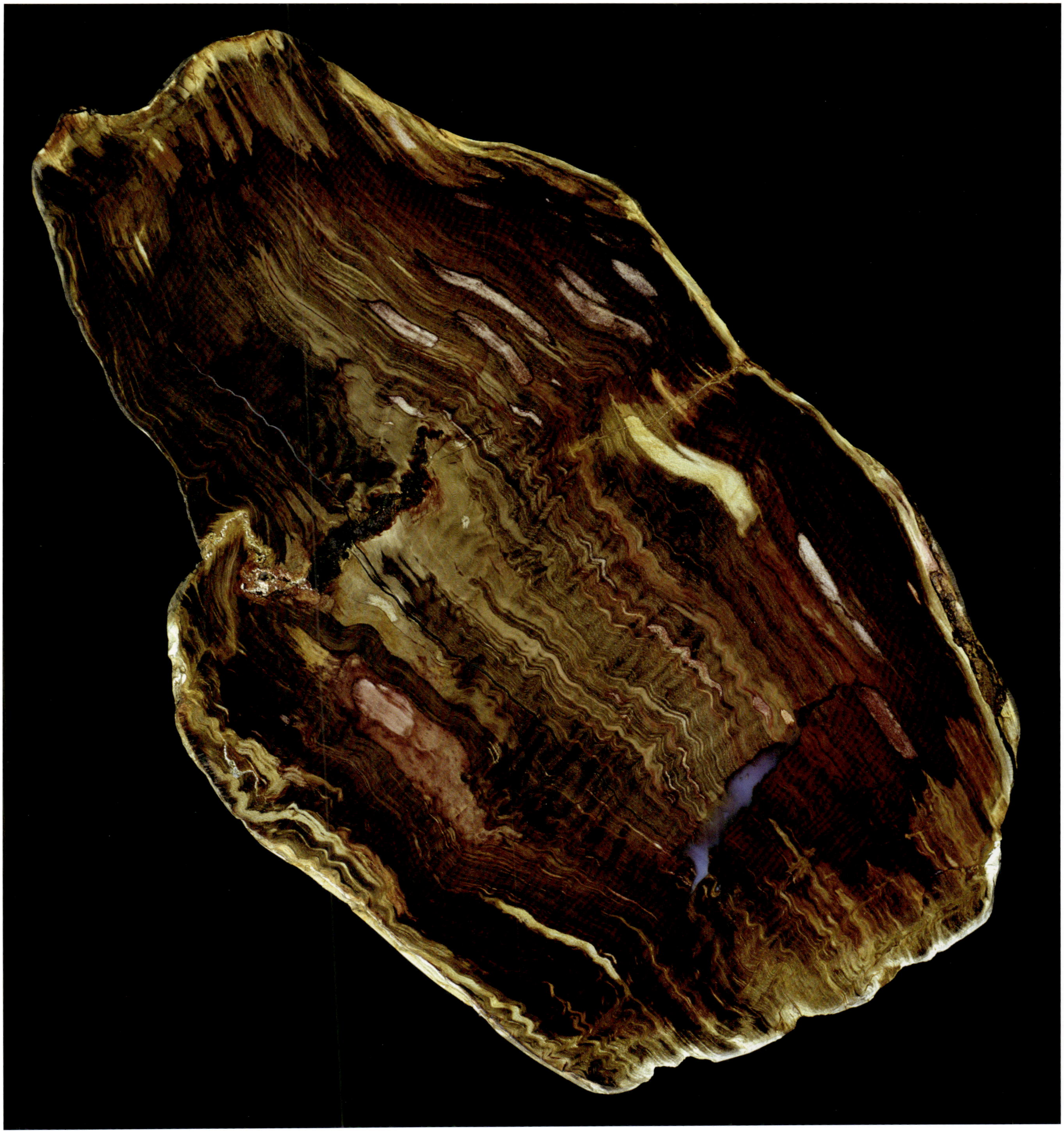

Unidentified conifer. Jurassic. Queensland, Australia. Daniels specimen, 27 cm.

Micro image showing cell structure in fossil wood. Jurassic. Lune River, Tasmania, Australia. Daniels specimen.

Unidentified conifer. Jurassic. Lune River, Tasmania, Australia. Daniels specimen, 22 cm.

Unidentified conifer. Jurassic. Tasmania, Australia. Daniels specimen, 16 cm.

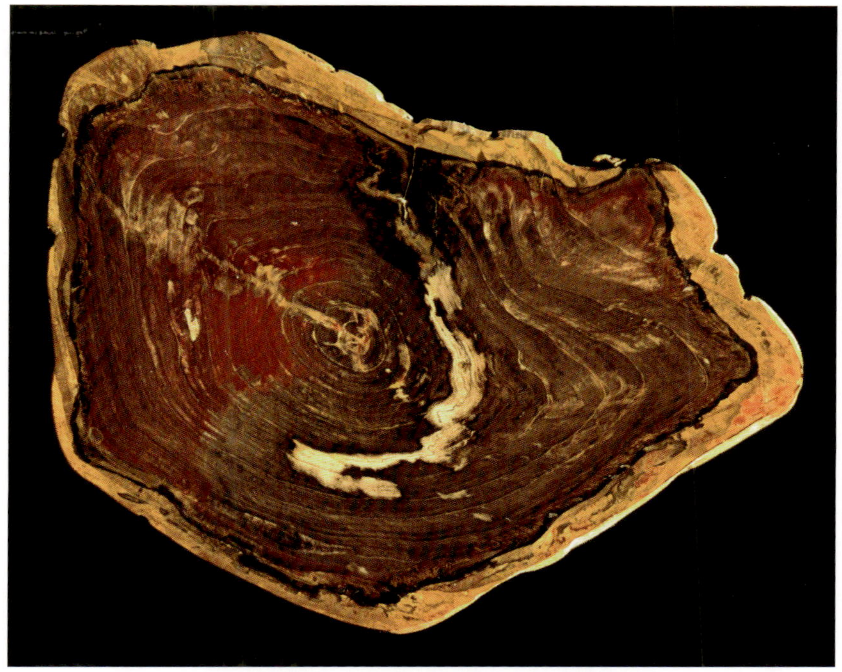

Unidentified conifer. Jurassic. Queensland, Australia. Daniels specimen, 10 cm.

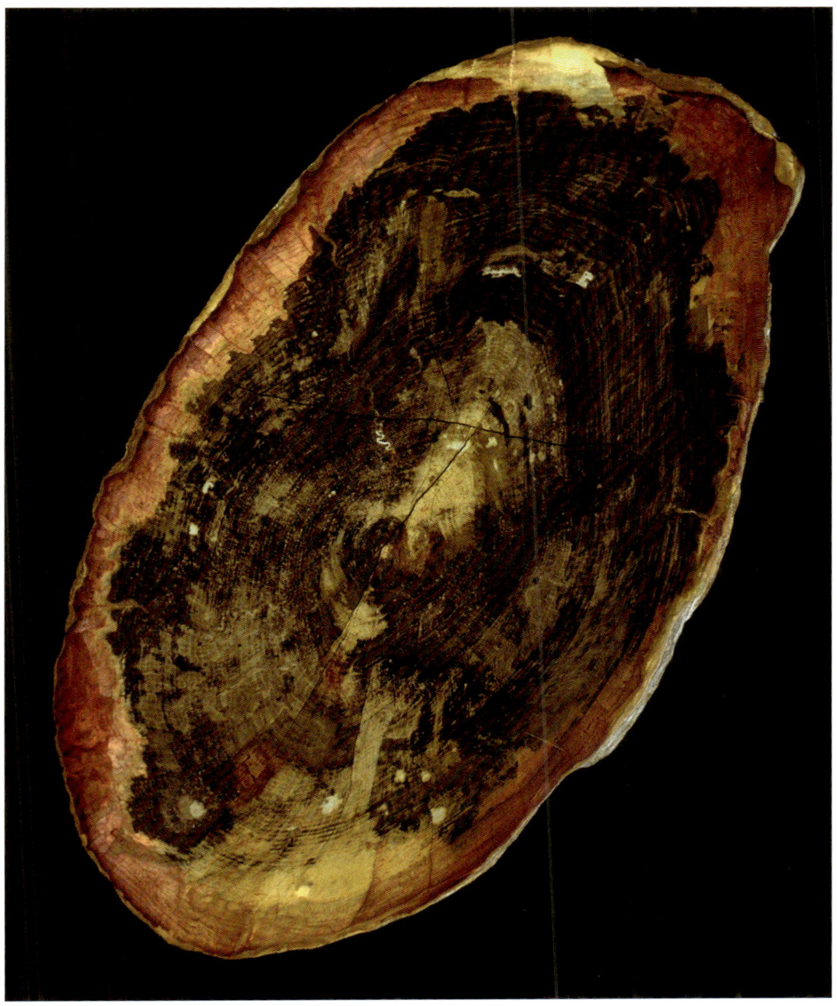

Unidentified conifer. Jurassic. Queensland, Australia. Daniels specimen, 20 cm.

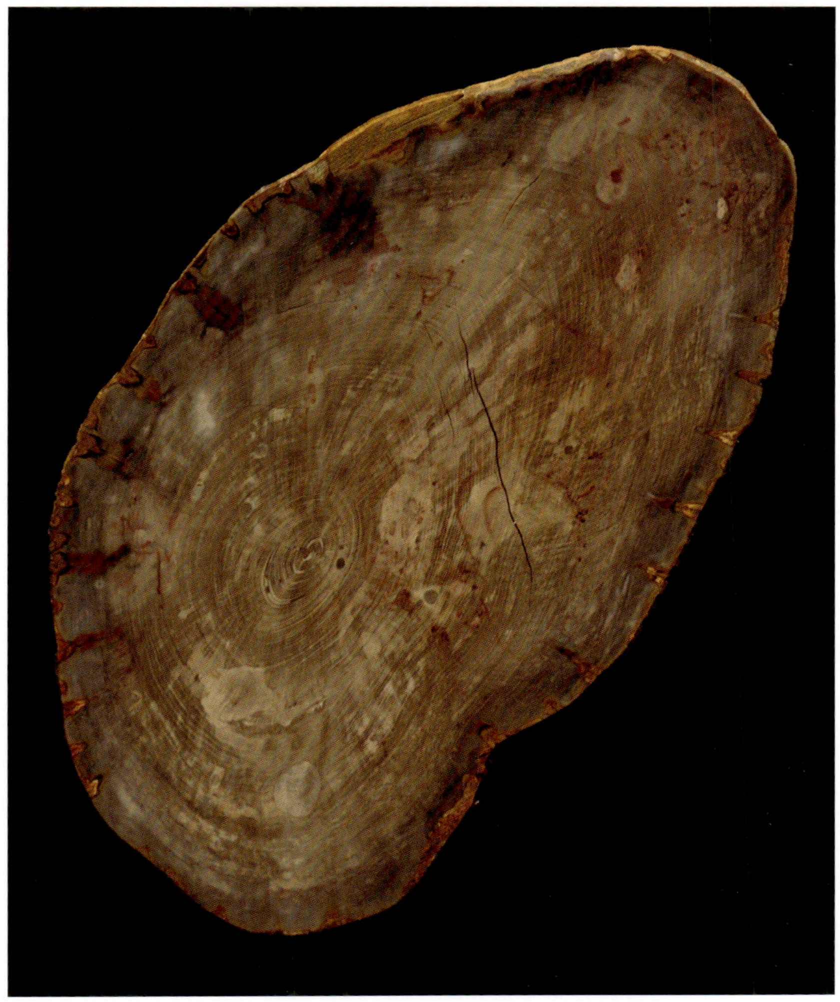

Unidentified conifer. Jurassic. Queensland, Australia. Daniels specimen, 19 cm.

Unidentified conifer. Jurassic. Queensland, Australia.
Daniels specimen, 8.5 cm.

Unidentified conifer. Jurassic. Queensland, Australia.
Daniels specimen, 9 cm.

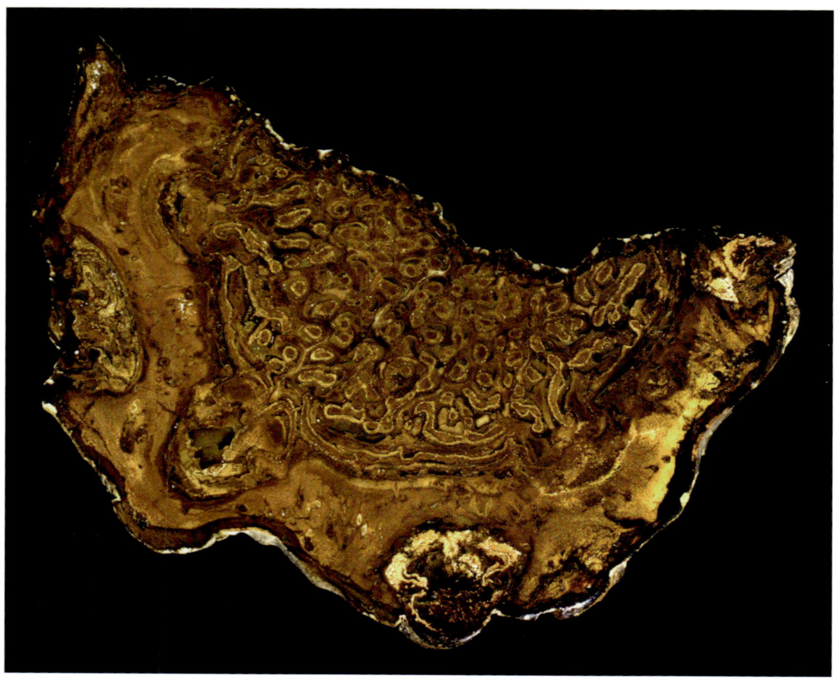

Fossil tree fern. Jurassic. Lune River, Tasmania, Australia.
Daniels specimen, 8 cm.

Unidentified conifer. Jurassic. Queensland, Australia.
Daniels specimen, 11 cm.

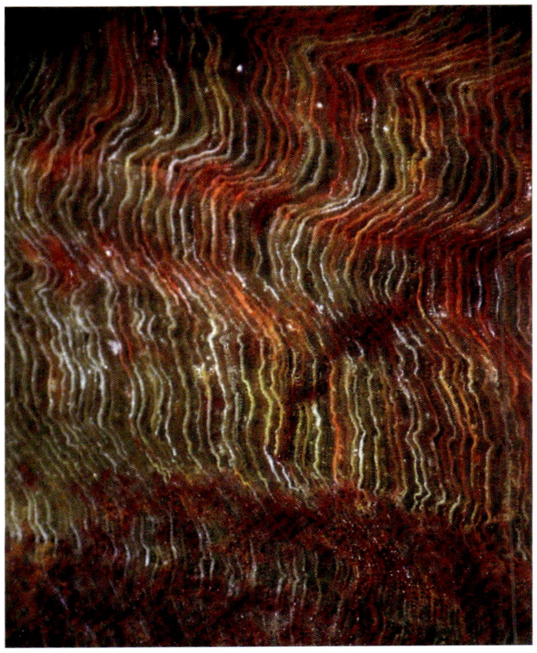

Micro image showing cell structure in fossil wood. Jurassic. Queensland, Australia. Daniels specimen.

Micro image showing cell structure in fossil wood. Jurassic. Queensland, Australia. Daniels specimen.

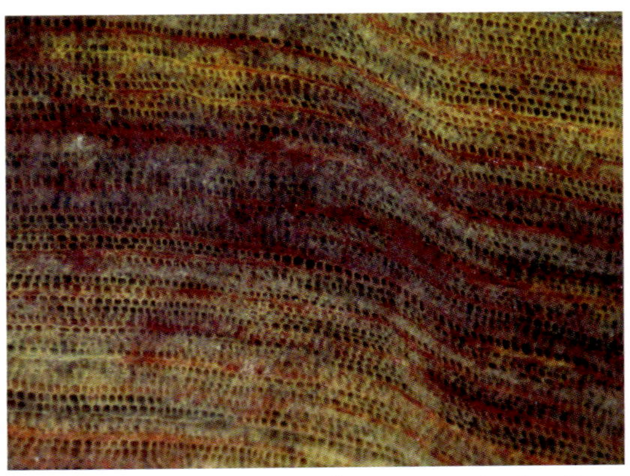

Micro image showing cell structure in fossil wood. Jurassic. Queensland, Australia. Daniels specimen.

Micro image showing cell structure in fossil wood. Jurassic. Queensland, Australia. Daniels specimen.

Micro image showing cell structure in fossil wood. Jurassic. Queensland, Australia. Daniels specimen.

Micro image showing cell structure in fossil wood. Jurassic. Queensland, Australia. Daniels specimen.

Micro image showing cell structure in fossil wood. Jurassic. Queensland, Australia. Daniels specimen.

Macro photograph of shipworm tunnels in fossil conifer. Windalia Radiolarite Formation. Lower Cretaceous. Kennedy Ranges, Western Australia, Australia. Daniels specimen.

Macro photograph of shipworm tunnels in fossil conifer. Windalia Radiolarite Formation. Lower Cretaceous. Kennedy Ranges, Western Australia, Australia. Daniels specimen.

Micro image showing structure in fossil shipworm wood. Windalia Radiolarite Formation, Lower Cretaceous. Kennedy Ranges, Western Australia, Australia. Daniels specimen.

Macro photograph of shipworm tunnels in fossil conifer. Windalia Radiolarite Formation. Lower Cretaceous. Kennedy Ranges, Western Australia, Australia. Daniels specimen.

Macro photograph of shipworm tunnels in fossil conifer. Windalia Radiolarite Formation. Lower Cretaceous. Kennedy Ranges, Western Australia, Australia. Daniels specimen.

Macro photograph of shipworm tunnels in fossil conifer. Windalia Radiolarite Formation. Lower Cretaceous. Kennedy Ranges, Western Australia, Australia. Daniels specimen.

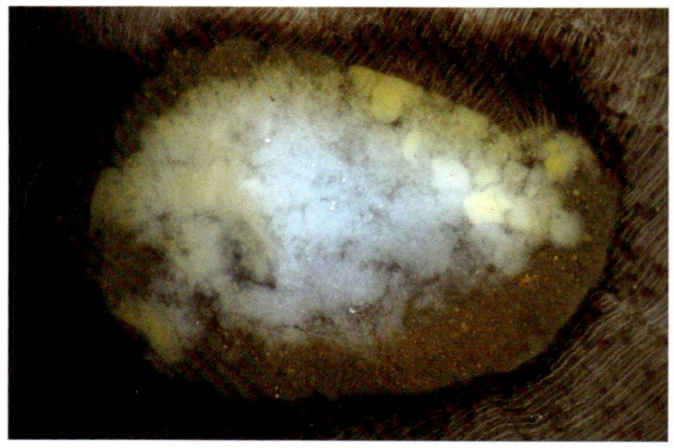

Micro image showing structure in fossil shipworm wood. Windalia Radiolarite Formation, Lower Cretaceous. Kennedy Ranges, Western Australia, Australia. Daniels specimen.

Micro image showing structure in fossil shipworm wood. Windalia Radiolarite Formation, Lower Cretaceous. Kennedy Ranges, Western Australia, Australia. Daniels specimen.

Micro image showing structure in fossil wood. Jurassic. Lune River, Tasmania, Australia. Daniels specimen.

True *Pentoxylon*, from Lower and Middle Jurassic sediments in India's Rajmahal Hills, has woody vascular segments arranged around the center. The number of segments is frequently 5 but ranges from 4 to 10. *Pentoxylon* has coniferous-like secondary wood and spirally arranged leaf bases. The leaves have characteristics of cycadophyte leaves. Describing Australian polyvascular fossil wood is difficult because most stems are heavily abraded as though river-worn or otherwise tumbled by water or glacial action, resulting in a lack of external structures. Specimens with a preserved center display multiple vascular segments arranged symmetrically as in *Pentoxylon*. In most specimens, multiple vascular segments multiply outward from the center in a wave-like pattern, sometimes numbering in hundreds of waves. As fossils, these stems can reveal beautiful patterns and colors when cut and polished.

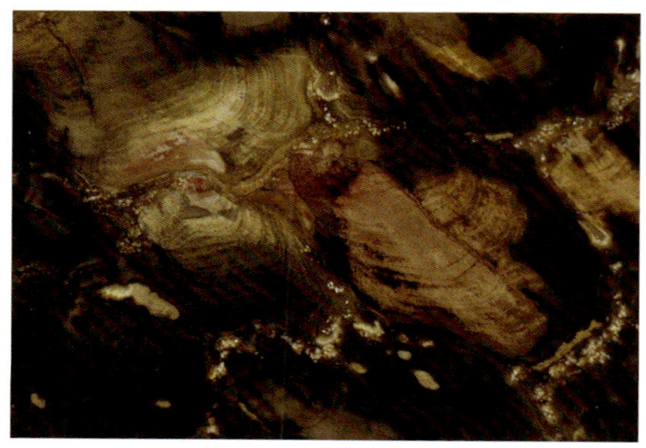

Macro photograph showing detail of polyvascular fossil wood. Jurassic. Queensland, Australia. Daniels specimen.

Polyvascular wood. Jurassic. Queensland, Australia. Daniels specimen, 18.5 cm.

Macro photograph showing detail of polyvascular fossil wood. Jurassic. Queensland, Australia. Daniels specimen.

Macro photograph showing detail of polyvascular fossil wood. Jurassic. Queensland, Australia. Daniels specimen.

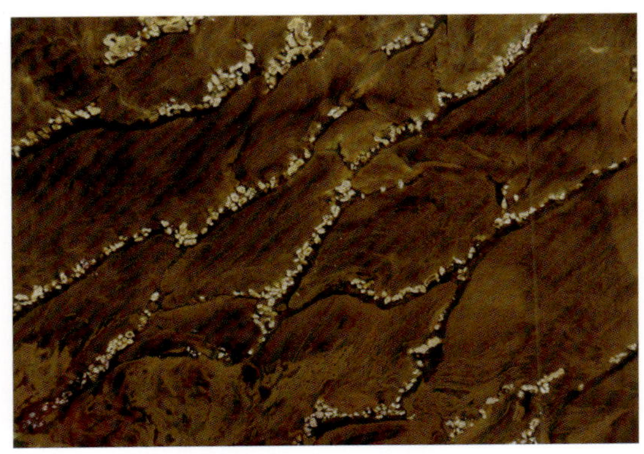

Macro photograph showing detail of polyvascular fossil wood. Jurassic. Queensland, Australia. Daniels specimen.

Polyvascular wood. Jurassic. Queensland, Australia. Daniels specimen, 19 cm.

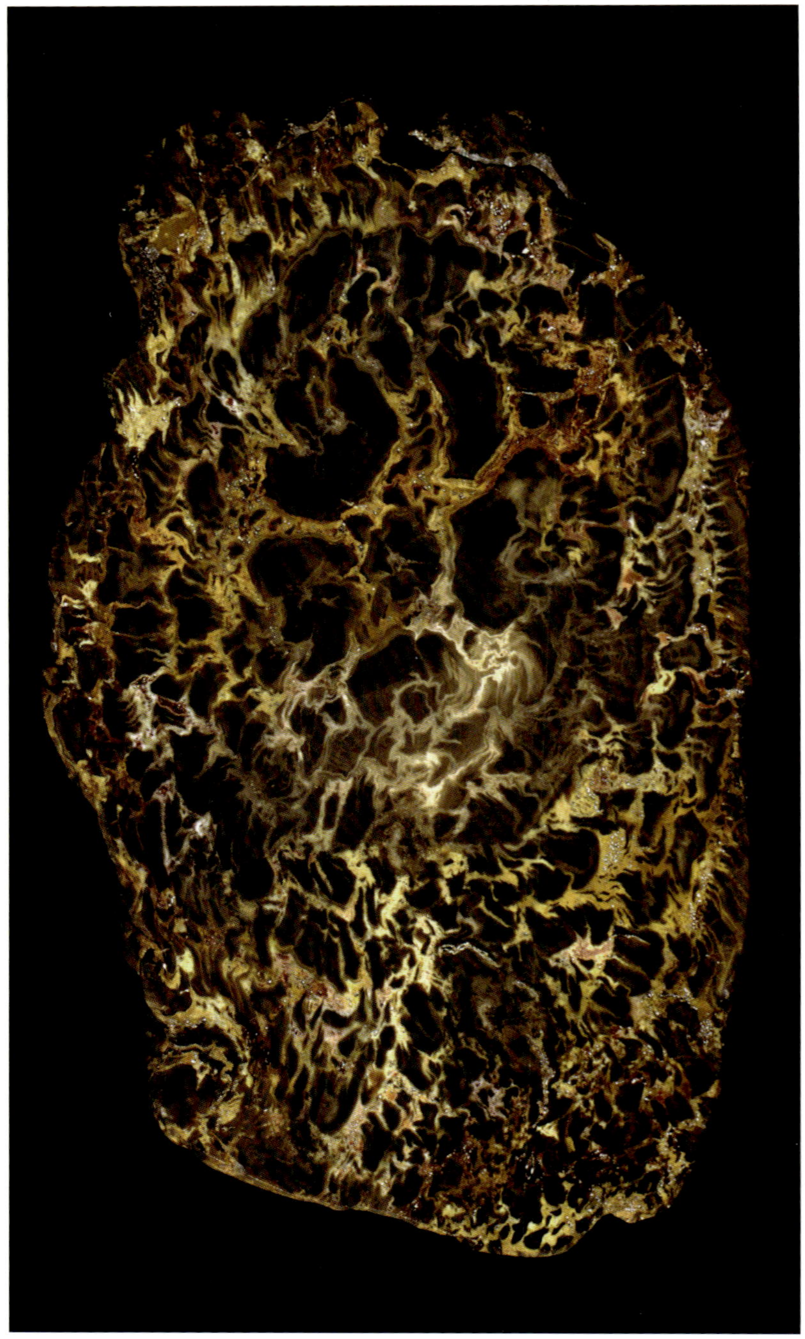

Polyvascular wood. Jurassic. Queensland, Australia. Daniels specimen, 13.5 cm.

People worldwide know fossil woods from the Triassic Chinle Formation of northern Arizona. Specimens of "rainbow wood," so described because of the range and brilliance of colors, reside in museums around the globe. The Chinle Formation is thick, expansive, and complex. It represents what was once a huge expanse of floodplains, deltas, and marshes at the western edge of Pangea. At Petrified Forest National Park in northern Arizona, the Chinle Formation has an approximate thickness of 400 meters (1,300 feet).

The mechanisms for preservation of plants through silicification were evidently quite perfect, resulting in a fascinating array of fossil plant remains to collect, study, and admire. Fossil woods from Chinle Formation exposures in Arizona and Utah yield distinctly differently mineralized specimens of the same few identified genera. The most prevalent fossil wood encountered (more than 80 percent) is *Araucarioxylon* (*Araucarioxylon arizonicum*), followed at a distance by *Woodworthia* (*Woodworthia arizonica*), then in far fewer numbers by *Schilderia* (*Schilderia adamanica*) and "star dadoxylon." *Woodworthia* and *Schilderia* specimens are found only in upper Chinle rocks. *Araucarioxylon* in this location were huge trees, some reaching 59 meters (192 feet) in height. *Woodworthia* were smaller than *Araucaria*, and *Schilderia* and "star dadoxylon" were smaller than *Woodworthia*. *Woodworthia*, a member of the Protopinaceae family, displays transitional characteristics bridging Paleozoic *Dadoxylon* and modern conifers with wood of the *Araucarioxylon* type.

Woodworthia. Chinle Formation, Triassic. Henry Mountains area, Utah, USA. Daniels specimen, 19 cm.

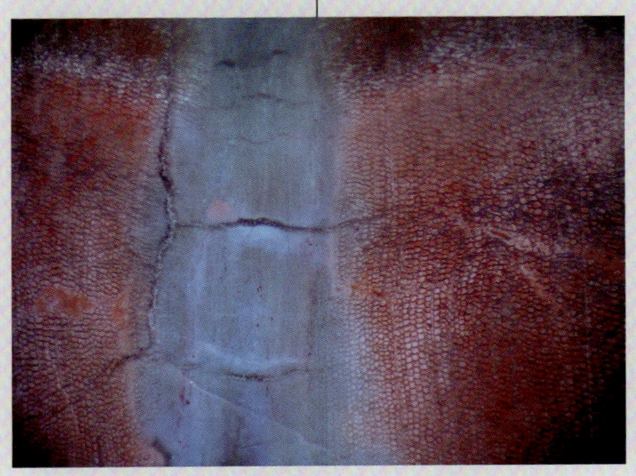

Micro image showing cell structure in *Woodworthia*. Chinle Formation, Triassic. Henry Mountains area, Utah, USA. Daniels specimen. Average tracheid size approximately 0.04 mm.

Chinle fossil woods designated informally as "star dadoxylon" are a distinct fossil from the recognized fossils of *Araucarioxylon*, *Woodworthia*, and *Schilderia*. In this book, we use the term *Dadoxylon* for fossil wood with coniferous, *Araucaria*-like tracheid structure from the Permian Period or earlier, and the term star dadoxylon for Triassic Period woods with *Araucaria*-like tracheid structure and a distinct star pattern at the pith. This star pattern structure is present throughout the life of the tree and is not the result of post-depositional decay or mineralization, although stems with these features are frequently confused as star dadoxylon. Immature stems of star dadoxylon have a growth pattern at the pith of tapered, interlocking, divided xylem sheaves that later blossom into the star. Triassic Chinle Formation exposures in both Arizona and Utah yield star dadoxylon, albeit rarely.

Other plant fossils of the 41 described genera preserved in the Chinle Formation include cycadeoids, ferns, and horsetails, all rarely preserved as well-silicified specimens. Because of distinct gross characteristic differences, the four major Triassic Chinle woods of Arizona and Utah are easy to identify with use of a hand lens. *Woodworthia* grew with persistent spike-like short shoots that left distinct, internal, broad ray-like patterns and prominent small external dimples or bumps. Not all transverse sections capture a portion of these short shoots, so the exterior merits inspection as well. The wide variety of *Woodworthia* types feature considerable range in size and distribution of short shoots. Some of the world's finest specimens come from the Chinle Formation of Arizona and Utah. *Schilderia* has an abundance of pronounced multiseriate rays and an undulating tracheid pattern. Star dadoxylon has a distinct star pattern at the pith. Almost every other large specimen of fossil wood from the area is *Araucarioxylon* and is easily identified with low-power magnification should sufficient structure be preserved. As the forests in which these trees grew were of long duration, the genera involved undoubtedly experienced numerous natural selection shifts, resulting in a variety of species. The fossil record preserves this evidence and patiently awaits further scientific analysis.

Star dadoxylon. Chinle Formation, Triassic. Northern Arizona, USA. Daniels specimen, 7 cm.

Star dadoxylon. Chinle Formation, Triassic. Circle Cliffs, Utah, USA. Daniels specimen, 5 cm.

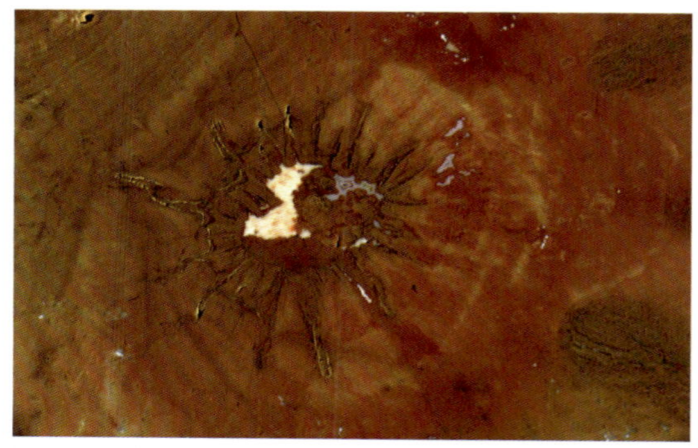

Macro photograph showing detail of star dadoxylon. Chinle Formation, Triassic. Northern Arizona, USA. Daniels specimen.

Star dadoxylon. Chinle Formation, Triassic. Northern Arizona, USA. Daniels specimen, 9 cm.

Star dadoxylon. Chinle Formation, Triassic. Northern Arizona, USA. Daniels specimen, 9 cm.

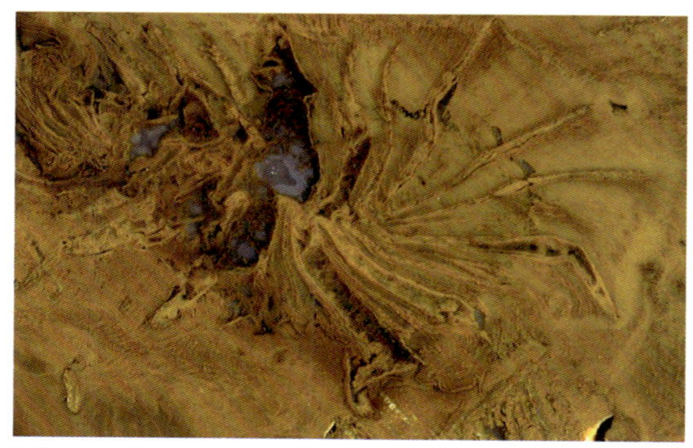

Macro photograph showing detail of star dadoxylon. Chinle Formation, Triassic. Northern Arizona, USA. Daniels specimen.

Star dadoxylon. Chinle Formation, Triassic. Henry Mountains area, Utah, USA. Mower specimen, 17 cm.

Schilderia. Chinle Formation, Triassic. Northern Arizona, USA. Gray specimen, 8 cm.

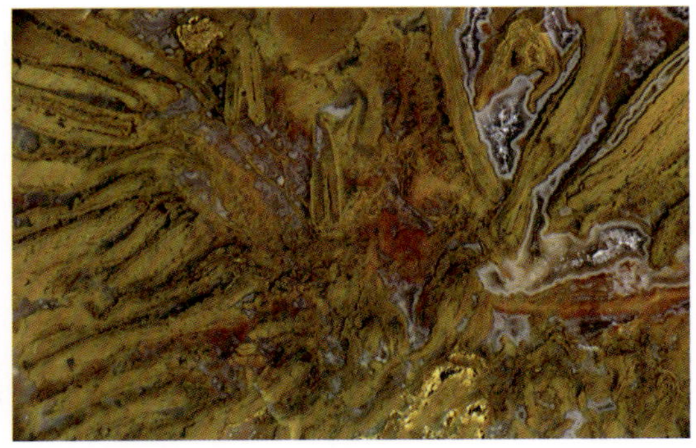

Macro photograph showing detail of star dadoxylon. Chinle Formation, Triassic. Northern Arizona, USA. Daniels specimen.

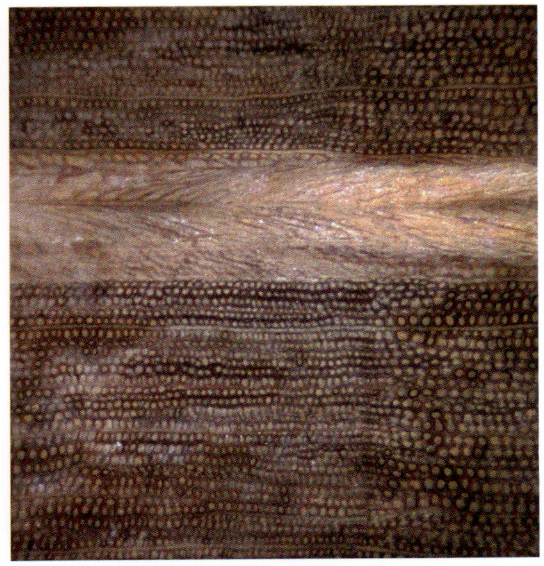

Micro image showing detail of *Schilderia* cell structure. Chinle Formation, Triassic. Northern Arizona, USA. Daniels specimen.

Micro image showing detail of *Schilderia* cell structure. Chinle Formation, Triassic. Northern Arizona, USA. Daniels specimen. Widest ray approximately 0.3 mm.

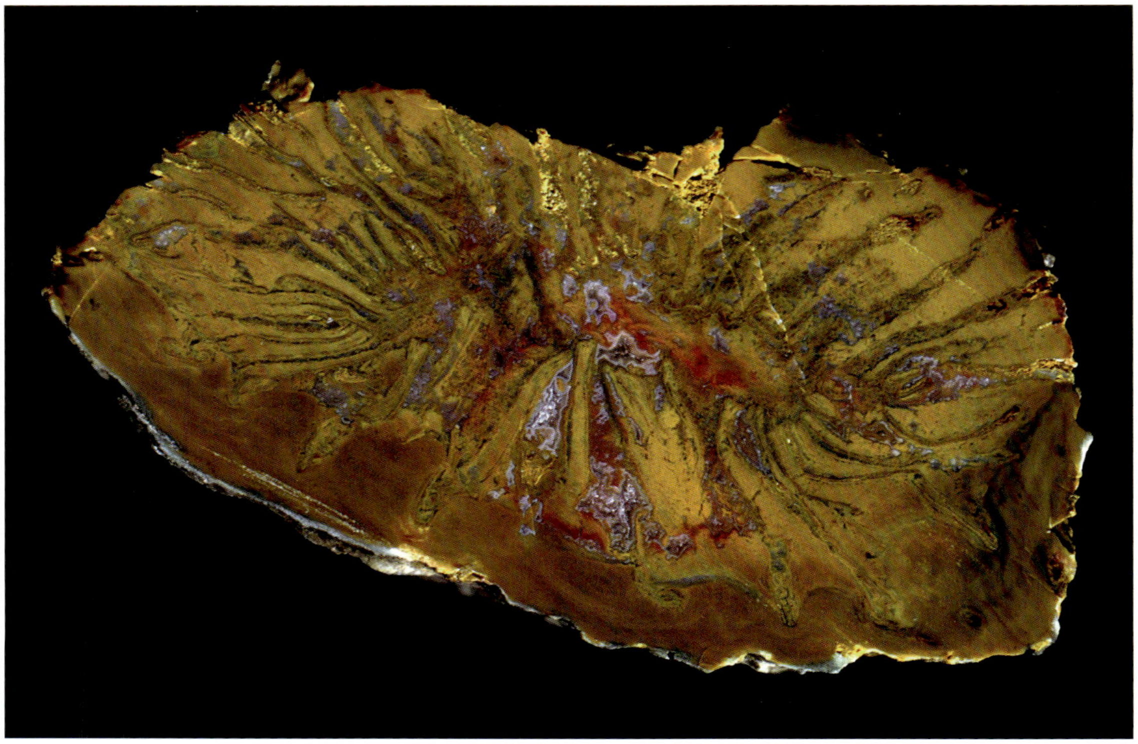

Star dadoxylon. Chinle Formation, Triassic. Northern Arizona, USA. Daniels specimen, 6.5 cm.

Micro image showing detail of star dadoxylon cell structure. Chinle Formation, Triassic. Henry Mountains area, Utah, USA. Mower specimen.

Macro photograph showing detail of star dadoxylon. Chinle Formation, Triassic. Northern Arizona, USA. Daniels specimen.

Micro image showing detail of star dadoxylon cell structure. Chinle Formation, Triassic. Northern Arizona, USA. Daniels specimen.

Micro image showing detail of star dadoxylon cell structure. Chinle Formation, Triassic. Henry Mountains area, Utah, USA. Mower specimen.

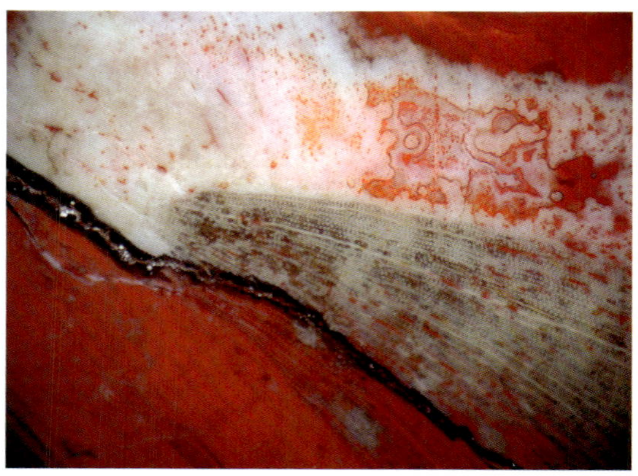

Micro image showing detail of star dadoxylon cell structure. Chinle Formation, Triassic. Northern Arizona, USA. Daniels specimen.

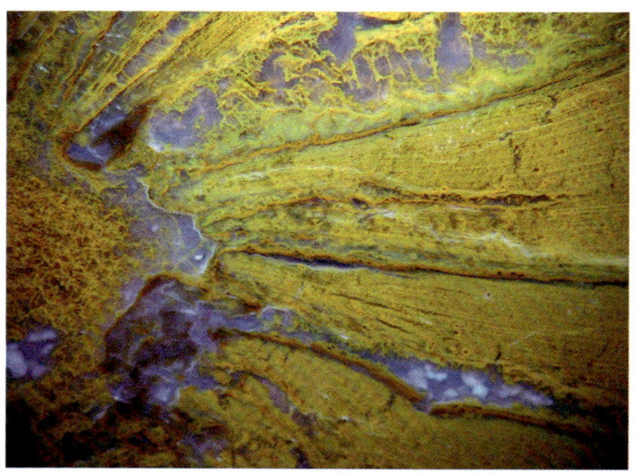

Micro image showing detail of star dadoxylon cell structure. Chinle Formation, Triassic. Northern Arizona, USA. Daniels specimen.

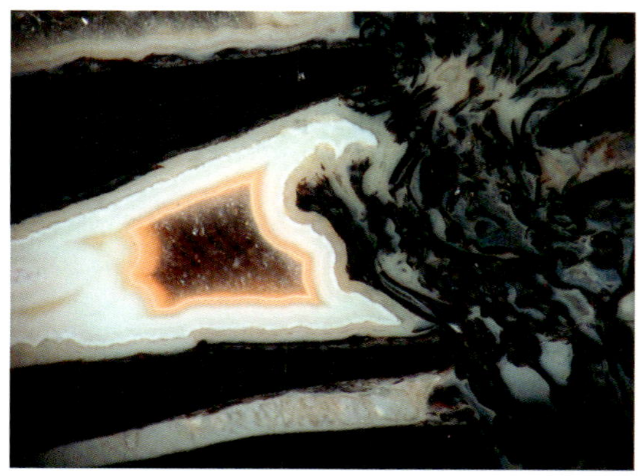

Micro image showing detail of star dadoxylon cell structure. Chinle Formation, Triassic. Henry Mountains area, Utah, USA. Mower specimen.

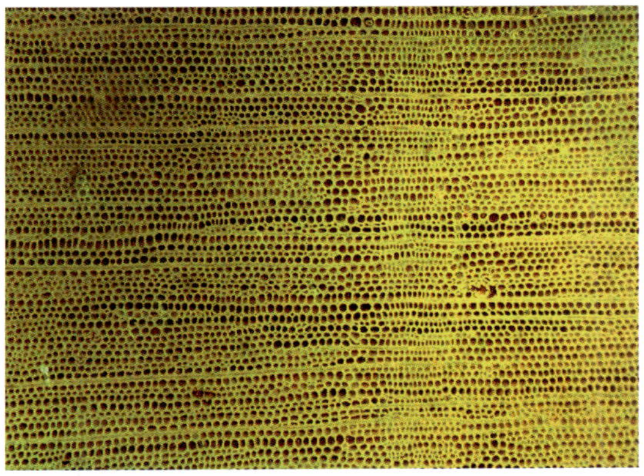

Micro image showing detail of coniferous cell structure. Morrison Formation, Jurassic. Henry Mountains area, Utah, USA. Daniels specimen.

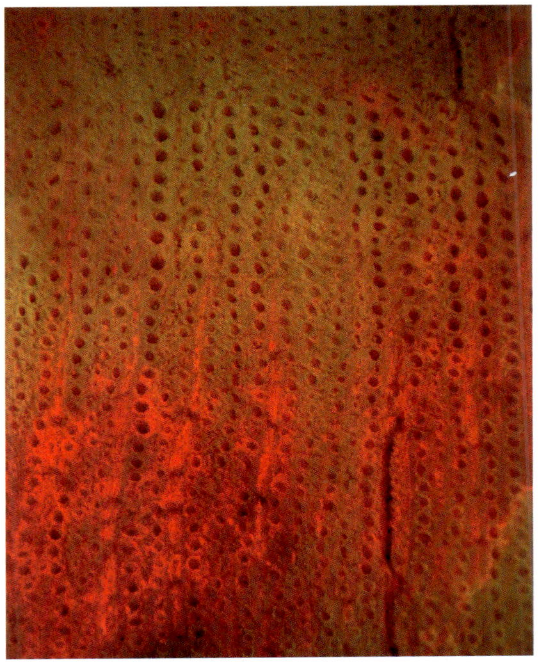

Micro image showing detail of coniferous cell structure. Morrison Formation, Jurassic. Henry Mountains area, Utah, USA. Daniels specimen.

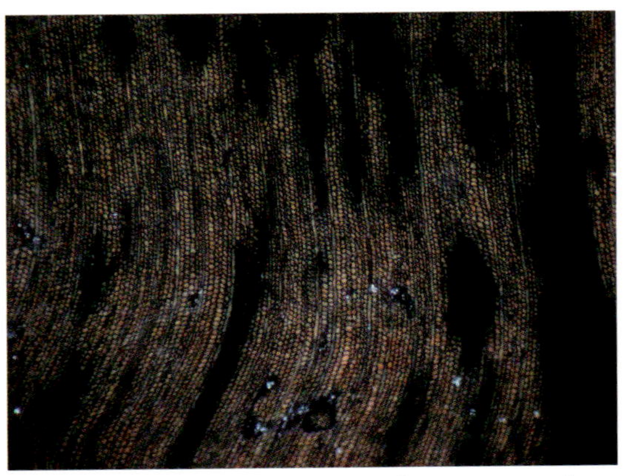

Micro image showing detail of *Araucarioxylon* cell structure. Chinle Formation, Triassic. Circle Cliffs, Utah, USA. Daniels specimen.

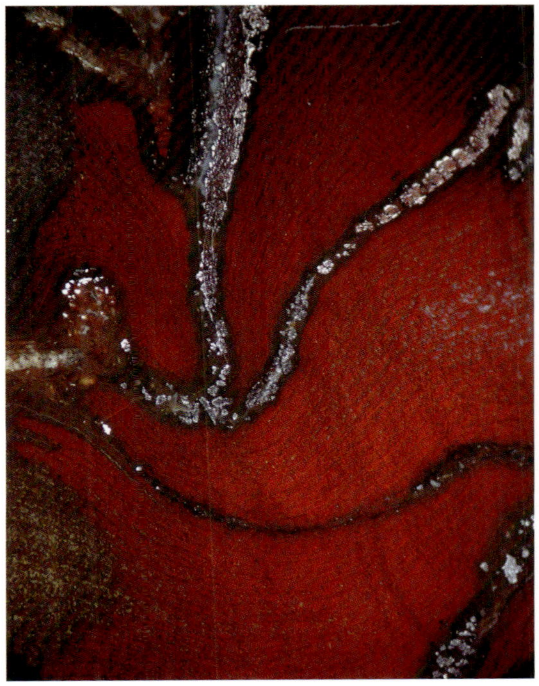

Micro image showing detail of star dadoxylon cell structure in an immature stem. Chinle Formation, Triassic. Circle Cliffs, Utah, USA. Daniels specimen. Average tracheid size approximately 0.03 mm.

Zimbabwe, Africa, is the source of many attractive and well-preserved Triassic Period fossil wood specimens, including *Araucarioxylon* and *Woodworthia*. Zimbabwean *Woodworthia* is among the most beautiful and well preserved in the world. Rare specimens of Zimbabwean wood are colored deep green by chromium. One especially interesting and enigmatic fossil wood from these horizons is the genus *Rhexoxylon*. These trees were fairly large with straight upright trunks. The structure consists of secondary wood in multiple vascular segments. The main identifying characteristic is the development of xylem in fan-like patterns, either solitary or in clusters. Similarities exist between *Rhexoxylon*, true *Pentoxylon* from India, the Australian Jurassic polyvascular gymnosperm previously discussed, and yet unidentified Jurassic fossil woods from McElmo Canyon in Colorado. *Rhexoxylon* has been suggested as a possible origin of *Pentoxylon* because of similarities in the secondary wood (Stewart and Rothwell).

***Woodworthia*. Triassic. Gokwe, Zimbabwe. Daniels specimen, 18.5 cm tall.**

Woodworthia group. Triassic. Gokwe, Zimbabwe. Daniels specimens, 18 cm tallest.

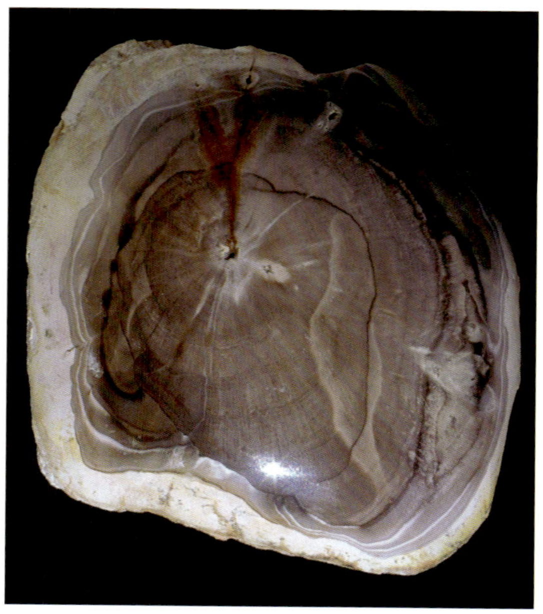

Araucarioxylon. Triassic. Gokwe, Zimbabwe. Daniels specimen, 6.5 cm.

Araucarioxylon. Triassic. Gokwe, Zimbabwe. Daniels specimen, 7.5 cm.

Araucarioxylon. Triassic. Gokwe, Zimbabwe. Daniels specimen, 13 cm.

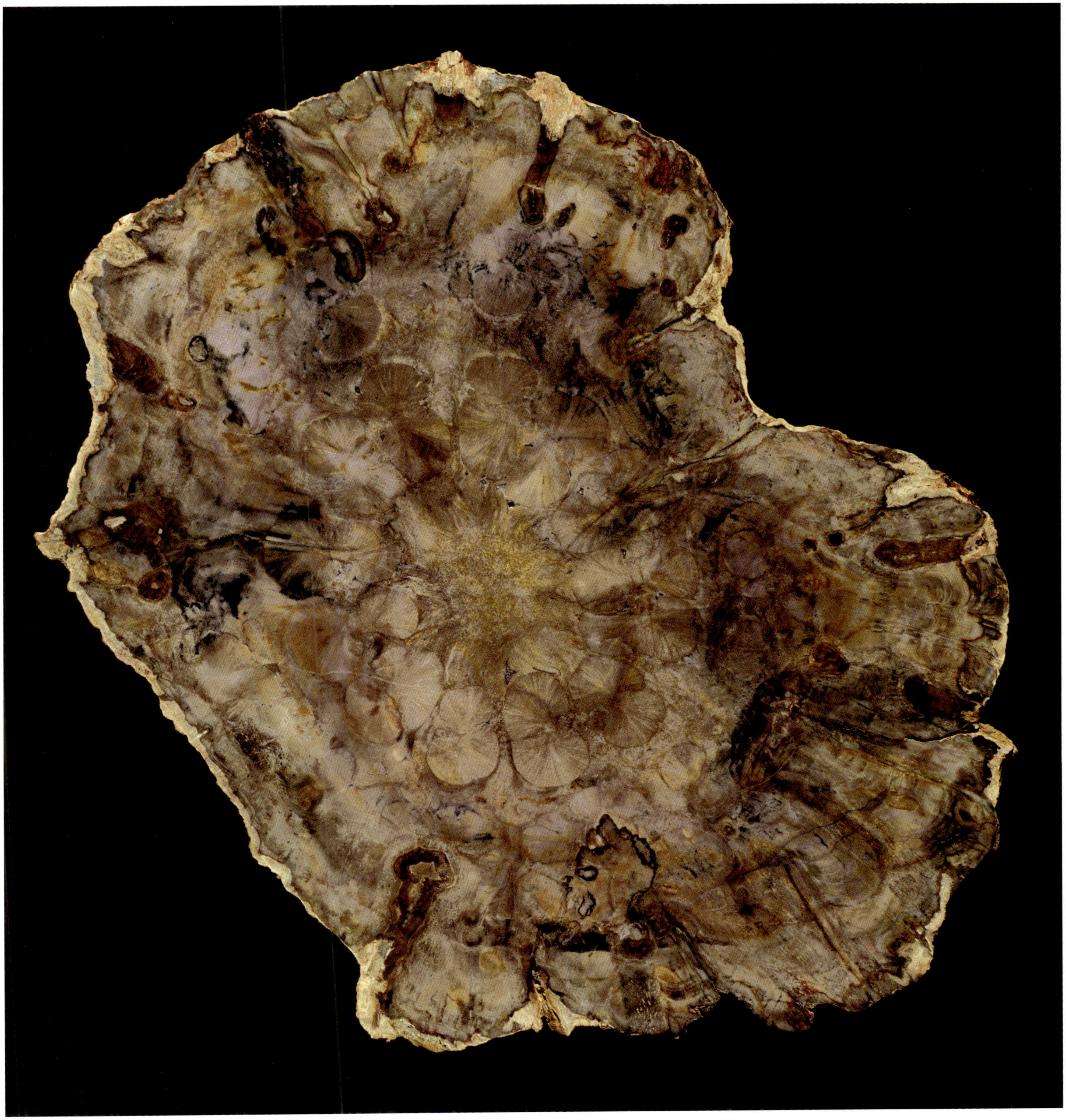

Rhexoxylon. Triassic. Gokwe, Zimbabwe. Ed Rigel specimen, 36 cm.

Rhexoxylon. Triassic. Zambezi River Gorge, Zimbabwe. Daniels specimen, 28 cm long.

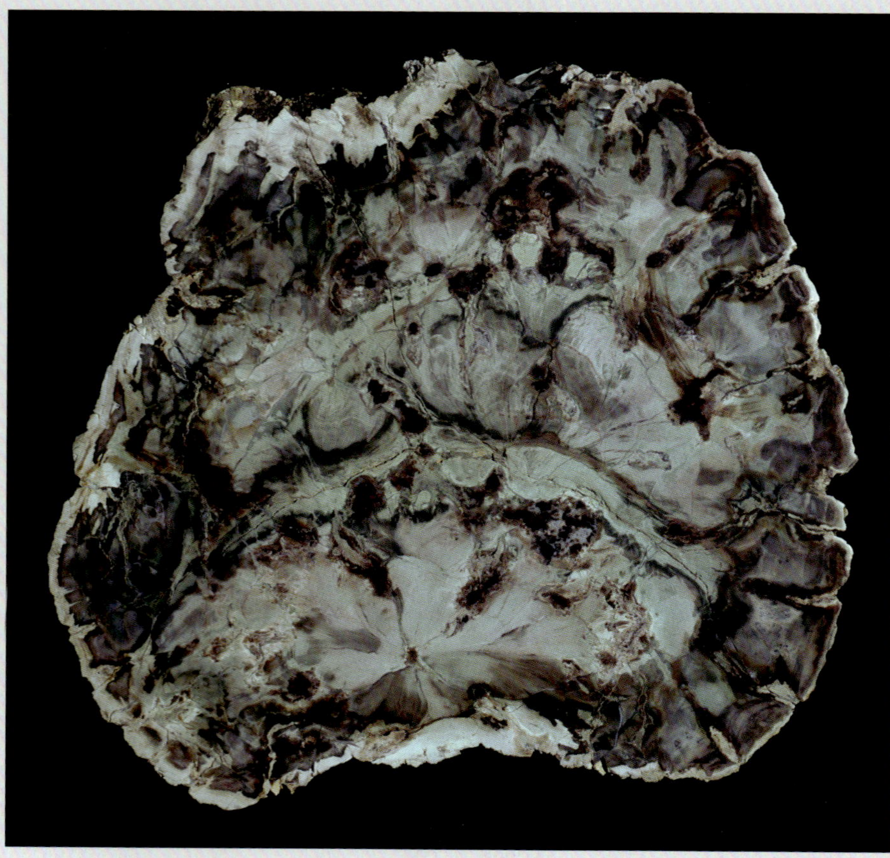

Macro photograph showing cell structure detail in *Rhexoxylon*. Triassic. Zambezi River Gorge, Zimbabwe. Daniels specimen.

Rhexoxylon. Triassic. Zambezi River Gorge, Zimbabwe. Daniels specimen, 26.5 cm.

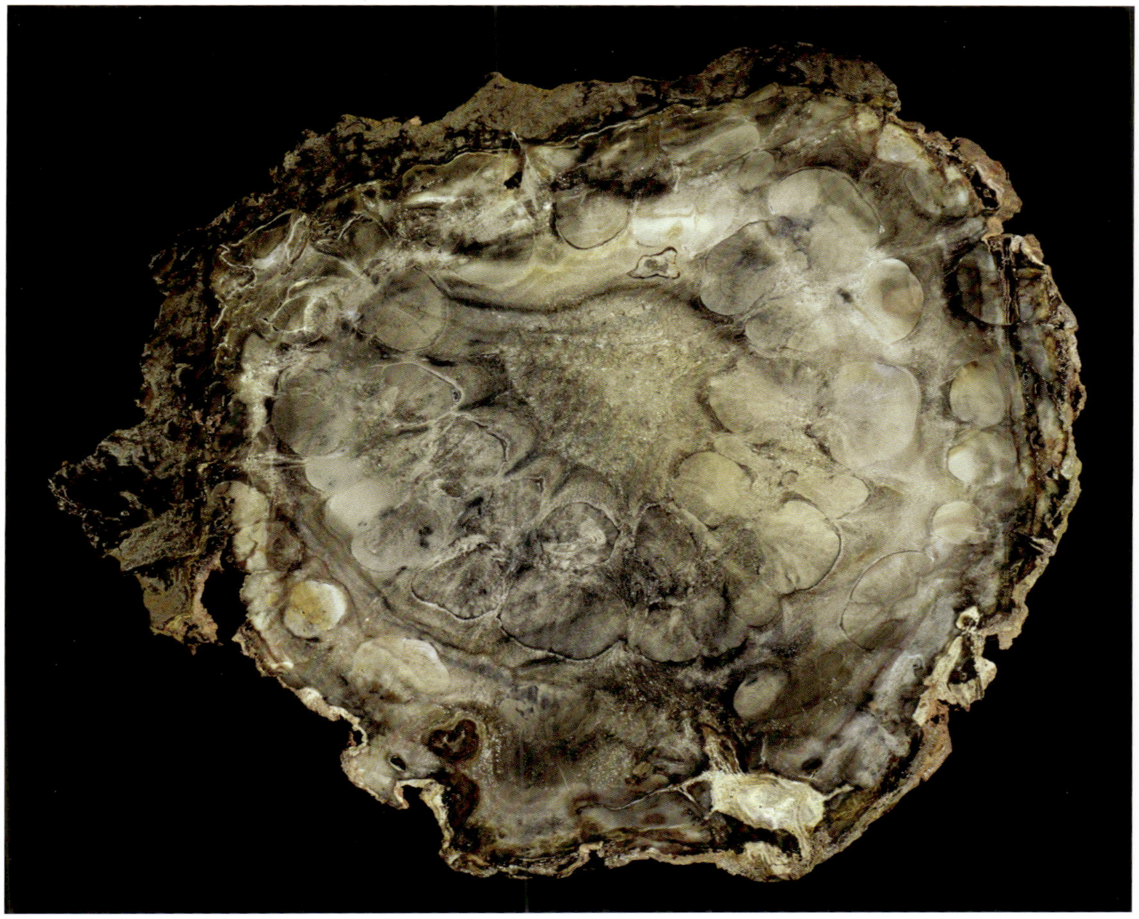

Rhexoxylon. Triassic. Gokwe, Zimbabwe. Daniels specimen, 22 cm.

Macro photograph showing cell structure in *Rhexoxylon*. Triassic. Gokwe, Zimbabwe. Daniels specimen.

Rhexoxylon. Triassic. Gokwe, Zimbabwe. Daniels specimen, 17 cm.

Macro photograph showing cell structure in *Rhexoxylon*. Triassic. Gokwe, Zimbabwe. Daniels specimen.

Rhexoxylon. Triassic. Gokwe, Zimbabwe. Daniels specimen, 23 cm.

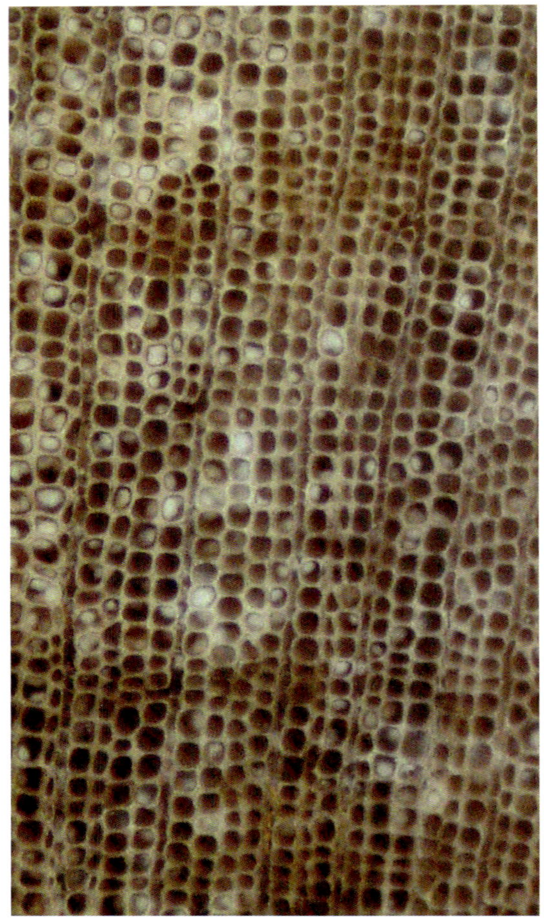

Micro image showing cell structure in *Rhexoxylon*. Triassic. Gokwe, Zimbabwe. Daniels specimen.

Micro image showing cell structure in *Rhexoxylon*. Triassic. Zambezi River Gorge, Zimbabwe. Daniels specimen.

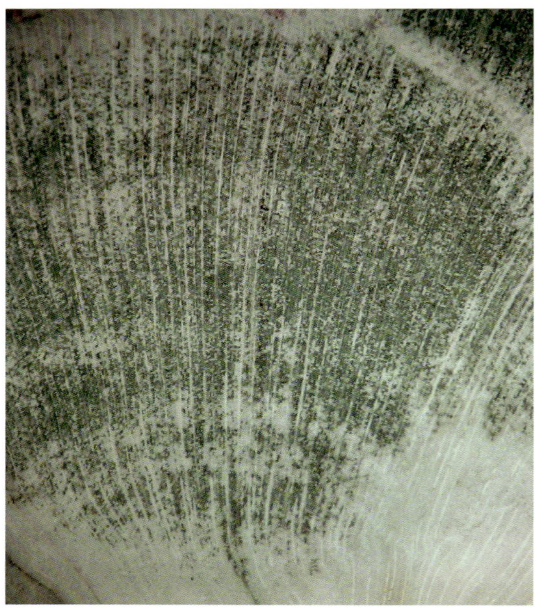

Micro image showing cell structure in *Rhexoxylon*. Triassic. Zambezi River Gorge, Zimbabwe. Daniels specimen.

Micro image showing cell structure in *Rhexoxylon*. Triassic. Gokwe, Zimbabwe. Daniels specimen.

Woodworthia. Triassic. Gokwe, Zimbabwe.
Daniels specimen, 8.5 cm.

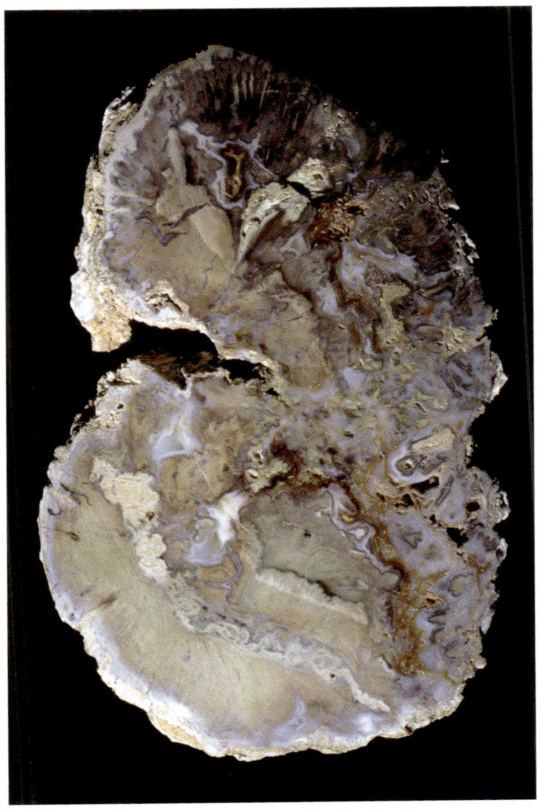

Woodworthia. Triassic. Gokwe, Zimbabwe.
Daniels specimen, 10 cm.

Woodworthia. Triassic. Gokwe, Zimbabwe.
Daniels specimen, 8.5 cm.

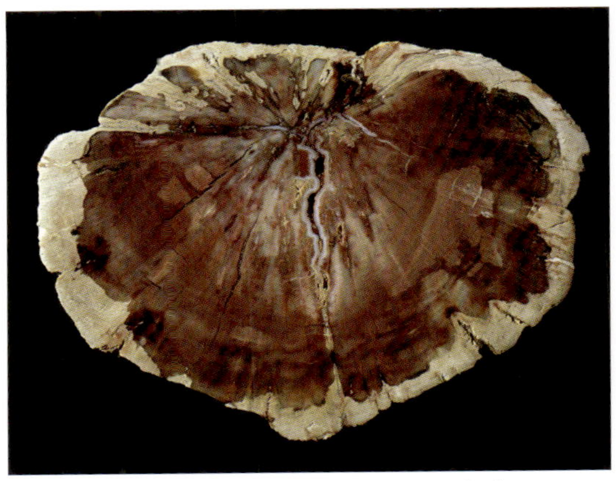

Woodworthia. Triassic. Gokwe, Zimbabwe.
Daniels specimen, 10 cm.

Woodworthia. Triassic. Gokwe, Zimbabwe.
Daniels specimen, 7.5 cm.

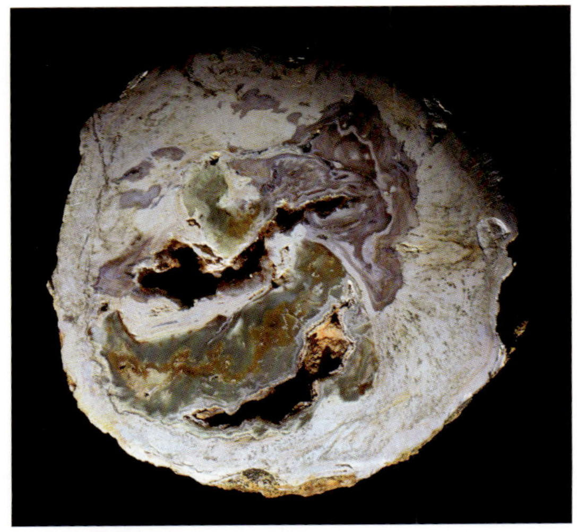

Woodworthia. Triassic. Gokwe, Zimbabwe.
Daniels specimen, 8 cm.

Woodworthia. Triassic. Gokwe, Zimbabwe.
Daniels specimen, 8.5 cm.

Woodworthia. Triassic. Gokwe, Zimbabwe.
Daniels specimen, 7 cm.

Macro photograph showing cell structure in *Araucarioxylon*. Triassic. Gokwe, Zimbabwe. Daniels specimen.

Micro image showing cell structure in *Woodworthia*. Triassic. Gokwe, Zimbabwe. Daniels specimen.

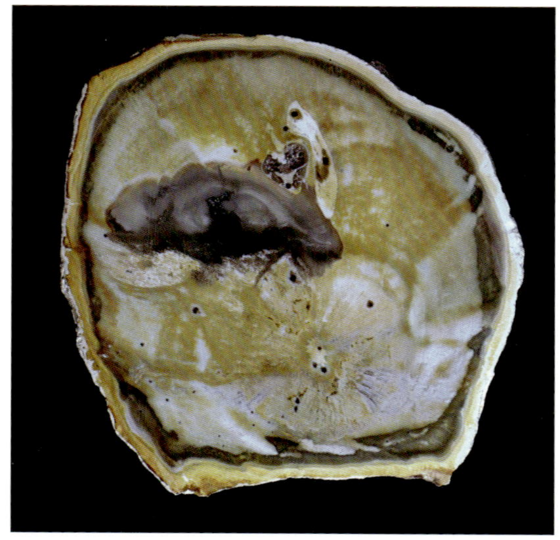

Unidentified conifer. Mesozoic. Malagasy Republic. Daniels specimen, 7 cm.

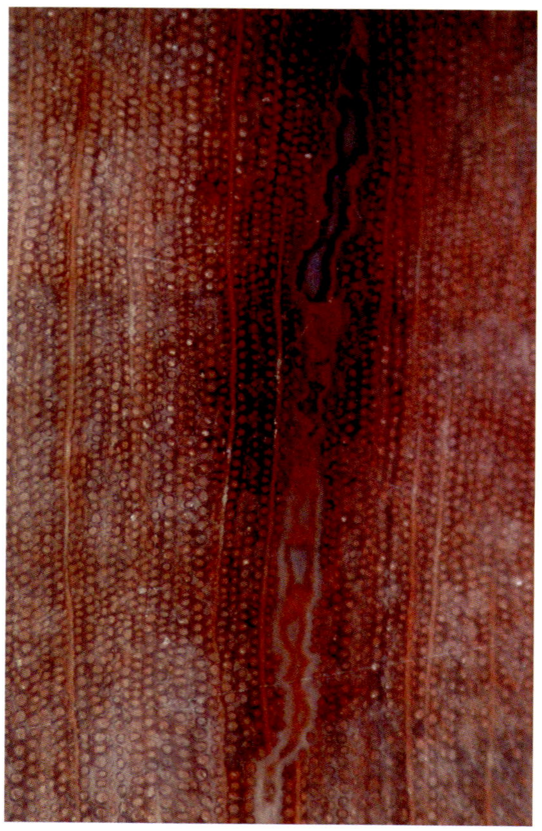

Micro image showing cell structure in *Woodworthia*. Triassic. Gokwe, Zimbabwe. Daniels specimen.

The island of Madagascar (Malagasy Republic) is home to many colorful and fine fossil wood specimens. Unfortunately, commercial fossil wood collectors who mine the material expend minimal effort preserving evidence of the source, frequently mixing specimens from various locations prior to marketing them to the West. The island's complex geology has received little professional attention. Although beautiful, many Malagasy fossil wood specimens are highly mineralized, erasing evidence necessary for identification. Others have extremely well-preserved cell structure. On the basis of transverse section observation, much of Madagascar's fossil wood has the appearance of *Araucarioxylon* and *Podocarpoxylon*. Various horizons yield occasional specimens of *Rhexoxylon* and *Woodworthia*. Madagascar occupied a central location in Gondwana, and significant expanses of exposed strata date to the Mesozoic Era. Certainly fossil woods from the island date to all Mesozoic periods, yet correlating fossil woods with distinct periods is problematic. One interesting type from near the village of Ikitsy, typically colored pink with swirls of red in intricate patterns, appears from the morphology to be a root or a vine.

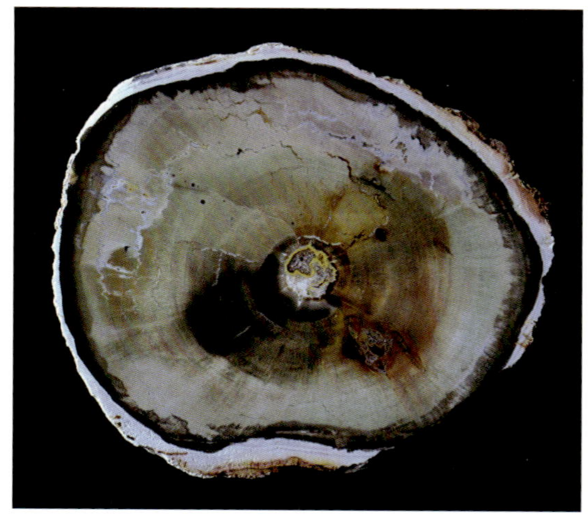

Unidentified conifer. Mesozoic. Malagasy Republic. Daniels specimen, 8.5 cm.

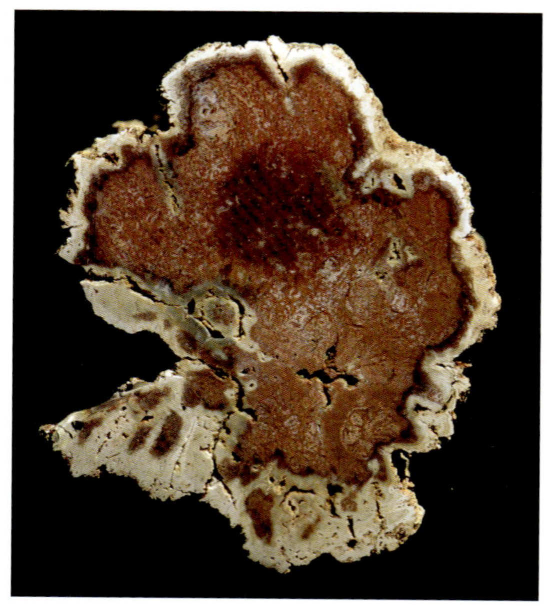

Unidentified Ikitsy pink. Mesozoic. Ikitsy, Malagasy Republic. Daniels specimen, 6 cm.

Unidentified set Ikitsy pink. Mesozoic. Ikitsy, Malagasy Republic. Daniels specimens, 14 cm largest.

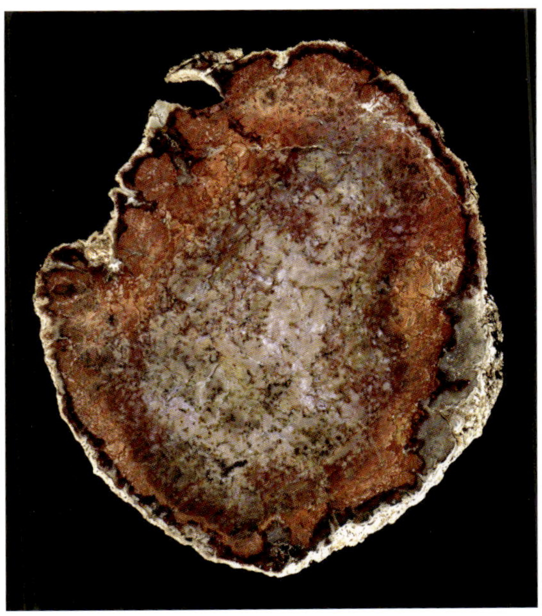

Unidentified Ikitsy pink. Mesozoic. Ikitsy, Malagasy Republic. Daniels specimen, 14 cm.

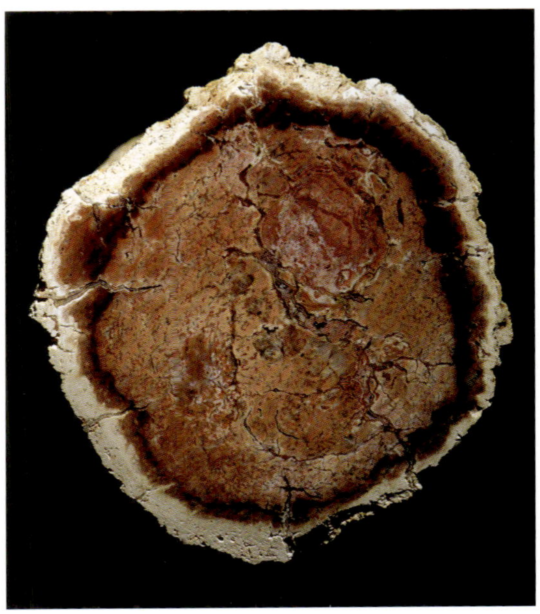

Unidentified Ikitsy pink. Mesozoic. Ikitsy, Malagasy Republic. Daniels specimen, 5 cm.

Unidentified Ikitsy pink. Mesozoic. Ikitsy, Malagasy Republic. Daniels specimen, 8 cm.

Unidentified conifer. Mesozoic. Malagasy Republic. Daniels specimen, 25.5 cm.

Unidentified conifer. Mesozoic. Malagasy Republic. Daniels specimen, 30 cm.

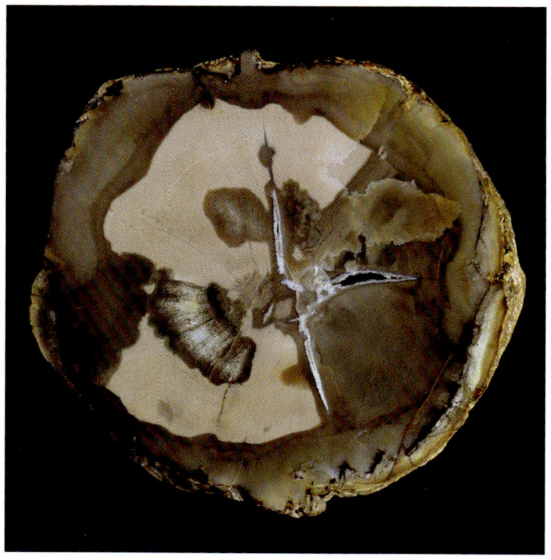

Podocarpoxylon. Mesozoic. Malagasy Republic. Daniels specimen, 7.5 cm.

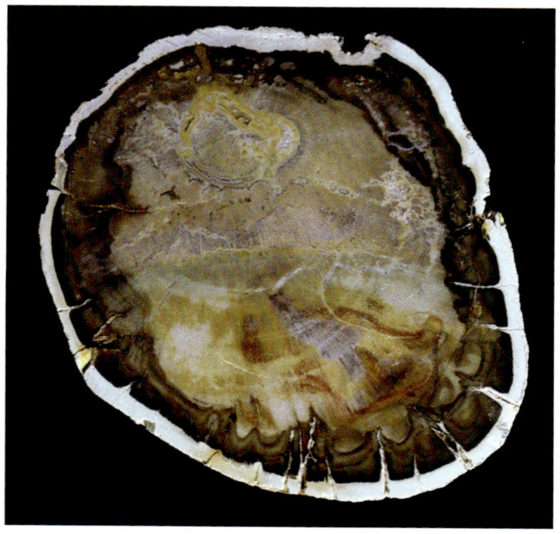

Unidentified conifer. Mesozoic. Malagasy Republic. Daniels specimen, 8 cm.

Podocarpoxylon. Mesozoic. Malagasy Republic. Daniels specimen, 7.5 cm.

Micro image showing cell structure in fossil wood. Mesozoic. Malagasy Republic. Daniels specimen.

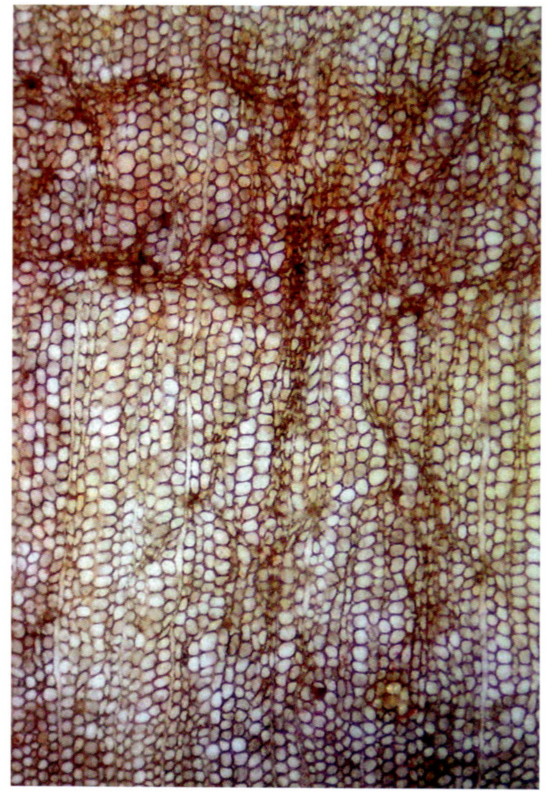

Micro image showing cell structure in *Podocarpoxylon*. Mesozoic. Malagasy Republic. Daniels specimen. Approximate average earlywood tracheid size 0.03 mm.

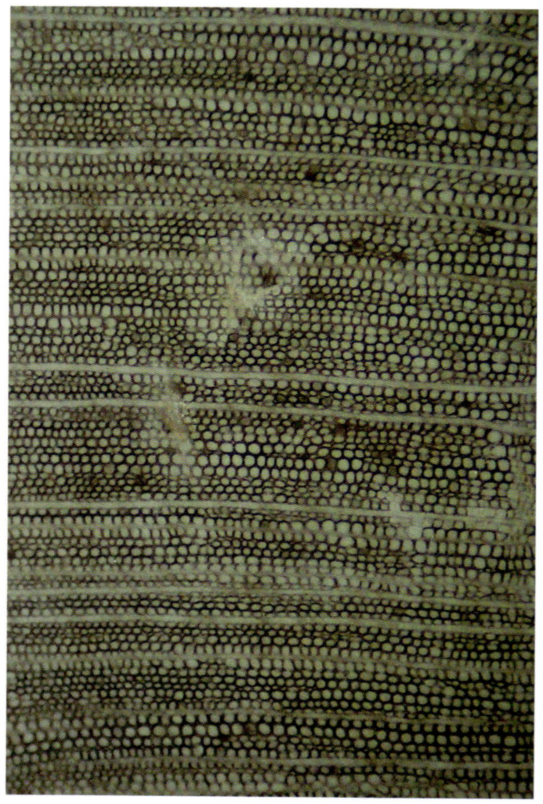

Micro image showing cell structure in *Podocarpoxylon*. Mesozoic. Malagasy Republic. Daniels specimen. Approximate average early-wood tracheid size 0.03 mm.

Micro image showing cell structure in *Podocarpoxylon*. Mesozoic. Malagasy Republic. Daniels specimen. Approximate average early-wood tracheid size 0.03 mm.

Rhexoxylon. Mesozoic. Malagasy Republic. Tidwell specimen, 24.5 cm.

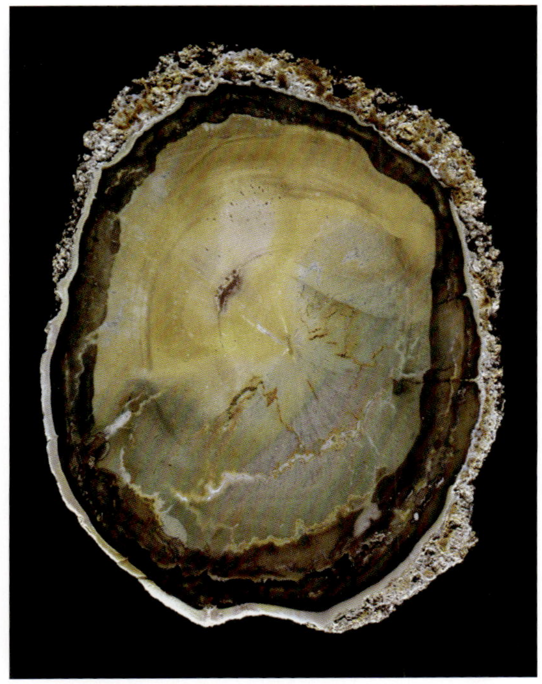

Unidentified conifer. Mesozoic. Malagasy Republic. Daniels specimen, 8.5 cm.

Podocarpoxylon. Mesozoic. Malagasy Republic. Daniels specimen, 10 cm.

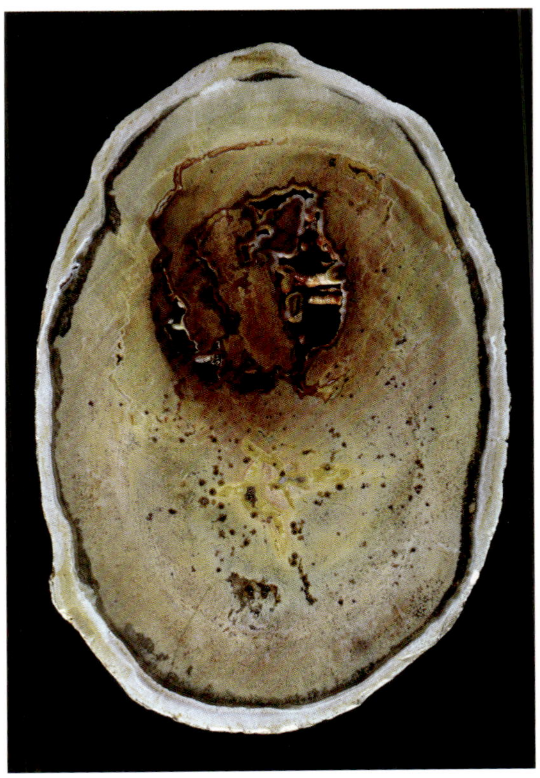

Unidentified conifer. Mesozoic. Malagasy Republic. Daniels specimen, 10 cm.

Unidentified Ikitsy pink. Mesozoic. Ikitsy, Malagasy Republic. Daniels specimen, 8 cm.

Unidentified conifer. Mesozoic. Malagasy Republic. Daniels specimen, 8 cm.

Unidentified Ikitsy pink. Mesozoic. Ikitsy, Malagasy Republic. Daniels specimen, 6 cm.

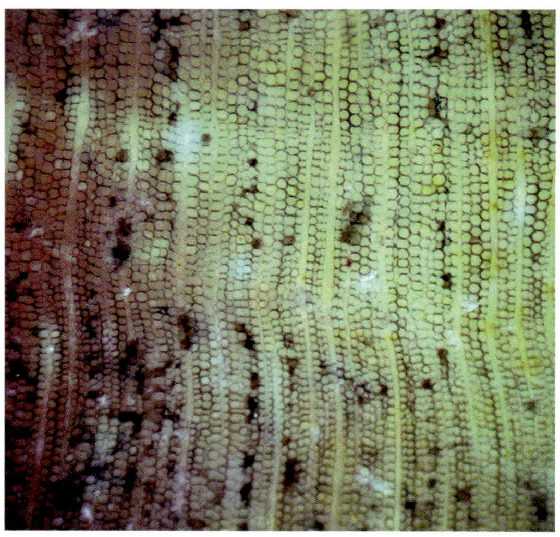

Micro image showing cell structure in fossil wood. Mesozoic. Malagasy Republic. Daniels specimen.

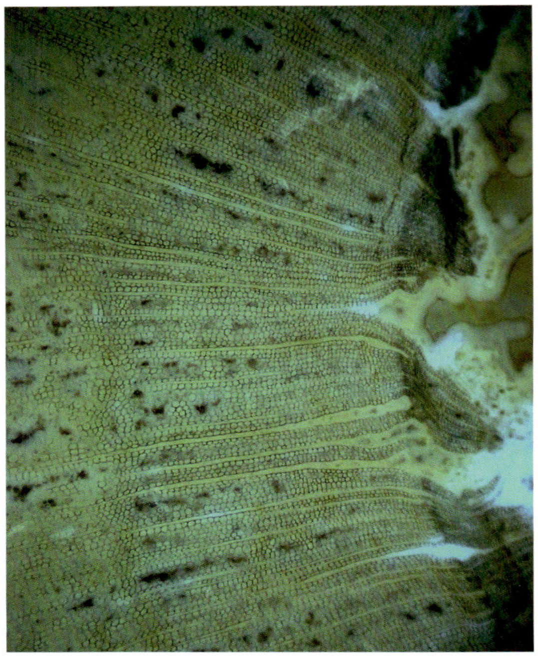

Micro image showing cell structure in fossil wood. Mesozoic. Malagasy Republic. Daniels specimen.

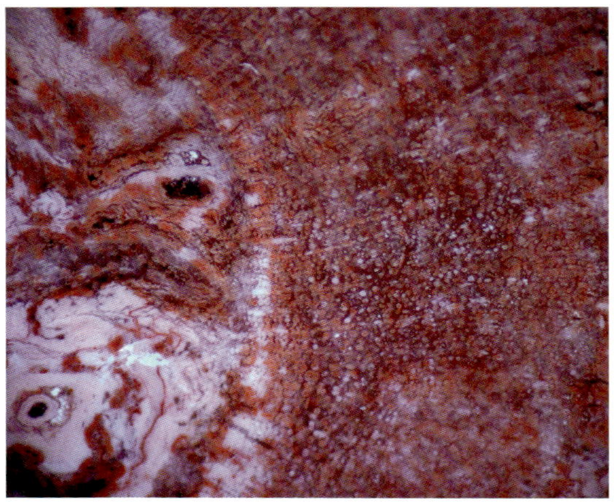

Micro image showing cell structure in fossil wood. Mesozoic. Ikitsy, Malagasy Republic. Daniels specimen.

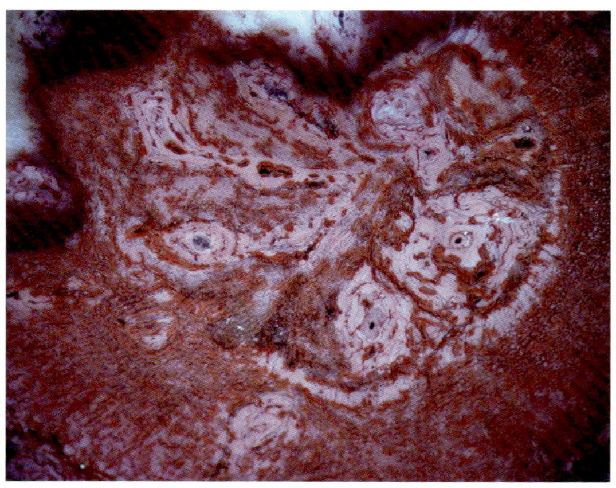

Micro image showing cell structure in fossil wood. Mesozoic. Ikitsy, Malagasy Republic. Daniels specimen.

Vast expanses of arid lands in the Patagonian provinces of Chubut, Rio Negro, and Santa Cruz expose Mesozoic Era formations rich in fossils. Repeated huge volcanic eruptions in the area allowed the preservation of some of the best-preserved fossil plant material in existence. The Bosques Petrificados Natural Monument in Santa Cruz is the Argentine answer to the Petrified Forest National Park in the United States. Jurassic Period formations in Santa Cruz yield perfectly preserved wood and cones. An enormous forest of predominately *Araucaria*-like trees was buried in hundreds of feet of volcanic ash, entombing the forest and preserving wood and cones as fossils. Even the bark of some trees was preserved, as well as entire limbs with cones still attached. Cut and polished *Araucaria mirabilis* cones display incredible detail, including well-preserved ovules. Cones of a wide range of sizes and shapes occur in Patagonia, with most of the cones coming from Jurassic Period deposits such as those in Santa Cruz. *Pararaucaria* wood and cones are also found in Santa Cruz Jurassic Period deposits. *Pararaucaria* wood combines characteristics of Taxodiaceae and other conifer families. Well-preserved short shoots or aerial lignotubers from *Araucaria*-like tees are found with the wood and cones. Cretaceous Period deposits preserve *Araucarioxylon*, cycadophytes, osmundaphytes, *Palmoxylon*, and *Tempskya*, as well as cycad seeds and dates.

Unidentified cones. Jurassic. Santa Cruz, Argentina. Daniels specimens, 3 cm.

Araucaria mirabilis cone group. Jurassic. Santa Cruz, Argentina. Daniels specimens, 9.5 cm largest.

Pararaucaria and other small cones. Jurassic. Santa Cruz, Argentina. Daniels specimens, 7 cm largest.

Pararaucaria cone. Jurassic. Santa Cruz, Argentina. Daniels specimen, 7 cm.

Unidentified cone on branch. Jurassic. Santa Cruz, Argentina. Dayvault specimen, 2.5 cm.

Unidentified cone. Jurassic. Santa Cruz, Argentina. Daniels specimen, 2.5 cm.

Pararaucaria cone. Jurassic. Santa Cruz, Argentina. Daniels specimen, 7 cm.

Araucaria mirabilis cone. Jurassic. Santa Cruz, Argentina. Daniels specimen, 7 cm.

Araucaria mirabilis cone. Jurassic. Santa Cruz, Argentina. Daniels specimen, 8 cm.

Araucaria mirabilis cone. Jurassic. Santa Cruz, Argentina. Daniels specimen, 7.5 cm.

Araucaria mirabilis cone. Jurassic. Santa Cruz, Argentina. Daniels specimen, 9 cm.

Araucaria mirabilis cone. Jurassic. Santa Cruz, Argentina. Daniels specimen, 9.5 cm.

Araucaria mirabilis pollen cone. Jurassic. Santa Cruz, Argentina. Daniels specimen, 10.5 cm.

Unidentified *Araucaria*-like cone. Jurassic. Patagonia, Argentina. Daniels specimen, 10 cm.

Araucaria mirabilis double cone. Jurassic. Santa Cruz, Argentina. Daniels specimen 7.5 cm.

Araucaria mirabilis cone. Jurassic. Santa Cruz, Argentina. Daniels specimen, 8.5 cm

Unidentified *Araucaria*-like cone. Jurassic. Chubut, Argentina. Daniels specimen, 11.5 cm.

Unidentified *Araucaria*-like cone. Jurassic. Chubut, Argentina. Dayvault specimen, 14.5 cm.

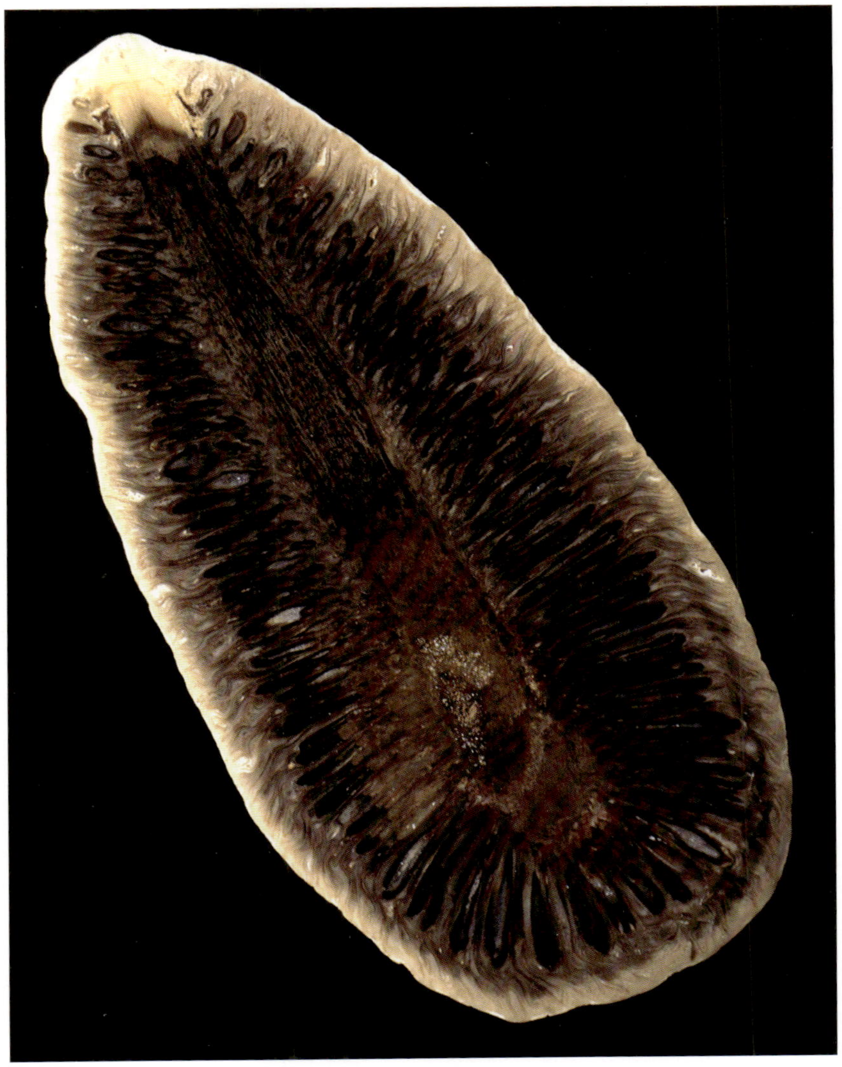

Unidentified *Araucaria*-like cone. Jurassic. Chubut, Argentina. Dayvault specimen, 14 cm.

Araucaria mirabilis cone. Jurassic. Santa Cruz, Argentina. Dayvault specimen, 6 cm.

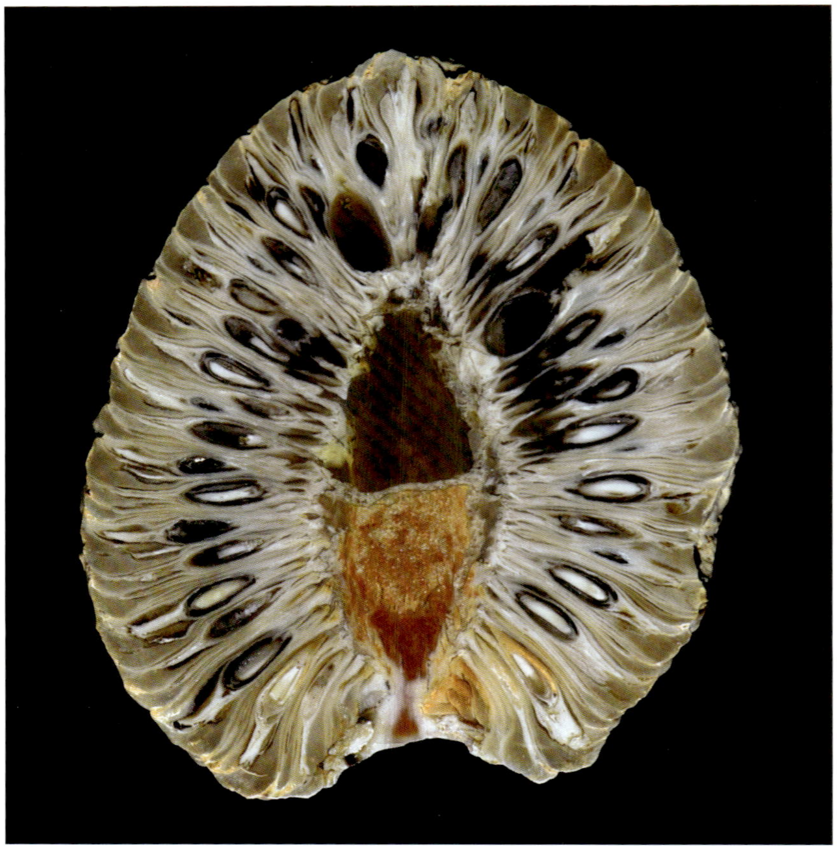

Araucaria mirabilis cone. Jurassic. Santa Cruz, Argentina. Daniels specimen, 8 cm.

Araucaria mirabilis cone. Jurassic. Santa Cruz, Argentina. Daniels specimen, 6.5 cm.

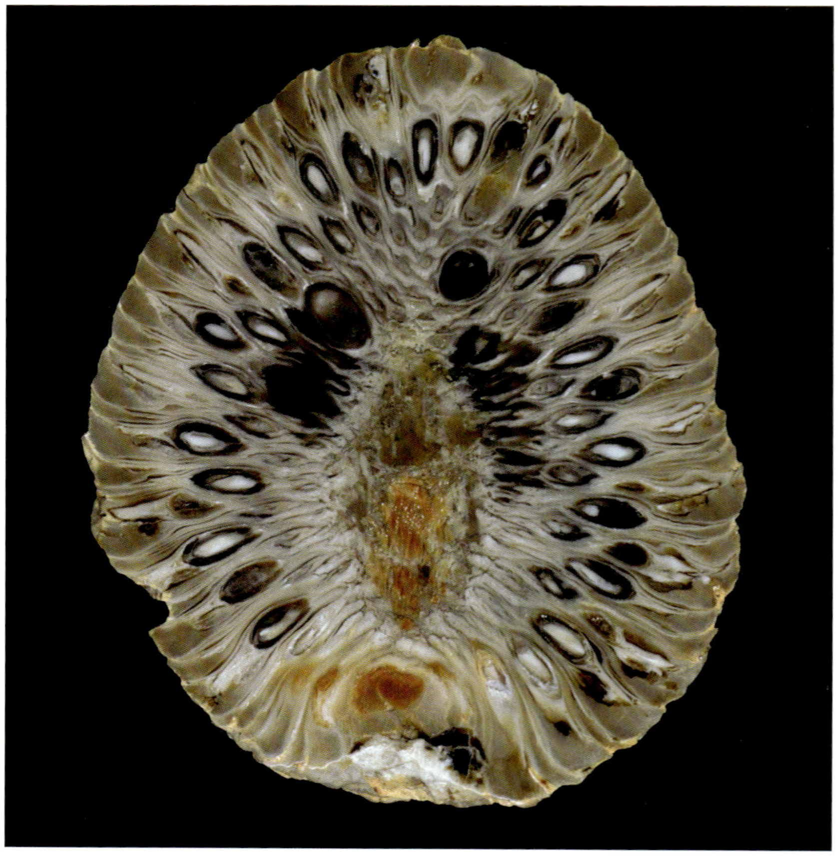

Araucaria mirabilis cone. Jurassic. Santa Cruz, Argentina. Daniels specimen, 8 cm.

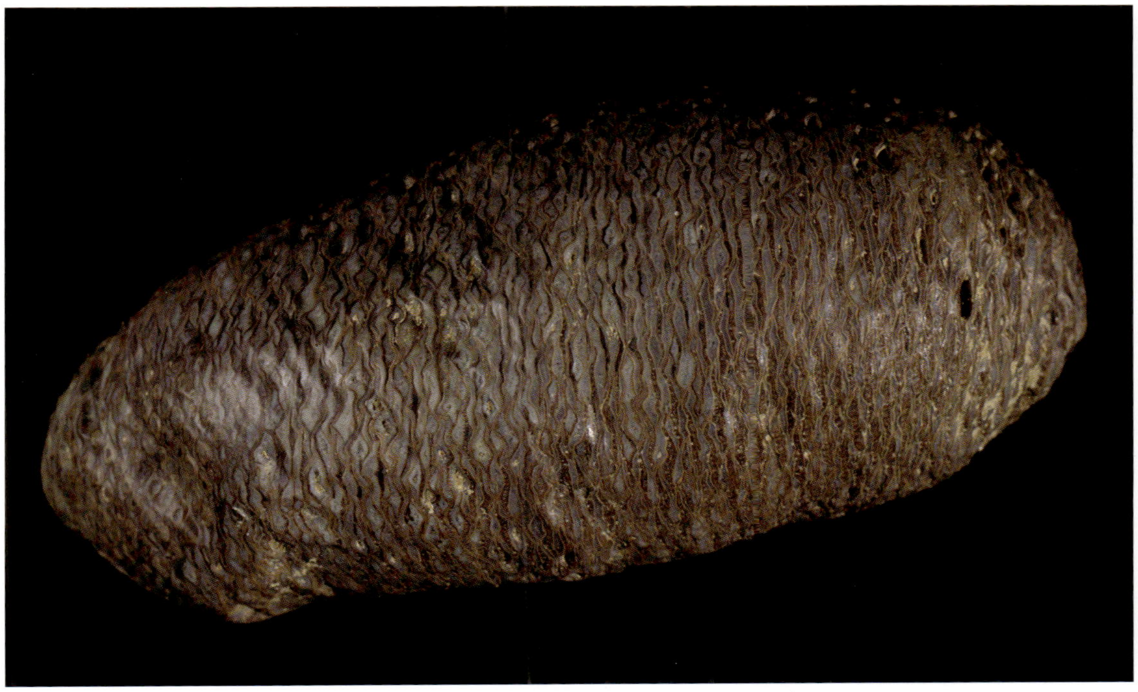

Unidentified *Araucaria*-like cone. Jurassic. Chubut, Argentina. Daniels specimen, 16 cm.

Araucaria mirabilis cone. Jurassic. Santa Cruz, Argentina. Daniels specimen, 7.5 cm.

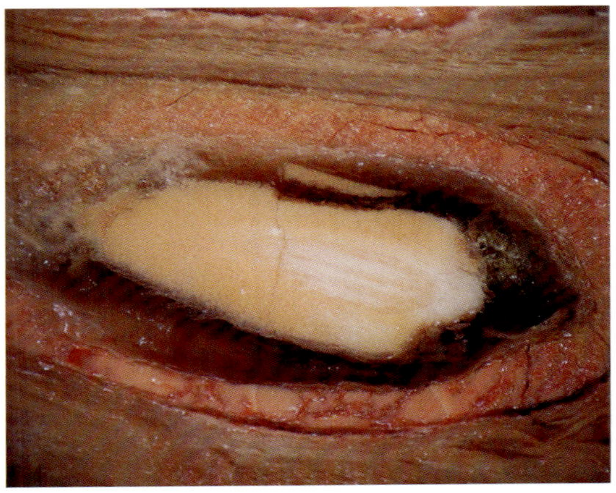

Micro image showing detail in *Araucaria mirabilis* cone. Jurassic. Santa Cruz, Argentina. Daniels specimen.

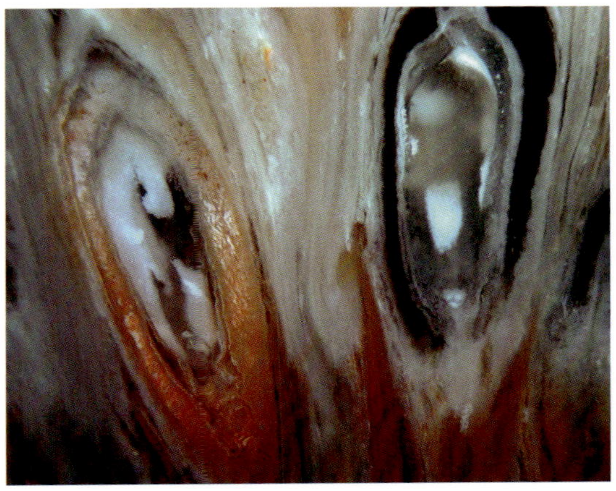

Micro image showing detail in *Araucaria mirabilis* cone. Jurassic. Santa Cruz, Argentina. Daniels specimen.

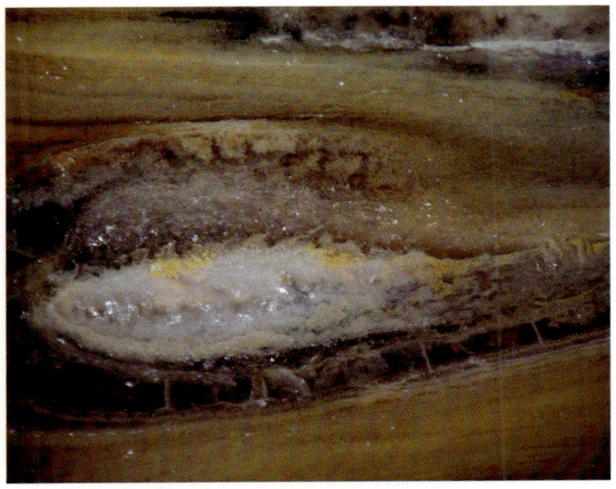

Micro image showing detail in *Araucaria mirabilis* cone. Jurassic. Santa Cruz, Argentina. Daniels specimen.

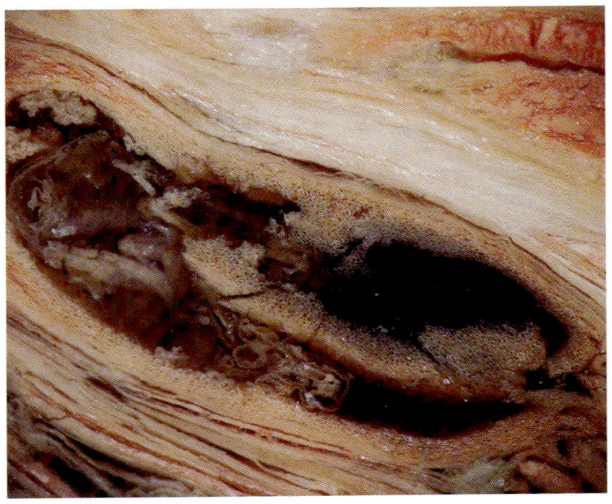

Micro image showing detail in *Araucaria mirabilis* cone. Jurassic. Santa Cruz, Argentina. Daniels specimen.

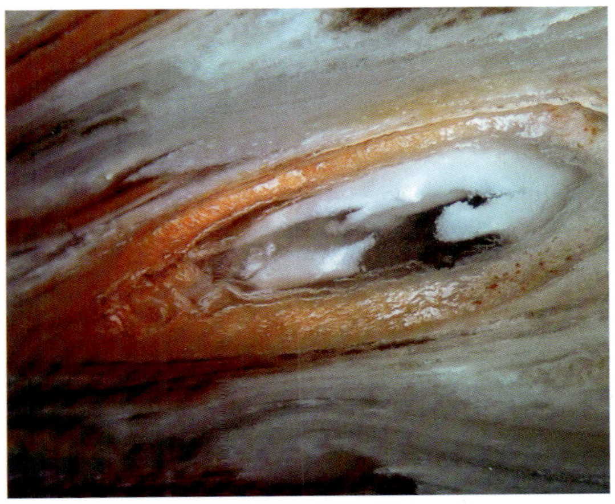

Micro image showing detail in *Araucaria mirabilis* cone. Jurassic. Santa Cruz, Argentina. Daniels specimen.

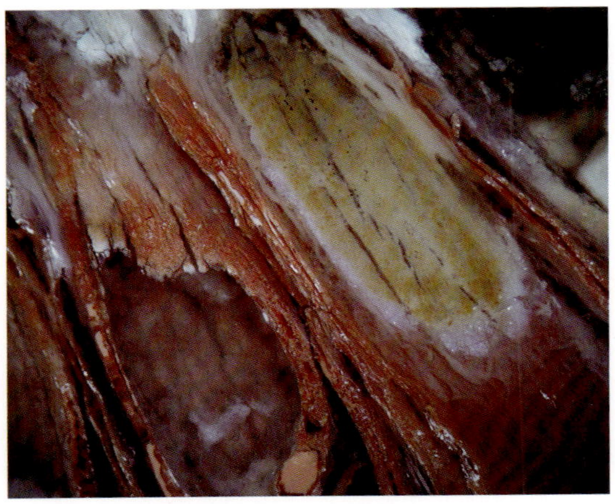

Micro image showing detail in *Araucaria mirabilis* cone. Jurassic. Santa Cruz, Argentina. Daniels specimen.

Araucaria mirabilis cone in matrix. Jurassic. Santa Cruz, Argentina. Korpusik specimen, cone 8 cm; branch 18 cm.

Pararaucaria cone. Jurassic. Santa Cruz, Argentina. Daniels specimen, 4 cm.

Unidentified cone. Jurassic. Santa Cruz, Argentina. Daniels specimen, 5 cm.

Pararaucaria cone. Jurassic. Santa Cruz, Argentina. Daniels specimen, 6 cm.

Pararaucaria cone. Jurassic. Santa Cruz, Argentina. Daniels specimen, 3.5 cm.

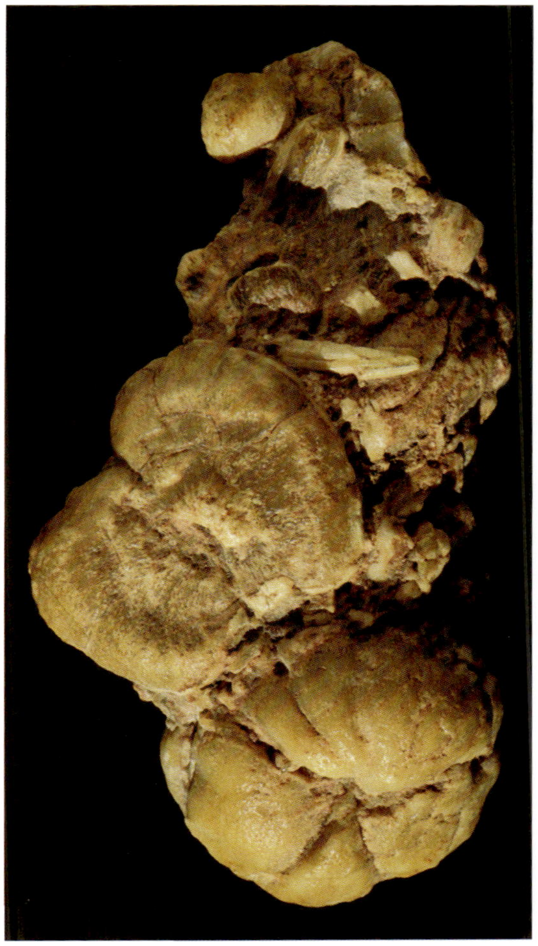

Short shoot or aerial lignotuber. Jurassic. Santa Cruz, Argentina. Daniels specimen, 8 cm.

Araucarixoylon. Jurassic. Santa Cruz, Argentina. Daniels specimen, 14.5 cm tall.

Short shoots or aerial lignotubers. Jurassic. Santa Cruz, Argentina. Daniels specimens, 4.5 cm largest.

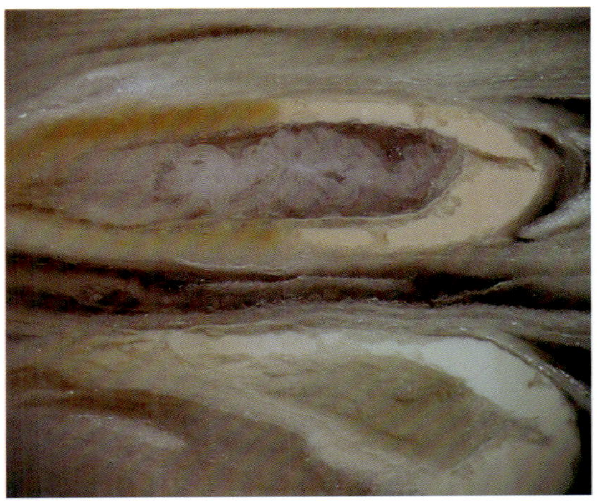

Micro image showing detail in *Araucaria mirabilis* cone. Jurassic. Santa Cruz, Argentina. Daniels specimen.

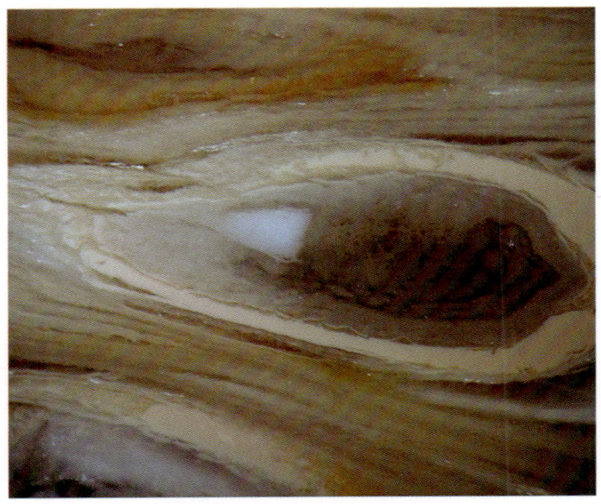

Micro image showing detail in *Araucaria mirabilis* cone. Jurassic. Santa Cruz, Argentina. Daniels specimen.

Assorted fossil woods from Patagonia. Mesozoic. Patagonia, Argentina. Daniels specimens, 16.5 cm tallest.

Araucarixoylon. Jurassic. Santa Cruz, Argentina. Daniels specimen, 14 cm.

Araucarixoylon. Jurassic. Santa Cruz, Argentina. Daniels specimen, 9.5 cm.

Micro image showing detail in *Araucarioxylon* fossil wood. Jurassic. Santa Cruz, Argentina. Daniels specimen.

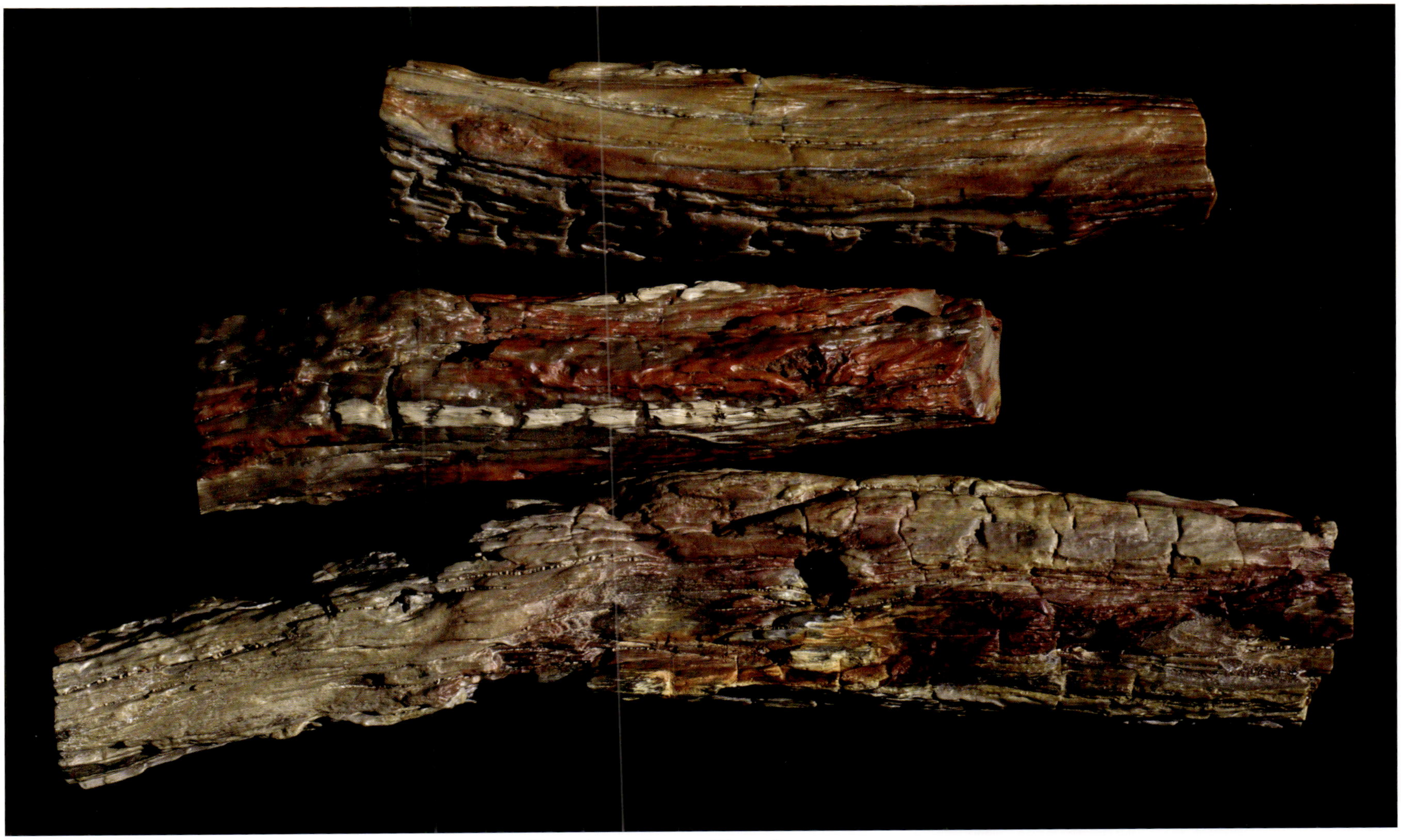

Unidentified conifers. Cretaceous. Rio Negro, Argentina. Daniels specimens, 51 cm longest.

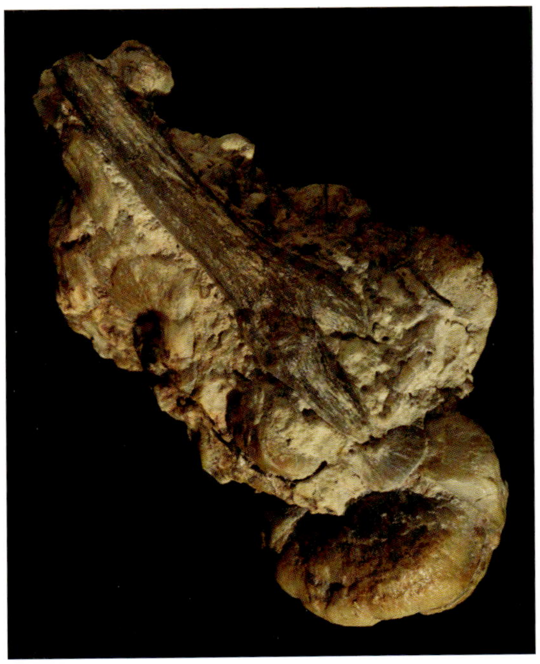

Short shoot or aerial lignotuber. Jurassic. Santa Cruz, Argentina. Daniels specimen, 8 cm.

Unidentified probable aerial lignotubers. Mesozoic. Chubut, Argentina. Daniels/Dayvault specimens, 7 cm largest.

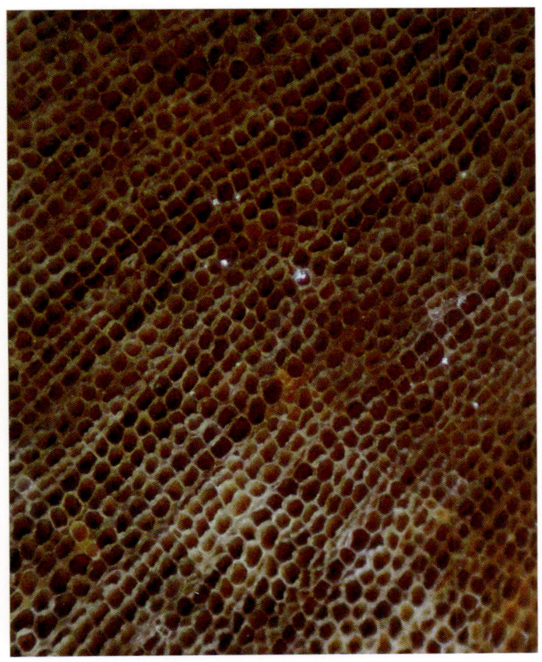

Micro image showing detail in *Araucarioxylon* fossil wood. Jurassic. Santa Cruz, Argentina. Daniels specimen.

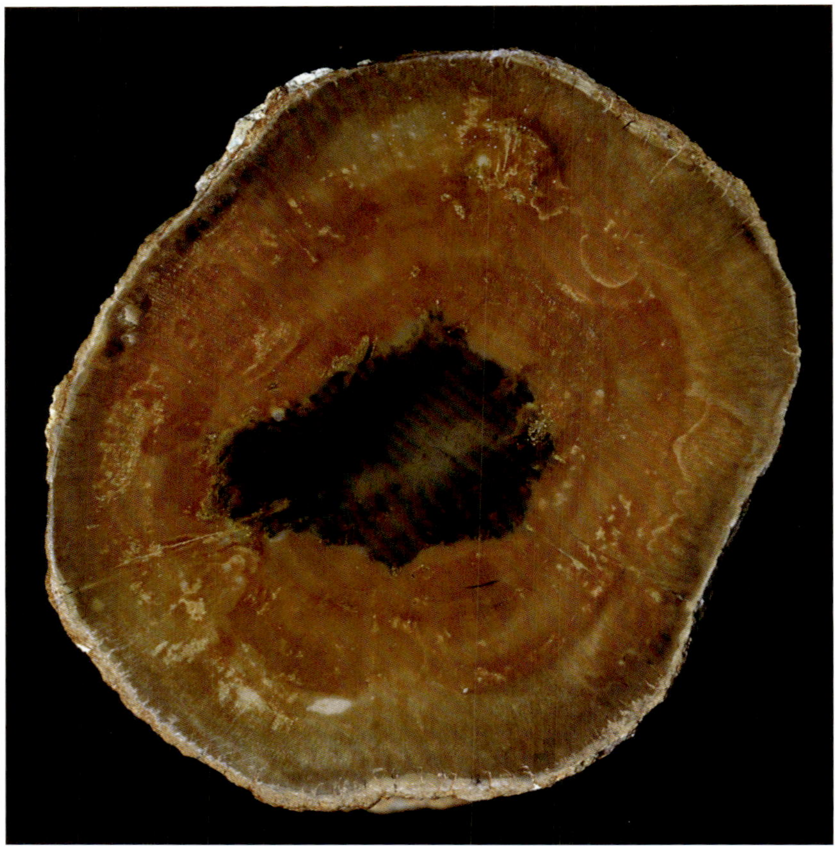

Araucarixoylon. Jurassic. Santa Cruz, Argentina. Daniels specimen, 6.5 cm.

Araucarixoylon. Jurassic. Santa Cruz, Argentina. Daniels specimen, 11.5 cm.

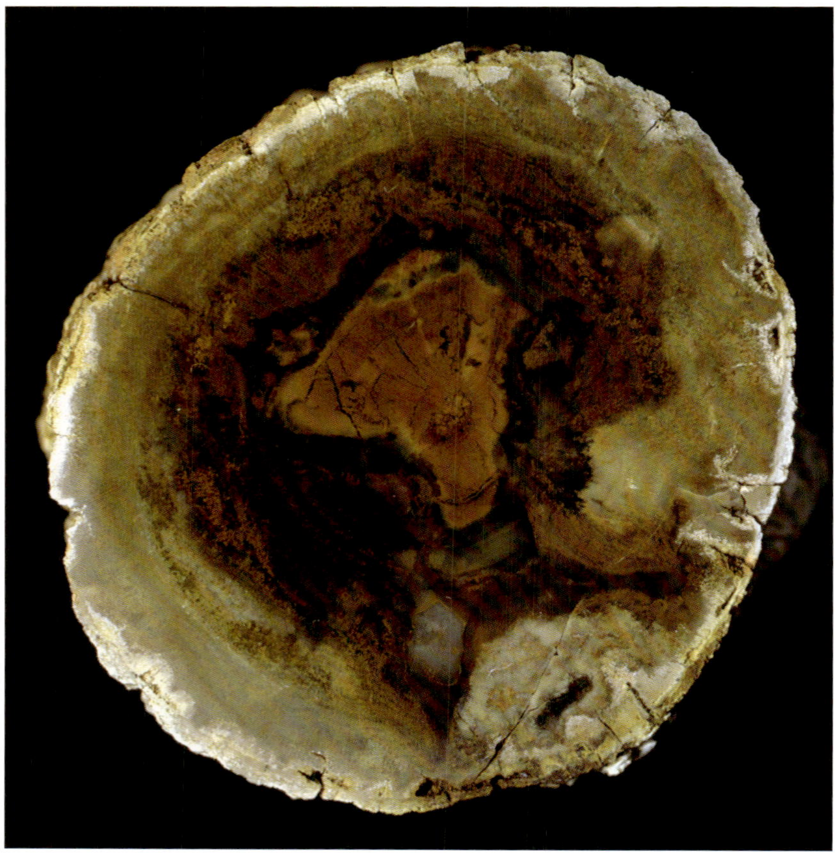

Araucarixoylon. Jurassic. Santa Cruz, Argentina. Daniels specimen, 5 cm.

Araucarixoylon. Jurassic. Santa Cruz, Argentina. Daniels specimen, 9.5 cm.

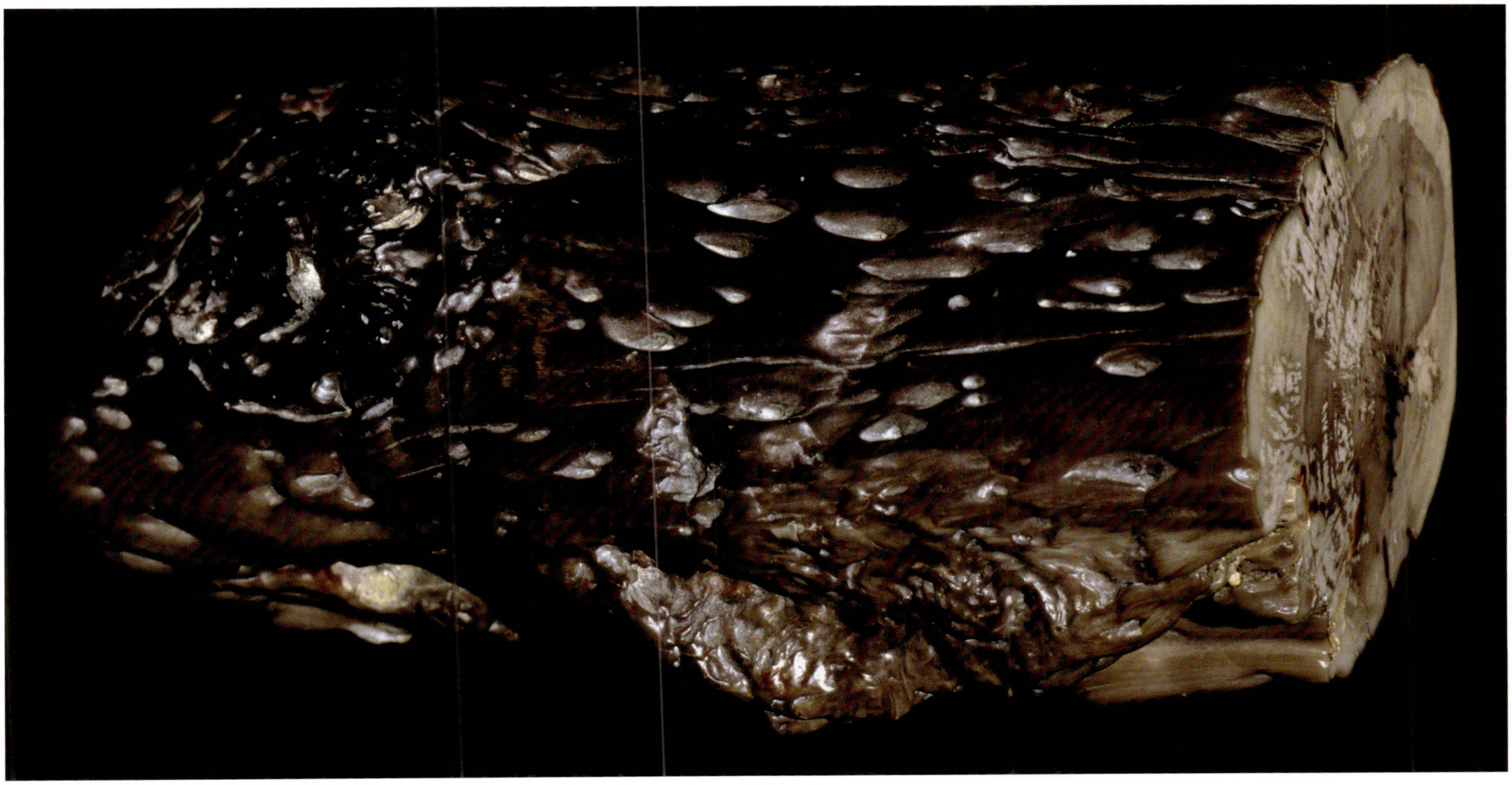

Unidentified conifer. Cretaceous. Rio Negro, Argentina. Daniels specimen, 20 cm long.

Araucarixoylon. Jurassic. Santa Cruz, Argentina.
Daniels specimen, 19 cm.

Araucarixoylon. Jurassic. Santa Cruz, Argentina.
Daniels specimen, 14 cm.

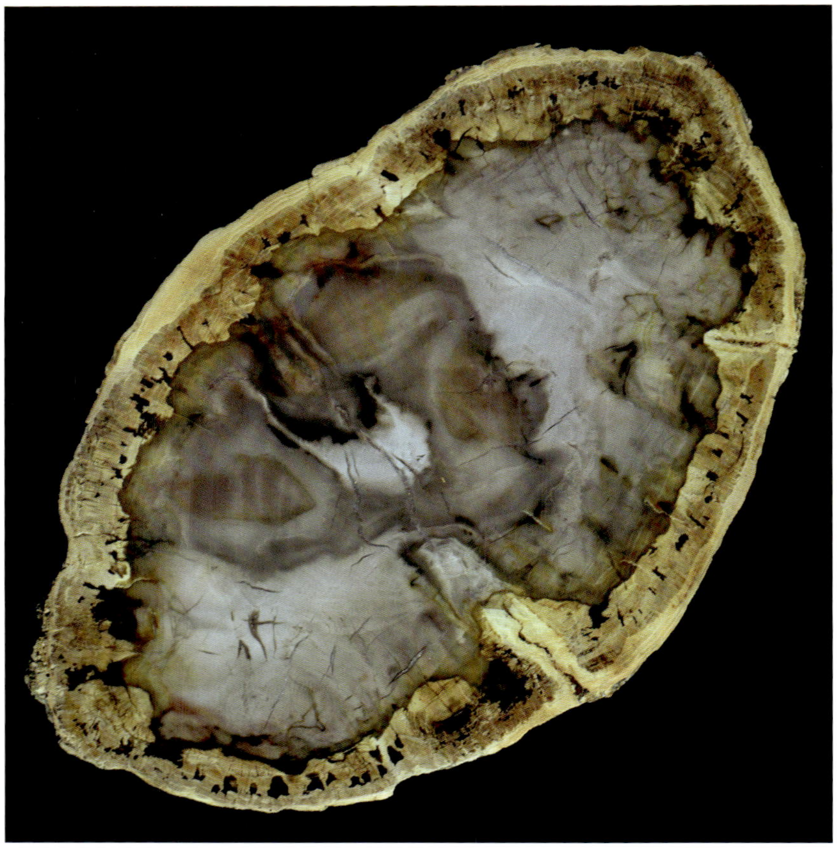

Araucarixoylon. Jurassic. Santa Cruz, Argentina.
Daniels specimen, 20.5 cm.

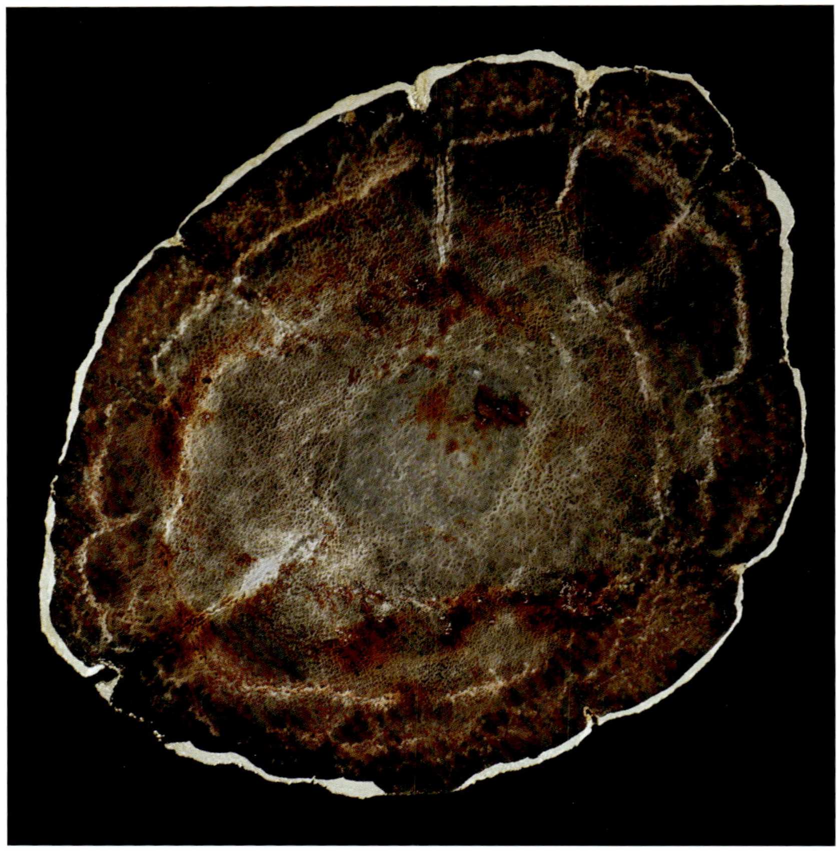

Palmoxylon. Cretaceous. Rio Negro, Argentina.
Daniels specimen, 15 cm.

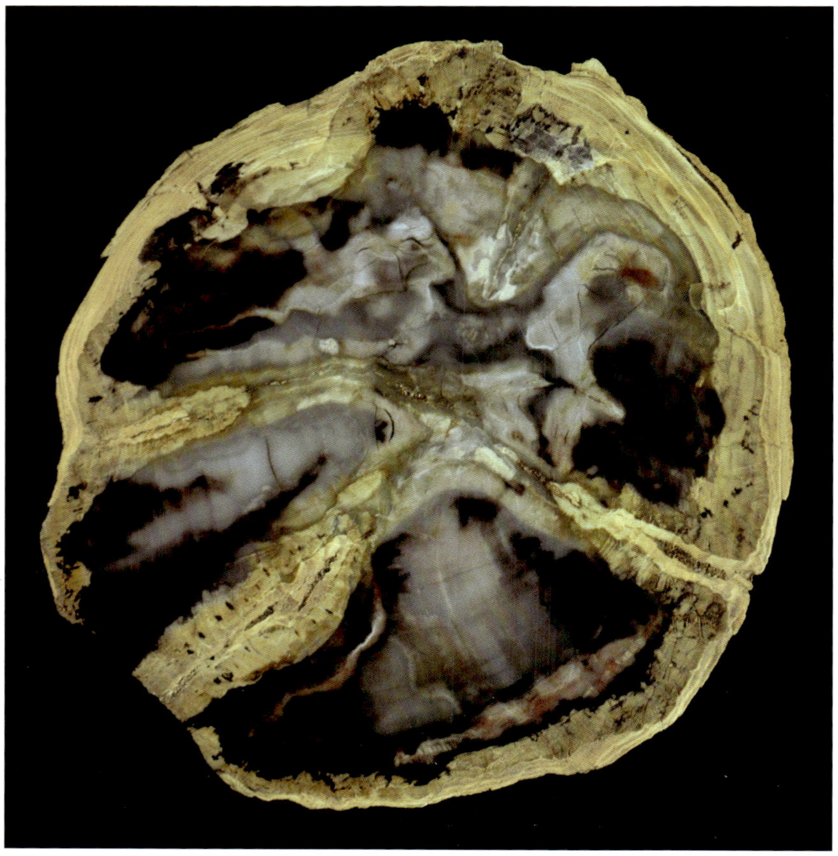

Araucarixoylon. Jurassic. Santa Cruz, Argentina.
Daniels specimen, 19 cm.

Unidentified conifer. Cretaceous. Rio Negro, Argentina.
Daniels specimen, 18 cm.

Carporichnus bertheorum [dates]. Cretaceous. Rio Negro, Argentina. Daniels specimens, 3 cm largest.

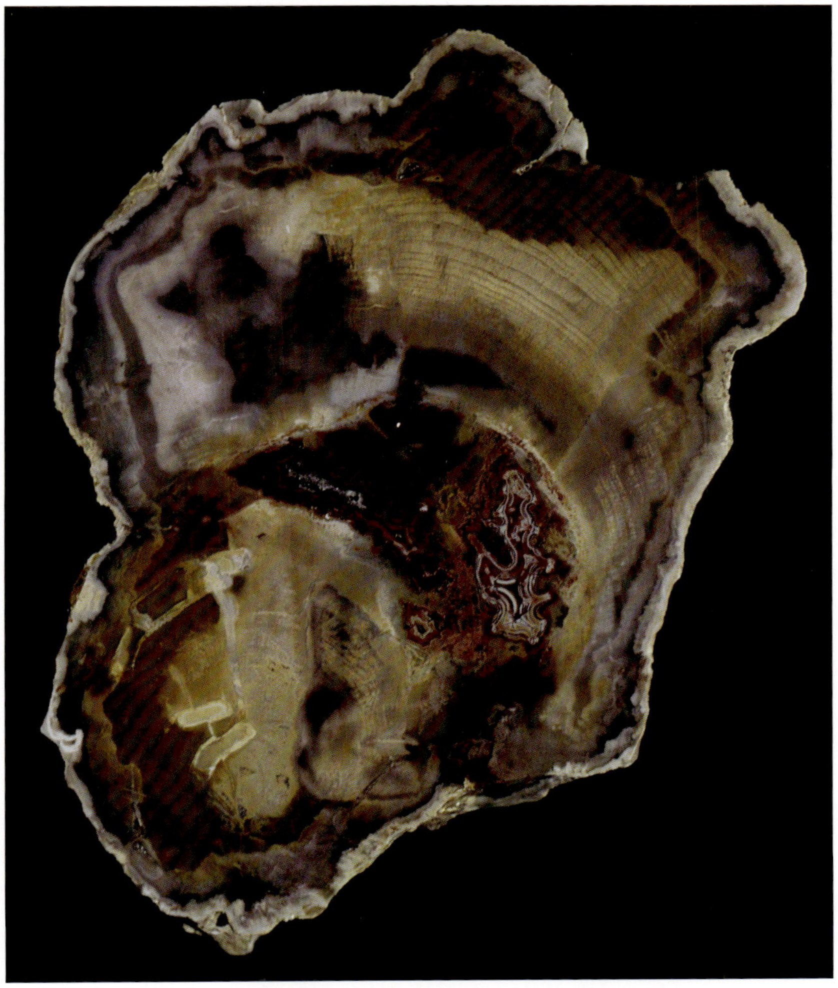

Unidentified conifer. Cretaceous. Rio Negro, Argentina. Daniels specimen, 14 cm.

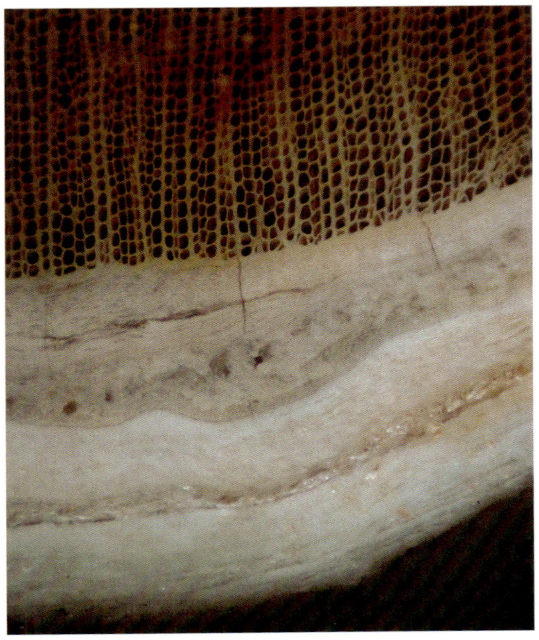

Micro image showing detail in *Araucarioxylon* fossil wood including secondary wood, vascular cambium, and cortex (bark). Jurassic. Santa Cruz, Argentina. Daniels specimen. Maximum bark thickness depicted 1.6 mm.

Micro image showing detail in unidentified conifer. Cretaceous. Rio Negro, Argentina. Daniels specimen.

Micro image showing detail in unidentified conifer. Cretaceous. Rio Negro, Argentina. Daniels specimen.

Borer evidence in fossil wood. Cretaceous. Rio Negro, Argentina. Daniels specimen, 10 cm.

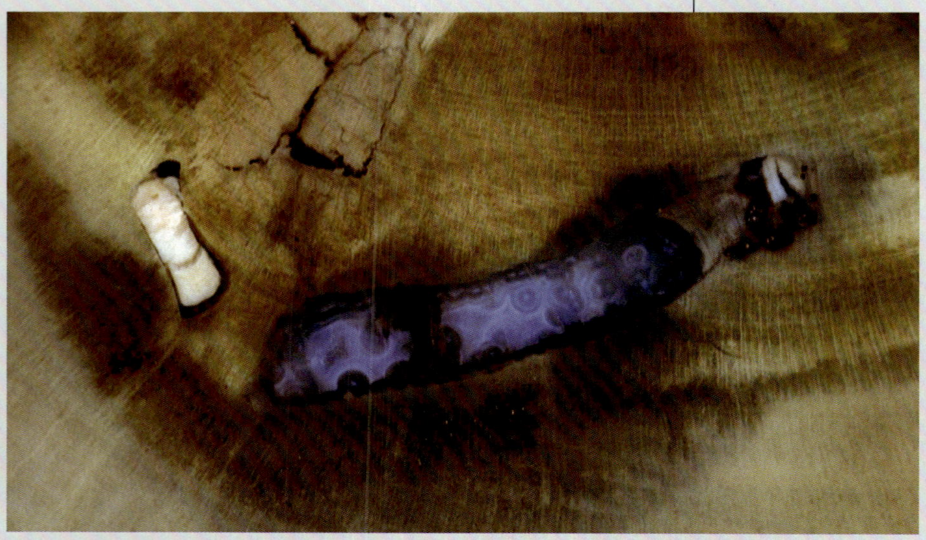

Macro photograph of borer evidence in fossil wood. Cretaceous. Rio Negro, Argentina. Daniels specimen.

Unidentified conifer. Mesozoic. Chubut, Argentina. Daniels specimen, 8 cm.

Unidentified conifer. Cretaceous. Rio Negro, Argentina. Daniels specimen, 5 cm.

Araucarixoylon. Jurassic. Santa Cruz, Argentina. Daniels specimen, 6 cm.

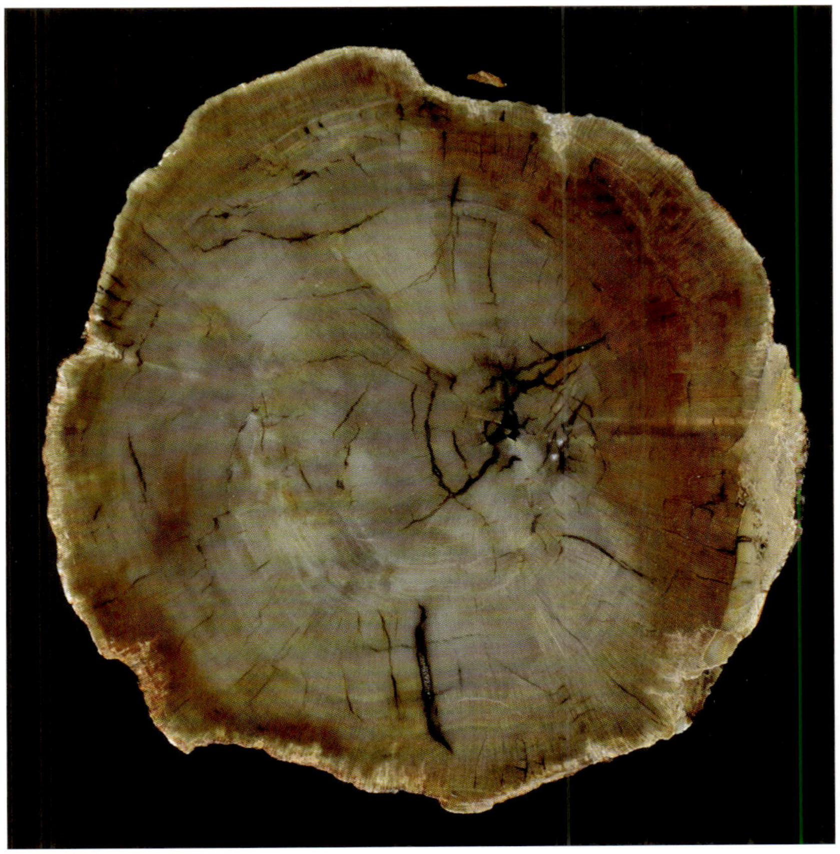

Araucarixoylon. Jurassic. Santa Cruz, Argentina.
Daniels specimen, 9 cm.

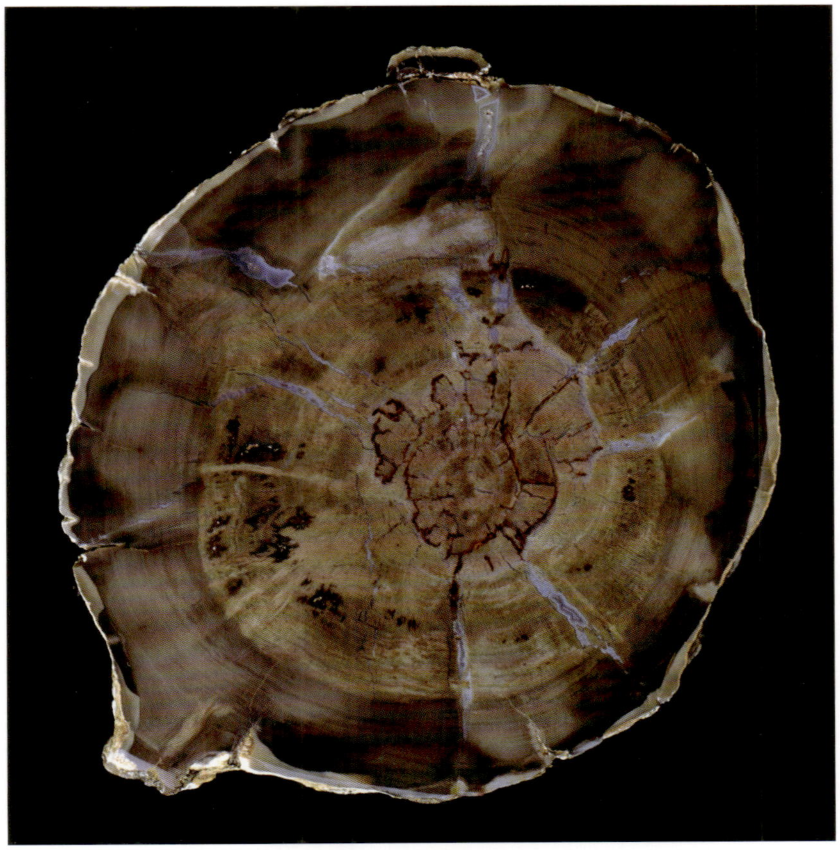

Unidentified conifer. Cretaceous. Rio Negro, Argentina.
Daniels specimen, 10.5 cm.

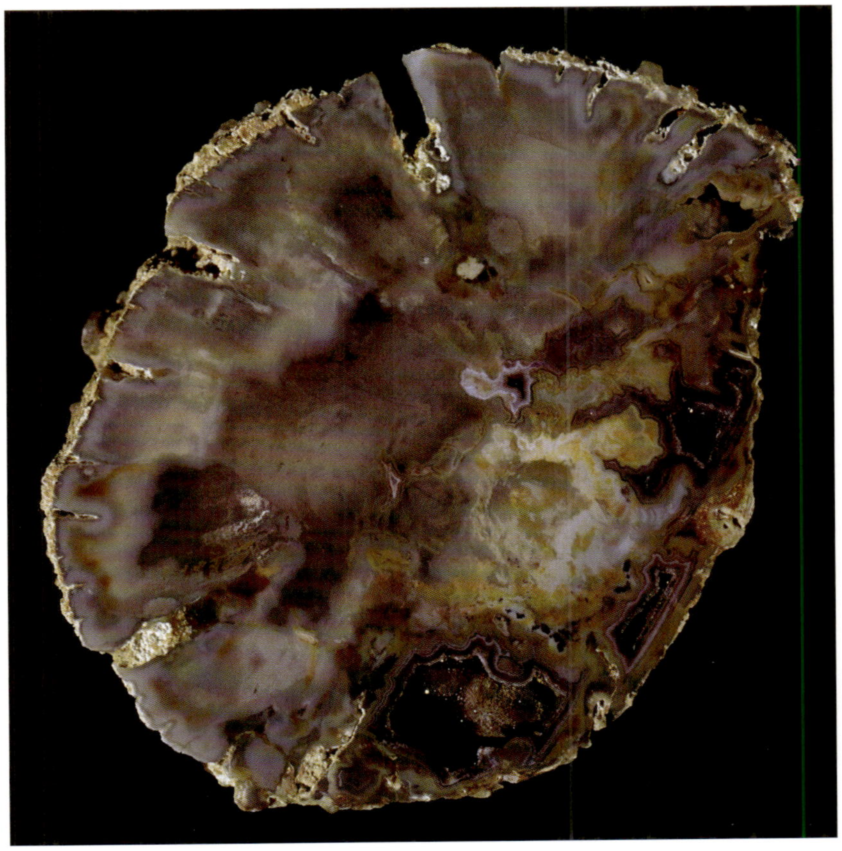

Unidentified conifer. Mesozoic. Chubut, Argentina.
Daniels specimen, 9 cm.

Unidentified conifer. Cretaceous. Patagonia, Argentina.
Daniels specimen, 10.5 cm.

Unidentified conifer. Mesozoic. Chubut, Argentina. Blankman specimen, 20.5 cm tall.

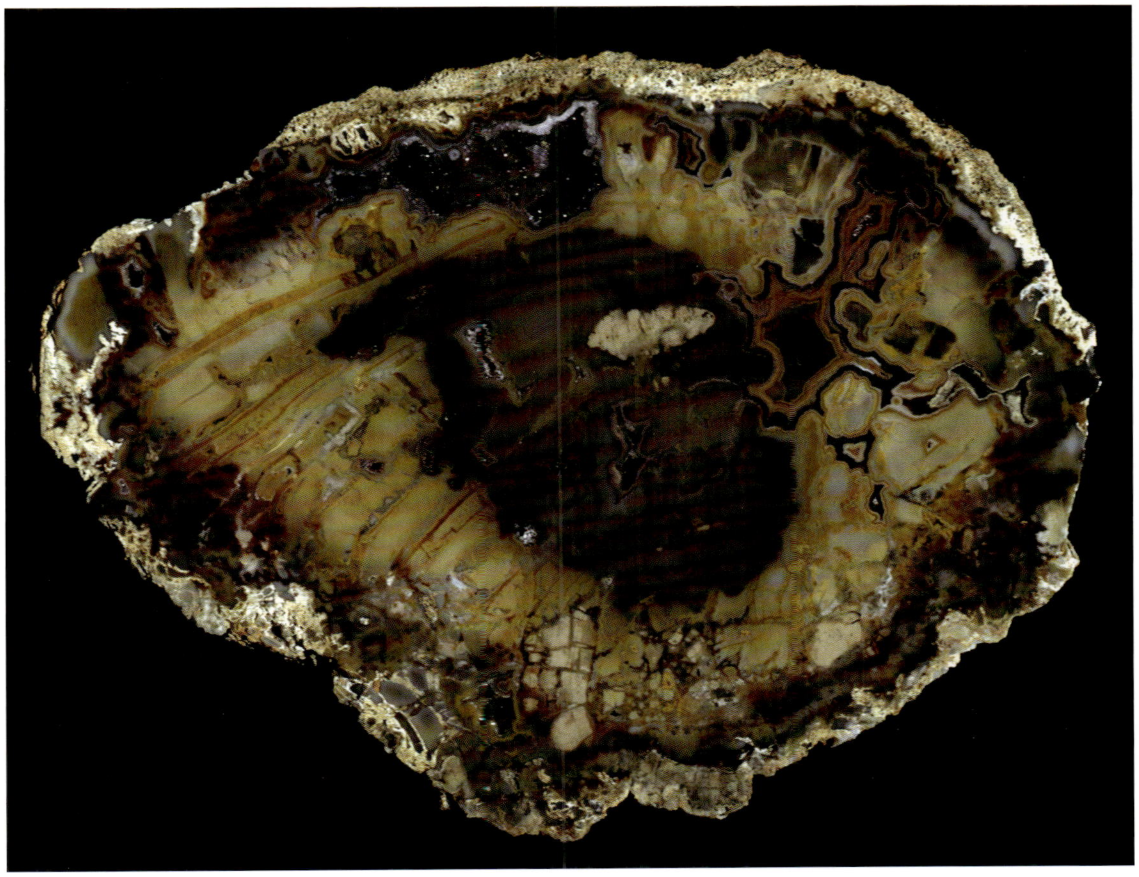

Unidentified conifer. Mesozoic. Chubut, Argentina. Daniels specimen. 13.5 cm.

Unidentified conifer. Mesozoic. Chubut, Argentina. Daniels specimen, 14 cm.

Modern families of conifers and ferns arose during the Triassic Period. Podocarps and araucarias were the predominant conifers during the Early Jurassic Period. Many Mesozoic Era conifer families have living members today. Conifers dating to the Jurassic and the Triassic Periods include Araucariaceae, Podocarpaceae (podocarps), Taxaceae (yews), and Taxodiaceae (bald cypress). The Cretaceous Period, known for the rise of flowering plants, saw forests of *Sequoia* and *Metasequoia* in northern latitudes, while araucarias and podocarps frequently populated southern climates.

A Cretaceous Period fern, *Tempskya*, left fossil remains in Lower Cretaceous rock throughout the Western United States, with particularly well-preserved specimens found in Colorado, Idaho, Oregon, Utah, and Wyoming. *Tempskya* is the form genus for the false trunks of a short-to-medium-height tree fern. The false trunk of *Tempskya* is composed of loosely organized leaf bases and branched stems encased in a dense mat of adventitious roots; the stem resembles a poorly preserved palm trunk. The roots formed a conical false stem that enabled it to assume an upright growth position. Some transverse section specimens of *Tempskya* consist almost entirely of roots, perhaps a circumstance resulting from stems dying and disintegrating as the false trunk grew outward and upward. Particularly glassy, attractive, well-preserved *Tempskya* specimens come from the Blue Mountains of Oregon. *Tempskya* specimens have been discovered in Lower Cretaceous deposits in Europe, North America, and South America.

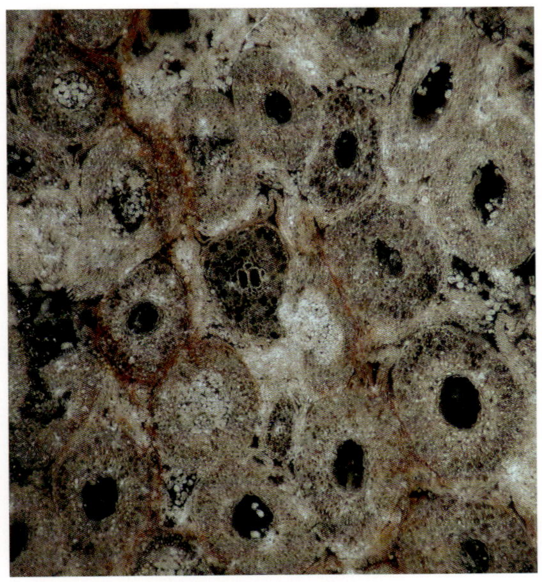

Micro image showing *Tempskya* detail. Lower Cretaceous. Idaho, USA. Daniels specimen.

Tempskya. Lower Cretaceous. Southeast Utah, USA. Daniels specimen, 42 cm long.

Tempskya. Lower Cretaceous. Blue Mountains, Oregon, USA. Daniels specimen, 23.5 cm long.

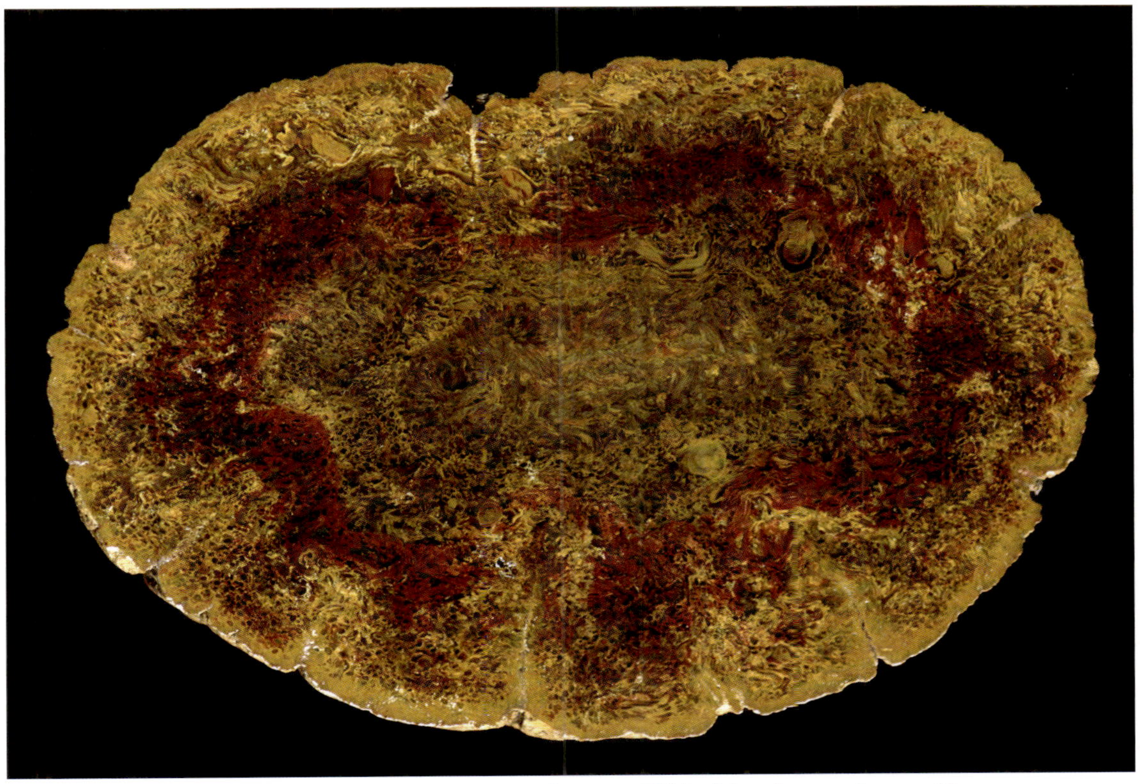

Tempskya. Lower Cretaceous. Blue Mountains, Oregon, USA. Murphy specimen, 13 cm.

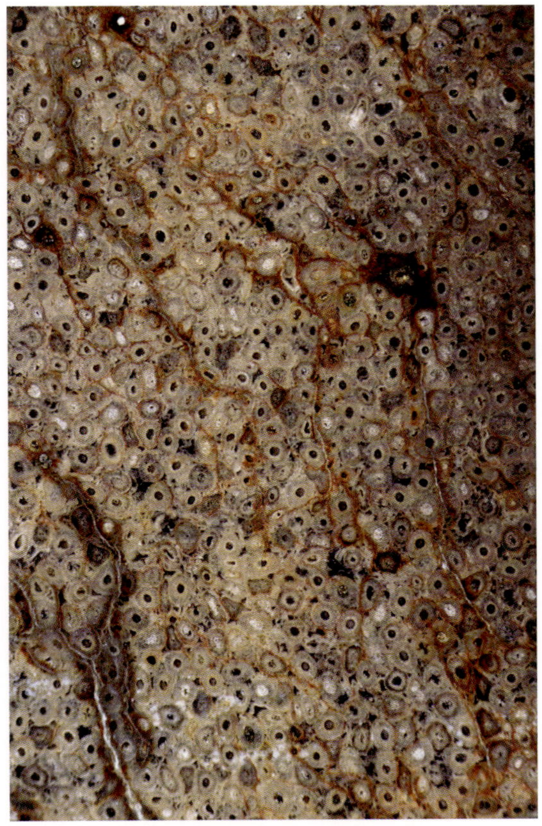

Macro photograph showing *Tempskya* detail. Lower Cretaceous. Idaho, USA. Daniels specimen.

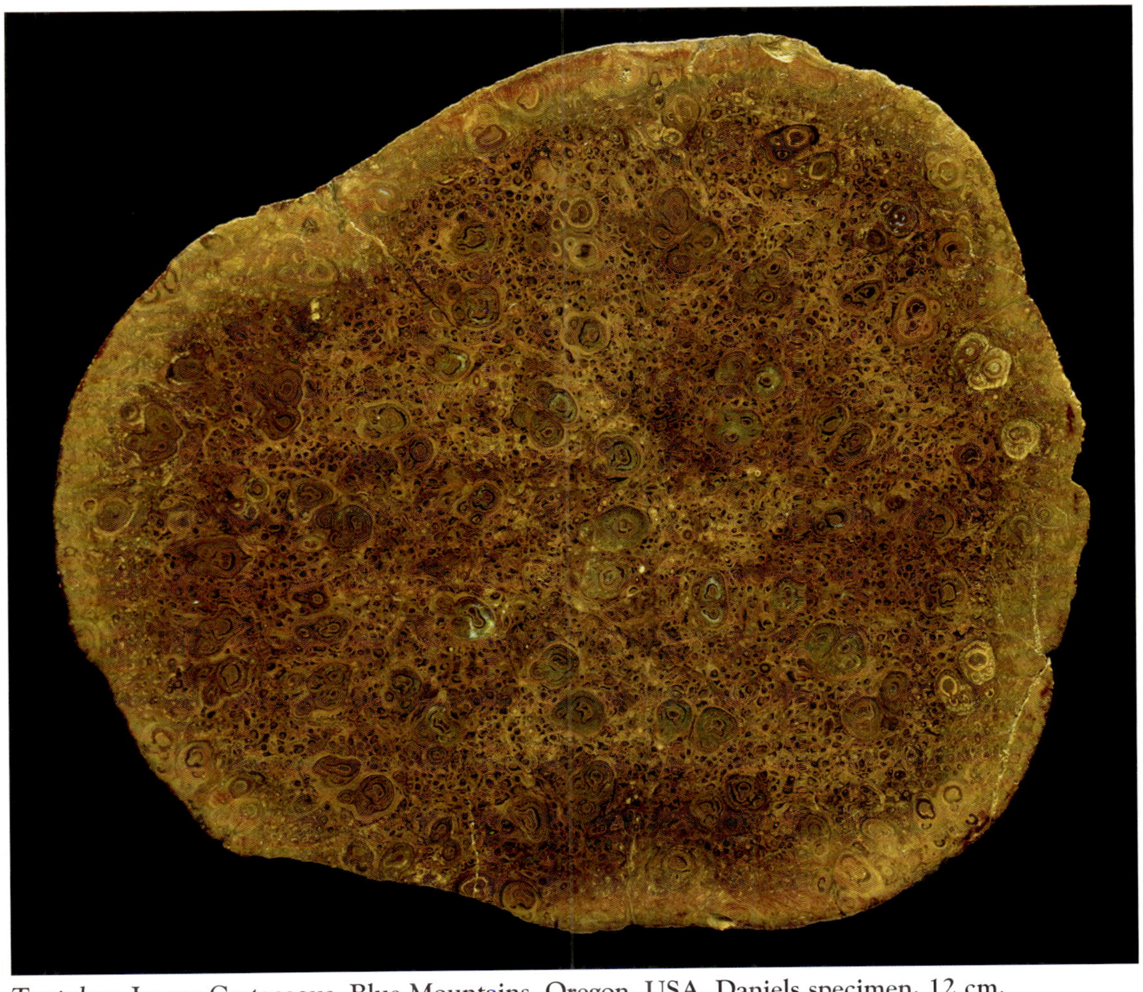

Tempskya. Lower Cretaceous. Blue Mountains, Oregon, USA. Daniels specimen, 12 cm.

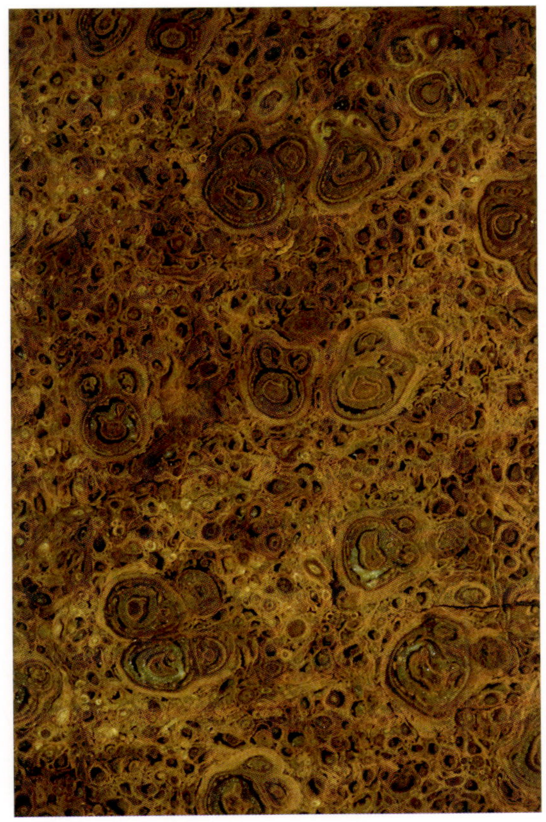

Macro photograph showing *Tempskya* detail. Lower Cretaceous. Blue Mountains, Oregon, USA. Daniels specimen.

Micro image showing *Tempskya* detail. Lower Cretaceous. Blue Mountains, Oregon, USA. Daniels specimen. Width of "monkey face" approximately 0.6 cm.

Micro image showing *Tempskya* detail. Lower Cretaceous. Blue Mountains, Oregon, USA. Daniels specimen.

At the end of the Mesozoic Era, something catastrophic happened, resulting in a time of darkness and cold temperatures. All the dinosaurs died. We call the stratigraphic boundary between the Cretaceous and the Tertiary periods the K/T boundary, and we call the reason the K/T event. Explanations include an asteroid impact and massive volcanic action. Most plant species apparently survived this event.

Identification of Paleozoic Era Fossil Woods

The fossil cliffs at Joggins in Nova Scotia, Canada, provide a stunning preserve of Upper Paleozoic Era plants. The mineralization of these fossils is poor, however, as is typical for Carboniferous Period fossils. It was during the Carboniferous that the first cycads and conifers evolved. Fossil wood from this age is rarely well silicified, colorful, or attractive, yet when we visit an area such as Joggins, the ancient forests come to life. Joggins is a small village near the head of the Bay of Fundy, where gargantuan tides of more than 16 meters (50 feet) are routine. In winter, ice-laden tides scour the ancient cliffs, exposing more fossil remains. The alternating beds of sandstone, shale, and siltstone formed 300 to 310 million years ago and now tilt approximated 20 degrees to the south. As you walk north along the shore, you travel back in time, thousands of years with each step, through abundant remains of ancient forests. The fossil wood remains at Joggins are typically bark and leaf impressions and pith casts, rather than petrified or silicified wood. Well-preserved fossils of animal life at Joggins, particularly small amphibians that evidently fell into disintegrating tree trunks, are of great interest to students of vertebrate paleontology.

The largest trees preserved at Joggins are *Lepidodendron* and *Sigillaria*. Both grew to some 30 meters (close to 100 feet) in height and left stumps at Joggins up to 1 meter (3 feet) in diameter. These two lycopods are differentiated by the spirally arranged leaf scar patterns on the bark. *Lepidodendron* bark is patterned with diamond-shaped scars, and *Sigillaria* leaf scars appear as vertical rows separated by ridges. Lycophytes have one of the longest fossil records of all plants, beginning in the Early Devonian Period. In the Late Carboniferous Period, they formed dense tropical forests and were one of the largest living organisms—some reaching 50 meters (more than 160 feet) in height. They were able to grow closely packed together—as many as 1,000 to 2,000 per hectare—because of their long slender trunks with leaves only at the top. The only lycophytes to survive the Jurassic were herbaceous forms. They now survive as "ground pine" or tiny club mosses that are not pine or mosses but are known as fern allies.

Other large fossil wood remains preserved at Joggins are *Calamites* (form genus for the stems) and the primitive conifer-like *Cordaites*, an extinct Paleozioc Era tree that had wood and seeds like a conifer and bore strap-shaped leaves up to a meter in length. *Cordaites* became extinct during the Permian, and the one extant apparent *Calamites* descendant, horsetail (*Equisetum*), survives with 25 species in a range of habitats as a small plant, frequently in marshy areas, along streambeds, and often in the shadow of modern giant trees such as the California redwood. The *Calamites* at Joggins grew to 5 meters (15 feet) in height with a stem diameter of 10 centimeters (4 inches) and formed thick undergrowth to the larger lycopods. *Calamites* grew as early as the Late Devonian and existed into the Permian Period. They reproduced by spores. Some reached heights of 10 meters (32 feet) or more. *Calamites* pith casts are identified by tight vertical ridges and frequent joints. As pith casts, they preserve just the shape of the central hollow pith of the original plant. Numerous locations throughout the world contain Carboniferous fossil wood remains, but none in as exciting a setting as Joggins. Another Carboniferous Period fossil wood deposit in Nova Scotia is located on the shore at Port Hood on Cape Breton Island where barite-replaced fossil wood erodes from limestone. Jurassic Period formations at Cape Split along the Scots Bay also yield fossil wood.

Macro photograph showing detail of fossil leaf impressions. Pennsylvanian. Joggins, Nova Scotia, Canada. Daniels specimen.

Minas Basin, Nova Scotia, Canada, from The Lookoff.

Fossil-bearing cliff at Joggins, Nova Scotia, Canada.

Fossil-bearing cliff at Joggins, Nova Scotia, Canada.

Fossil wood eroding from cliff at Joggins, Nova Scotia, Canada.

Shore at Port Hood, Nova Scotia, with fossil wood eroding from limestone.

Modern horsetail (*Equisetum*).

Shore along Minas Channel near Scots Bay, Nova Scotia, Canada.

Cliff at Port Hood, Nova Scotia, with fossil wood eroding from limestone.

Calamites pith cast. Pennsylvanian. Joggins, Nova Scotia, Canada. Daniels specimen, 12.5 cm.

As interesting and stimulating as the fossil cliffs of Joggins are, the fossil wood is of little interest to collectors of silicified wood specimens. Fortunately, any museum or private collector wanting Paleozoic Era fossil wood representatives can turn to a petrified forest in Brazil where beautiful specimens with amazing cell preservation from the Permian Period are abundant and well preserved. This petrified forest is located in the Permian Pedra do Fogo Formation in the Maranhao Basin of northeastern Brazil, near the city of Araguaina. Some of the wood is conifer in the nature of *Dadoxylon*. Much is wood of the tree ferns *Psaronius* and *Tietea singularis*. A far fewer number of stems are *Calamites*, and far fewer still are the rare fossil fern *Grammatopteris*. Fossil wood of Brazilian *Dadoxylon* often has well-preserved fine rows of tracheids.

Psaronius stems contain complex structure similar to modern tree ferns. They grew to 8 meters (26 feet) in height with a crown of large pinnately compound leaves. The stem structure is complicated by reliance on a massive mantle of adventitious roots for support. As the stem increases in diameter with height, the size of the root mantle decreases. The center of the *Psaronius* stem contains vascular bundles in long, worm-like strands that are generally symmetrical and, depending on the location in the tree, can be in a star-like pattern. In sections with complex structure, the appearance is reminiscent of a maze.

The complexity of the vascular system increases higher in the trunk. *Tietea singularis* transverse sections reveal central vascular bundles in smaller shapes; they can be O-shaped, C-shaped, omega-shaped, or short, fat, rounded wavy segments with occasional longer ribbons. *Grammatopteris* transverse sections include a distinct complex stele central to a mass of roots and large frond bases. As with all tree fern-like fossils, the structure alters with the position of the transverse section in the trunk. When the frond bases are cut at a level in which fronds were still vital when preserved, the appearance is similar to the fern *Osmunda*, with distinct frond bases, more rounded toward the center and wing-like toward the perimeter. *Calamites* identification is not difficult with specimens from mature plants because of the unique structure. Permineralized *Calamites* fossils are distinguishable by the empty, scalloped pith. A thin diaphragm of tissue stretched across the pith only at the nodes, leaving the remainder of the stem tubular. Primary xylem projects inward, and the rays between the xylem sheaves in the stem create the outer margins, giving the pith its shape. The pith is hollow, although often infilled with chalcedony, with a finely scalloped edge. On immature plants, and perhaps higher in the trunk where not yet fully developed, the pith can be closed, yet the serrated edges of the developing pith opening are apparent under close inspection. In many fossil *Calamites* specimens, compression collapsed the pith prior to fossilization.

Unidentified fossil wood cast. Pennsylvanian. Port Hood, Nova Scotia, Canada. Daniels specimen, 9 cm.

Calamites. Permian. Araguaina, Tocantins, Brazil. Daniels specimen, 11.5 cm.

Calamites in opalized sand cast with a stem or appendage of some sort. Permian. Araguaina, Tocantins, Brazil. Daniels specimen, 21 cm.

Calamites. Permian. Araguaina, Tocantins, Brazil. Daniels specimen, 11.5 cm.

Macro photograph showing *Calamites* detail. Permian. Araguaina, Tocantins, Brazil. Daniels specimen.

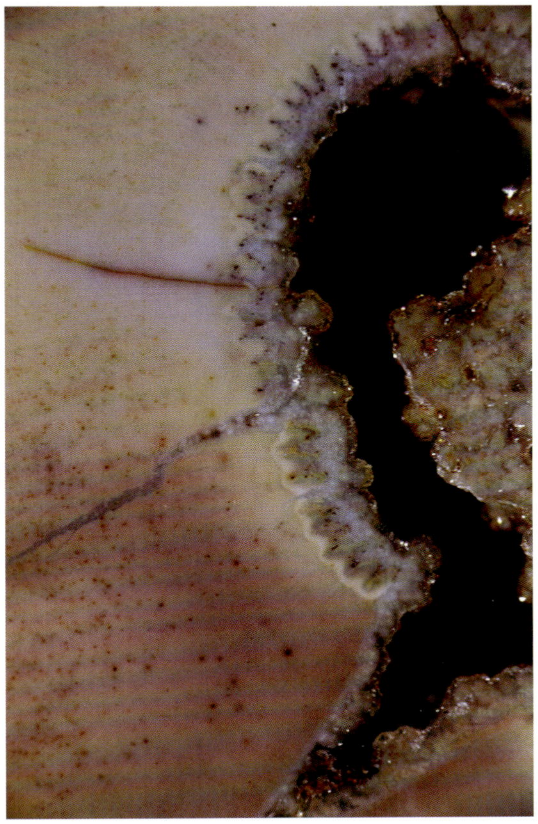

Macro photograph showing *Calamites* detail. Permian. Araguaina, Tocantins, Brazil. Daniels specimen.

Macro photograph showing *Calamites* detail. Permian. Araguaina, Tocantins, Brazil. Daniels specimen. Length of view approximately 5 cm.

Micro image showing *Calamites* detail. Permian. Araguaina, Tocantins, Brazil. Daniels specimen. Width of view approximately 1.5 cm.

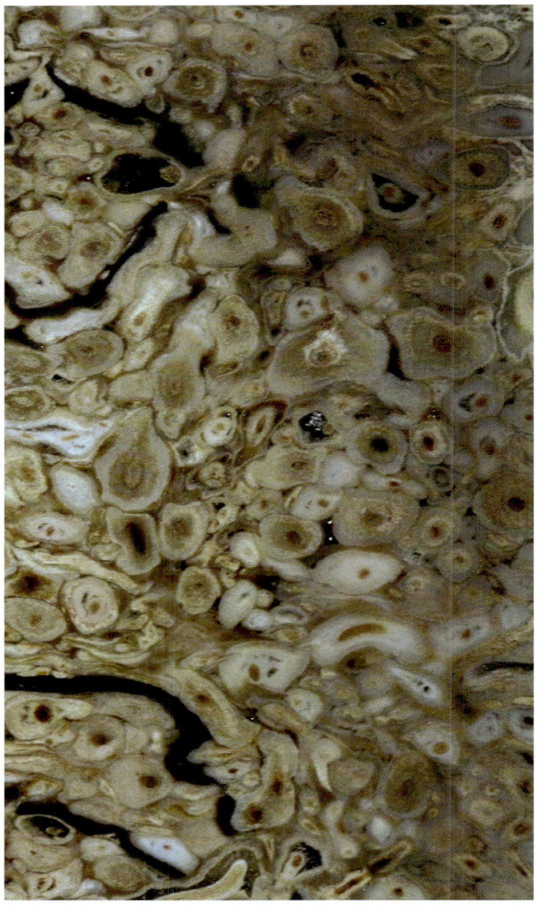

Macro photograph showing *Grammatopteris* detail. Permian. Araguaina, Tocantins, Brazil. Daniels specimen.

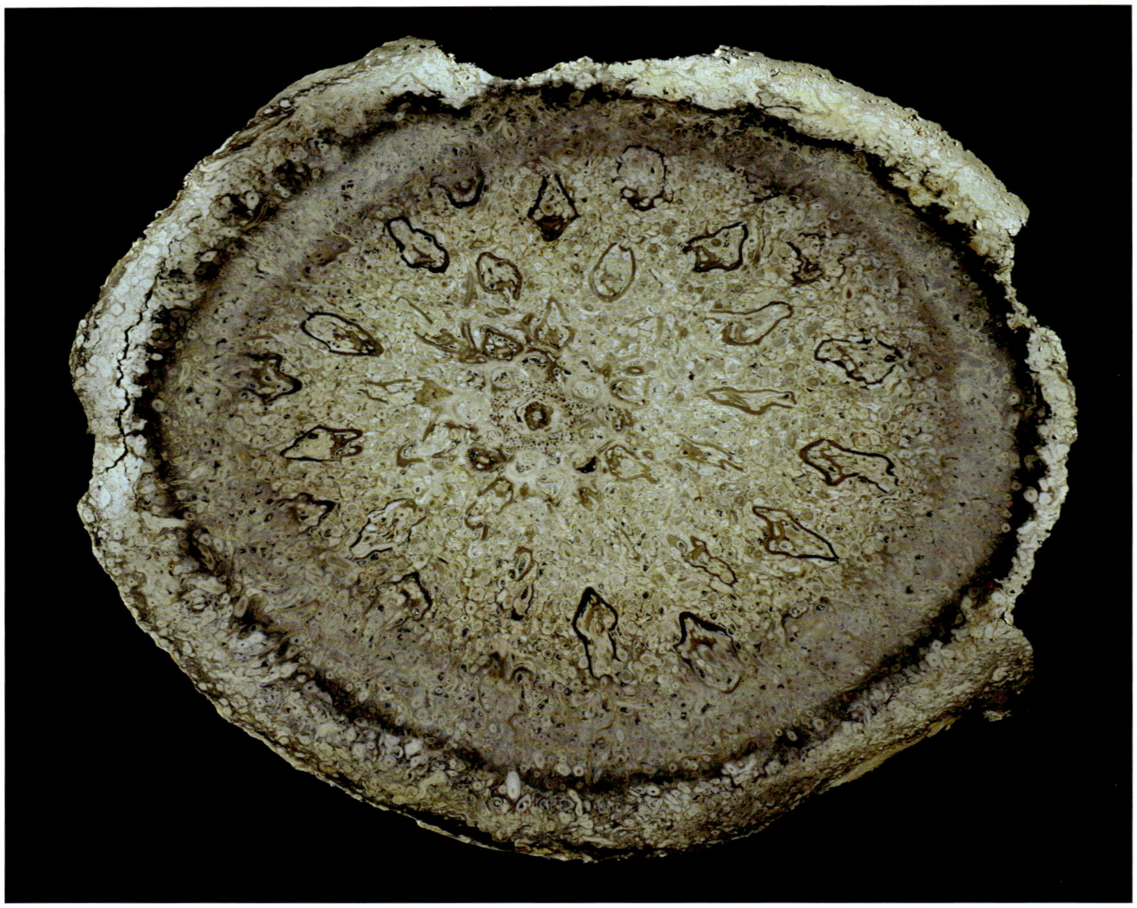

Grammatopteris. Permian. Araguaina, Tocantins, Brazil. Haggerty specimen, 26.5 cm.

Macro photograph showing *Psaronius* detail. Permian. Araguaina, Tocantins, Brazil. Daniels specimen.

Psaronius. Permian. Araguaina, Tocantins, Brazil. Daniels specimen, 15 cm.

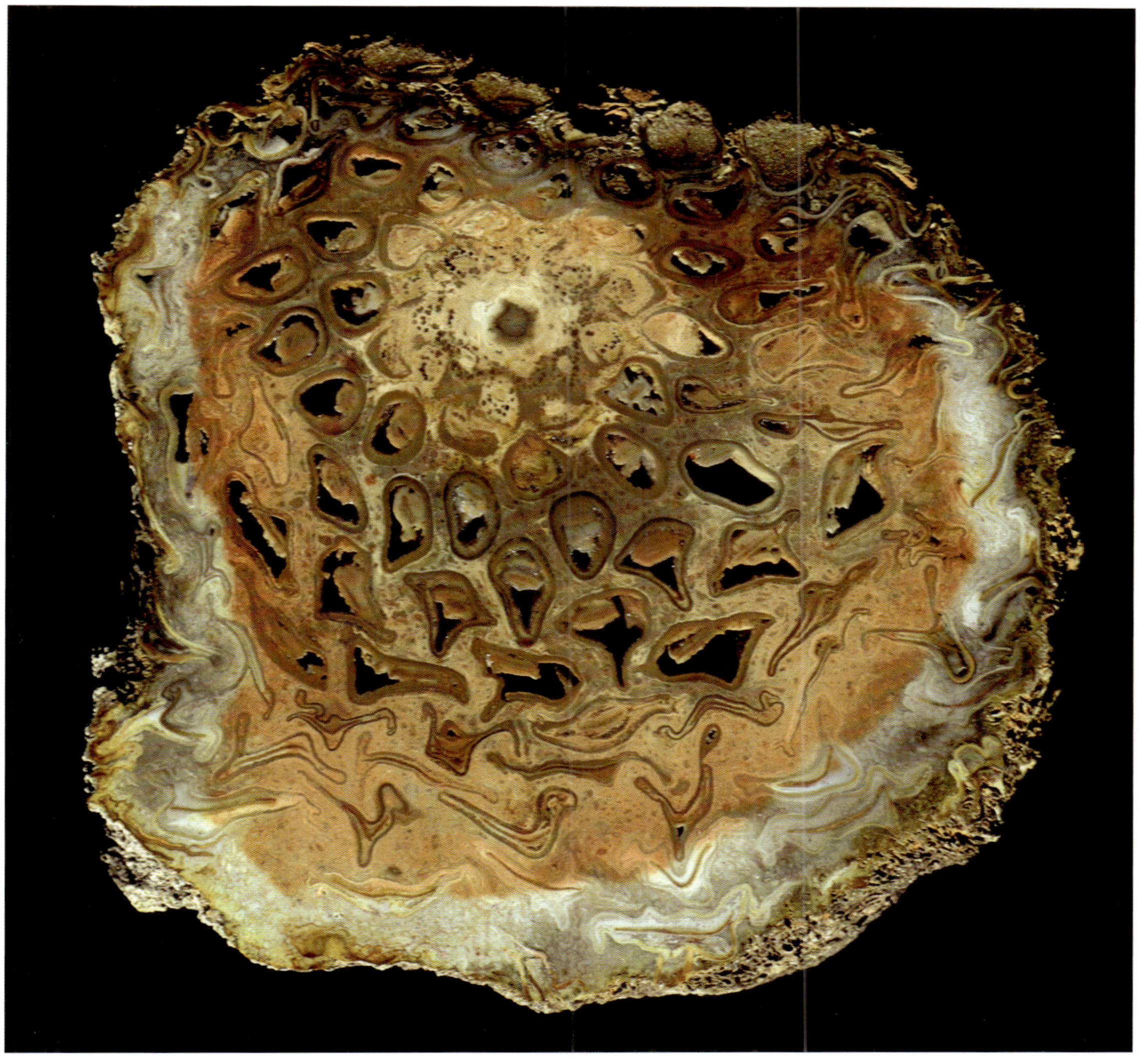

Grammatopteris. Permian. Araguaina, Tocantins, Brazil. Daniels specimen, 16.5 cm.

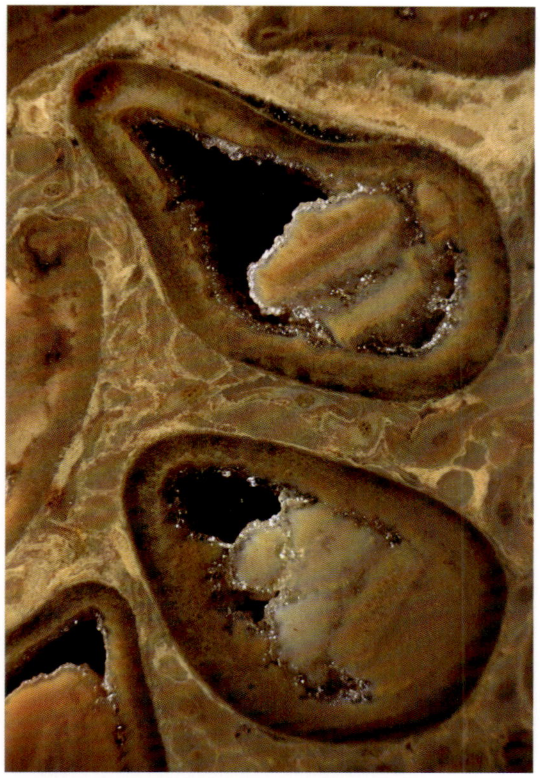

Macro photograph showing *Grammatopteris* detail. Permian. Araguaina, Tocantins, Brazil. Daniels specimen.

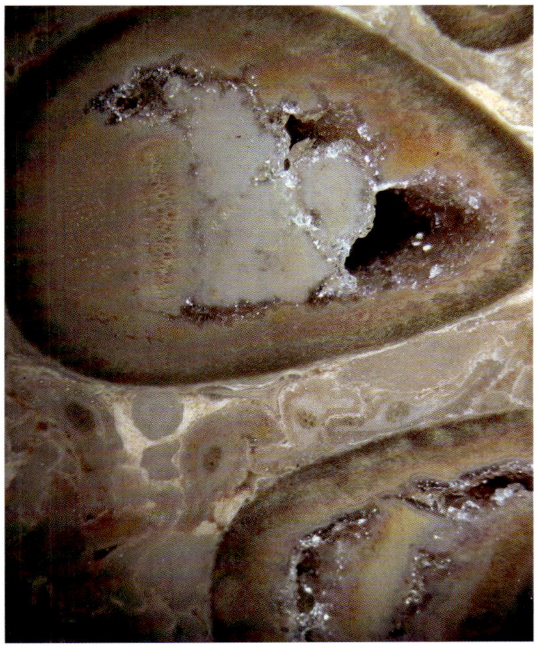

Micro image showing *Grammatopteris* detail. Permian. Araguaina, Tocantins, Brazil. Daniels specimen. Width of view 2 cm.

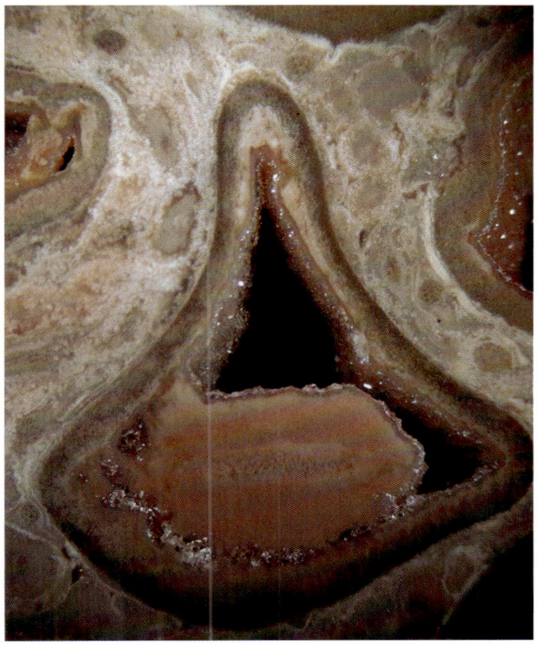

Micro image showing *Grammatopteris* detail. Permian. Araguaina, Tocantins, Brazil. Daniels specimen. Width of frond base 1.5 cm.

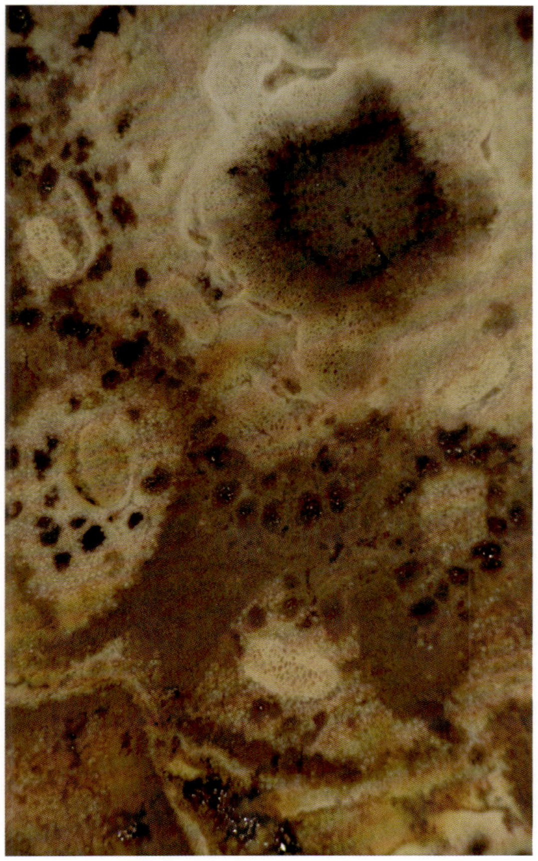

Macro photograph showing *Grammatopteris* detail. Permian. Araguaina, Tocantins, Brazil. Daniels specimen.

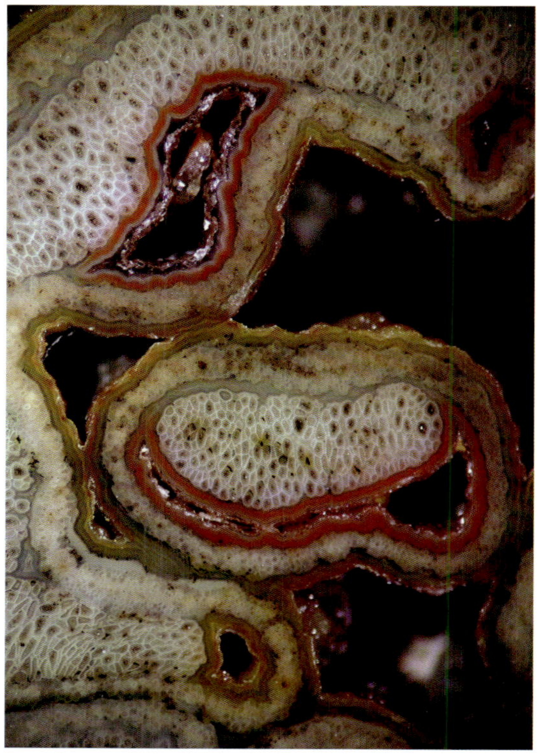

Micro image showing *Psaronius* detail. Permian. Araguaina, Tocantins, Brazil. Daniels specimen. Width of view approximately 1.2 cm.

Psaronius. Permian. Araguaina, Tocantins, Brazil. Daniels specimen, 11.5 cm.

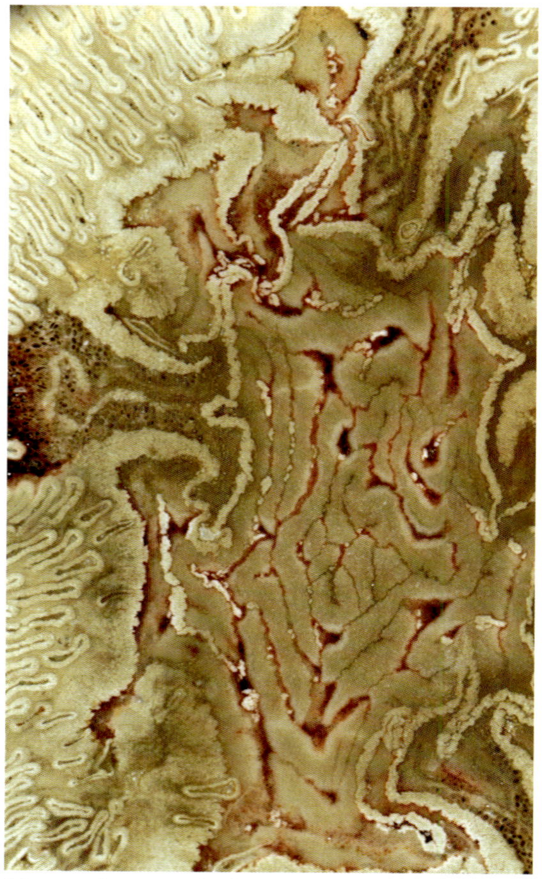

Macro photograph showing *Tietea singularis* detail. Permian. Araguaina, Tocantins, Brazil. Daniels specimen.

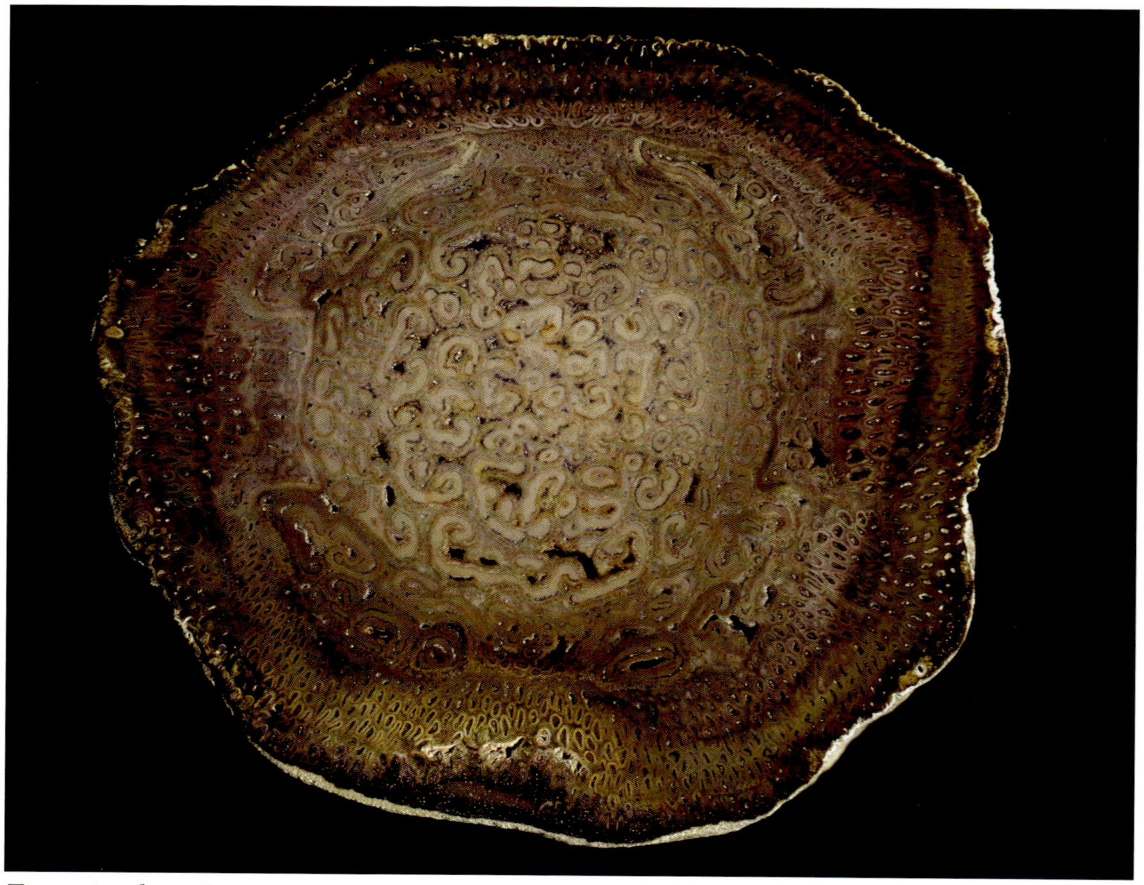

Tietea singularis. Permian. Araguaina, Tocantins, Brazil. Daniels specimen, 15 cm.

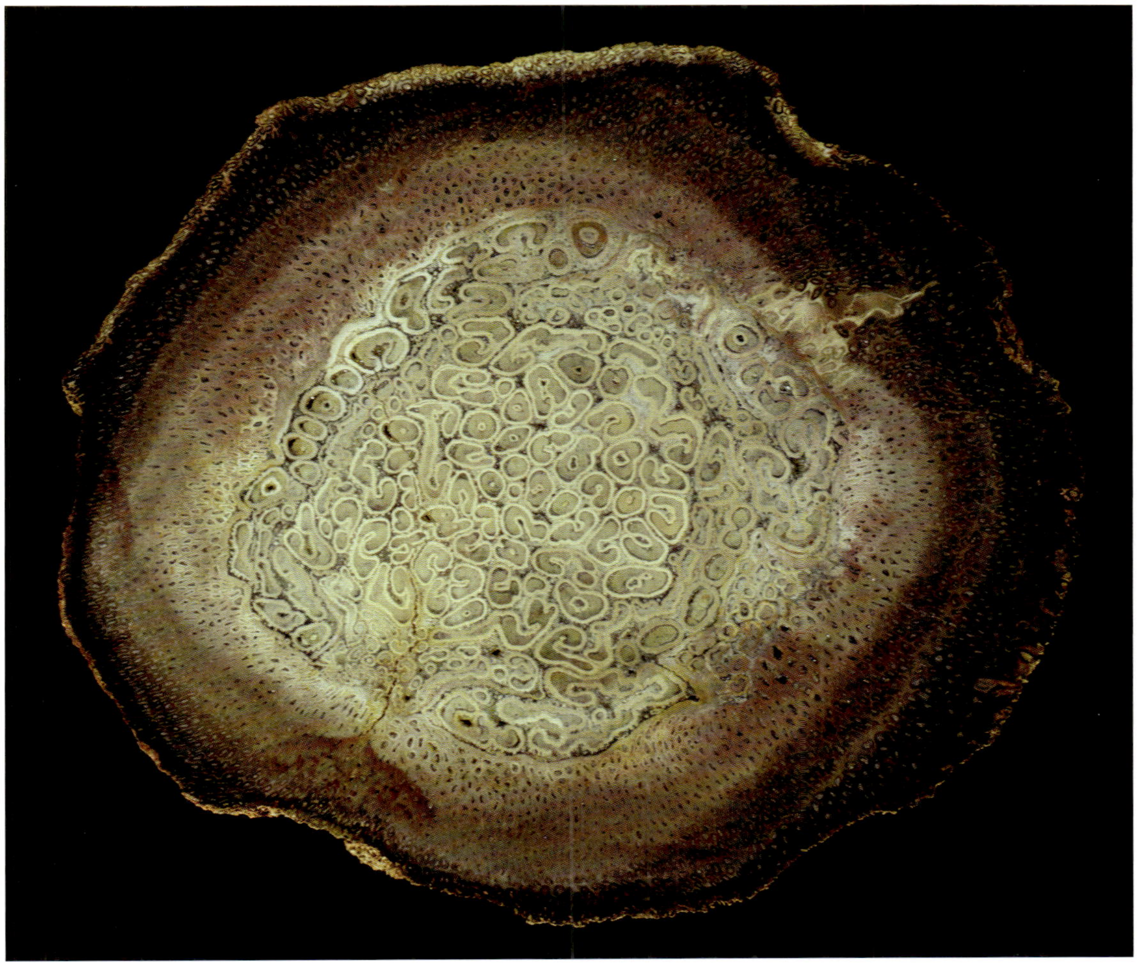

Tietea singularis. Permian. Araguaina, Tocantins, Brazil. Daniels specimen, 18 cm.

Macro photograph showing *Tietea singularis* detail. Permian. Araguaina, Tocantins, Brazil. Daniels specimen.

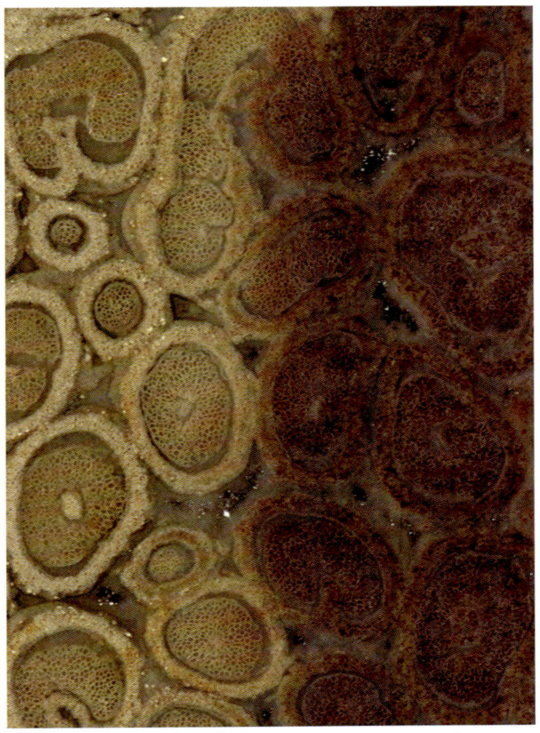

Macro photograph showing *Tietea singularis* detail. Permian. Araguaina, Tocantins, Brazil. Daniels specimen.

Micro image showing *Tietea singularis* detail. Permian. Araguaina, Tocantins, Brazil. Daniels specimen.

Macro photograph showing *Tietea singularis* detail. Permian. Araguaina, Tocantins, Brazil. Daniels specimen.

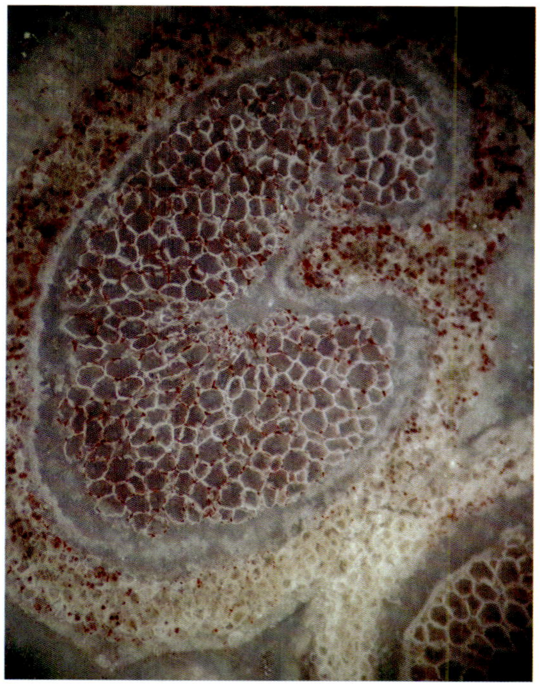

Micro image showing *Tietea singularis* detail. Permian. Araguaina, Tocantins, Brazil. Daniels specimen.

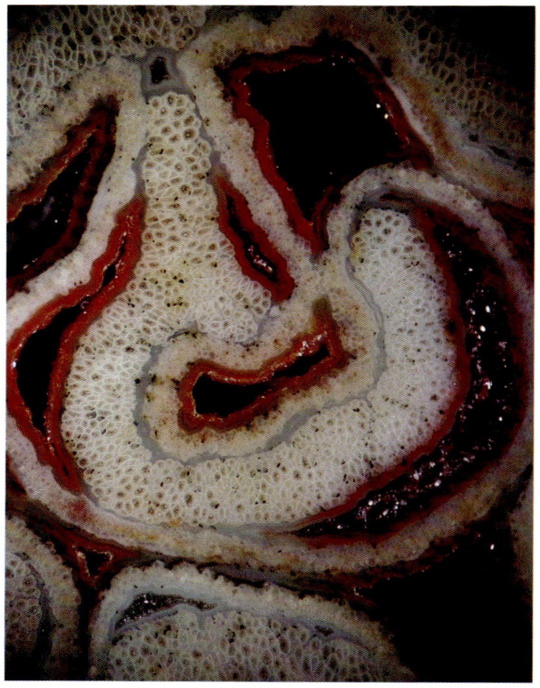

Micro image showing *Psaronius* detail. Permian. Araguaina, Tocantins, Brazil. Daniels specimen. Width of view approximately 1.2 cm.

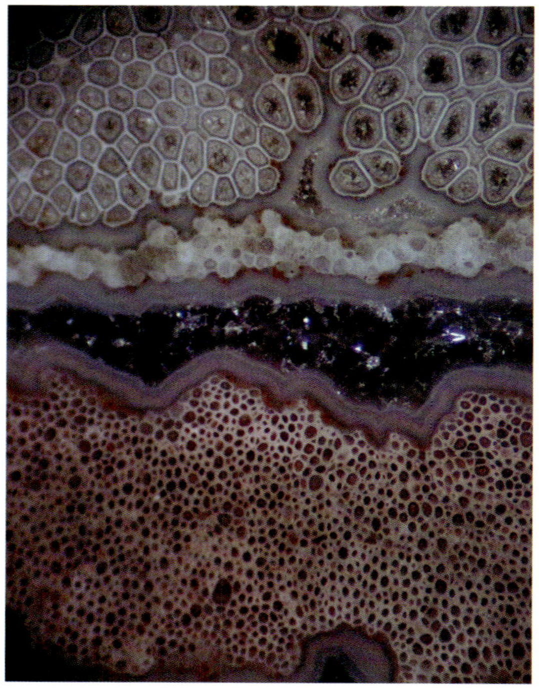

Micro image showing *Psaronius* detail. Permian. Araguaina, Tocantins, Brazil. Daniels specimen.

Micro image showing *Tietea singularis* detail. Permian. Araguaina, Tocantins, Brazil. Daniels specimen.

Micro image showing *Psaronius* detail. Permian. Araguaina, Tocantins, Brazil. Daniels specimen.

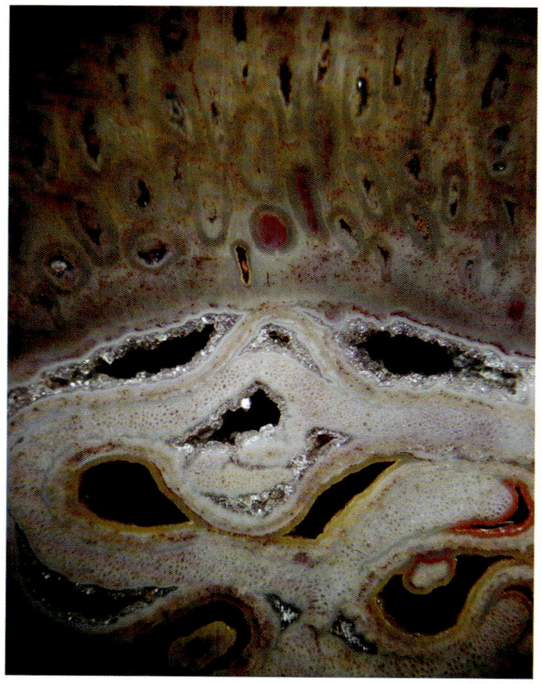

Micro image showing *Psaronius* detail. Permian. Araguaina, Tocantins, Brazil. Daniels specimen.

Macro photograph showing *Dadoxylon* detail. Permian. Araguaina, Tocantins, Brazil. Daniels specimen.

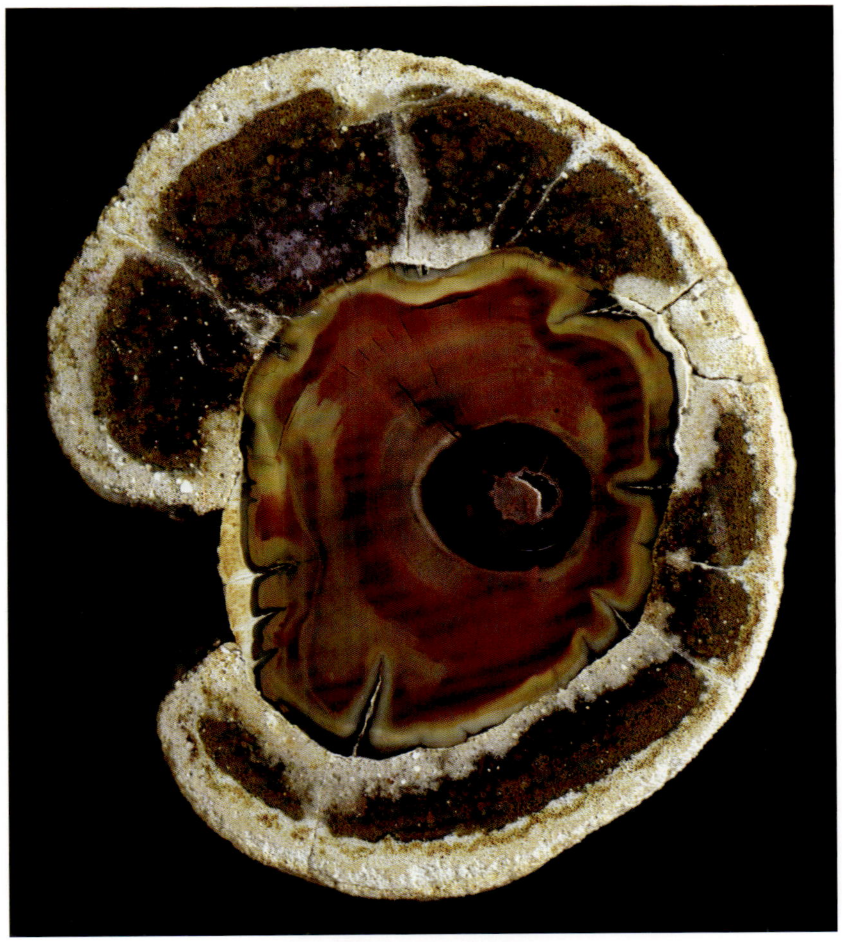

Dadoxylon in opalized sand cast. Permian. Araguaina, Tocantins, Brazil. Daniels specimen, 10.5 cm.

Micro image showing *Dadoxylon* detail. Permian. Araguaina, Tocantins, Brazil. Daniels specimen. Average tracheid size 0.04 mm.

Dadoxylon. Permian. Araguaina, Tocantins, Brazil. Daniels specimen, 16 cm.

Dadoxylon. Permian. Araguaina, Tocantins, Brazil. Daniels specimen, 25 cm.

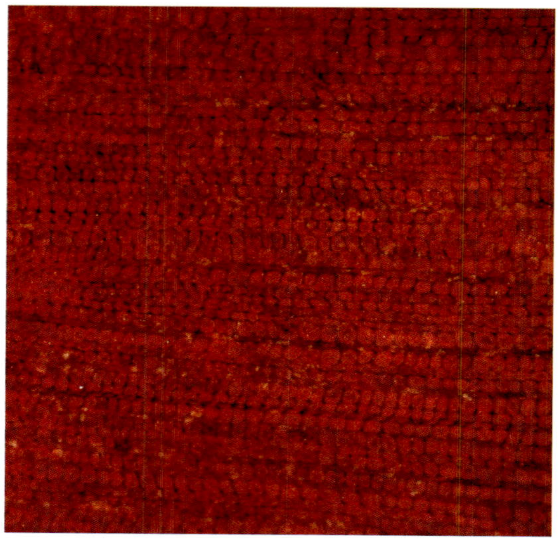

Micro image showing *Dadoxylon* detail. Permian. Araguaina, Tocantins, Brazil. Daniels specimen. Average tracheid size 0.025 mm.

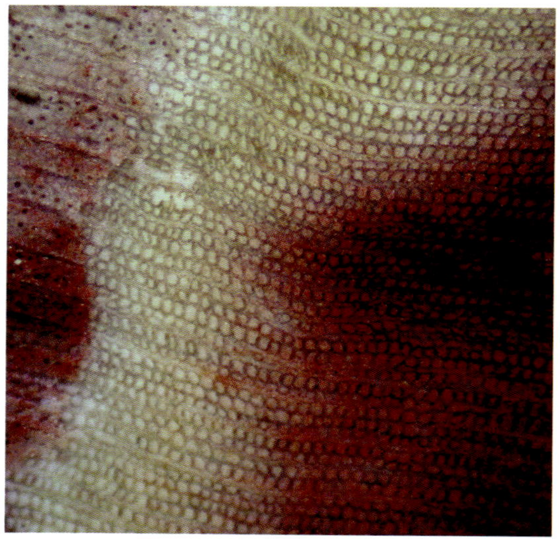

Micro image showing *Dadoxylon* detail. Permian. Araguaina, Tocantins, Brazil. Daniels specimen. Average tracheid size 0.025 mm.

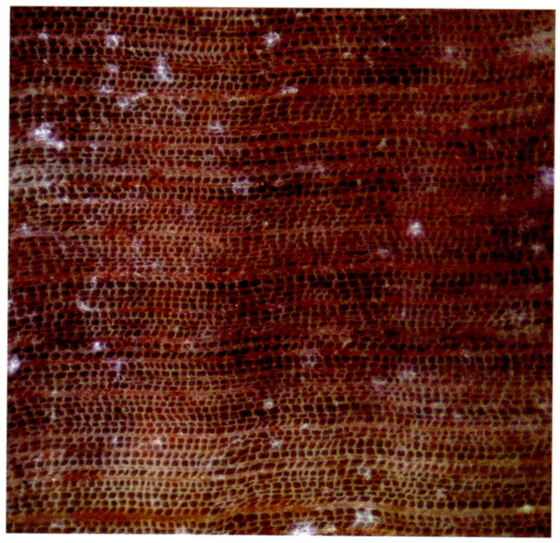

Micro image showing *Dadoxylon* detail. Permian. Araguaina, Tocantins, Brazil. Daniels specimen. Average tracheid size 0.035 mm.

Macro photograph showing *Dadoxylon* detail. Permian. Araguaina, Tocantins, Brazil. Daniels specimen.

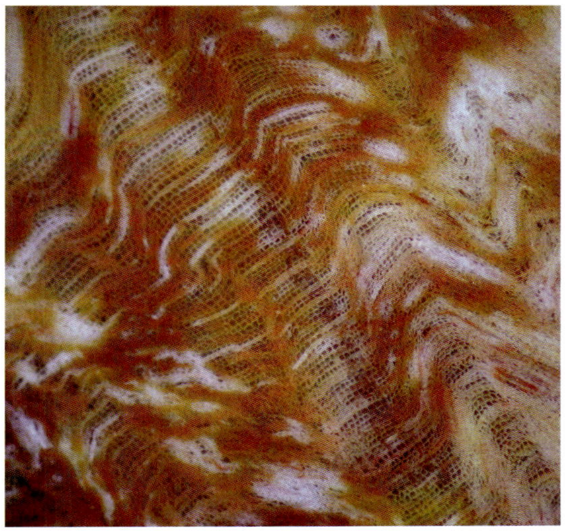

Micro image showing *Dadoxylon* detail. Permian. Araguaina, Tocantins, Brazil. Daniels specimen. Average tracheid size 0.035 mm.

Micro image showing *Calamites* detail. Permian. Araguaina, Tocantins, Brazil. Daniels specimen. Average tracheid size 0.0625 mm.

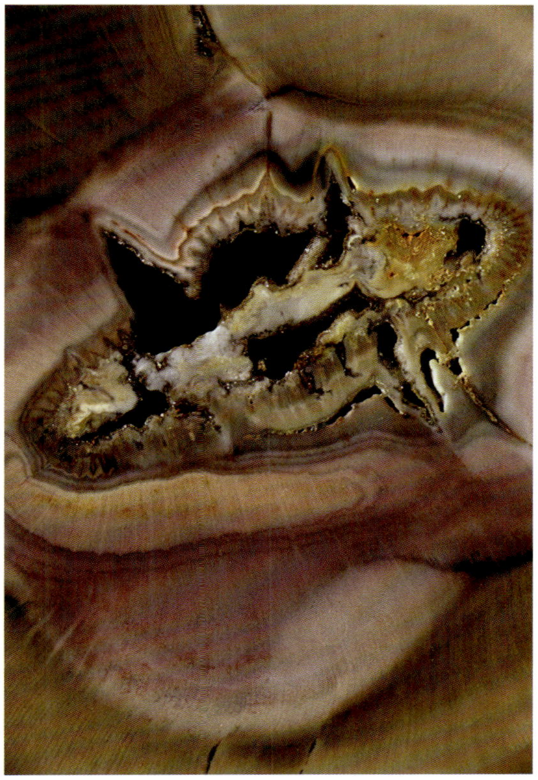

Macro photograph showing *Calamites* detail. Permian. Araguaina, Tocantins, Brazil. Daniels specimen.

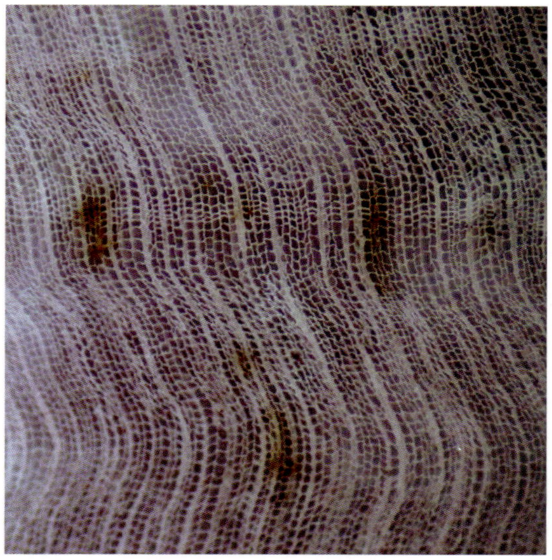

Micro image showing *Calamites* detail. Permian. Araguaina, Tocantins, Brazil. Daniels specimen. Average tracheid size 0.0625 mm.

Micro image showing *Calamites* detail. Permian. Araguaina, Tocantins, Brazil. Daniels specimen. Average tracheid size 0.0625 mm.

A little understood cataclysmic event of an enormity we cannot comprehend marks the end of the Permian Period, when seed plants first became dominant. Whatever happened caused the most devastating change of all time to the Earth's vegetation, resulting in the extinction of numerous plant and animal species, including many lycopods, ferns, and conifers. We know the extinction as the P/T event, for Permian/Triassic; the stratigraphic level is the P/T boundary.

II

CHAPTER TWO

Wood to Stone

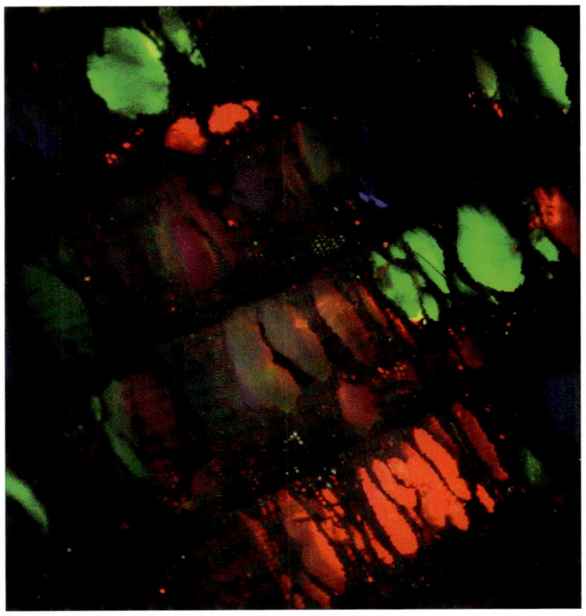

FEW OF US HAVE HELD A PIECE OF PETRIFIED WOOD IN OUR HANDS without wondering how the wood turned into stone. And what is the type of stone? How long did it take to form? Why can you still see minute woody textures? What happened to the original wood? What causes the colors? Fossil wood collectors and scientists have been puzzling over these questions for centuries. This chapter presents ideas that researchers have formulated in an attempt to answer these questions. As might be expected, the actual physio-chemical mechanisms causing petrifaction are complex and only recently have we started to understand the processes.

What is petrified wood? Jackson suggests that the general term petrifaction means a process whereby organic matter is converted into a stony substance by the infiltration of water containing dissolved inorganic elements or compounds that substitute for the original organic materials, sometimes retaining the original structure. Collectors quickly learn that petrified wood usually comes in two general forms. Either it looks like wood on the outside and shows detailed preservation of cellular structure on the inside, or it looks like wood on the outside and exhibits various textures on the inside, none of which resemble wood. In some cases, a single specimen of petrified wood can exhibit both of these characteristics.

Petrified wood that shows detailed internal woody texture is what most collectors think of as petrified wood and is described by such terms as permineralized or replaced. These terms are often used as interchangeable equivalents, but they do not have the same meaning. Most importantly, replacement is now considered improper terminology when discussing the formation of petrified wood and should not be used. The commonly used term permineralization simply means mineral matter has been deposited in the original pore spaces. A good example of permineralization is the spectacular infilling of the open parts of cells by precious opal that occurs in specimens from Virgin Valley, Nevada.

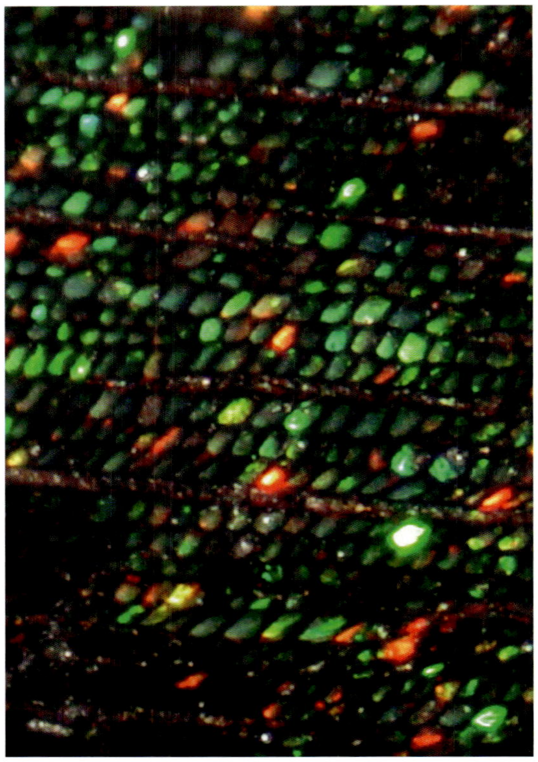

Transverse view of individual cell lumina that are filled with areas of precious opal; cellular walls are dark-colored common opal, producing a stained glass window effect with magnification. Virgin Valley Formation, Miocene. Royal Peacock Opal Mine, Virgin Valley, Nevada, USA. Dayvault specimen, Individual cell lumen about 0.025 mm.

Old texts and publications describe how inorganic molecules such as silica replace individual carbon molecules in the original wood. This is the so-called atom-by-atom replacement process, which is no longer considered accurate but is still occasionally used. Instead, the modern concept supports initial replication of the wood by a process called templating. This is accomplished on a supramolecular (consisting of more than one molecule) scale by infiltration and emplacement, whereby silica dissolved in ground water infiltrates and impregnates host wood and begins to replicate detailed textures of the woody fabric. Additional silica eventually infills any remaining openings and fortifies the wood as described for permineralization.

Petrified wood that contains no relic internal texture is usually some sort of cast, although permineralization and organization of microcrystalline silica can proceed to the extreme and obliterate original cellular textures. Limb casts are formed by the loss of the original wood, which leaves a relatively large void in the sediment, and by later infilling of the void by mineral matter. Casts can result in beautiful and interesting fossil woods, although detailed cellular characteristics of the original wood are lost. Both petrifaction of wood that leaves a detailed histological record of the precursor and cast formation that involves deposition of mineral matter in a large void are complicated processes and subjects of continuing research.

CONDITIONS FOR PETRIFACTION

How does a living tree meet such a glorious finale and become a stony replica of its former self? The plant must first die and become buried in a sedimentary depression, such as a river channel or lake, where water saturates the enclosing sediment and the wood. Without this saturated environment, wood ordinarily decomposes in a matter of a few years by the action of insects, bacteria, and especially fungi.

Leo and Barghoorn in their landmark paper of 1976, along with others, list necessary requirements for wood to petrify, and, just as importantly, detrimental conditions under which the original wood will not retain microscopic textures important to subsequent detailed preservation. These parameters affect plant tissue decomposition and allow preservation of microscopic detail to proceed or fail.

- **Burial.** Rapid burial and deposition below the water table are important for preservation of wood. These processes slow down exposure to oxygen and fungi, both agents causing decomposition of wood after the tree dies. Timbers retrieved from stagnant lake bottoms with even slightly reducing bottom sediments were intact even after hundreds of years. Wood retrieved from anaerobic or extremely oxygen-deficient lake mud retains much of its microtexture even after thousands of years.

- **pH.** This value is a measure of acidic or basic quality of an aqueous solution. In nature, a pH value may range from 2 to 10 or higher. A pH value of 7 is a neutral solution; less than 7 is acidic; greater than 7 is basic. Wood-attacking fungi can withstand a wide range of pH values from about 4 to 9 but operate best in the range from about 4.5 to 7; therefore, pH values in the range of 7 to 9 inhibit fungal attack. Solutions with pH values greater than 9, as found in alkaline or very basic solutions, can destroy the molecules that form wood. Detailed preservation of microscopic woody textures is unlikely if the original wood falls into a highly alkaline lake, such as those sometimes found in desert climates. A pH of 8 is most favorable for development of colloidal silica and 5 to 6 for development of solutions containing silicic acid. Therefore, a solution with a pH of 6 to 9 is probably the most favorable for both the polymerization of silicic acid, which replaces wood, and for preventing initial chemical breakdown of cellulose. Fortunately, this situation is not uncommon in nature.

- **Silica source.** An abundant source of dissolved silica is needed. The most abundant sources of dissolved silica are from the breakdown of volcanic glass (from siliceous tuffs or lava flows) and from siliceous organisms such as diatoms.

- **Temperature.** Fungal activity causes much of the early rotting of wood and, thereby, removes any possibility of detailed petrifaction. Fungal activity nearly ceases at temperatures less than 39° Fahrenheit (F) [4° Celsius (C)] and more than 104° F [40° C]. Plant tissue begins to break down above 212° F [100° C]. Temperature is not too important in normal near-surface sedimentary environments but becomes important in hot springs deposits.

- **Sedimentary setting.** The immediate sedimentary environment that entombs the wood is primarily responsible for optimizing the preceding criteria. Rapid burial in muddy or fine-grained sediment, preferably submerged in water, is ideal. Many fossil wood localities are associated with volcanic ash or volcaniclastic (water-transported ash) sediments resulting from ash fall eruptions or with flood events that drained volcanic ash terrain. Many petrified wood beds originate between successive lava flows where a forest had time to establish and grow before being covered by the next eruption.

Petrified forests at Yellowstone National Park are a famous example of petrified wood in an ash bed environment. Princeton University professor Ehrling Dorf identified at least 27 individual forests preserved in a 2,000-foot-thick (610 meters) sedimentary deposit that formed when ash or ash-flow tuffs buried mature forests, preserving many trees in upright positions. Volcanic-related flood events that extract, transport, and bury trees are more common. The Petrified Forest Member of the Triassic Chinle Formation in Arizona and the Brushy Basin Member of the Jurassic Morrison Formation in Utah include numerous deposits of volcaniclastic sediments that contain petrified wood. In these examples, wood was protected from oxygen and the ash provided a source of silica.

WHAT MINERALS ARE DOING THE PETRIFICATION?

St. John reported in 1927 at least 40 minerals that replace wood to some extent. The 1994 *extraLapis* No. 7 issue published a list of at least 58 minerals and other substances that petrify wood. By far the most common are hydrous silica (opaline varieties) and microcrystalline quartz (such as chalcedony), followed by calcium carbonate (calcite), iron oxides, hydrous iron oxides (particularly goethite), and iron sulfides (pyrite and marcasite). Many metals are concentrated in fossil wood because of the local reducing environment produced by decomposition of organic carbon compounds in the wood. Some fairly common minerals that permineralize wood or produce wood casts include copper minerals such as chrysocolla, bornite, chalcopyrite, covellite, tenorite, and even native copper. Woodward and others report that copper-mineralized wood provided the primary ore for the Nacimiento Mining District of north central New Mexico. The deposit occurs in sandstones of the Chinle Formation and yielded logs to 60 feet in length that had been mineralized with chalcocite, covellite, bornite, chalcopyrite, and other oxidized copper minerals, as well as small amounts of native silver. Between 1881 and 1960, 7,500,000 pounds of copper and 75,000 ounces of silver were produced from this deposit. Old mining timbers containing native copper have been recovered from Cyprus to Arizona. On the Colorado Plateau and at other locations, minerals containing uranium and vanadium, including carnotite, uraninite, coffinite, tyuyamunite, volborthite, and others, are sometimes sufficiently concentrated in fossil wood to produce high-grade ore deposits. Carbonate minerals other than calcite that mineralize wood include aragonite, dolomite, hydromagnesite, siderite, cerrusite, azurite, and malachite. Sulfate minerals that mineralize wood include barite, gypsum, langite, and schulenbergite. A number of organic substances have been identified in fossil wood, including amber, "succinite," jet, hartite, humboldtine, mellite, simonellite, and stepanovite. Fluorite, galena, sphalerite, zeolites, vivianite, apatite, and even gold have been detected in fossil wood. The opal varieties and microcrystalline quartz preserve the greatest detail of original vascular tissue structure because of the method of petrifaction.

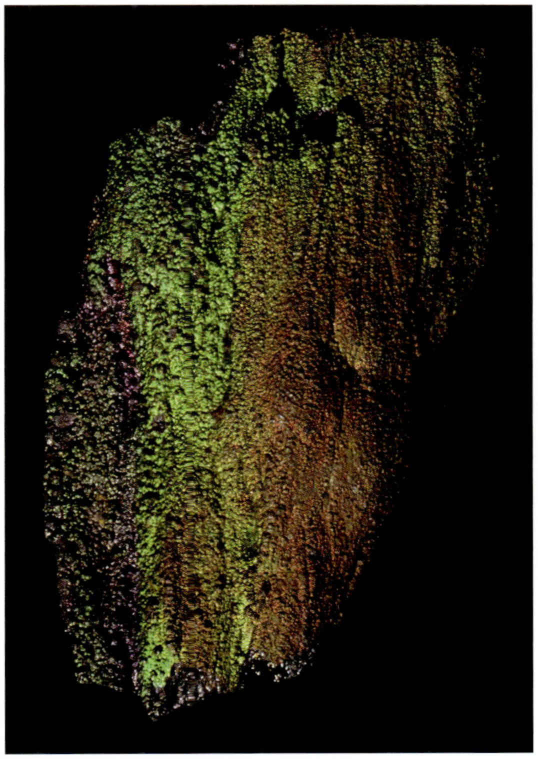

Hematite wood. Limb cast with iridescent hematite replacing wood. Specimen is from the old Tintic Mining District, Tintic, Utah, USA. Daniels specimen, 5 cm.

Malachite after wood. Copper minerals frequently replace wood usually by permineralization. Reducing environments produced by the carbon in wood cause the precipitation of metal ions such as copper. These may initially form sulfides that can oxidize and combine with carbonate in solution to produce other minerals. Chinle Formation, Triassic. Nacimiento Mining District, northern New Mexico, USA. Daniels specimen.

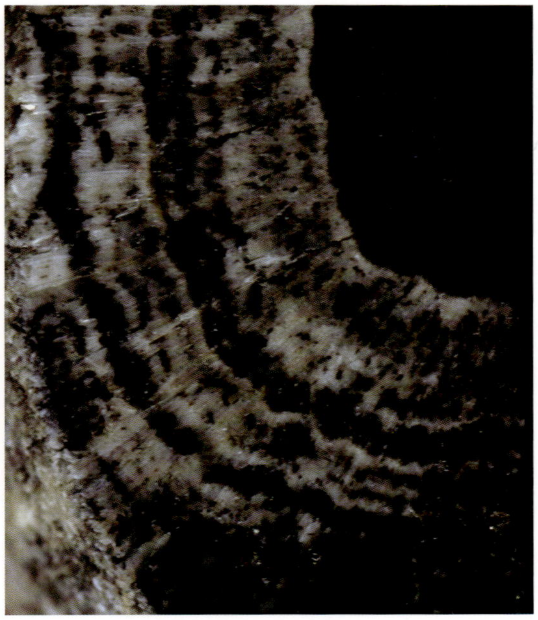

Barite limb cast. Banded structure is from barite crystallization. Carboniferous, Port Hood, Nova Scotia, Canada. Daniels specimen.

Fluorite replacing *Dadoxylon* with vestiges of ring structure remaining. Permian. Chemniz-Hilbersdorf, Germany. Daniels specimen.

Chalcocite with malachite, covellite, bornite, pyrite, chalcopyrite, silver, and other copper secondary minerals replacing wood from the Nacimiento Mining District, northern New Mexico, USA. Chinle Formation, Triassic, Daniels specimens, 6.5 and 8 cm.

Concentrically banded red wood limb cast showing minor oxidation of agate in several areas to produce vivid orange highlights. Morrison Formation, Late Jurassic. Moab, Utah, USA. J. Tibbitts specimen, 3.5 cm.

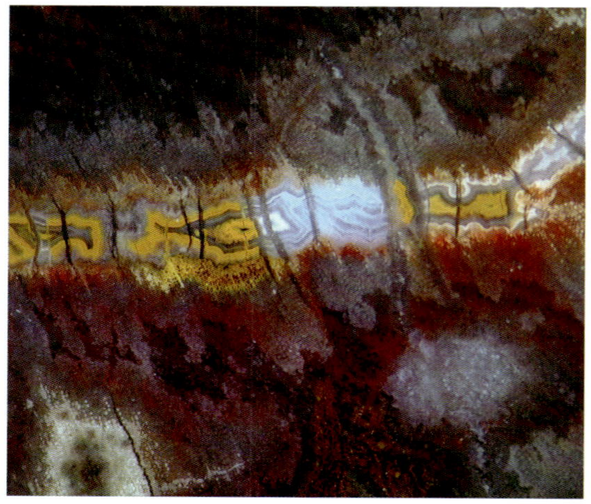

Agate fortification in petrified wood. The different colors of agate reflect different degrees of oxidation that are partitioned by fractures. Petrified Forest Member, Chinle Formation, Triassic. Northern Arizona, USA. Daniels specimen. Width of view about 15 mm.

Opalized wood with areas of precious opal. This specimen demonstrates the detailed templating of woody tissue by opal. Virgin Valley Formation, Miocene. Virgin Valley, Nevada, USA. Hodson specimen, 6.5 cm.

COLORS OF WOOD

Collectors are interested in the aesthetic qualities of petrified wood, especially brightly colored varieties. Most colors result from traces of transition metal oxides or hydroxides or the way light interacts with colloidal-size materials in translucent sections of the petrified wood. Colors are produced by iron, manganese, chromium, nickel, copper, and uranium, but remnant carbon can also produce dark colored to black petrified wood. By far the most important coloring agent is ferric iron (iron in the +3 valence state written as Fe^{3+}) that produces many shades of yellow, orange, red, brown, and black. Variables in oxidation of iron can initiate aesthetic changes in wood or agate banding in limb casts.

Oxidation of silicified wood and agate fortification in slightly brecciated specimen. The green and orange areas of petrified wood are surrounded by bluish to red chalcedony, producing a colorful wood specimen. Clarno Formation, Eocene/Oligocene. Hampton Butte, Oregon, USA. Daniels specimen. Width of view about 12 mm.

Differential oxidation produces dramatic color changes along agate banding in this red wood specimen. Brushy Basin Member Morrison Formation, Late Jurassic. Moab, Utah, USA. Dayvault specimen. Width of view about 4 mm.

In most cases, the lower the concentrations of trace impurities, the lighter colored the wood. Light blues in agates are produced when metal impurities are very low and the chalcedony contains high concentrations of micron-to-smaller-sized bubbles containing water. These bubbles scatter incident light similar to the way smoke scatters light and produces a bluish haze, sometimes called Rayleigh scattering. This same phenomenon probably accounts for bluish regions of petrified wood. Lavender is traditionally attributed to traces of manganese oxides, but more recent research suggests this desirable shade is due to iron. One mechanism is a combination of microscopic orange-to-red hydrous iron oxide particles disseminated in bluish translucent chalcedony. The eye adjusts for the combination of background color and particles and tells the brain to see lavender. Oxidized uranium (U^{6+}) and vanadium (V^{5+}) can form minerals that impregnate carbonized wood and produce bright yellow and orange colors. Uranium (U^{4+}) as pitchblende or coffinite produces sooty black wood that is much sought after as high-grade uranium ore. Bluish opaque wood is often produced by copper minerals such as azurite or chalcocite but can also be produced by vivianite, an iron phosphate. Manganese causes dark colors in wood from dark brown to dark gray and black. The flashes of colors seen in precious opal are an entirely a different process, which is described later. Green colors can be caused by traces of ferrous iron (Fe^{2+}), chromium, copper, nickel, and perhaps vanadium, probably in that order of abundance.

The cause of the beautiful and much sought after green-colored woods from the northwestern United States, Zimbabwe, Arizona, and Czech Republic have been of interest to collectors and researchers for some time. To help answer "why is this green wood green," Maxim Boyanov and Ken Kemner at the Advanced Proton Source at the Argonne National Laboratory (ANL) near Chicago, Illinois, agreed to analyze samples of wood from a number of locations. Ten 1-centimeter polished cubes of wood were sent to ANL for elemental analysis using synchrotron X-ray fluorescence (XRF). Four samples were analyzed using iron K-edge X-ray adsorption near-edge spectroscopy (XANES) to identify the average valence state of the iron atoms in those samples. Extended X-ray adsorption fine structure (EXAFS) techniques were used for these samples to identify the average local chemical environment about the iron atoms in these samples and to determine the crystalline phase with which the iron atoms are associated. Eight varieties of green-colored wood from a number of localities, one typical sample of red-colored wood from Arizona, and one bright yellow-colored wood sample from Arizona were submitted for analyses. Descriptions and analytical results are:

- **Red:** Arizona: typical red even-colored wood from Arizona. Iron, uranium, and possible gallium identified by XRF. XANES and EXAFS show iron dominating as +3 valence, similar to a hematite standard, estimated 90 percent in this form.
- **Yellow:** Winslow, Arizona: even-colored yellow with small dark spots. XRF spectra indicate iron dominating with some strontium and possibly some vanadium. XANES and EXAFS data are identical and show 90 percent of the iron is in the form of goethite.
- **Green:** Zimbabwe 1: relatively even-colored light green wood. XRF analysis shows chromium, iron, and strontium dominating; minor uranium. XANES spectrum shows that iron is in +3 valence state but does not match available standards.
- **Green:** Zimbabwe 2: somewhat variegated light green and brown. XRF data show similar results to Zimbabwe 1 with chromium, iron, and strontium dominating but with much higher chromium and slightly higher iron and uranium contents than Zimbabwe 1.
- **Green:** Tom Miner Basin, Montana: green with many dark streaks and spots. The XRF spectra show high iron, about 15 times as high as the Zimbabwean samples. Rubidium and strontium are significant. No chromium was detected. Dark spots appear to contain higher concentrations of iron and rubidium, and lighter areas have higher strontium concentrations.
- **Green:** Winslow, Arizona 1: even-colored light green. XRF spectra show strontium and chromium dominating with some iron. XANES and EXAFS spectra indicate iron is in +3 valence state that shows similarity to the lepidocrosite mineral standard.
- **Green:** Winslow, Arizona 2: light green with dark bands. XRF spectra indicate chromium, iron, and strontium dominating. Light green areas have higher iron; dark parts have higher strontium; chromium appears higher in one of the lighter areas but concentrations vary.
- **Green:** Winslow, Arizona 3: dark grayish green with light green band. XRF spectra show chromium and strontium dominating with minor iron. Vanadium and rubidium may also be present.
- **Green:** Hampton Butte, Oregon: even-colored dark green. XRF spectra indicate rubidium, iron, and strontium dominating; no chromium was detected.
- **Green:** Budweis, Czech Republic: fairly even-colored dark green. XRF spectra show yttrium and chromium dominating; iron and strontium are also present.

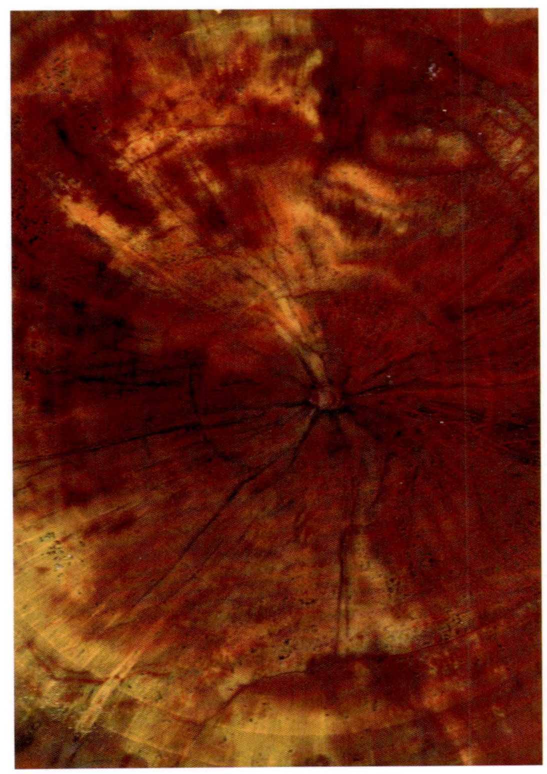

***Araucarioxylon*. Chinle Formation, Triassic. Northern Arizona, USA. Daniels specimen, 12 cm.**

Unidentified fossil wood. Eocene. Tom Miner Basin, Montana, USA.
Daniels specimen, 24 cm long.

Unidentified fossil wood. Tertiary. Budweis, Czech Republic. Daniels specimen, 8 cm.

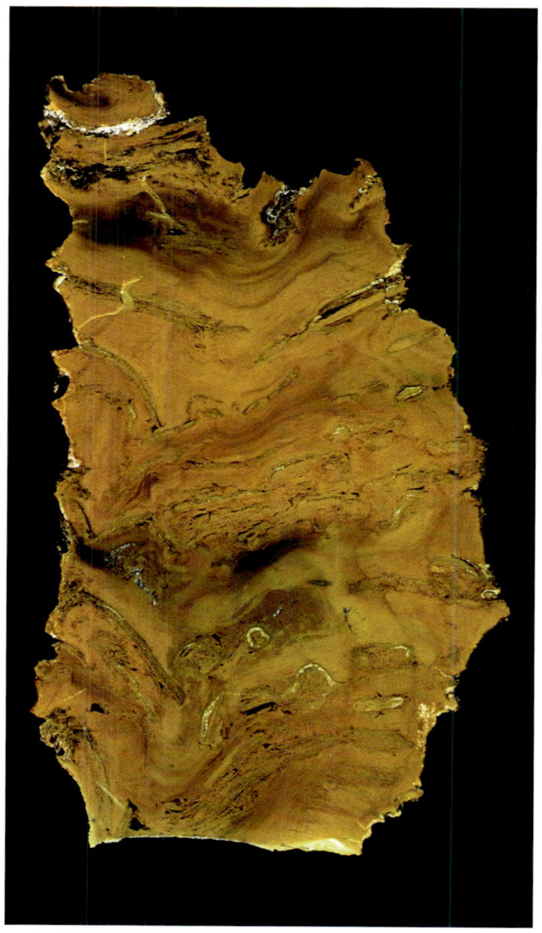

Unidentified fossil wood. Chinle Formation, Triassic. Northern Arizona, USA.
Daniels specimen, 11.5 cm.

Unidentified fossil wood. Chinle Formation, Triassic. Northern Arizona, USA.
Daniels specimen, 15 cm.

Unidentified fossil wood. Clarno Formation, Eocene. Hampton Butte, Oregon, USA. Daniels specimen, 17 cm.

Araucarioxylon. Triassic. Gokwe, Zimbabwe. Daniels specimen, 7.5 cm.

ssil wood. Clarno Formation, on Butte, Oregon, USA. n, 8.5 cm.

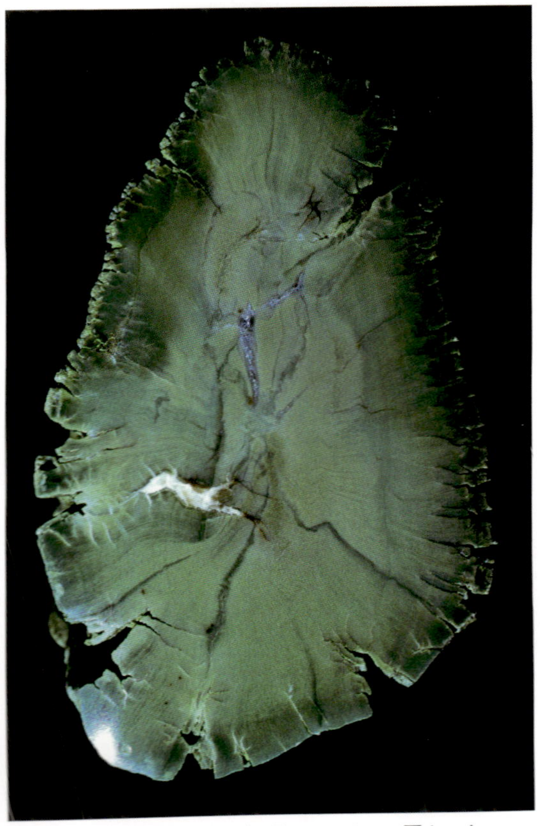

Araucarioxylon. Chinle Formation, Triassic. Northern Arizona, USA. Daniels specimen, 9.5 cm.

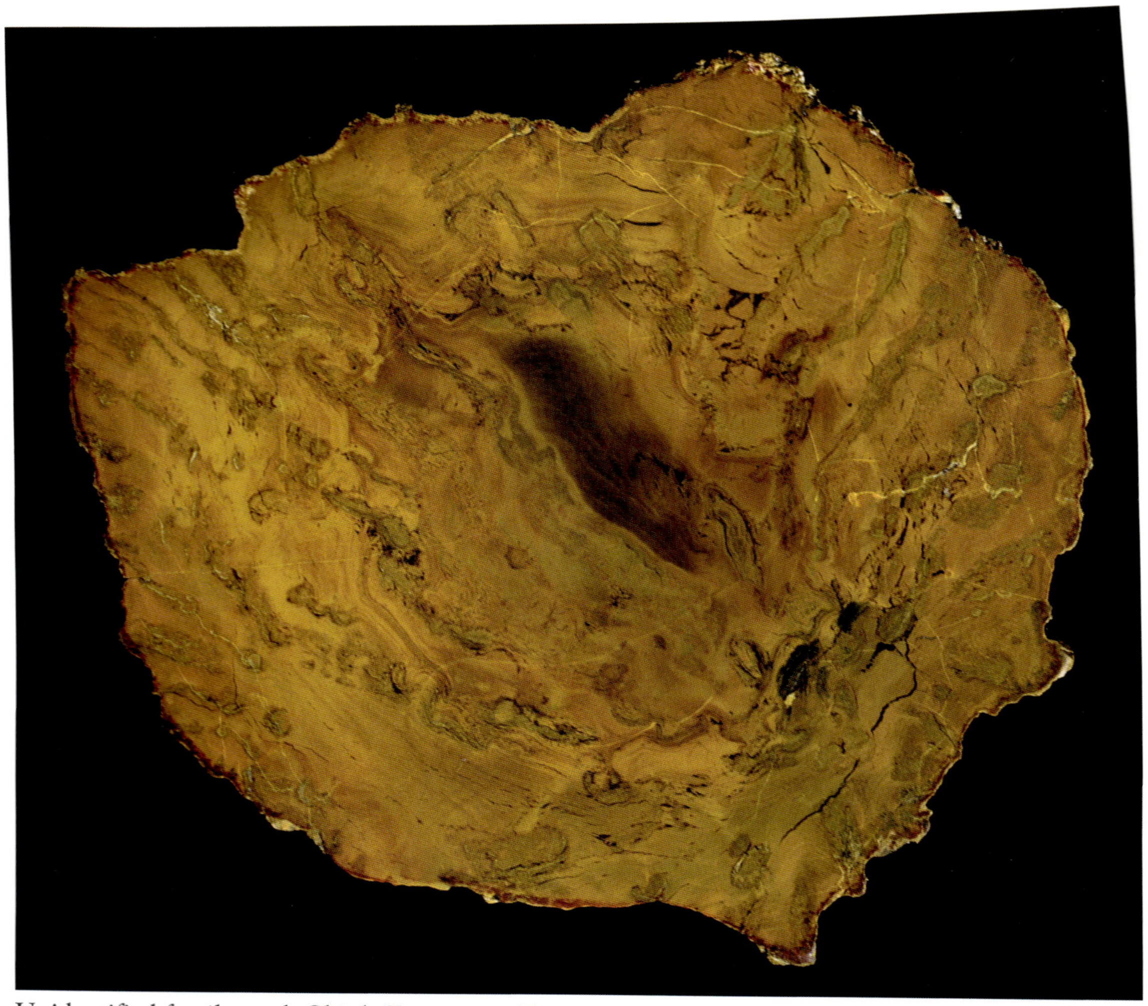

Unidentified fossil wood. Chinle Formation, Triassic. Northern Arizona, USA. Daniels specimen, 16 cm.

Woodworthia. Chinle Formation, Triassic. Norhern Arizona, USA. Daniels specimen, 10 cm.

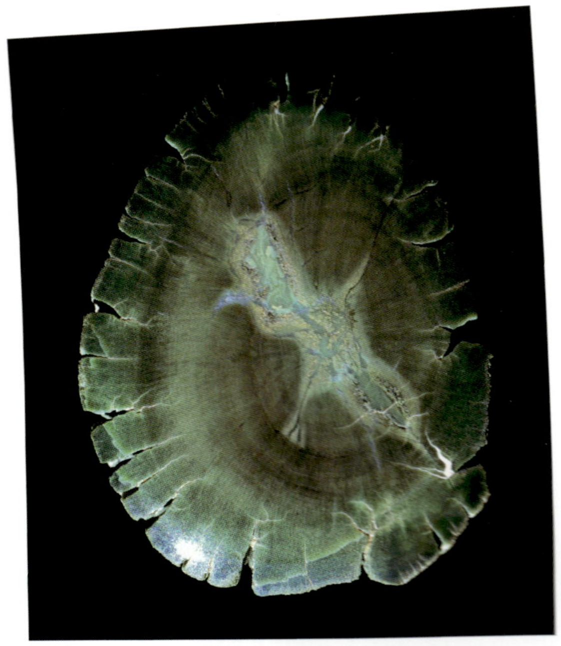

Woodworthia. Chinle Formation, Triassic. Northern Arizona, USA. Daniels specimen, 5.5 cm.

These analytical results suggest that so
green woods, in particular the lighter gr
samples from Winslow and Zimbabwe and
darker green sample from Budweis, con
substantial amount of chromium, which i
bly the coloring agent. The dark gree
from Hampton Butte and the lively gr
from Tom Miner Basin contain iron
nant trace element. Chromium is al
samples, thus iron is probably the
The red and yellow colors of th
are dominated by iron in the
which closely matches spe
hematite (red) and goethite
respectively. Therefore, t
occurs as mineral phase
Contributions to woo
rubidium, gallium, and

Two samples of
submitted for X-
determine if the g
observed as a crystalline
Diffractometer a
Junction, Colorad
Hood analyzed s
only peaks for qu
as standard clay
mineral phases w
centrated sufficie
spectra for identi
green wood fro
mixed-layered
percent illite

identified f
ene. Hamp
iels specim

Moab, Utah, also contained mixed-layer illite/smectite with about 90 percent illite. These minerals are not good candidates for affecting green color. No other minerals were identified by this method. Unless a mineral is highly ordered, this method is usually not sensitive enough to detect concentrations less than 5 percent of the total mass. Therefore, this method of analysis did not provide evidence for a cause of green color in petrified wood.

ATTEMPTS TO MAKE PETRIFIED WOOD

One of the first individuals who tried to convert wood into a rock was an alchemist named Basil Valentine in 1520. He apparently used some form of powered silica and a derivative of grapes called subtartrate of potash. Results were equivocal, but already silica had been recognized as a key ingredient. At about the same time, a man named Kentmann produced a recipe for making stony wood of suitable quality for use as whetstones or flintstones. Conveniently, it was a by-product of beer production. The stony wood material was made by taking alder wood that had been simmered with hops and burying it for 3 years in sand or gravel in a cellar. He claimed success. As recently as the 1940s, Longyear published a recipe for petrified wood that called for mixing equal amounts of gum salt, rock alum, white vinegar, and pebble powder. After the concoction settles down, place any type of wood or porous organic material in the mixture and it will soon petrify. Modern investigators discount the veracity of these early methods, but admit accurate information about procedures and ingredients is limited.

In the past half-century, researchers have at least partially synthesized silicified wood in the laboratory. In the 1960s, Drum replicated vascular tissue in birch twigs by soaking them in a sodium silicate solution for 12 to 14 hours and transferring them to chromic acid for 2 to 3 days. Siliceous casts of some vascular cellular tissue resulted. Leo and Barghoorn made siliceous wood "lithomorphs" in late 1970s by boiling wood pieces in water until they were waterlogged and gas free. Then the wood was alternately placed in two different jars, one containing water and the other ethyl silicate, for days to months at a time and kept at 150° F [70° C]. The specimens were then treated with acid to remove any residual organic carbon and prepared for electron microscope study. Although the material was still quite fragile, it did show incipient silicification and replication for microscopic woody textures.

Early in 2005, material scientists at the U.S. Department of Energy Pacific Northwest National Laboratory demonstrated that petrified wood could be produced quickly using artificial conditions. They made silicon carbide wood by soaking small cubes of ordinary pine or poplar in an acid bath, then in a silica solution for a couple days, heating the sample to 2520° F [1400° C] for 2 hours in an argon atmosphere, and then letting it cool. The resulting material replicated the microscopic structure of the wood and produced porous silicon carbide that may be useful as an extremely tough microfiltering medium.

SILICA VARIETIES AND GENESIS

Most petrified wood is available to collectors because it is harder than the sedimentary rocks from which it weathers, thereby leaving lag deposits on the ground surface. This same hardness also provides suitable lapidary material that can be polished. By far the most common substance throughout the world that replicates wood is some form of silica, which is simply one silicon atom and two oxygen atoms; the formula is written as SiO_2. Silica molecules may combine in a variety of forms, ranging from amorphous silica or opal with 10 percent or more water in its interstices to various minerals with the same chemical formula but different internal structures such as cristobalite, tridymite, and quartz. Chalcedony, the most common type of silica that replaces fossil wood, is a microcrystalline fibrous form of the mineral quartz that also forms most agates in the world. Although chalcedony is the most common type of silica in petrified wood, it probably did not start out in that form.

The sequence of silica genesis in petrified wood generally, but not always, follows the pathway expressed by

Silicic Acid ➔ Polymerized Silicic Acid ➔ Opal-A (amorphous) ➔ Opal-CT (cristobalite tridymite) ➔ Microcrystalline Fibrous Quartz (chalcedony) or Microcrystalline Granular Quartz (chert or flint)

This sequence of chemical and mineralogical events is similar to those that have been ascribed to the production of chert or chalcedony in marine biogenic sediments or silica deposits in siliceous hot springs.

SILICIC ACID, POLYMERIZATION, AND EARLY TEMPLATING

It is no coincidence that petrified wood is commonly associated with volcanic ash deposits or in rocks that resulted from ash that has been mobilized and redeposited by water. Years ago, Murata and, somewhat later, Siever recognized that when volcanic glass begins to take on water or hydrate and crystallize or devitrify, it readily releases silicon, aluminum, sodium, potassium, calcium, and other elements to ground water. These elements ionize and recombine to form silica, silicate compounds such as sodium silicate, or silicate minerals such as clays. This process begins within years after deposition of the glass and is the main source of monosilicic acid [formula: $Si(OH)_4$], which means four hydroxyl (OH) molecules combine with one silicon atom (Si). Monosilicic acid tends to form larger molecules by combining with itself and losing water, a process called polymerization as shown in the diagram:

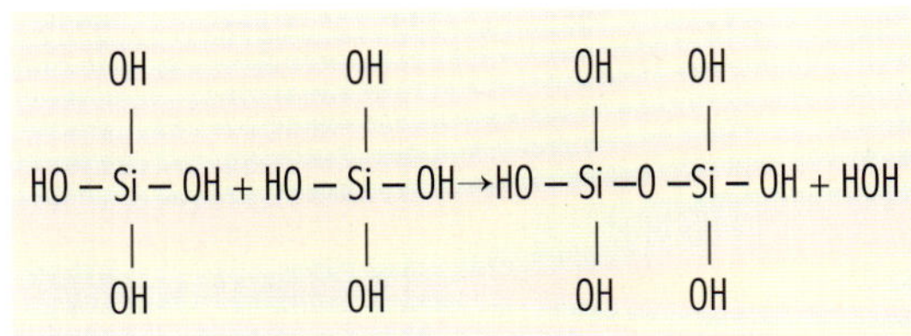

The resulting process produces a large repeating silica molecule that contains hydroxyl bonding sites, which can combine with hydroxyl sites on carbon molecules. Even in its polymerized form, silicic acid molecules are relatively small, allowing them to penetrate woody tissue easily. Silica is preferentially removed from ground water by orienting itself and attaching to carbon molecules in the cellular components of the wood by means of hydrogen bond formation. The diagram shows hydrogen bonds between polymerized silicic acid and vascular woody tissue:

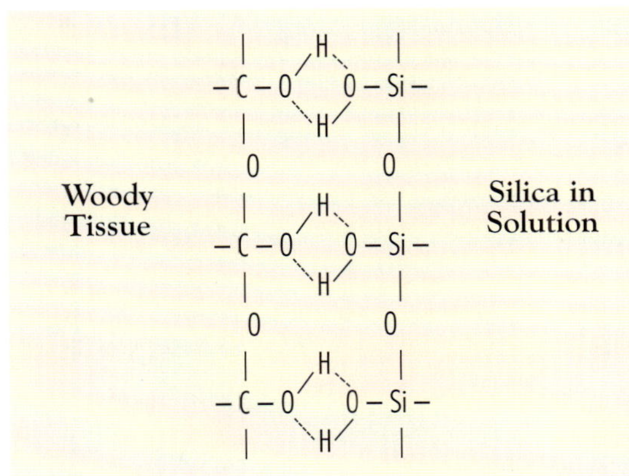

Hydrogen bonds are not particularly strong, but the number of bonding sites in molecules composing wood is enormous, which provides integrity to the silicification process by strength in numbers. At least twofold evidence exists to substantiate hydrogen bonding. Silicified wood often occurs in matrix or host rocks that are not consolidated by silica cement, suggesting a preferential affinity of silica for wood via hydrogen bonding. The concentrations of silica required to accomplish initial templating are lower than the concentrations required to form colloidal silica. The only other mineral substances that preserve detailed plant structures are goethite (hydrous iron oxide) and hydroxyl apatite (calcium phosphate). Both species probably achieve hydrogen bonding by sharing the hydroxyl molecules with plant tissues. Wieland discussed fine preservation of cycadeoids from Maryland by iron mineralization (probably goethite) about 100 years ago, and Buurman published electron microscope images showing detailed preservation in the 1970s of phosphatized wood from Bruneau, Idaho.

As the silica-cellulose bonding chemical process advances, silica bonds to other silica atoms instead of carbon compounds. With the availability of more silica, the surfaces consist of layer after layer of large silica molecules, but a replica of the original texture, termed a lithomorph, is preserved. Further deposition by the silica produces a film that continues to replicate the fine details of the woody structures in a process termed templating. Whether the original molecules causing this templating are simple silicic acid or an already polymerized version of the acid is still debated. Organic material remaining during this process is subject to deterioration during the addition of silica, which may be beneficial by offering more sites for templating, but the original replicated structures or lithomorph may continue to exist.

Further polymerization of the silica may cause the formation of silica gel, which can crystallize by a different process described later in this chapter, and infill lumina (openings in the center of cells), thereby further fortifying the wood. This infilling of open voids by mineral matter is called permineralization. Other porous organic materials, such as bones, are petrified by a similar infilling of the cellular structure. Permineralization can preserve details of woody tissue, but it does not preserve the finest details of plant structures. Colloidal silica, usually as micron-sized spheres, may be too large to penetrate the finer interstices of the organic tissues and does not readily form hydroxyl bonds with the complex molecules constituting plant tissue. However, permineralization of wood is common and is often observed as quartz crystals or agate fortifying cell lumina or other microscopic areas in wood in conjunction with templated walls of cellular material.

WOOD COMPOSITION

To understand the complexity and probable sequence of the templating process, a rudimentary understanding of the principal woody tissues—cellulose, hemicellulose, and lignin—and their general molecular composition is useful. These tissues are composed of polysaccharides, giant organic molecules that account for more than 95 percent of the dry weight in wood. The remaining weight of wood consists of proteins, starch, inorganic salts, and a few other substances. Cellulose is composed of thousands of glucose sugar molecules and is about 1,000 times larger than hemicellulose. Cellulose forms linear molecules that organize into bundles called microfibrils that coil around one another like strands of a cable. Each microfibrils strand may contain a billion cellulose molecules laid down parallel to one another in a layer. The next layer of microfibrils is oriented approximately at right angles to the layer below it, making an extremely tough tissue that imparts structural integrity to the plant. Hemicellulose polymers are branched polysaccharides that provide interpenetrating matrix support to the cellulose.

Lignin is another complex and high molecular-weight polymer that has many forms and is responsible for providing rigidity to cell walls. It is

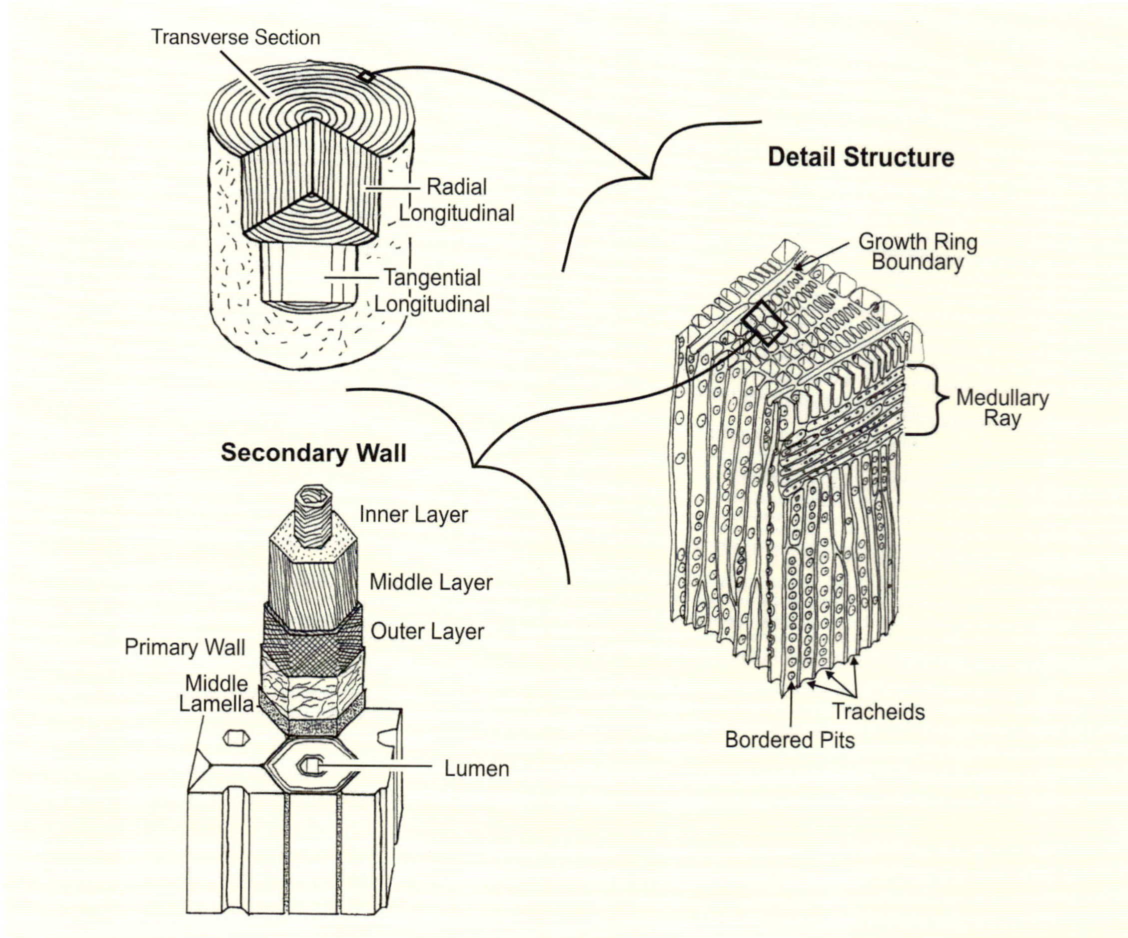

Top view shows the three planes that fossil wood is sliced — transverse, radial, and tangential — to study structures. The detail structure shows some structures in coniferous wood and several anatomical features discussed in the text. Lower enlargement of middle diagram depicts layering of cell walls and opening in center of cell called the lumen. Templating of polymerized silicic acid begins with the most accessible region, the lumen, and progresses via bordered pits and tiny cracks formed by decay and desiccation to inner cell walls and beyond. Permineralization is most prominently developed in relatively large openings such as lumen or in larger regions such as desiccation cracks or hollow centers. Modified from Wardrop and Bland.

hydrophobic (will not bond with or absorb water) and is more resistant to decay than the other polysaccharide wood components. This material helps preserve wood from early decay that can destroy detailed structures and textures.

After a plant dies and is buried, decomposition occurs initially and more completely from the microbial and fungal metabolism of cellulose and hemicellulose. All three tissues (cellulose, hemicellulose, and lignin) are more resistant to decay under reducing and anaerobic (oxygen-deficient) conditions. Cells walls are multilayered and contain both cellulose and lignin. If the cell walls decompose before silicification, details of histology are lost and, subsequently, even the most faithful templating by silicia can preserve only a vestige of the former tissue. Conversely, Leo and Barghoorn speculate that slight degradation or decay of the original wood might actually enhance petrifaction by increasing permeability of cell walls, thereby creating more void space for infiltration of silica. They also suggest that fungal decomposition might also provide more active hydroxyl bonding sites by the incipient breakdown of cellulose and lignin. Jefferson studied Cretaceous conifers from Antarctica and suggested that the increased size of cell wall openings in the microfibrils was due to enzymatic activity of white rot fungus (which removes lignin) before silicification.

If wood degrades too rapidly, the templating process could have little original fabric to replicate, or the wood may be completely consumed and only a cast could form. If too little decay occurs, the remaining woody products might convert to some type of carbonized remains. A balance must be achieved between degradation and templating. Optimal conditions for petrifaction are met by the slow and steady degradation of wood accompanied by equally slow and steady mineral deposition. As minerals precipitate by templating of the cell walls and by permineralization of lumina (plural of lumen), the organic cellular constituents simultaneously break down. This process offers an explanation of how so much of the original carbon tissue disappears and yet a high-fidelity siliceous replica remains.

CONTINUED TEMPLATING AND BEYOND

Initial silica templating usually occurs on the most easily accessible cellular surfaces, such as the inner cell walls and the lining of the pit chamber where the bordered pits are found, as shown in the diagram. Silicification builds outward away from the inner cell walls until the lumina are filled. Silica templating may continue through the pits or portals that form the connective tissue between the tracheids. During these early stages of templating, the polymerized silicic acid molecules are small enough to penetrate cell inner walls into the inner and middle layers and deposit silica in intimate association with the microfibrils. Ideally, other structural units of the wall composed of decay-resistant lignin remain intact during this time and help maintain the shape of the cell wall and, thus, the anatomical woody texture until more advanced silicification occurs. Lignin is one of the last molecules to be replaced and may in fact never be replaced.

Leo and Barghoorn report that the amount of organic carbon remaining in silicified wood samples dissolved in hydrofluoric acid is 1 to 5 percent by weight. Lignin residue is the suspect relic hydrocarbon. George Mustoe of Western Washington University in Bellingham, Washington, approached the problem of measuring the amount of original carbon remaining in the petrified wood with a different method. A known volume of powdered silicified wood is heated to burn off relict organic matter, and the results are compared to the amount of carbon that would be present in an equal volume of modern wood. By this quantitative technique, he has shown that relict organic matter seldom exceeds 10 percent and usually constitutes much less of the total volume.

Through these early phases of templating, the nascent silicified wood may be friable or even fragile, although lithomorphs of the original woody textures may exist to some degree. The initial emplacement of silica can occur extremely rapidly in geologic terms, especially in volcanic or hot spring settings. Trees buried in ash from an 1886 eruption in New Zealand were partially templated by silica, according to Leo and Barghoorn. In another example, Allen and Day reported a lodgepole pine located in Yellowstone National Park that was partially silicified in only 13 years after it was inundated by hot spring solutions. The tree was alive until the hot spring shifted its location in 1959. Under normal conditions of ambient ground water temperatures, this initial templating process requires a much longer period of time.

This initial emplacement of silica does not produce what collectors think of as a good specimen of petrified wood, but it is an important step in the preservation of internal textures. If conditions remain stable in the ground and a steady source of silica continues to exist, the process of templating by polymerized silicic acid and/or permineralization by colloidal silica will continue. After the initial stages of mineralization, cellular walls may be disrupted by the molecular strength of the new silica megamolecules, allowing access by additional silica inside the cell walls and more complete templating. As silicification progresses, it systemically reduces the porosity of the host wood, thus requiring ever increasing time for the deposition of any addition silica. The final production of hard, solid petrified wood requires not only the nearly complete infilling of large and microscopic voids but also the ultimate conversion of original opaline silica to a harder and stronger variety of silica called chalcedony. Sometimes the tiny lumina develop microagates formed by the subsequent organization and crystallization of the silica. Frequently, desiccation cracks or breaks in the original wood will contain fortification agate that is formed by processes that will be discussed later. Many interesting and exceptionally beautiful examples of silicified wood from Utah and other western states contain areas of relic wood associated with colorful fortification agate. But this is jumping ahead in our story.

FORMATION OF AMORPHOUS OPAL (OPAL-A) INCLUDING PRECIOUS OPAL

The original silica that templates woody tissue continues to lose water and organize internally to form a harder and more durable material called amorphous opal or opal-A. (Note: opal-A was the term originally used to describe biogenic amorphous opal, and opal-A' was the term for inorganic amorphous opal. Opal-A will be used in this book to describe amorphous opal regardless of origin). The most frequently found variety of opal-A is common opal, hence its name. Common opal is usually a dull waxy material, transparent to opaque, and light to dark in color with excellent conchoidal fracture. Precious opal is rare by comparison and exhibits bold flashes of colors when turned from side to side. Both common and precious opal consist of primary transparent silica spheres that range from 20 to 50 nanometers (0.0000008 to 0.000002 inch) in diameter. Features this small can only be seen with an electron microscope, which was first used for this purpose in 1964. The tiny spheres tend to organize themselves into larger spheres by nucleating around a primary opal sphere and subsequently forming one to several shells around it.

Fortunately, when specimens of opal are broken, they often break through the spheres instead of around them so that their internal structure can be observed. When viewed under high magnification, the surface shows the internal concentric arrangement of the tiny opal spheres that form larger composite spheres as shown in the diagram.

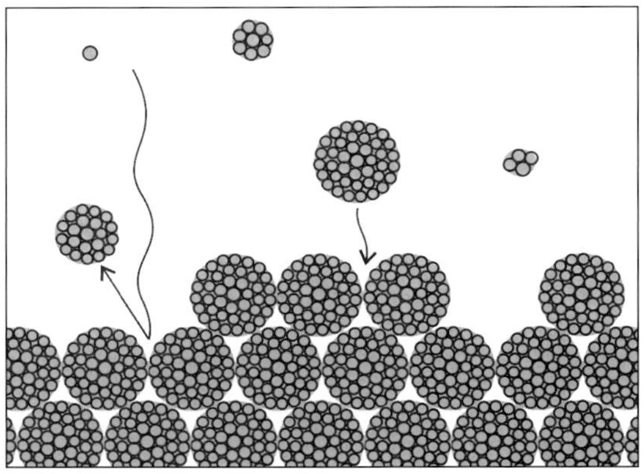

Composite silica spheres of even size deposit in an orderly fashion on the bottom of a cavity to form precious opal.

Precious opal, which sometimes petrifies wood and produces arguably the most beautiful and certainly the most valuable specimens of fossil wood in the world, is most often a variety of opal-A (opal-CT can also form precious opal). Precious opalized wood is a combination of templating and permineralization. Darragh, Gaskin, and Sanders describe the internal arrangement of silica spheres in opal-A, and, in particular, how this arrangement causes the incredible colors of precious opal.

The difference between common opal, sometimes called potch opal, with no play of colors and precious opal with its stunning display of colors is linked with the packing and size of these composite opal spheres. Scanning electron micrograph images of common opal show that the composite spheres are irregularly shaped, are of many sizes, and are not packed together in a regular grid. In this case, the opal might be thought of as a mixture of microspheres, some not even spheres but irregular rounded shapes that range in relative size from marbles to oranges.

Precious opals, on the other hand, consist of uniform-size composite spheres highly organized into a tightly packed grid with all spheres touching. This might be thought of as similar to a well-packed crate of oranges. Researchers believe the silica spheres acquire this orderly arrangement by slowly developing in a silica solution. They are possibly held in suspension by their similar surface charges, but the spheres eventually settle extremely slowly to the bottom of an open space in a near-surface environment during prolonged periods of evaporation. The exact mechanism that allows the almost perfect packing is not understood. Iler reports that it may be related to uniform negative surface charge on the spheres that repulse each other until a minute concentration of electrolyte is added to the solution, which reduces the surface charge and allows orderly settling. However it happens, opal spheres build up successive layers and achieve their requisite uniform-size, three-dimensional array. Atoms in crystals can be packed in a similar fashion, but the silica spheres are billions of times larger. Additional transparent opal cements the spheres together, giving it a hardness of 5.5 to 6.5 and greater strength, which provides the needed physical requirements of a gemstone.

Voids that result from the intersections of the composite spheres contain air, water, or a combination of both. Opal contains at least 1 percent water, and often has between 3 to 9 percent water, but can have to 15 or even 20 percent water. The water occurs in the intersections of the spheres and as hydrogen-bonded water films on inner surfaces, according to Gains and others. Castor and Ferdock report that Virgin Valley opal has to 13 percent water in these traps. This large water content is probably the reason that much of the Virgin Valley opalized wood crazes upon desiccation and must be kept hydrated or slowly allowed to hydrate after extraction from the soil for better preservation.

The voids are the key to constructing a three-dimensional diffraction grating that produces the colors of opal. If light enters the opal, it is scattered at these void intersections because the material in the voids has a slightly different refractive index than the opal spheres. Refractive index is the degree of deflection for a light ray as it passes from one medium to another of a different density (as from air to opal), which changes its velocity. Precious opal contains randomly oriented internal faults (tiny fractures) that cut across these closely packed and ordered microspheres, providing the "grains" or flashes of color. Light rays break into colors that we can see only if the size of the spacing between the voids is about half the wavelength of visible light, which is in the 150 to 300 nanometer size range. Diffraction gratings nearer the 150 nanometer range will produce shorter wavelength light in the violet, blue, and green range visable in some of the finer opalized wood from Virgin Valley, Nevada, and Lightning Ridge, New South Wales. Larger gratings nearer the 300 nanometer size produce longer wavelength light until a red color is produced. Opals that produce red diffraction colors can produce lower wavelength colors (violets and blues) when viewed at low angles. The converse is not true. Individual single "grains" of opals seen as infillings of cell lumen can resemble stained glass windows. As opalization continues and if original templating is only partially complete, leaving open voids, larger areas of opal may become continuous and the cellular structure becomes less apparent, producing larger flashes of color. Some of the finest opalized wood results from complete loss of original wood during burial and direct fortification of the void by precious opal (a cast) or the advanced destruction of finest textures by the continued formation and infiltration by opal.

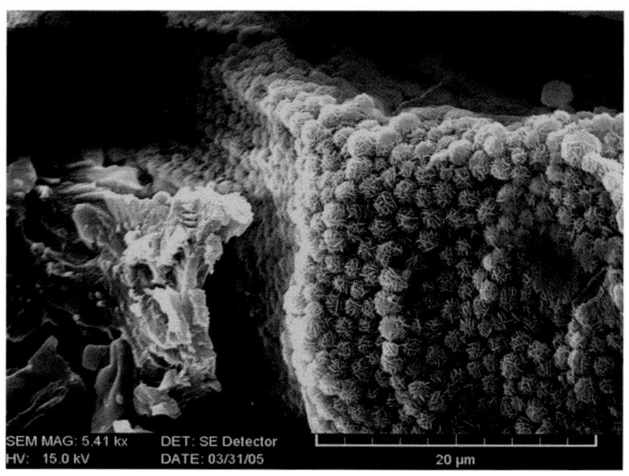

Precious opal silica spheres. This striking precious opal specimen from Rainbow Ridge Opal Mine in Virgin Valley, Nevada, USA, contains colorful opals infilling each lumen separated by dark cell walls. Microspheres of opal demonstrate the orderly packing required to produce the color pattern for precious opal. The platy exterior texture suggests this is opal-CT. Virgin Valley Formation, Miocene. Mustoe specimen and image. Scanning electron microscope (SEM) image. Magnification approximately x 2,200.

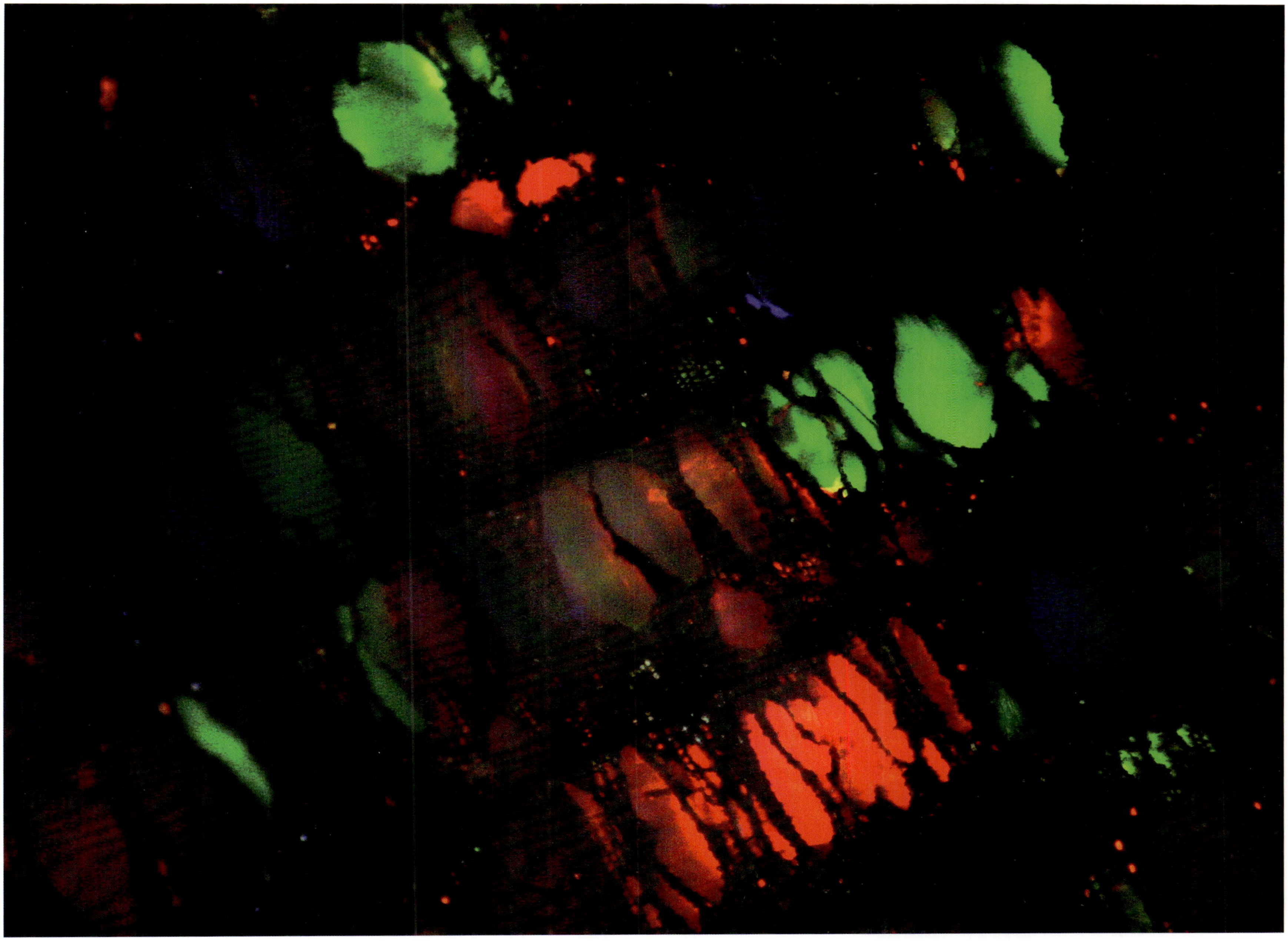

Transverse view of precious opal showing larger areas of opal separated by growth bands. Individual cell lumen are still visible in most growth bands. Virgin Valley Formation, Miocene. Rainbow Ridge Opal Mine, Virgin Valley, Nevada, USA. Hodson specimen.

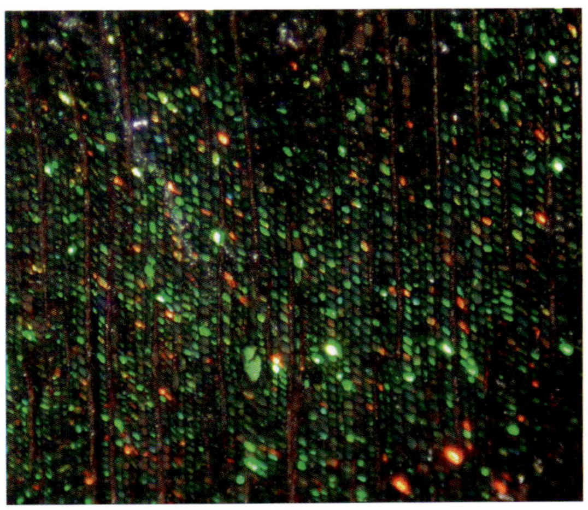

Transverse view of individual cell lumina that are filled with areas of precious opal. Cellular walls are dark-colored common opal, producing a stained glass window effect with magnification. Virgin Valley Formation, Miocene. Virgin Valley, Nevada, USA. Dayvault specimen, individual cell lumen about 0.025 mm.

Transverse view of precious opal showing opal expanding beyond cellular boundaries. In some cases, fractures contain larger areas of continuous precious opal. Virgin Valley Formation, Miocene. Rainbow Ridge Opal Mine, Virgin Valley, Nevada, USA. Hodson specimen.

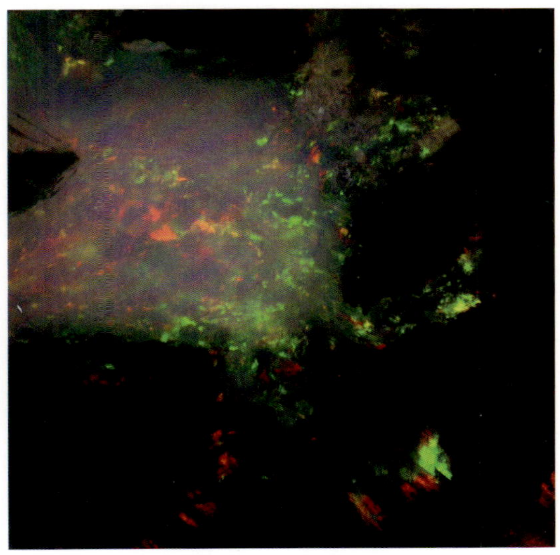

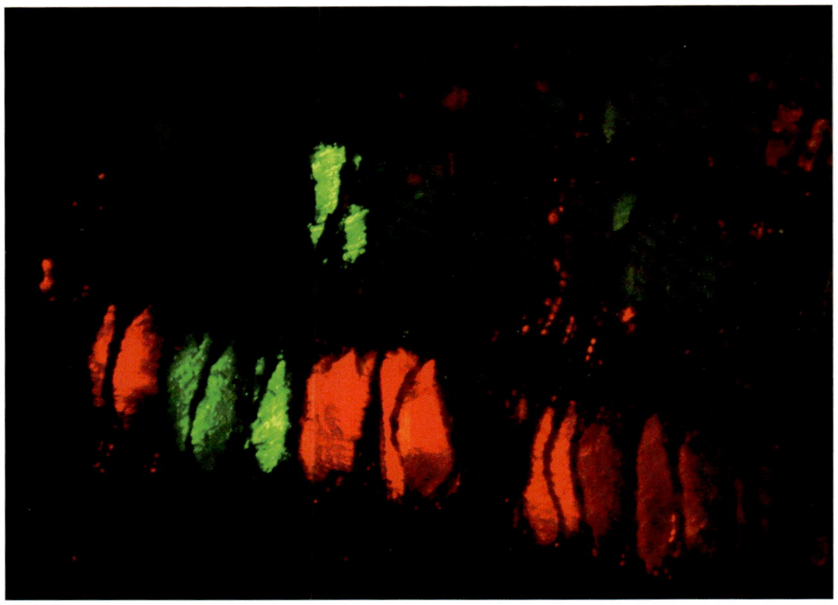

Another example from Virgin Valley. Virgin Valley Formation, Miocene. Rainbow Ridge Opal Mine, Virgin Valley, Nevada, USA. Hodson specimen.

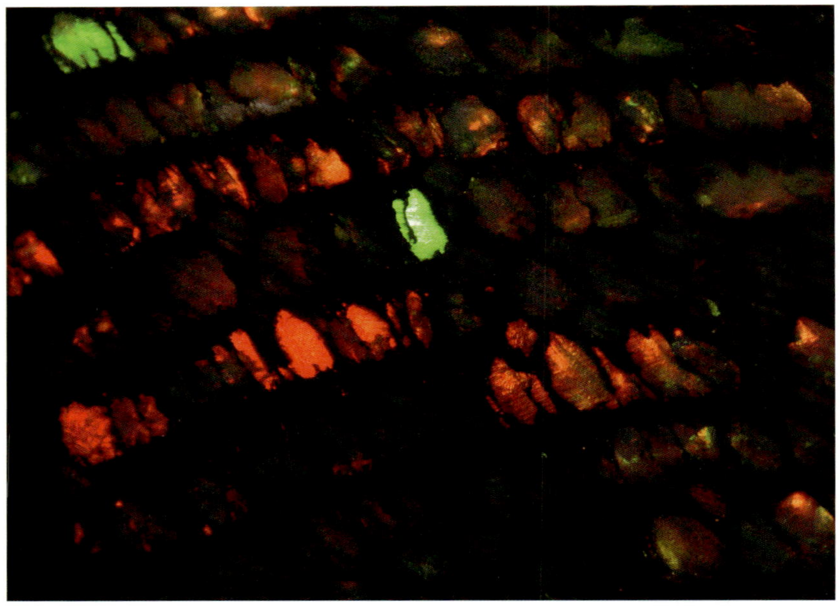

Transverse view of precious opal showing larger areas of opal separated by growth bands. Opals displaying larger areas of color separated by dark areas of common opal are of great interest for jewelry. Virgin Valley Formation, Miocene. Rainbow Ridge Opal Mine, Virgin Valley, Nevada, USA. Hodson specimen.

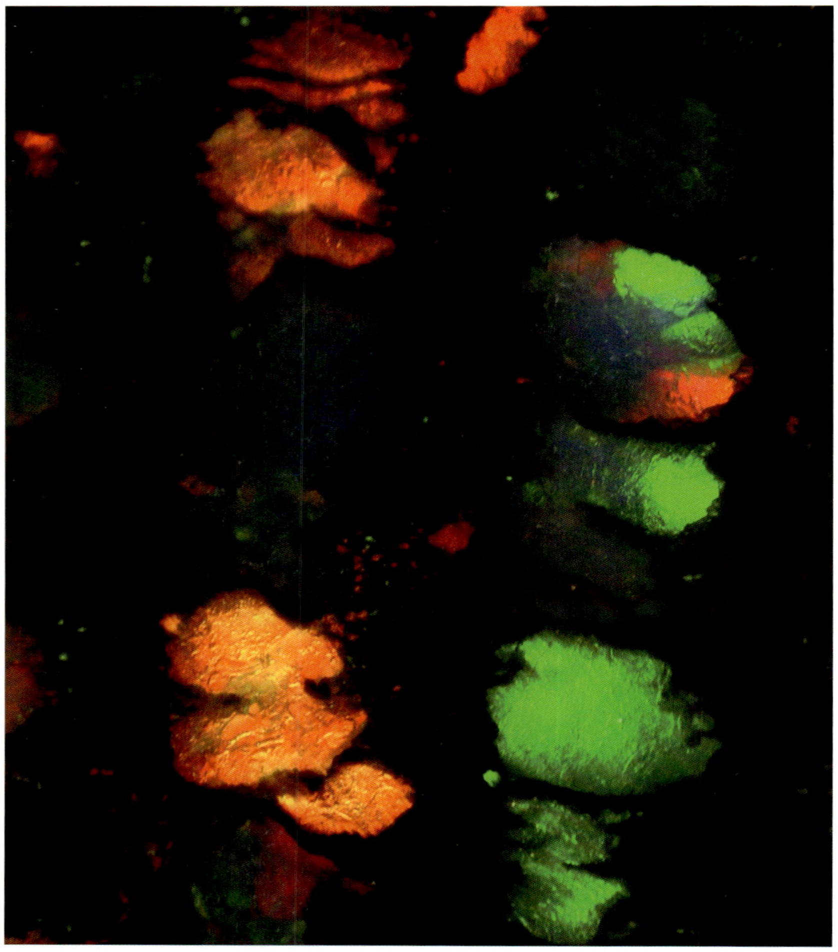

Transverse view of precious opal showing larger areas of opal separated by growth bands. Individual cell lumen are still visible in most growth bands. Virgin Valley Formation, Miocene. Rainbow Ridge Opal Mine, Virgin Valley, Nevada, USA. Hodson specimen.

Transverse view of precious opal showing larger areas of opal separated by growth bands. Individual cell lumen are still visible in most growth bands. Virgin Valley Formation, Miocene. Rainbow Ridge Opal Mine, Virgin Valley, Nevada, USA. Hodson specimen.

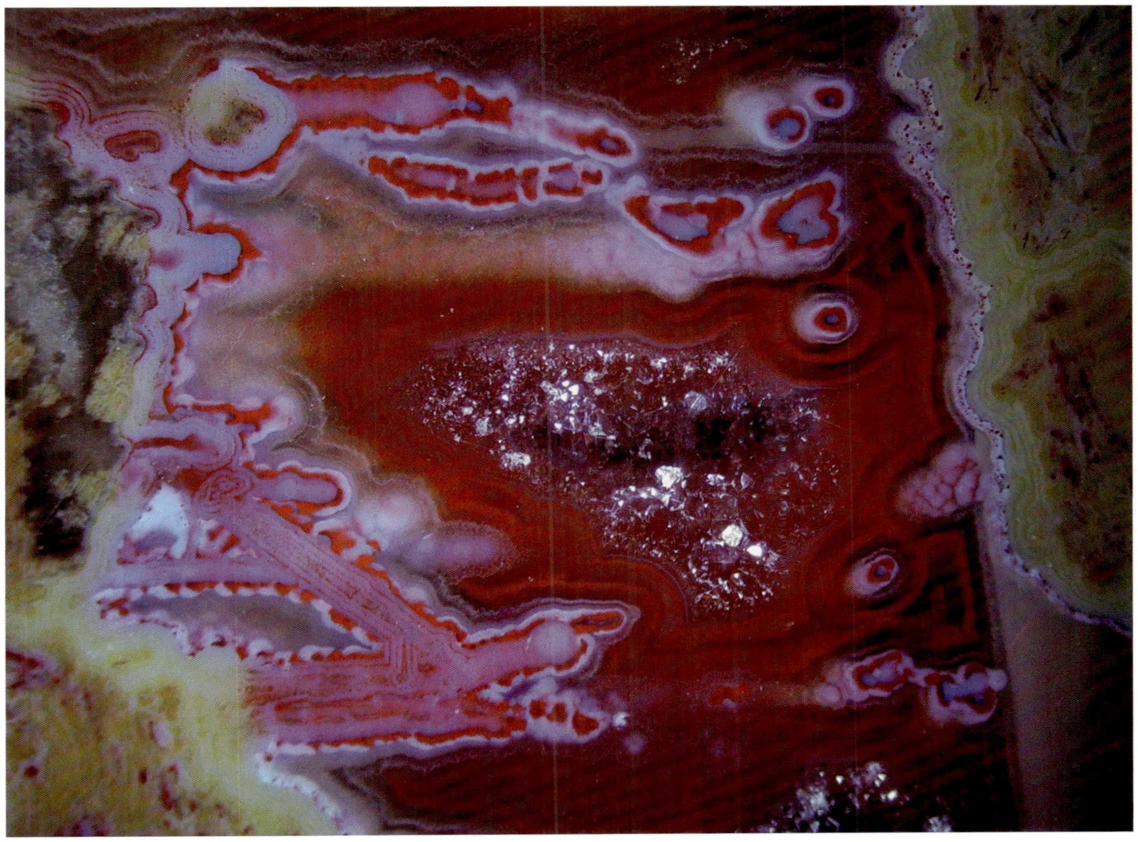

Close view of colorful agate fortification in petrified wood. Quartz center occurs on one section. Unidentified conifer. Mesozoic. Malagasy Republic. Daniels specimen. Width of view about 18 mm.

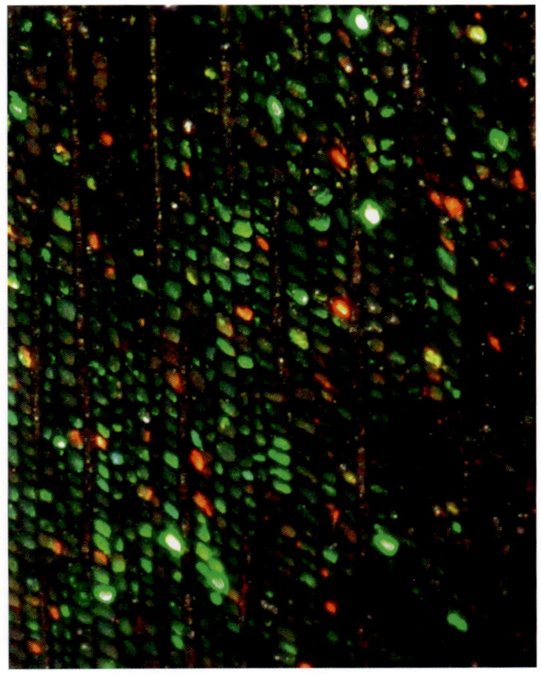

Precious opal fills cell lumina, and cellular walls are replaced by dark-colored common opal. The result is a micro "cathedral opal," which is unfortunately too small to be appreciated without magnification. Virgin Valley Formation, Miocene. Virgin Valley, Nevada, USA. Dayvault specimen, individual cell lumen about 0.025 mm.

Tangential view of tracheids filled with precious opal. Virgin Valley Formation, Miocene. Virgin Valley, Nevada, USA. Hodson specimen.

Precious opal limb cast with light-colored fragments of templated silicified wood floating in opal. Griman Creek Formation, Lower Cretaceous, Lightning Ridge, New South Wales, Australia. Dayvault specimen, 5 cm.

"Black opal" is desirable because the background opal containing the color flashes is less milky and, therefore, produces greater contrasts of colors. The darker color of the parent opal may result from inclusions of carbon or oxides of iron or titanium or from other light-interference structural phenomenon. These substances or structures cause the absorption of some undiffracted light, which produces a higher contrast between incident and refracted rays. The organic content of petrified wood sometimes provides this darker background color to the opal, creating an exceptional gemstone. The famous Roebling opal dug from Virgin Valley in 1917, now on display in the National Museum of Natural History (The Smithsonian), is an example of a wood cast black opal. Large limb casts weighing as much as 130 pounds with regions of precious black opal have been discovered during the past 20 years at the same location. This isolated area of northwestern Nevada still produces fine opal. Although hundreds of mining claims have been staked in Virgin Valley for opal, Castor and Ferdock report that three mines on patented ground—Bonanza Opal Mine, Rainbow Ridge Opal Mine, and Royal Peacock Opal Mine—continue to produce commercial quantities of precious opal and offer fee digging to the public.

Ordinary common opal provides some of the most detailed reproductions of plant physiology. These specimens usually occur in fairly young formations, ranging from Miocene to recent age. As opal-A is transformed into opal-CT, and eventually into chalcedony, details of plant physiology tend to be lost.

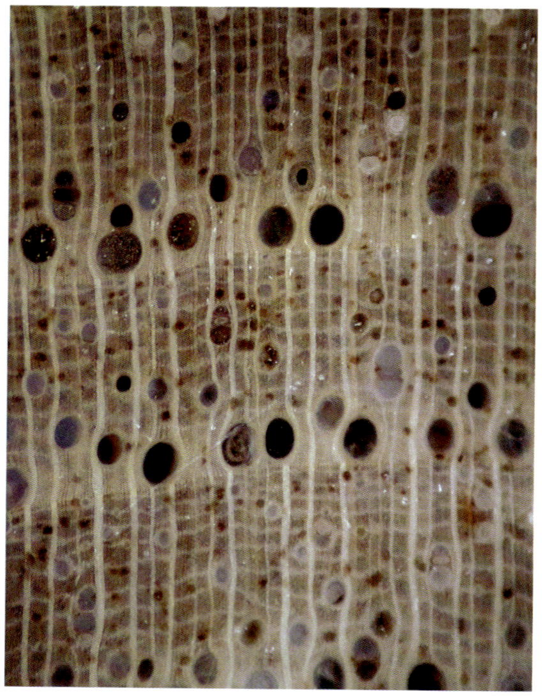

Microscopic image of hickory-like fossil wood from Deschutes, Oregon, USA, preserved as common opal. Pliocene. Daniels specimen.

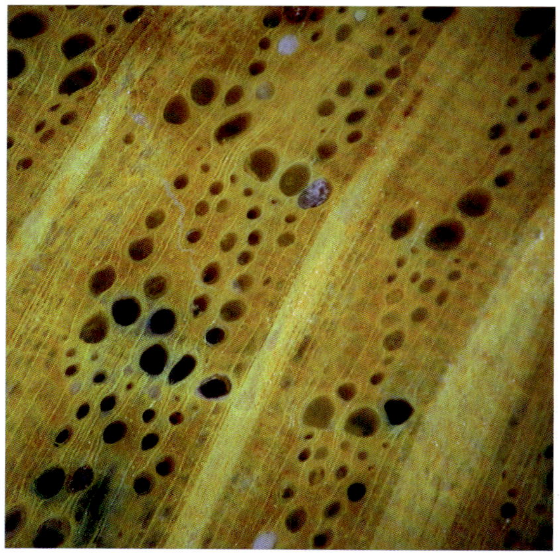

Microscopic image of oak-like fossil wood from Deschutes, Oregon, USA, preserved as common opal. Pliocene. Daniels specimen.

In fact, all opal is geologically young and, at surface temperatures and pressures, tends to disintegrate because of desiccation or transform into the more organized opal-CT. It is not uncommon to find a combination of opal-CT and chalcedony in the same specimen. The future of opal-A, if it persists, is to lose water and become more organized internally.

George Mustoe employed X-ray microanalytical techniques to observe this early opal-A formation on and in woody tissue. The accompanying image shows secondary electron images of opal-A that also forms botryoidal-to-spherical masses soon after the templating process begins. Wood composed exclusively of opal-A is uncommon but examples have been discovered at Virgin Valley.

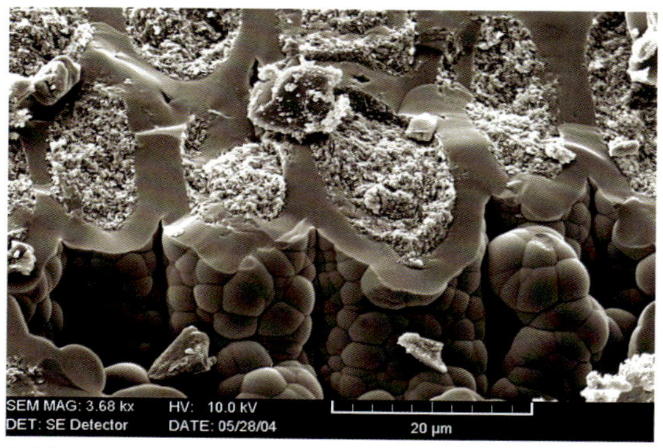

Oblique/transverse view of opalized wood. Cell walls are composed of opal-A; lumens are filled with porous opal, probably opal-CT. Washoe County, Nevada, USA. Miocene. Mustoe specimen and image.

ABIOTIC VERSUS BIOTIC PRODUCTION OF SILICA SPHERES

The formation of micron-sized silica spheres has always been presumed to be a purely chemical or abiotic process. Indeed, artificial opal is produced abiotically. But the importance of microbes in the formation of ore deposits, weathering of rocks to soils, and corrosion of metals is only beginning to be understood and appreciated. Robert Folk, professor emeritus at the University of Texas, Austin, and a well-known scientist in sedimentary rock formation and weathering processes, discusses the possibility that nanobacteria, sometimes called dwarf bacteria, in the 0.05 to 0.2 micron range might contribute to processes previously considered purely chemical. Although this theory has not received widespread acceptance among microbiologists, some cultured nanobacteria produced colonies of 30 to 70 nanometer silica spheroids—the same size range as precious opal spheres. The regular stacking pattern of precious opal spheres is characteristic of some viruses. As far back as the 1950s, Xeros observed a virus that constituted about 25 percent of a crane fly larvae and produced a brilliant display of colors similar to those in precious opal. Whether a relationship exists between the nearly perfect stacking of silica spheres in precious opal and the growth of viruses, nanobacteria, or some other bacteria is unknown but is worth further consideration.

FORMATION OF CRISTOBALITE TRIDYMITE OPAL (OPAL-CT)

Amorphous materials such as volcanic glass or opal-A tend to transform internally into increasingly organized materials, depending on the initial composition. As opal-A continues to lose water and become more ordered, it tends to form poorly crystalline interlayers of the minerals cristobalite and tridymite, called opal-CT. Both cristobalite and tridymite are polymorphs of quartz, meaning they have the same composition but the atoms are organized differently, resulting in different internal structures. Opal-CT looks very much like opal-A, and visually distinguishing the difference between the two is difficult if not impossible. Although much precious opal is opal-A, opal-CT infrequently forms precious opal. Opal-CT also contains up to 20 percent water held in its interstices, according to Gains and others.

Geologists usually think of cristobalite and tridymite as high-temperature forms of silica, commonly found in silica-rich volcanic rocks, but these minerals can also form at surface temperatures. So we have a type of opal that is no longer amorphous but instead shows the first indications of achieving internal structure.

The best clue regarding which opal variety exists in a piece of wood is probably the age. If a specimen of opalized wood is older than mid-Miocene, it probably contains more opal-CT than opal-A and also considerable chalcedony. Moxon

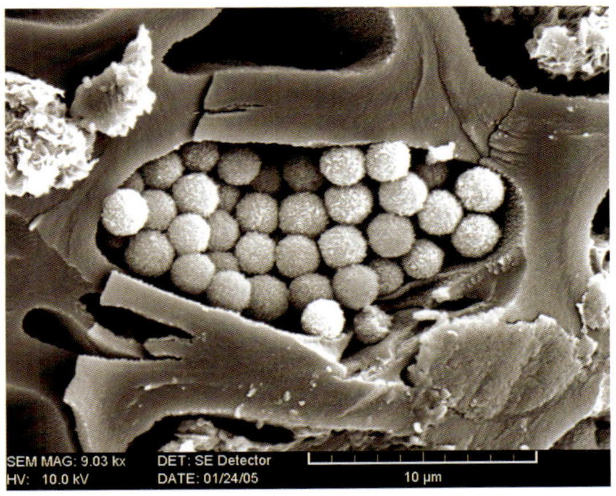

SEM image of opalized cell walls, probably opal-A, with opal-CT microspheres partially filling cell lumen. These spheres are too large and disordered to produce color patterns. Royal Peacock Opal Mine, Virgin Valley. Nevada, USA. Virgin Valley Formation, Miocene. Mustoe specimen and image.

and many others think this progression of opal varieties may occur through a process of dissolution and reprecipitation. However, opal-CT may be the primary form of opal that forms in woods without an opal-A precursor. Scurfield and Segnit, in their study of 75 woods, arrived at this conclusion, and Mustoe confirms this observation from studies of numerous fossilized wood specimens, mostly from the northwestern United States.

When viewed with an electron microscope, opal-CT resembles opal-A, except the microsphere surfaces begin to appear a bit platy. The accompanying scanning electron microscope (SEM) image shows an example of these platy botryoidal structures, sometimes called lepispheres, that grew in a cell lumen of Virgin Valley wood. Jones and Segnit, Stein, Buurman, Kuczumow and others, and Jefferson, among many other researchers, studied petrified wood from throughout the world and describe conversion of silica to opal varieties and eventually to chalcedony. They determined that common varieties of opalized wood are usually composed of opal-CT or a combination of opal-CT in association with chalcedony. These investigators thought that open cracks caused by desiccation of the dead wood before or during burial presented initial pathways for silica solutions to enter the wood. In the northwestern United States, petrified wood exhibiting a waxy or opaline appearance with a low density usually consists of opal-CT along with some chalcedony.

FORMATION OF CHALCEDONY

Chalcedony (pronounced "kal-sed′-ne") is an interesting variety of quartz that constitutes most petrified woods worldwide. Its internal properties have been studied for decades, and yet this material remains fascinating to mineralogists. Chalcedony is distinguished by its composition of microscopic twisted fibers of quartz, a property that makes it both hard (hardness 6.5 to 7) and unusually tough. Water is still present but in small quantities, usually less than 0.5 percent. Chalcedony forms banded agates, onyx, moss agate, sard, carnelian, chrysoprase, bloodstone, and petrified wood.

Flint, chert, and jasper differ from chalcedony. They also consist of microcrystalline varieties of silica but have a more granular microtexture instead of being fibrous. Flint usually refers to dark-colored microcrystalline silica produced by the transformation of microscopic siliceous marine organisms Chert is properly a rock that contains microcrystalline quartz but may also contain opal varieties, calcite, clays, and iron oxide minerals and is usually found in beds or lenses. Jasper is usually considered a variety of chert with higher concentrations of iron or manganese oxides, giving it darker colors. Massive beds of red jasper are associated with Precambrian Era bedded iron deposits. Some fossil wood limb casts contain granular microcrystalline quartz that has a jasper appearance.

As opal-CT continues to lose water and organize internally, quartz begins to become the dominant species. Buurman, Scurfield and Segnit, and Stein report that older specimens generally contain chalcedony as the dominant form of silica. With increasing age, quartz appears to grow either as microcrystalline chalcedony or as submicroscopic granular quartz. Well-developed quartz crystals, even microscopic ones, almost always seem to be limited to voids such as fractures and cell lumen where three-dimensional spaces exist for the crystals to grow. Chemical conditions for this occurrence of quartz are different from that of chalcedony.

The formation of chalcedony is complex and has been studied by many scientists including Clifford Frondel from Harvard University, Yifeng Wang of Sandia National Laboratories, and Enrique Merino of Indiana University. Quartz crystals are normally elongated along the "c" axis as is exemplified by single crystals from anywhere in the world. In cross section, the crystal is a hexagon with three "a" axes in one plane oriented perpendicular to the "c" axis and intersecting the centers of the prism faces. Chalcedony is also quartz, but it forms very differently. Instead of being elongated along the "c" axis, it is elongated along an "a" axis. This structure is related to the speed of crystallization from a supersaturated solution of silica or, in this case, the continued crystallization of an amorphous (mostly) silica solid. If a slab of Brazilian agate is placed in front of a light, the fibrous nature of chalcedony is apparent. Chalcedony that forms from the progressive organization of opal-CT in petrified wood is not as obviously fibrous to the unaided eye, but its fibrous texture is apparent when viewed with a petrographic microscope. Chalcedony fibers that originate from opal-CT may be much shorter and narrower, and the orientations may be numerous because of the nature of the material being petrified. Chalcedony that fortifies cracks in limbs or forms limb casts is texturally similar to other agates.

As chalcedonic quartz crystallizes, dislocations (which are something like holes or imperfections) in the crystal lattice allow aluminum ions (which are close in ionic radius or size and charge to silicon ions and common in ground water solutions) to substitute for silicon atoms. This irregularity in the crystal lattice causes twisting along the advancing growth front. As these fibers of quartz grow, other impurities in the opal (or silica gel for agates) such as iron, manganese, or other metal oxides are frequently not incorporated in the lattice. Instead, they are pushed along the growth front and eventually concentrate sufficiently to interrupt crystallization and cause a hiatus in growth. This process then repeats itself by initiating another advancing growth front beyond the impurities, thereby producing a growth band in the chalcedony. This banding is more pronounced in the formation of fortification agates, but a subtle version of this growth also occurs during the transition of opal-CT to chalcedony. Multiple growth

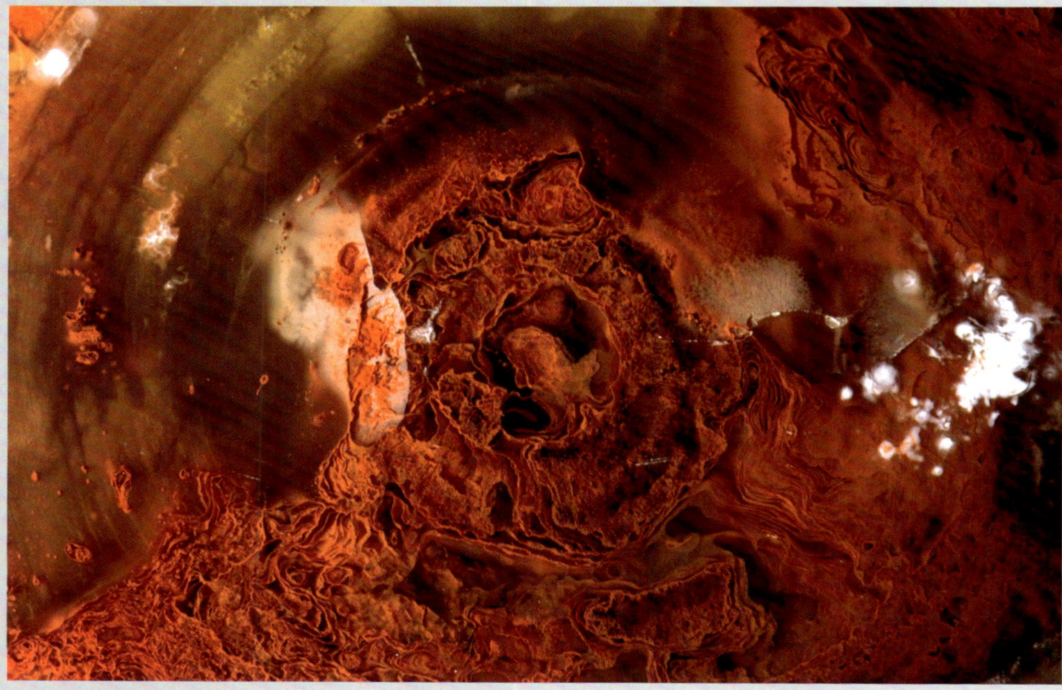

Advanced chalcedony formation. Vestiges of cellular structure are seen as very fine textures in the more translucent areas. Leafy red chalcedony textures are replacing and vaguely preserving woody ring structures. Cedar Mountain Formation, Lower Cretaceous. Castle Dale, Utah, USA. Dayvault specimen, width of view 3.5 cm.

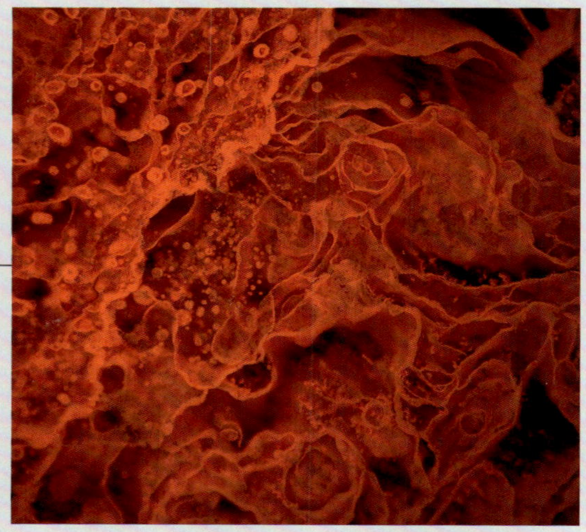

Micro image of specimen from Cedar Mountain Formation. The central part of the specimen contains the leafy texture often observed as chalcedony continues to crystallize and destroy cellular structure. Cedar Mountain Formation, Lower Cretaceous. Castle Dale, Utah, USA. Dayvault specimen, width of view 14 mm.

fronts in petrified wood are often microscopic and visually imperceptible. This continuing process of converting opal-CT to chalcedony in silicified wood decreases the permeability and increases the refractive index, hardness, and specific gravity.

Ideally, we have elongated twisted strands of microcrystalline quartz organizing from the slightly crystallized opal-CT that developed from amorphous opal. This process is somewhat analogous to, but not the same as, a sedimentary rock containing fossil shells that undergoes changes produced by heat and pressure as it is buried in the earth and recrystallizes into a metamorphic rock. As this physical process proceeds, microscopic minerals start to form and enlarge in the newly forming rock, and the original fossils and textures progressively disappear. The crystallization process overwhelms and obliterates antecedent components. So too do the microscopic details of the original wood diminish as the quartz crystallizes. It is remarkable that any original wood textures remain, much less that many microscopic cellular structures persist. As chalcedony formation progresses, however, the details often become increasingly ghostly. George Mustoe reports that, on occasion, microscopic structures still discernable in chalcedony with a petrographic microscope are sometimes obscured when the same specimen is viewed with an electron microscope. Although the crystallizing quartz has grown beyond its original cell dimensions, relic cellular architecture is still visible because of iron oxide pigments or vestiges of organic carbon that outline cell walls or other features. Therefore, some specimens of highly silicified wood, which in hand specimens and even under the electron microscope appear to have no cellular structure, really have subtle relic textures when optically magnified.

In other instances, once prominent growth bands may be the only vestige of original wood texture remaining in the specimen as silicification progresses. The example of wood from the Lower Cretaceous Cedar Mountain Formation of Utah shows relic annual growth bands in a matrix of translucent chalcedony. Sometimes only a vestige of the original ring tree structure remains, which can be confused with agate banding. Silicification can continue until all organic substance is lost and the crystallizing quartz erases all but traces of former organic structures. Triassic woods from Utah and Arizona often show the transition of preserved vascular tissue, which had been so carefully replicated, to leafy or nodular and rod-like textures.

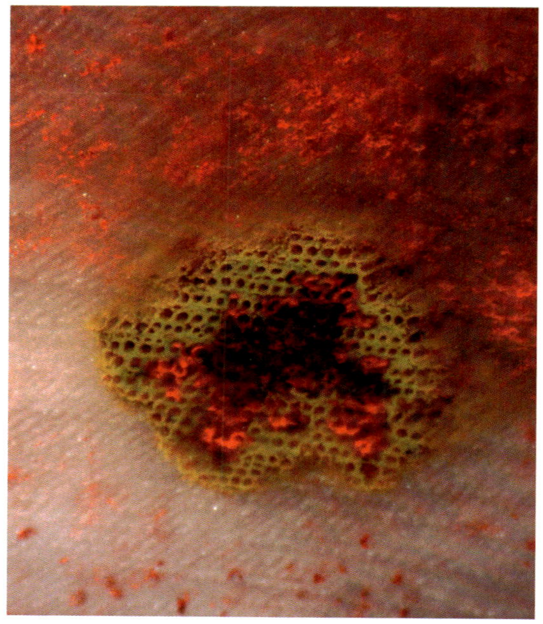

Cellular structure that is losing identity by altering to chalcedony. Petrified Forest Member, Chinle Formation, Triassic. Northern Arizona, USA. Daniels specimen. Width of view about 2 mm.

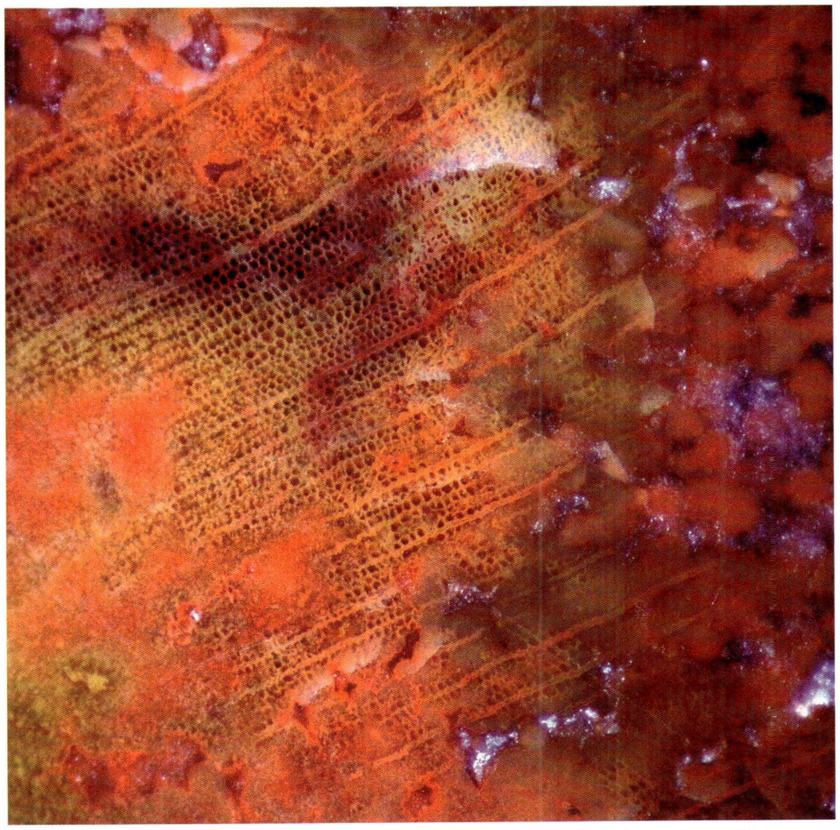

Destruction of cellular structure with chalcedony development. This specimen shows tracheids on the left altering to blebs and rounded packets of colorful chalcedony. Growth structures are easily visible on the left and mostly obliterated on the right of the specimen. Star dadoxylon. Chinle Formation, Triassic. Henry Mountains area, Utah, USA. Mower specimen. Width of view about 2 mm.

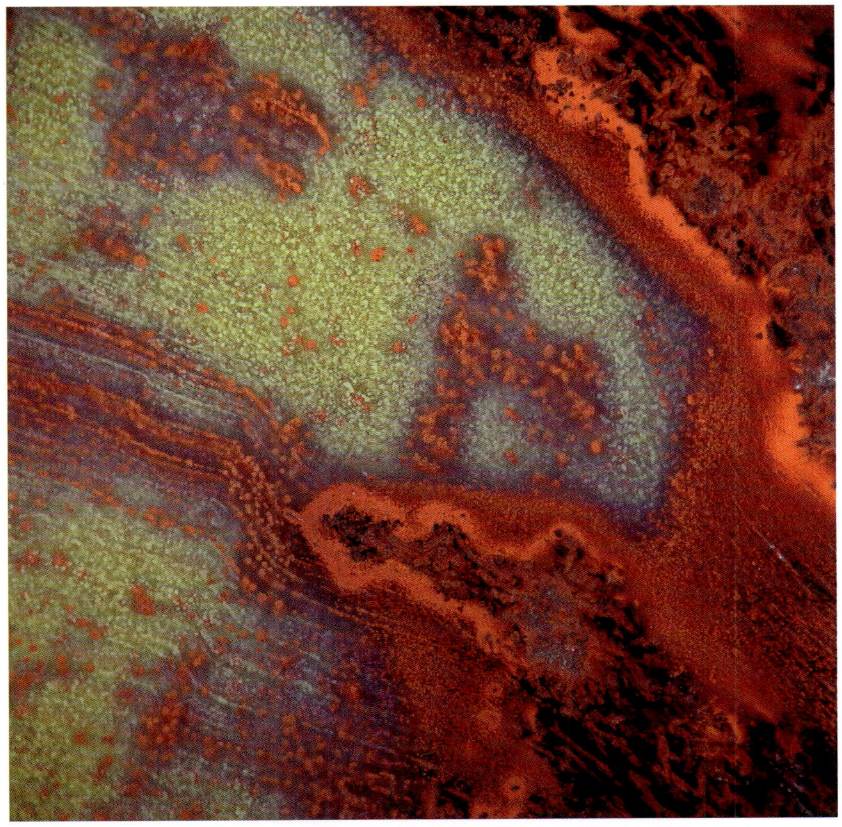

Destruction of cellular structure with chalcedony development. This specimen shows cell structures on the upper right side that are missing toward the lower left. Cellular textures disintegrate into blobs of yellow chalcedony. Star dadoxylon. Chinle Formation, Triassic. Henry Mountains area, Utah, USA. Mower specimen. Width of view about 8mm.

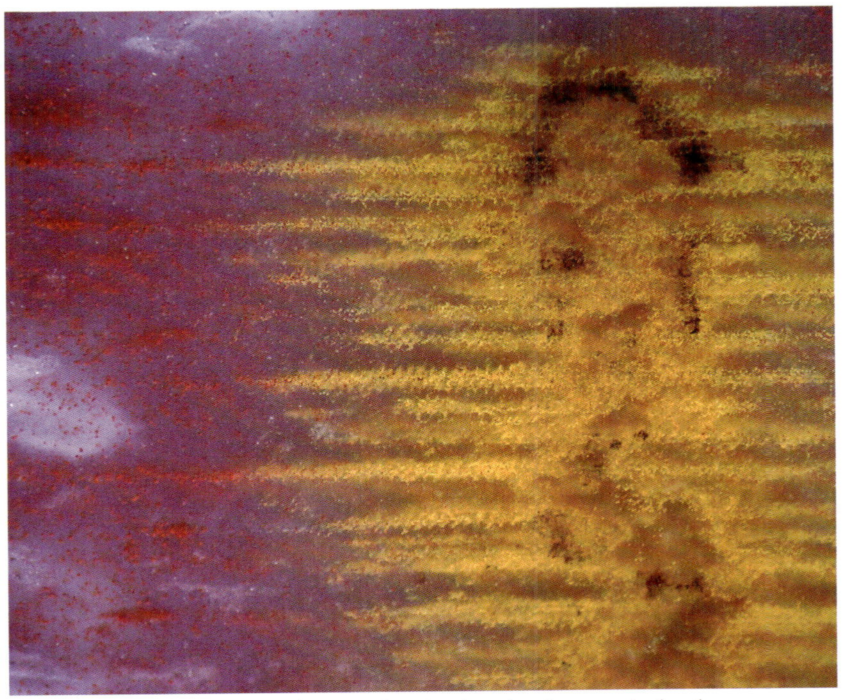

Cellular structure that is losing identity by altering to chalcedony. Differential oxidation produces contrasting colors. Petrified Forest Member, Chinle Formation, Triassic. Northern Arizona, USA. Daniels specimen. Width of view about 8 mm.

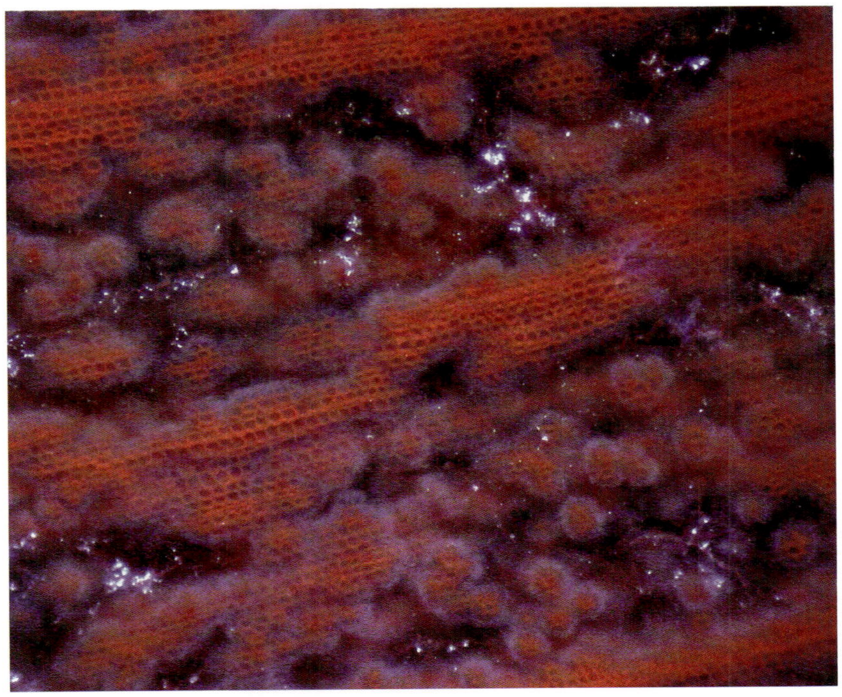

Cellular structure that is losing identity by altering to rounded packets of chalcedony. Petrified Forest Member, Chinle Formation, Triassic. Northern Arizona, USA. Daniels specimen. Width of view about 4 mm.

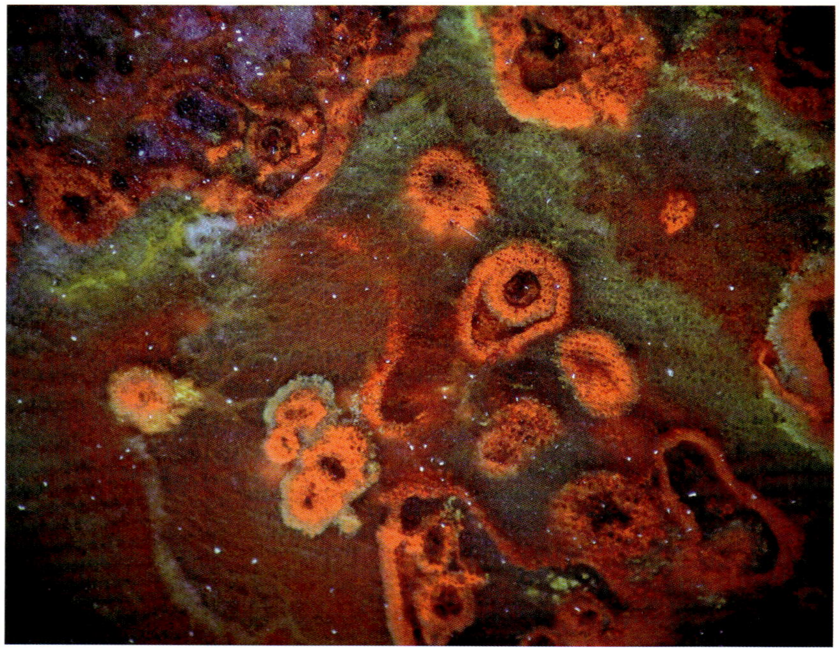

Destruction of cellular structure with chalcedony development. This specimen shows tracheids altering to rounded areas of colorful chalcedony. The colors apparently reflect degrees of oxidation. Morrison Formation, Jurassic. Henry Mountains area, Utah, USA. Daniels specimen. Width of view about 15 mm.

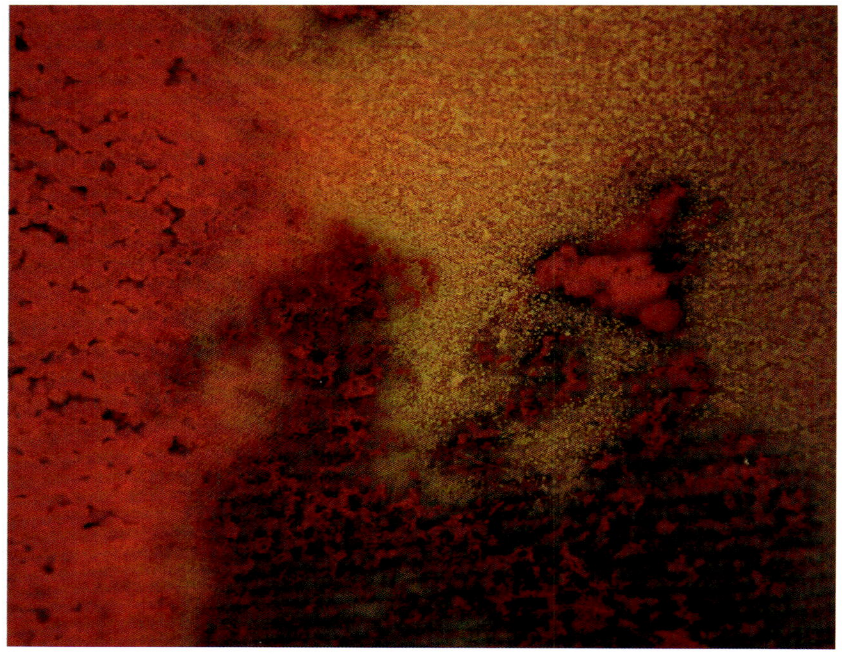

This specimen shows cell lumina altering to blebs of colorful chalcedony. Star dadoxylon. Chinle Formation, Triassic. Henry Mountains area, Utah, USA. Mower specimen. Width of view about 3 mm.

This specimen shows tracheid vestiges as ghostly white patches in chalcedony. Red banding is all that remains of original cellular textures. Unidentified conifer. Morrison Formation, Jurassic. Henry Mountains area, Utah, USA. Daniels specimen. Width of view about 8 mm.

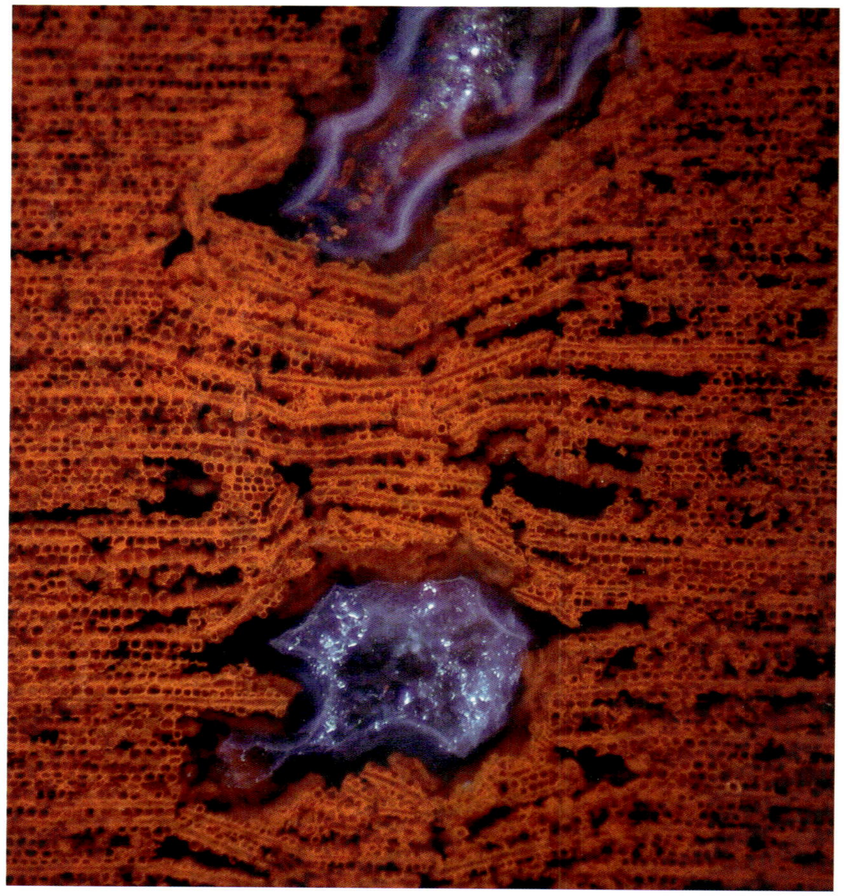

This specimen shows tracheids beginning to become separated and floating in clear chalcedony. Morrison Formation. Jurassic. Henry Mountains area, Utah, USA. Daniels specimen. Width of view about 4 mm.

Destruction of cellular structure with chalcedony development. This specimen shows cell walls being assimilated into the chalcedony and cell lumina vestiges remaining. Differential oxidation provides contrasting colors. Unidentified conifer. Mesozoic. Henry Mountains area, Utah, USA. Daniels specimen. Width of view about 10 mm.

This specimen shows tracheids altering to rounded packets of colorful chalcedony. Star dadoxylon. Chinle Formation, Triassic. Henry Mountains area, Utah, USA. Mower specimen. Width of view about 4 mm.

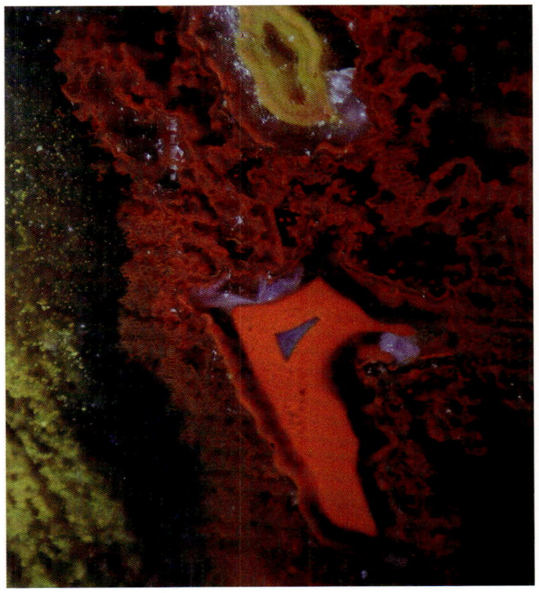

This specimen shows tracheid islands floating in clear chalcedony and the development of colorful lacy silica. Fortification agates are also present in this specimen. Star dadoxylon. Chinle Formation, Triassic. Henry Mountains area, Utah, USA. Mower specimen. Width of view about 3 mm.

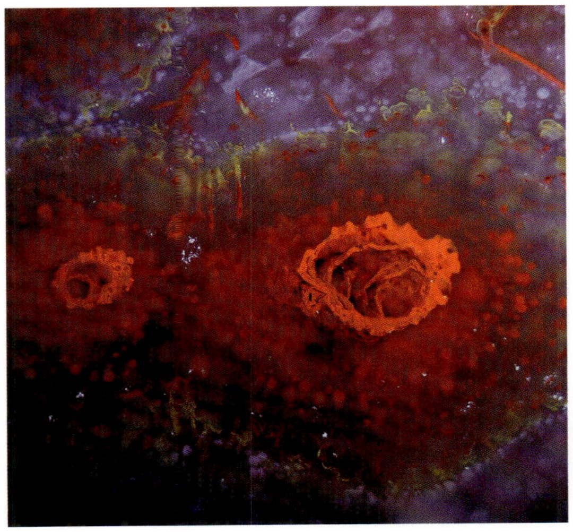

Multi-colors and textures of chalcedony in limb cast. Reddish silica aggregates float in translucent chalcedony. Explanation for this texture is unknown. Brushy Basin Member, Morrison Formation, Late Jurassic. Moab, Utah, USA. Daniels specimen. Width of view about 2 mm.

Some Jurassic petrified wood specimens found near Moab, Utah, thought to be limb casts may instead be examples of advanced silicification. Rod-and-cup or rod-and-ellipsoid radiating textures occur in many of these specimens. Whether these textures actually represent a vestige of cellular structures or some form of spherulitic growth is unknown.

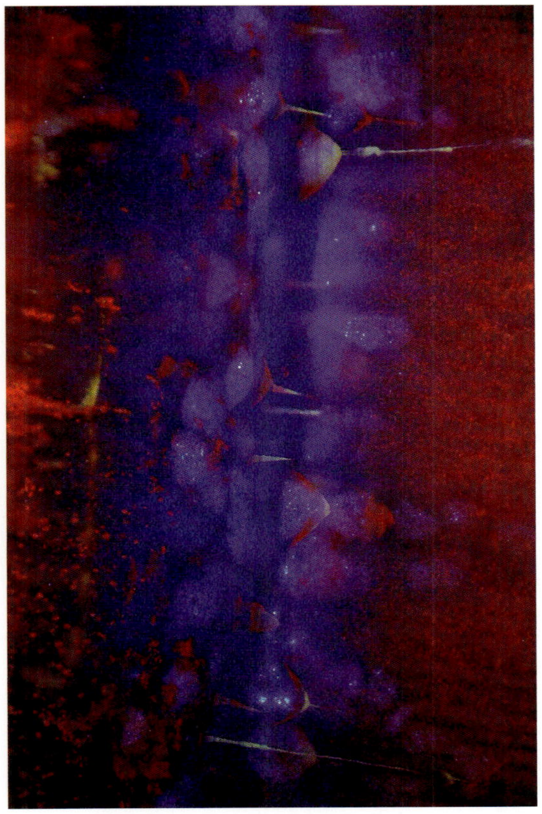

Rods and cones textures are seen in a number of red wood specimens from eastern Utah. This band of translucent-to-transparent chalcedony near the border of a limb cast(?) shows delicate acicular features that have oblate-to-prolate ellipsoids or part of this structure (cups) distributed along their length. The spears are always oriented radially from the center of the cast. Whether these features are related to original wood textures or to the formation of agate is unclear. Brushy Basin Member, Morrison Formation, Late Jurassic. Moab, Utah, USA. Dayvault specimen. Width of view about 4 mm.

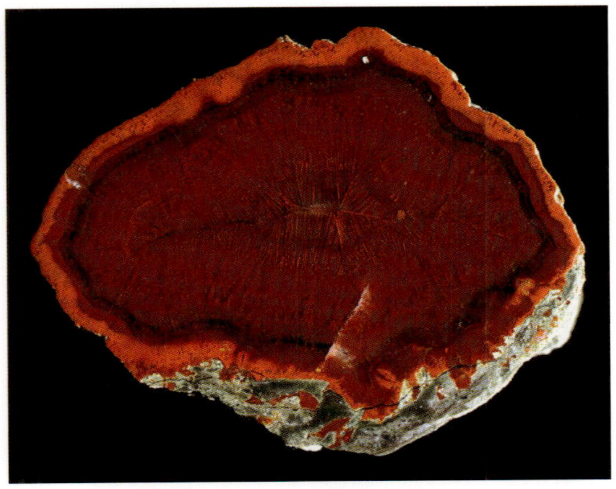

Limb cast that exhibits unusual rod and cone micro textures along the arcuate center of the specimen. Brushy Basin Member, Morrison Formation, Late Jurassic. Yellow Cat area, Utah, USA. Dayvault specimen, 5 cm.

Limb cast that exhibits unusual rod and cone micro textures in the clear darker band near the perimeter of the specimen. Brushy Basin Member, Morrison Formation, Late Jurassic. Yellow Cat area, Utah, USA. Dayvault specimen, 4.5 cm.

Rods and cones in a red wood limb cast from eastern Utah. This unusual texture, found in the central portion of a limb cast, shows opposing ellipsoids situated on either side of a central line highlighted as yellow chalcedony. Some ellipsoids contain quartz centers. Brushy Basin Member, Morrison Formation, Late Jurassic. Moab, Utah, USA. Dayvault specimen. Ellipsoids are about 1 mm.

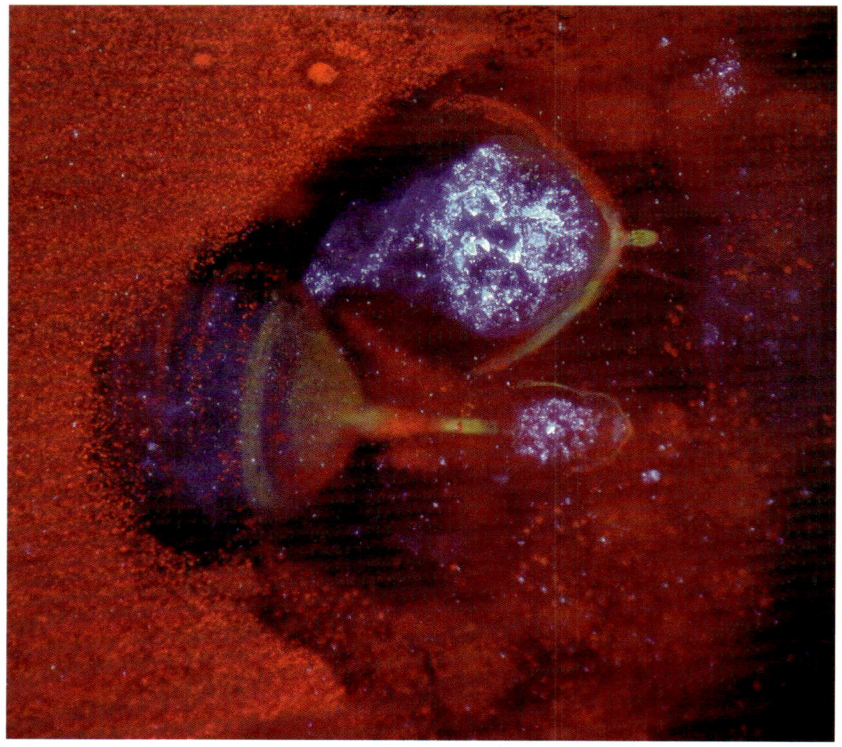

In this specimen, the cones show only vestiges of the rods. Points of attachment to rods were to the right of the photograph. Color zones are visible along the outside of the cone structures. The central portion of one cone is quartz. Brushy Basin Member, Morrison Formation, Late Jurassic. Moab, Utah, USA. Dayvault specimen. Ellipsoids are about 1 mm.

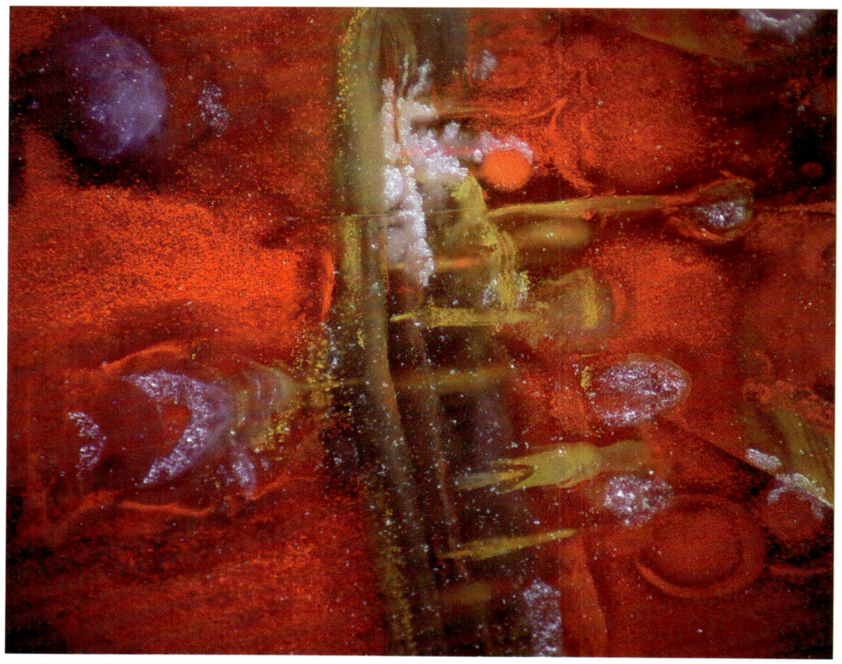

Rods and cones in a red wood limb cast from eastern Utah. Rods appear as shells in this specimen. Brushy Basin Member, Morrison Formation, Late Jurassic. Moab, Utah, USA. Dayvault specimen. Width of view about 3.15 mm.

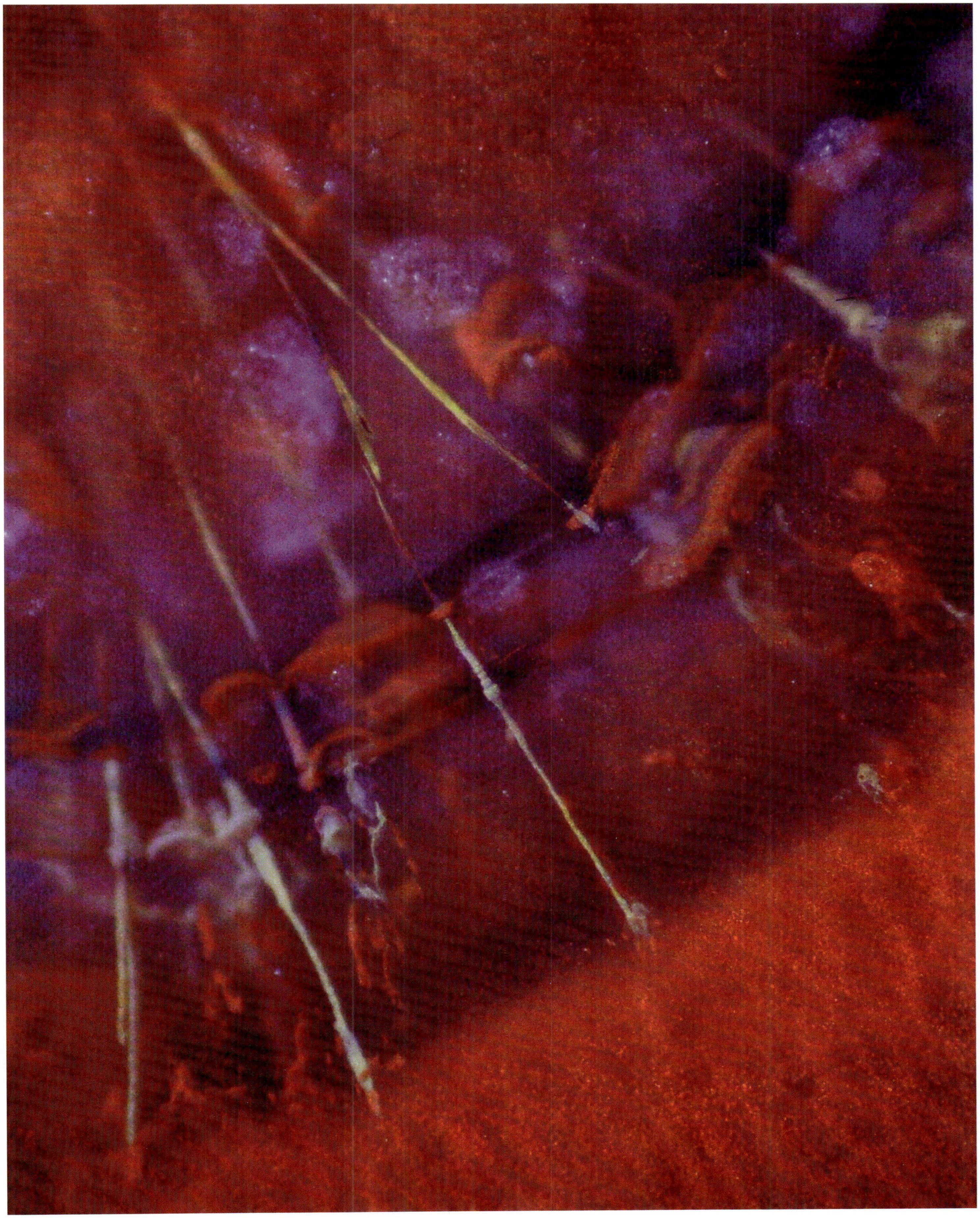

Rods and cones in a red wood limb cast from eastern Utah. This is another view near the perimeter of a limb cast along a translucent-to-transparent band of chalcedony that contains these interesting textures. Brushy Basin Member, Morrison Formation, Late Jurassic. Moab, Utah, USA. Dayvault specimen. Width of view about 4 mm.

Rods and cones in a red wood limb cast from eastern Utah. In this specimen the rods penetrate through opaque red chalcedony. Quartz often develops in the ellipsoids. Brushy Basin Member, Morrison Formation, Late Jurassic. Moab, Utah, USA. Dayvault specimen. Width of view about 1.75 mm.

Rods and cones in a red wood limb cast from eastern Utah. This is a view of the central part of an agate that produces rods or bundles of rods radiating from the center line of the specimen. Ellipsoids and other orb-like features are aligned along the rods or elsewhere in the chalcedony. Brushy Basin Member, Morrison Formation, Late Jurassic. Moab, Utah, USA. Dayvault specimen. Width of view about 7 mm.

Another unusual and unexplained texture is the microscopic coiled red (hematitic?) inclusions that "float" in colorless, translucent faintly banded limb casts. Occasionally, the corkscrew inclusions are concentrated along preferred bands. The helical shapes suggest an organic precursor, possibly vestiges of bacteria or algae.

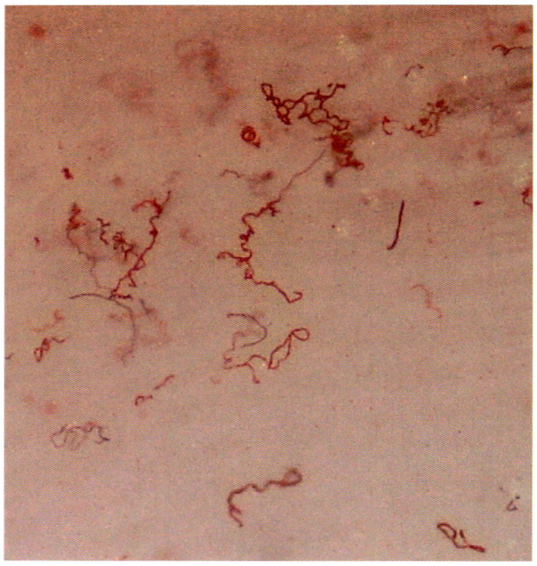

View of the twisted filaments in translucent agate in fossil wood from eastern Utah. A number of these specimens have been found at this locality. Brushy Basin Member, Morrison Formation, Late Jurassic. Moab, Utah, USA. Dayvault specimen. Maximum length of filaments about 0.2 mm.

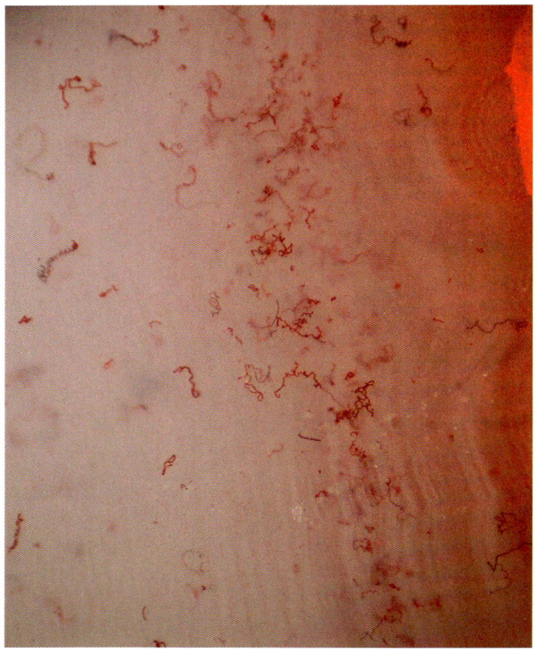

Small milky translucent agate section of fossil wood from a red wood locality in eastern Utah. Inclusions of minute red hematitic-looking twisted filaments float in the chalcedony. These features are sometimes concentrated along certain bands of the agate. The origin of these features is unknown, but the twisted appearance suggests a possible organic precursor. Brushy Basin Member, Morrison Formation, Late Jurassic. Moab, Utah, USA. Dayvault specimen. Width of view about 9 mm.

Microcrystalline quartz may not always be in the form of fibrous chalcedony but may instead be granular. This type of transformation can proceed from the same process that produced chalcedony. It results in generally less transparent, more earthy looking, and less pure silica such as jasper or chert. If vestiges of original wood textures are visible, they may have resulted from the silica templating process. If no relic structures of the wood remain, the microcrystalline granular silica may have infilled a cast. A problem with the opal-to-chalcedony transformation process is the nearly 20 percent difference in density between opal and chalcedony. How can you completely fill a cavity with chalcedony that evolved from opal without adding silica mass? Landmesser proposed that the diffusion of additional silicic acid fills the remaining space.

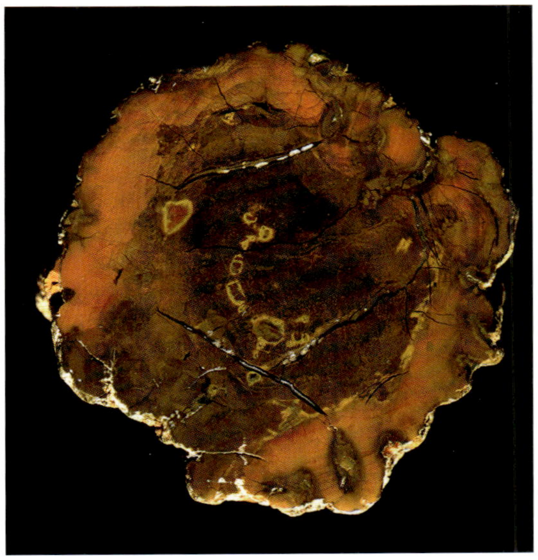

This unusual specimen is probably an example of advanced chalcedony formation in a silica-templated limb. Vestiges of cellular structure are visible around the perimeter, but the general texture appears to be more of a jasper. Morrison Formation, Late Jurassic. Ticaboo, Utah, USA. Hatch specimen, 7 cm.

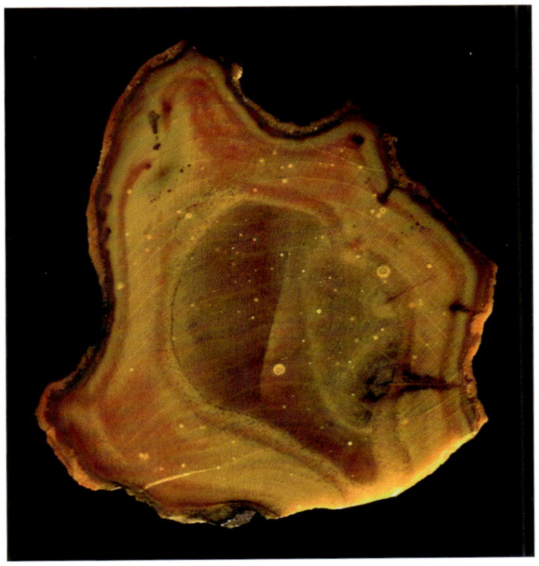

This specimen probably represents advanced chalcedony formation in a silica-templated limb. Cellular structure is visible, but an overprint of color banding crosses the original woody textures. Morrison Formation, Late Jurassic. Ticaboo, Utah, USA. Dayvault specimen, 5 cm.

EXCEPTIONS TO THE SILICIFICATION SEQUENCE

Dead wood buried in solutions containing silica does not always know it should follow the silicification processes just described. Silicification processes may differ, depending on the changing silica concentrations in the ground water, ambient temperatures, and pressures where the wood was buried. A good example would be the fairly common situation where fossil wood becomes silicified by the templating process, then the ground water chemical composition changes or disappears and the process stops. This leaves crusty friable wood full of holes where lumina, conductive vessels, or spaces between adjacent cells are not infilled by the permineralization process. Such fossil wood is usually of little interest to collectors. In another common example, cell walls are templated by the usual process and opal-CT is formed, but the chemistry of the ground water changes. In this case, silica concentrations decrease, and other openings in the wood are filled by euhedral (well-formed) quartz crystals rather than chalcedony.

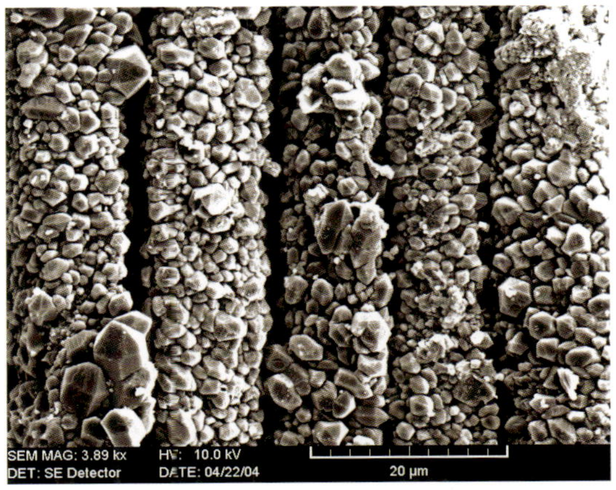

SEM of quartz crystals encrusting tracheids of *Cupressinoxylon*. Trinity Formation, Lower Cretaceous, Montague County, Texas, USA. Mustoe specimen and image.

FORMATION OF LIMB CASTS

INTRODUCTION

Limb casts are among the most beautiful and collectable of all fossil woods. Some of the finest limb cast specimens in the United States are collected in the Miocene Virgin Valley Formation of continental origin in the Virgin Valley, located in northwestern Nevada. At this location, precious opal permineralizes and templates limbs and cones and sometimes occurs as fracture fillings or along bedding planes. The most famous locality in the world for precious opal limbs, known also for many other fossils including marine animals, is the Lower Cretaceous Griman Creek Formation in Lightning Ridge, New South Wales, Australia. Elizabeth Smith reports that opalization at Lightning Ridge occurred between 10 and 60 million years ago, much more recently than the age of the fossils found there. Both localities have been mined for gemstones for more than 100 years.

Other notable localities for collectible limb casts in the United States include the so called "red wood" from the Yellow Cat, Moab, and Henry Mountains areas of eastern Utah; pink, green, and blue-gray casts from the South Fork of the Crooked River, Oregon; pink casts from Texas Springs, Nevada; colorless to bluish casts from Wiggins Fork and Dubois, Wyoming; and a variety of gray, yellow, and green casts from Tom Miner Basin, Montana, some of which are hollow and remarkably lined with amethyst crystals. These casts are all examples of chalcedony fortifications, usually as agate. Agate, in its most common form, is the distinctly banded variety of chalcedony. Onyx is also banded, but the layers are in horizontal straight lines. Carnelian refers to red translucent chalcedony that often is inter-banded with other colors of chalcedony.

THE BEGINNING

Limb casts begin their history by the same processes that produce templated and permineralized woods. A branch or tree is buried in sediment produced by an ash fall or by a flood event. But now a different transformation evolves. Instead of the original wood maintaining some degree of structural and histological integrity, accelerated processes of decay and rot remove most or all of the wood, while the enclosing sediment concurrently hardens sufficiently to maintain the shape of the original cavity or void. Often limb casts are elliptical or collapsed, indicating that the sediment was still soft while the wood was disappearing or that pressures from overlying consolidated sediments compressed the original void. If the enclosing sediment is fine grained, a detailed replica of the limb surface may be impressed in the rock, which results in a cast of the original limb. Some time after the original wood decomposes and is removed, solutions in the ground water begin depositing minerals in the open space. These minerals harden into a solid that preserves the outer surface impressed in the surrounding rock. As with other wood petrifactions, the most important mineralizing medium is silica. The result is usually a chalcedony limb cast or less commonly a beautiful carnelian agate; rarely, precious opal fills the void. A condition that can provide the chemical conditions to satisfy these requirements might exist in an oxygenated and often alkaline environment, such as alkaline lakes or river systems in arid regions.

Chemical and physical processes that produce limb casts are similar to those that produce most agates around the world. The theories presented in this chapter are borrowed from studies of agate formation, not specifically from studies of petrified wood limb-cast formation.

MINERALOGY OF AGATES

As previously mentioned, decomposition of volcanic rocks provides silica for most large deposits of agate or silicified wood. Zarins discusses how glassy fragments in these rocks rapidly hydrate and devitrify, yielding highly soluble sodium silicates that can eventually result in the formation of silica gels. In addition to silica, other silicate minerals are formed, such as the hydrous silicates of the zeolite group and the swelling bentonitic clays that produce, upon weathering, a distinctive popcorn appearance.

The next step is the deposition of aqueous silica into voids left by decomposed wood. A large number of researchers have puzzled over the genesis of chalcedony and agate for more than 100 years. Most recently, Clifford Frondel in the United States, Michael Landmesser in Germany, and Terry Moxon in Great Britain have spearheaded studies of this subject. Much of the discussion that follows derives from their work.

Interestingly, the formation of chalcedony is thought to follow a process called the spherulitic theory. Keith and Padden first put forth this premise for the formation of round or spherical mineral aggregrates in volcanic glass in the early 1960s. Snowflake obsidian owes its snowflakes to this type of crystallization. Spherulites are spherical mineral growths formed by initial crystallization of a seed crystal within a melt or solution that grows outward in all directions. In volcanic glass, this process begins with a single microscopic crystal fiber or crystallite of feldspar (sanidine) or of a silica variety (cristobalite) that begins to grow additional crystals outward from the center of the crystallizing nucleus. Early spherulites might look like a tiny bow tie floating in a glassy matrix, but the final product, if unimpeded, is a radiating spherical crystalline aggregate. Typically, this spherical growth is disrupted when two or more spherulites impinge on one another like soap bubbles and share boundaries. In the mid-1970s, Oehler expanded the theory of spherulitic formation in volcanic glass to explain similar textures that are produced in silica gel. The early impingement of growing spherulites and subsequent change in growth direction are important extrapolations to the formation of agates.

In the early 1990s, Pabian and Zarins and, somewhat later, Moxon expanded on this idea and described the physiochemical processes that might proceed in a low-temperature, low-pressure setting such as the interior of a limb cast or void in volcanic rocks. The crystallization of chalcedony in a void filled with a supersaturated silica solution, possibly as a gel, would begin at the outer margins and progress inward. The four diagrams illustrate this progression. Spherulitic growth may begin when small hemispherical groups of crystal fibers form near the edge of the void, as shown in Diagram A. As growth continues, expanding surfaces impinge on one another to produce a fairly smooth planar surface or sector boundary as demonstrated in the inset of Diagram B. This planar interface, produced by intersecting sub-parallel fibers, is oriented approximately perpendicular to the original surface of the cavity. The actual process can be more complex than the idealized diagram suggests, depending on the number of nucleation sites, growth directions of crystal bundles along the perimeter of an irregular cavity, changes in silica concentration, and other factors. With continued growth, the sector boundaries proceed toward the center of the cavity and produce the prominent radial texture common to agates, as shown in the diagram.

Breaks in the growth may be caused by impurities, such as iron oxides or other metal oxides, that are not incorporated into the crystallizing silica phase and build up at the crystal and liquid interface, eventually coating the inside of the cavity. These usually colloidal-size aggregates

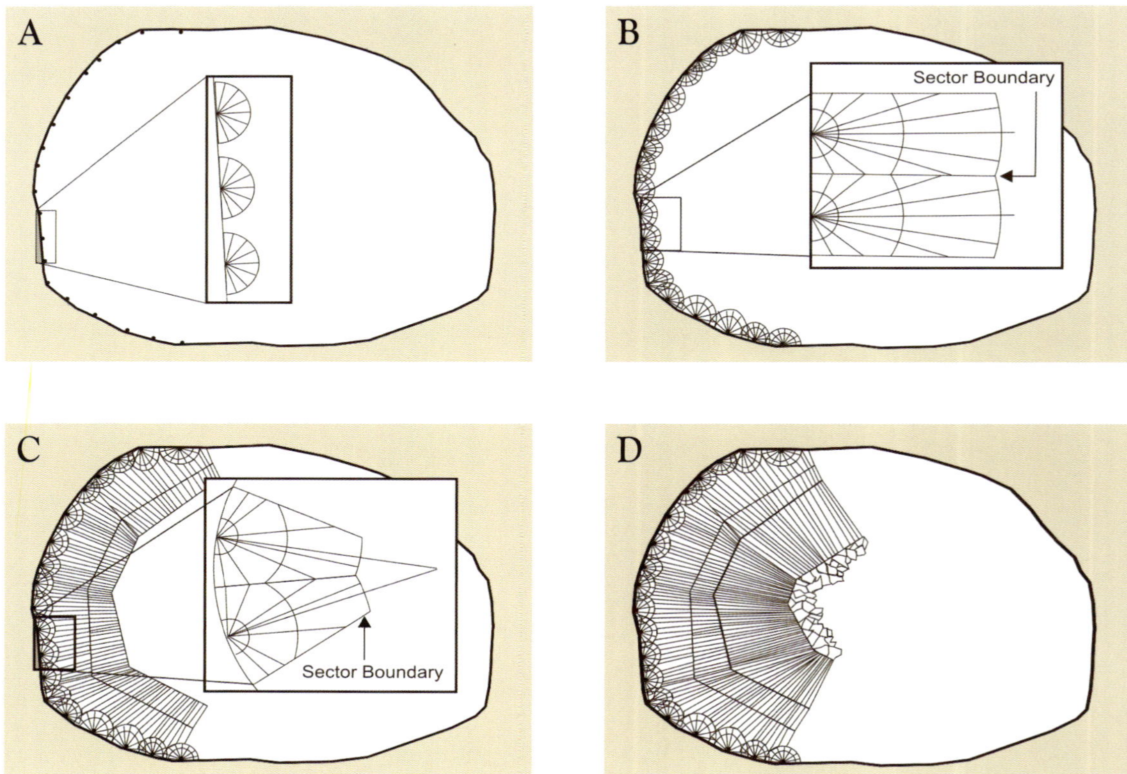

Genesis of Agate in a Cavity. A. Cross section of a limb cavity filled with supersaturated silica probably in the form of a gel. Nucleation of radiating quartz crystallites forms spherulites near the perimeter of the void. In reality, the void may contain some vestiges of silicified woody material and nucleation could begin in multiple locations and directions. B. Continued radial growth of quartz crystallites produces fibrous chalcedony bundles that impinge on one another (inset) to form a growth boundary or sector boundary. C. Further growth produces an inwardly converging radial fibrous structure as shown in the inset. The process may repeat if metal oxide impurities build up on the inwardly crystallizing front until crystallization is halted. Spherulitic growth may begin again and progress toward the center of the cavity, producing a discrete band. D. If the concentration of silica decreases sufficiently, coarse crystalline quartz may grow in the center of the cavity. Modified from Moxon and from Frondel.

hinder further growth. At this point, the formation of spherulites may simply begin again on the inside of the impurity surface and start pushing another ever increasing concentration of similar impurities toward the center of the cavity until the process is repeated. Concentric banding, often of different colors and varying transparencies, can be produced in this way. This banding is similar to the processes discussed in the transformation of opal-CT to chalcedony. Diagram C shows a second band that has formed by this process. Thin sections of chalcedony from Del Norte, Colorado, thundereggs show some of these progressive textures. Other processes that produce banding may result from pauses or changes in the rate of chalcedony crystallization for as yet unclear reasons. Concentric bands may be thick or extremely thin. At first, the continuous banding reproduces the contours on the outer margins of the cavity, but the banding becomes smoother and less representative of the original cavity as the agate formation proceeds inward. Other processes, such as a change to parallel banding, may partition future concentric banding in the remaining interior of the void.

Thin section of chalcedony showing demarcation between fine layer of spherulitic growth and adjacent coarser layer. Sector boundaries between individual spherulites in the coarser layer are similar to Diagram B in the agate genesis figure. Cross-polarized light with 1 lambda quartz plate. Rio Grande do Sul, Brazil. Kile specimen and photograph. Width of view 0.2 mm.

Red wood limb cast concentrically banded, subsequently horizontally banded, and finally concentrically banded. This demonstrates the change of agate formation from one type to another within the same specimen. Morrison Formation, Late Jurassic. Yellow Cat, Utah, USA. Dayvault specimen, 5 cm.

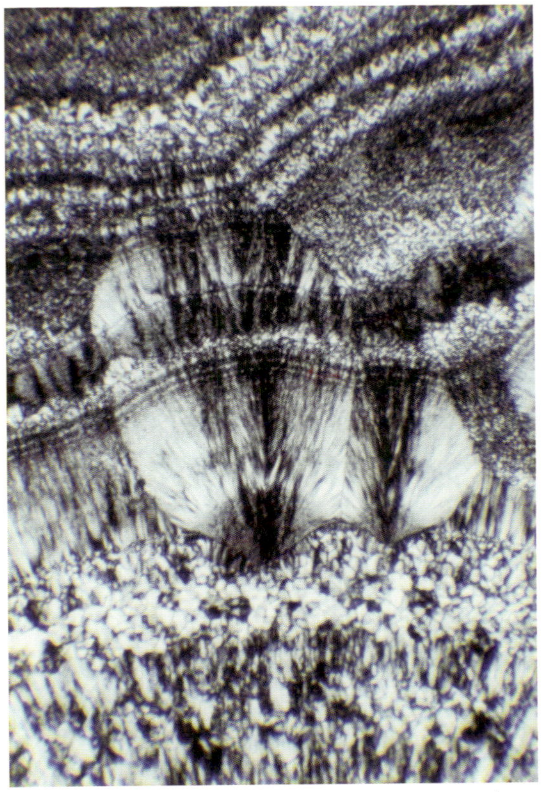

Thin section of chalcedony showing strongly developed spherulitic plume that crosses traditional banding in places. Changes from light to dark along chalcedony fibers are caused by light reacting to the twisting of the fiber as it grew. Cross-polarized light. Del Norte, Colorado, USA. Kile specimen and photograph. Width of view 3.9 mm.

Agate fortification in limb cast from eastern Utah. This specimen demonstrates how progressive bands of agate become less angular toward the center. Brushy Basin Member, Morrison Formation, Late Jurassic. Moab, Utah, USA. Dayvault specimen. Width of view about 10 mm.

Concentrically banded red wood limb cast of unusual yellow agate. Dendrites occur in the central portion. Morrison Formation, Late Jurassic. Moab, Utah, USA. Dayvault specimen, 5.5 cm.

Concentrically banded red wood limb cast. Morrison Formation, Late Jurassic. Moab, Utah, USA. Dayvault specimen, 3.5 cm.

Chalcedony continues to crystallize until the ground water solution becomes undersaturated with respect to silica, whereupon coarse crystalline quartz crystals may then form as illustrated in Diagram D. This change in silica concentration is often observed in the quartz cores of limb casts or as changes in the mode of silica crystallization from chalcedony to quartz and back to chalcedony. Other interesting chalcedony textures form in limb casts. Delicate tiny blebs of silica float within bands of chalcedony, strong radiating spikes often occur in or across bands of chalcedony, and contrasting colors sometimes juxtapose. Agate formation is opportunistic and can fill any open space, no matter how small. Wood from Oregon contains minute desiccation cracks that have been filled with small fortification agates. Numerous examples exist of partial agate fortification occurring in conjunction with templated silicified wood.

Concentrically banded red wood limb cast with quartz core. Morrison Formation, Late Jurassic. Moab, Utah, USA. Bittle specimen, 4 cm.

Thin section of chalcedony demonstrating concentric banding of fibrous chalcedony inward with a quartz center as shown in Diagram D of the agate genesis figure. Cross-polarized light. Del Norte, Colorado, USA. Kile specimen and photograph. Width of view 9 mm.

Thin section of chalcedony that begins growth at the boundary of the cavity. Well-developed spherulitic growth shows impingement of individual growth centers to form sector boundaries as demonstrated in Diagrams B and C of the agate genesis figure. A second band of spherulitic growth begins and shows similar impingement and so on. Cross-polarized light. Del Norte, Colorado, USA. Kile specimen and photograph. Width of view 4.8 mm.

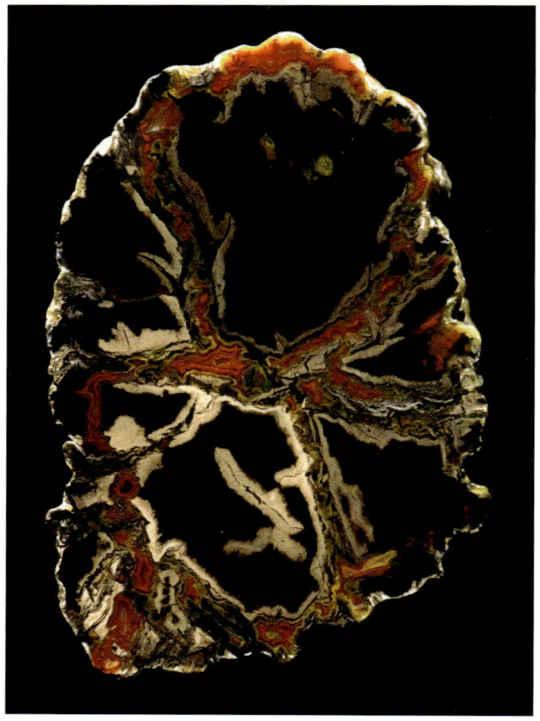

An example of fossil wood preserved by chalcedony through the templating process with areas of agate. Agate has fortified radial cracks in the original limb. Morrison Formation, Late Jurassic. Ticaboo, Utah, USA. Hatch specimen, 6.5 cm.

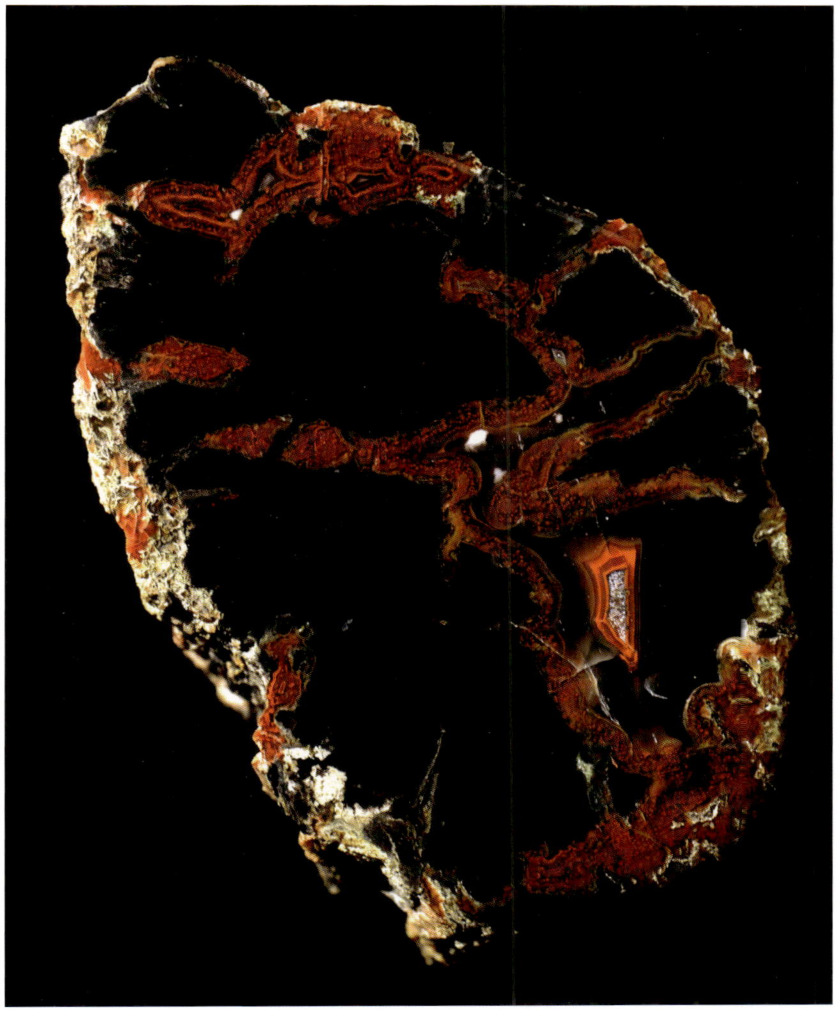

Red agate fortification highlights black silicified wood. Brushy Basin Member, Morrison Formation, Late Jurassic. Yellow Cat, Utah, USA. Dayvault specimen. Width of view 12 cm.

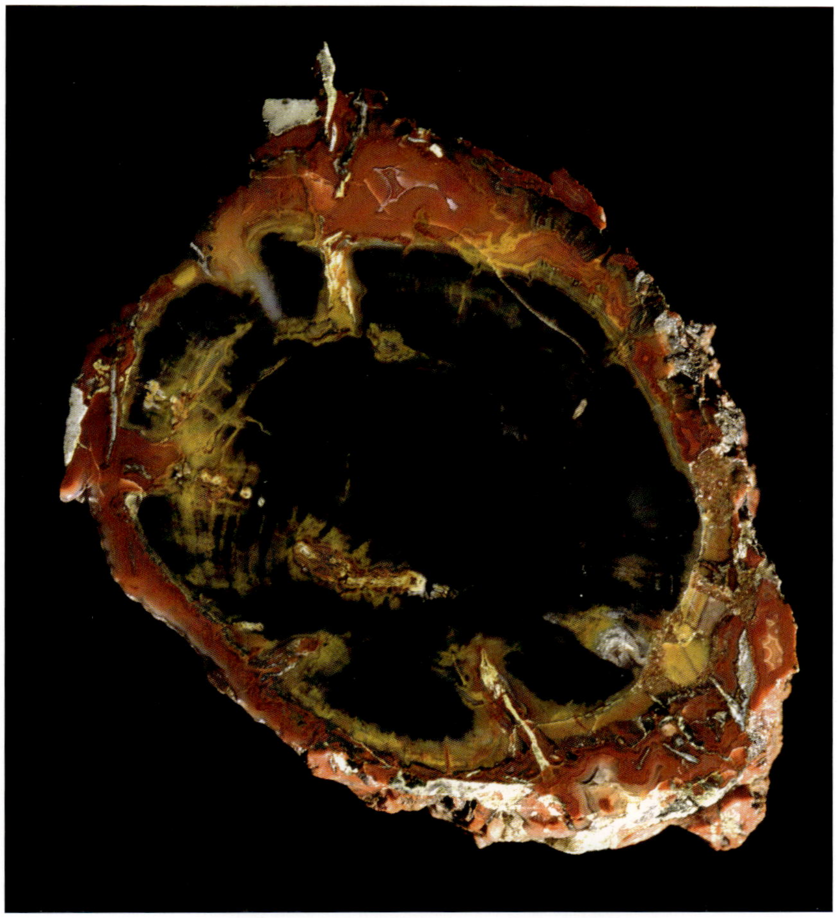

Fossil wood that has been preserved by chalcedony through the templating process sometimes contains areas of agate. Agate formed around the perimeter of this limb, probably as a result of the original limb drying out and leaving an initial void inside the cast. Morrison Formation, Late Jurassic. Ticaboo, Utah, USA. Dayvault specimen, 9 cm.

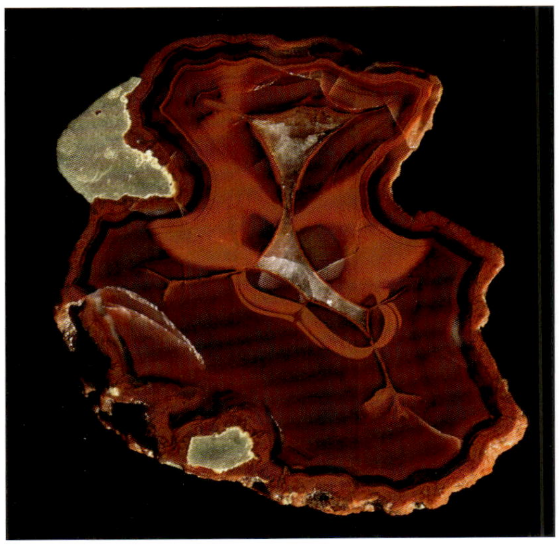

Concentrically banded red wood limb cast with quartz core. Morrison Formation, Late Jurassic. Moab, Utah, USA. Dayvault specimen, 4 cm.

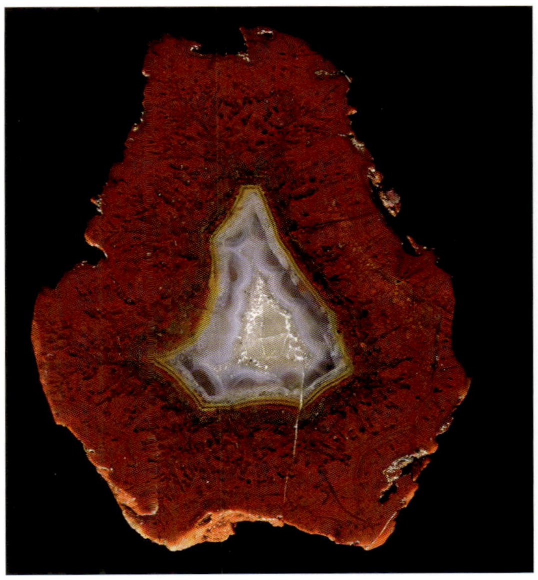

Concentrically banded red wood limb cast with quartz core. Lacy spherulitic textures occur in the banding. Morrison Formation, Late Jurassic. Moab, Utah, USA. Branson specimen, 7 cm.

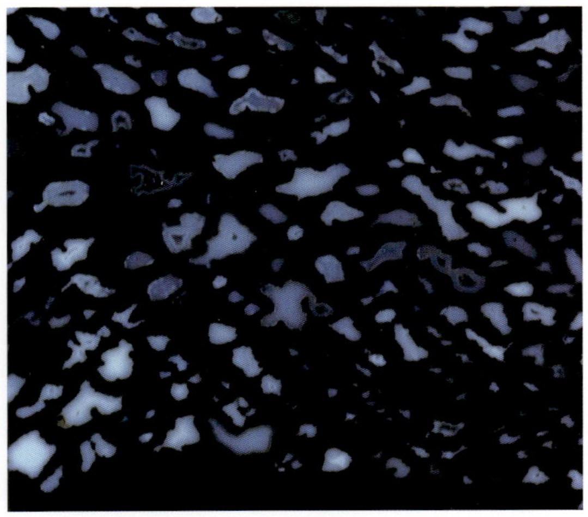

Petrified wood showing minute desiccation cracks that have received opportunistic infilled agate. Silica solutions follow openings in the wood and readily permineralize these areas. The cracks follow the coarse textures of the wood grain. Little Butte Volcanic Series, Oligocene/Miocene. Sweet Home, Oregon, USA. Daniels specimen. Width of view 2 cm.

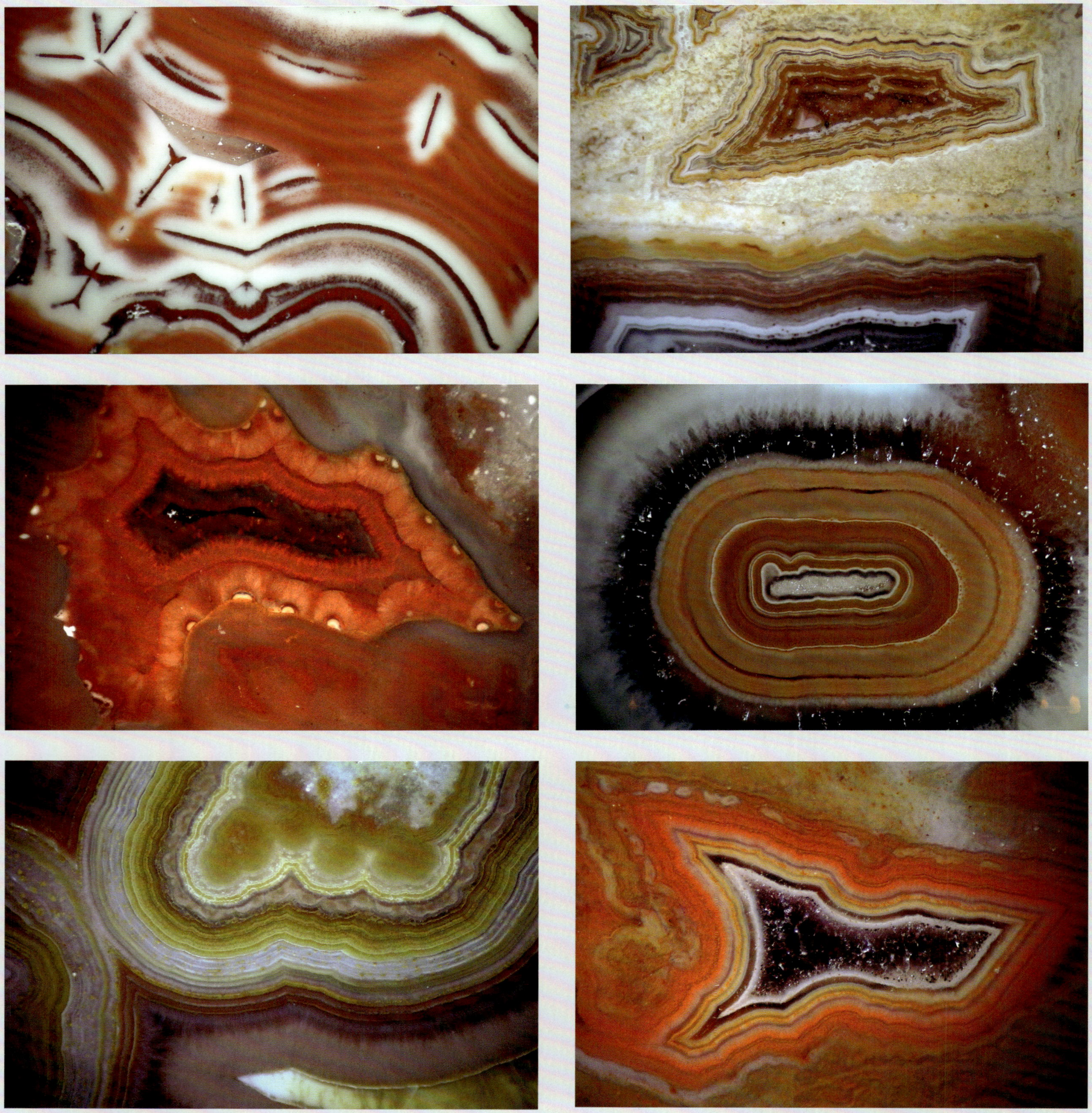

Micro image examples of colorful agate fortifications in fossil woods from Mesozoic formations in Chubut and Rio Negro provinces, Argentina. Daniels specimens.

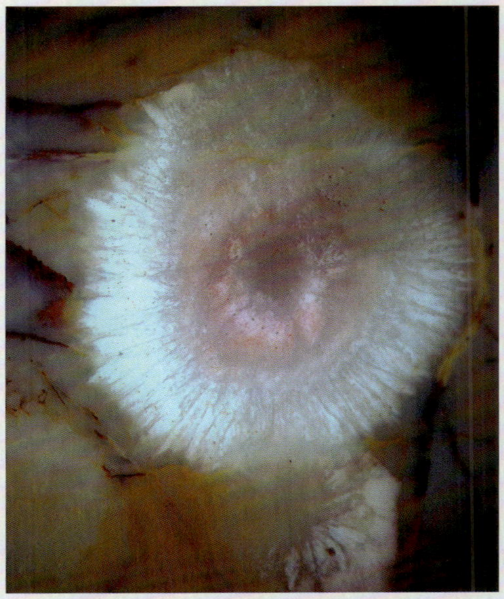
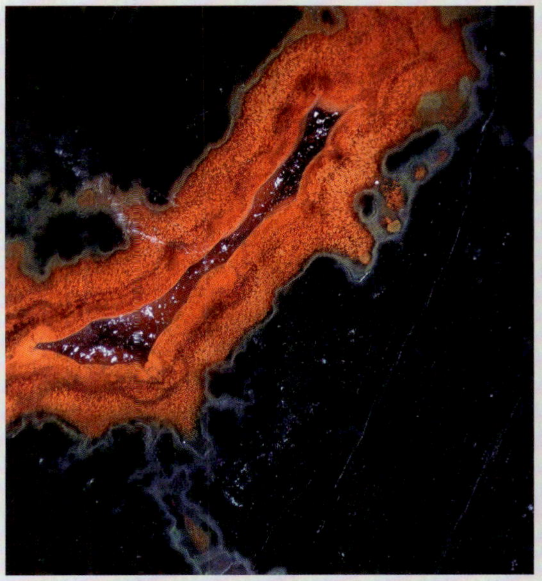
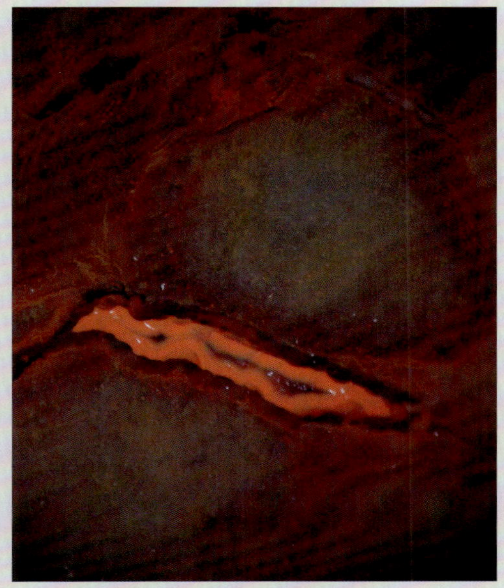

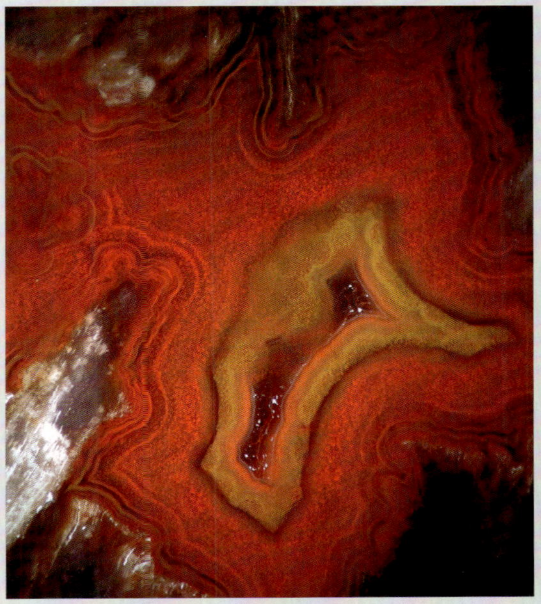

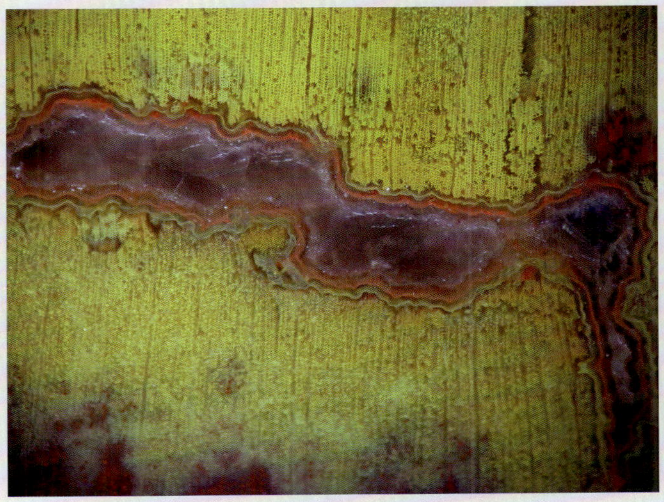
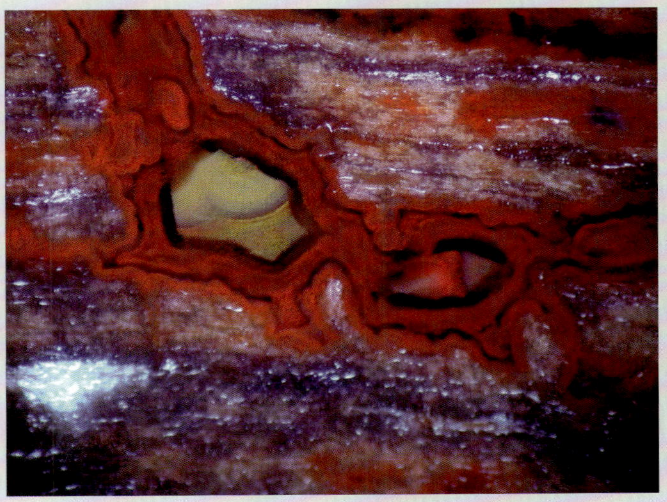

Micro image examples of colorful agate fortifications in fossil woods from the Jurassic Brushy Basin Member of the Morrison Formation in the Henry Mountains area, Utah, USA. Daniels specimens.

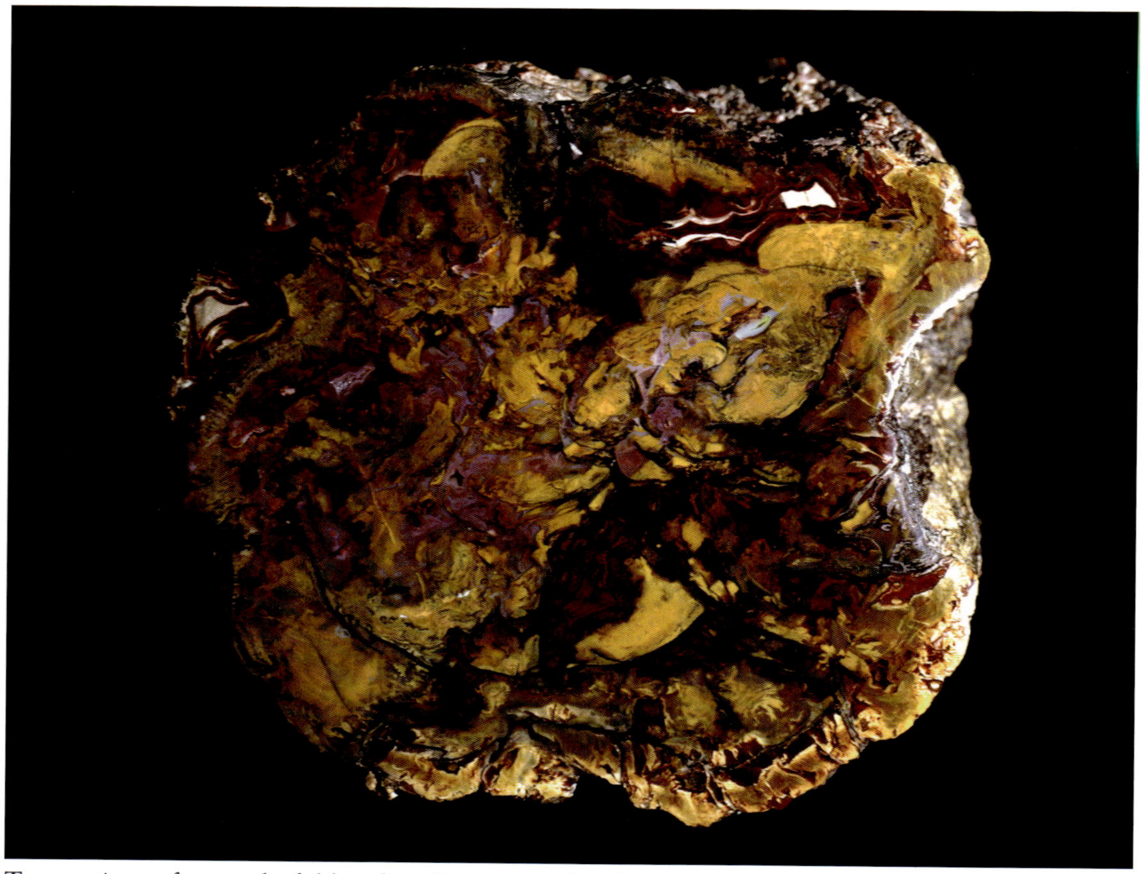

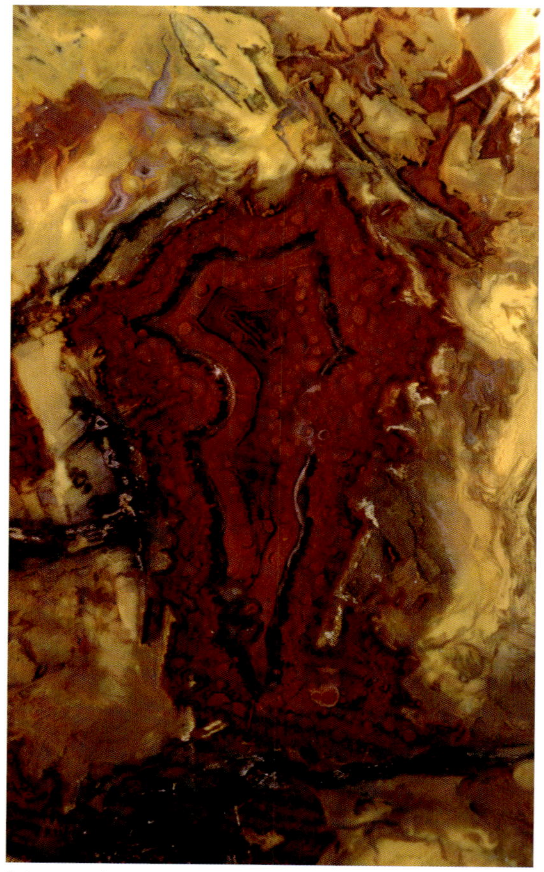

Two sections of one colorful log that shows agate fortification occurring in conjunction with templated silicified wood. Morrison Formation, Late Jurassic. Eastern Utah, USA. Daniels specimens, 13 and 14 cm.

Macro photographs of log to the left showing closer views of templated wood and agate structures. Morrison Formation, Late Jurassic. Eastern Utah, USA. Daniels specimen.

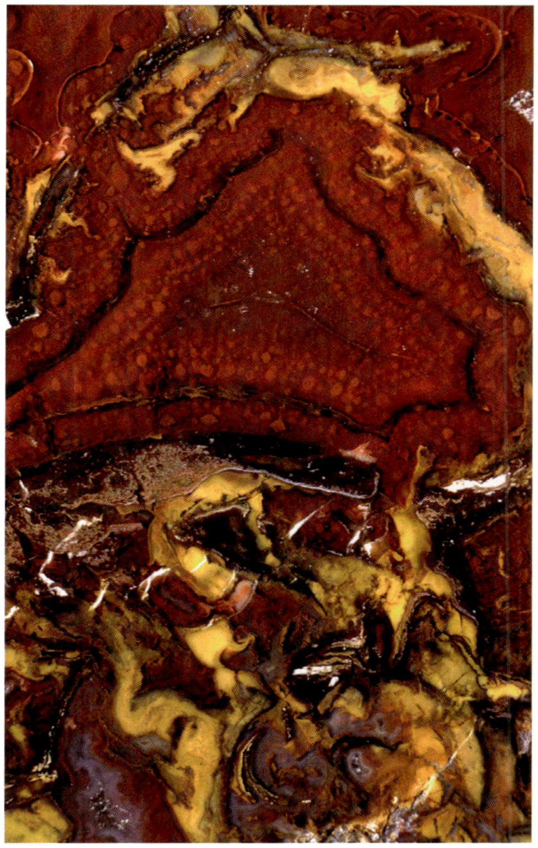

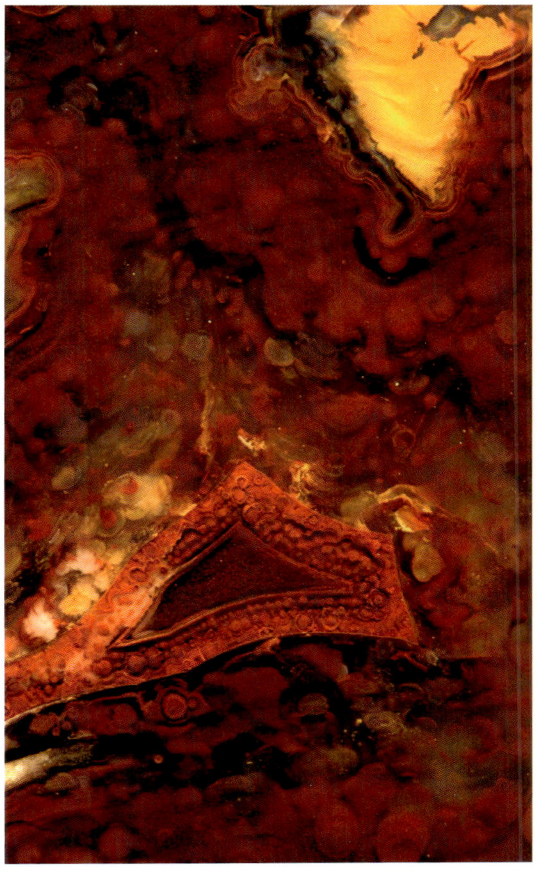

Macro photographs of log to the left showing closer views of templated wood and agate structures. Morrison Formation, Late Jurassic. Eastern Utah, USA. Daniels specimen.

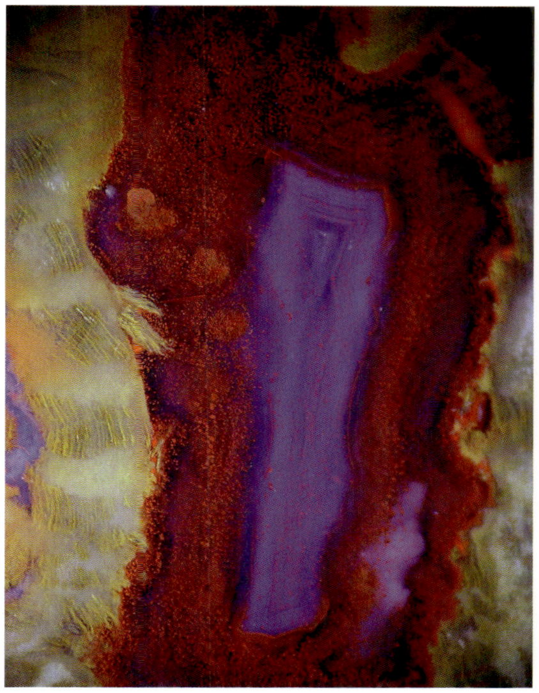

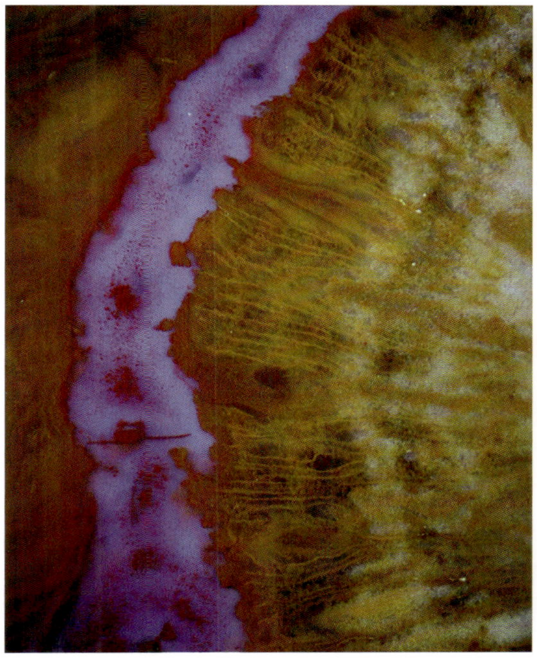

Micro images of log to left showing minute detail of templated wood and agate structures. Morrison Formation, Late Jurassic. Eastern Utah, USA. Daniels specimen.

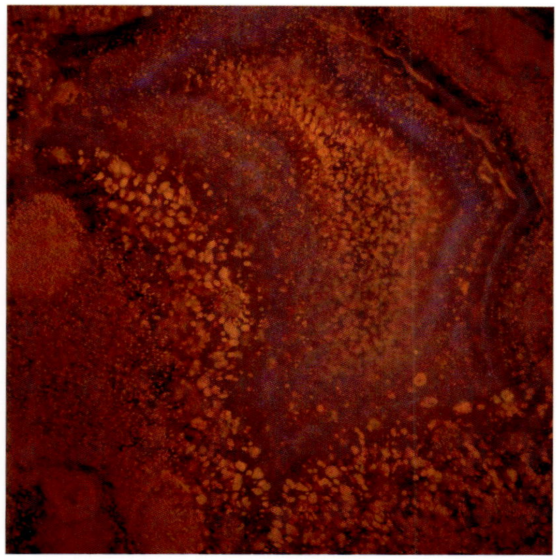

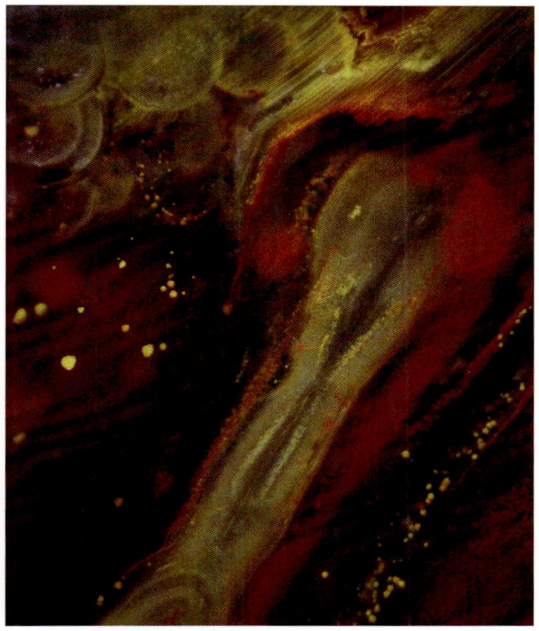

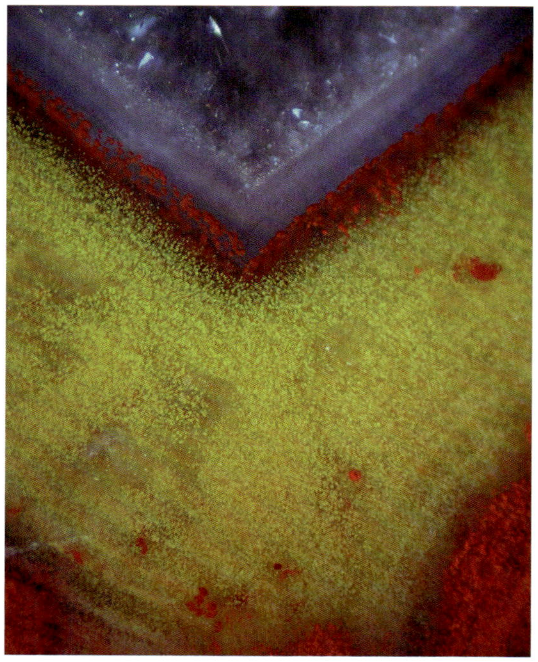

Multi-colors and textures of chalcedony producing agate banding. Blebs of yellow silica floating in chalcedony are common in agate limb casts from Utah. Quartz fills the central cavity on the top. Brushy Basin Member, Morrison Formation, Late Jurassic. Yellow Cat, Utah, USA. Daniels specimen. Width of view about 7 mm.

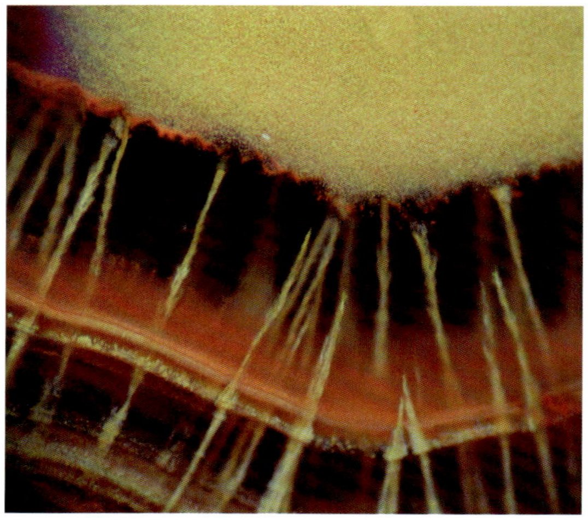

Close view of agate banding in a limb cast. A common texture observed in red wood agates from Utah is the development of radiating structures that are sometimes contained in individual bands and sometime cross two or more bands. Brushy Basin Member, Morrison Formation, Late Jurassic. Moab, Utah, USA. Dayvault specimen. Width of view about 3 mm.

When silica becomes undersaturated in the cavity solution, other minerals such as barite and calcite that can dissolve in a silica gel are also able to crystallize. Calcite and barite may crystallize long after the formation of chalcedony and quartz cease, and the force of crystallization by the later minerals may cause fractures and displacement of the agate.

Petrified wood showing minute desiccation cracks that have received opportunistic infilled agate. Silica solutions follow openings in the wood and readily permineralize these areas. The cracks follow the coarse textures of the wood grain. Little Butte Volcanic Series, Oligocene/Miocene. Sweet Home, Oregon, USA. Daniels specimen. Width of view 2 cm.

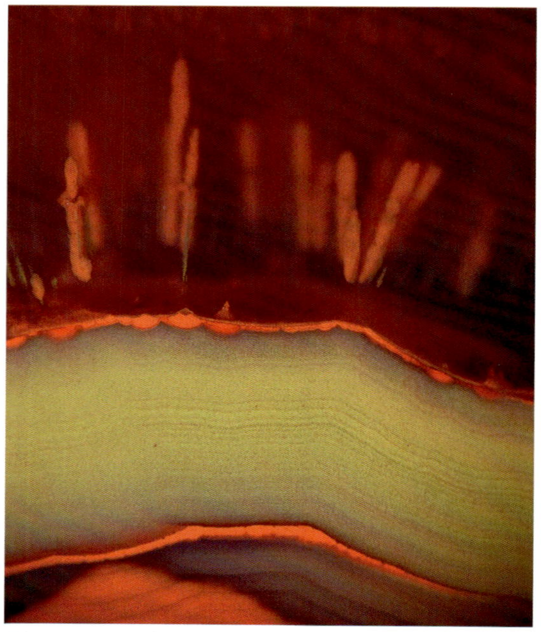

Close view of agate banding in a red wood limb cast from eastern Utah. Chalcedony changes abruptly from thin orange bands to translucent bands containing yellow silica blebs to another orange band and radiating spikes of silica in translucent chalcedony. Brushy Basin Member, Morrison Formation, Late Jurassic. Moab, Utah, USA. Dayvault specimen. Width of view about 10 mm.

Red wood limb cast displaying concentric banding with a vestige of petrified wood probably situated in the bottom of the cavity. Incipient failure of the agate occurs with sections floating in central barite area. Morrison Formation, Late Jurassic. Moab, Utah, USA. Dayvault specimen, 10 cm.

Same specimen of red wood limb cast, but 8 cm farther along limb section. Agate shows much greater failure and fragments float in central barite area. Barite is presumed to have formed much later than the agate and may have displaced sections of broken agate. 11 cm.

During chalcedony growth, the agate is in a semisolid state and can be plastically deformed. Examples of this deformation are observed in nature. Because chalcedony occupies more space than an equal volume of silica gel, outward pressure increases in the remaining core solution as crystallization proceeds. If internal pressure becomes greater than the strength of the enclosing and still-plastic agate, the lining of the agate can rupture and form an escape tube. This presumes that the enclosing rock is not completely hardened or is sufficiently porous to allow ejection of the silica gel.

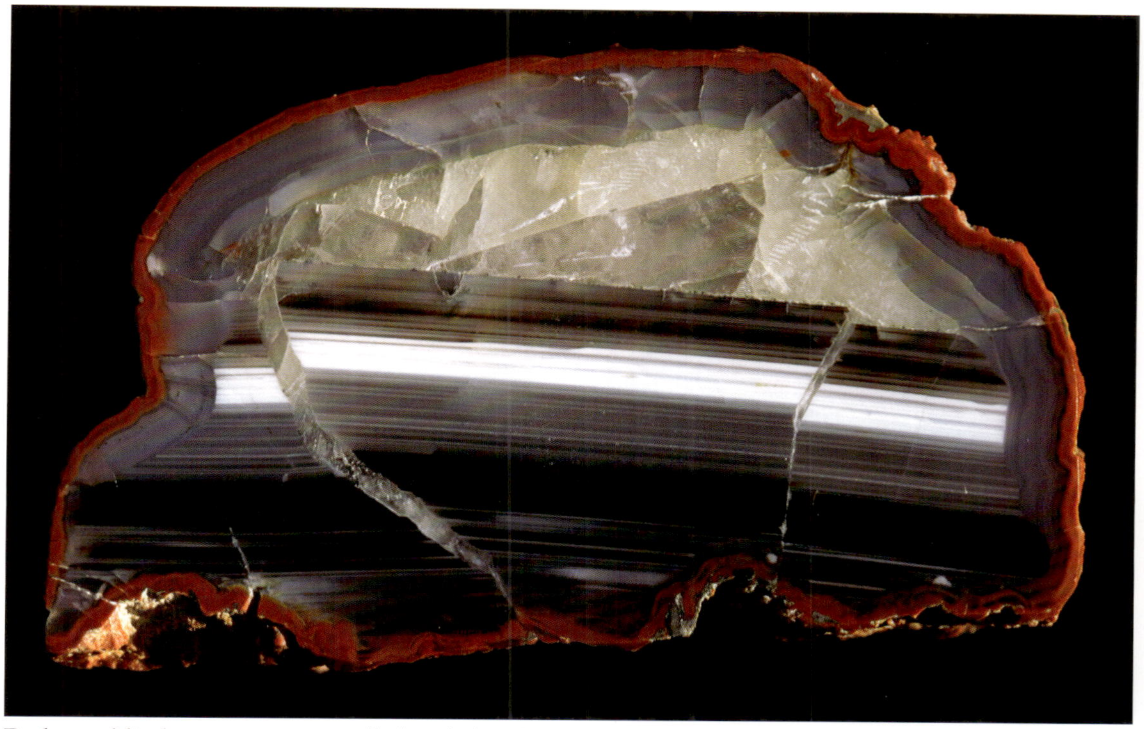

Red wood limb cast concentrically banded, subsequently horizontally banded, and finally barite filled the upper portion. Its crystallization may have contributed to the internal failure. Morrison Formation, Late Jurassic. Moab, Utah, USA. J. Walker specimen, 11 cm.

Red wood limb cast concentrically banded, subsequently horizontally banded, and finally concentrically banded with a quartz core. This demonstrates the change of agate formation from one type to another within the same specimen. Morrison Formation, Late Jurassic. Moab, Utah, USA. J. Walker specimen, 5 cm.

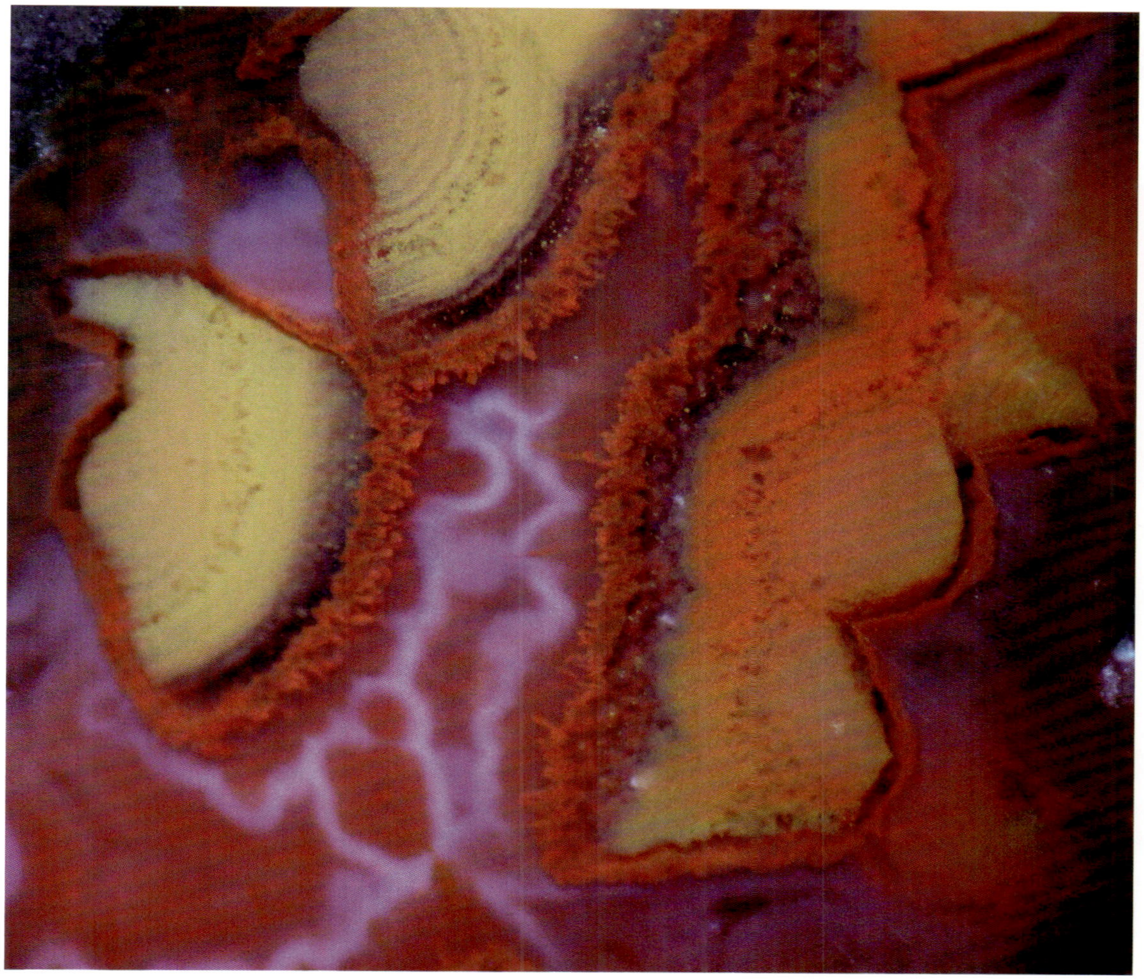

Section of a red wood limb cast showing sections of agate that were fractured and cemented by additional chalcedony. Interesting lacy textures surround the brecciated fragments. Brushy Basin Member, Morrison Formation, Late Jurassic. Moab, Utah, USA. Daniels specimen. Width of view about 9 mm.

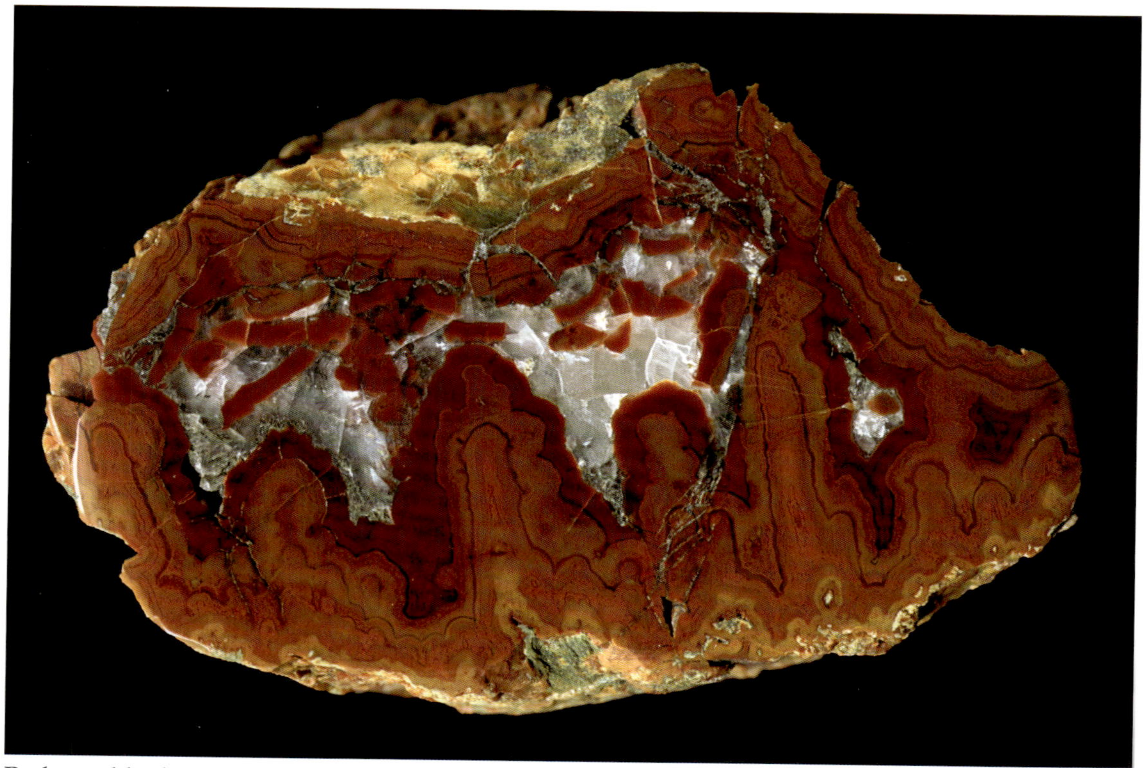

Red wood limb cast demonstrating initial concentric banding followed by tubular internal structures. Brushy Basin Member, Morrison Formation, Late Jurassic, Yellow Cat area, Utah, USA. Daniels specimen. Width 8 cm.

A variety of other minerals reside in the internal open parts of limb casts. Quartz commonly occupies the centers of limb casts. It is most commonly colorless to milky, but examples of amethystine or smoky quartz lining cavities are not uncommon. Additional minerals crystallized in open cavities after the agate formation is completed, and subsequent solutions become saturated with other compounds. Calcite, barite, and gypsum commonly either form crystals that line the center of the cavity or completely fill the remainder of the cavity. Hydrated iron or manganese oxides produce goethite or groutite needles that sometimes perch on or include other crystals. Pyrite or other sulfides are infrequently found in open centers of limb casts.

Microscopic crystals of goethite include quartz crystals in the center of a fortification agate. Conifer. Mesozoic. Malagasy Republic. Daniels specimen. Goethite crystals about 1 mm maximum.

Microscopic crystals of pyrite include quartz crystals in the center of a fortification agate. Conifer. Mesozoic. Malagasy Republic. Daniels specimen. Pyrite crystal about 0.8 mm maximum.

A former carbonate(?) mineral grew in an opening in petrified wood and later had silica growing over it. The carbonate eventually dissolved away leaving only casts as white chalcedony. Conifer. Chinle Formation, Triassic. Henry Mountains area, Utah, USA. Mower specimen. Width of view about 6 mm.

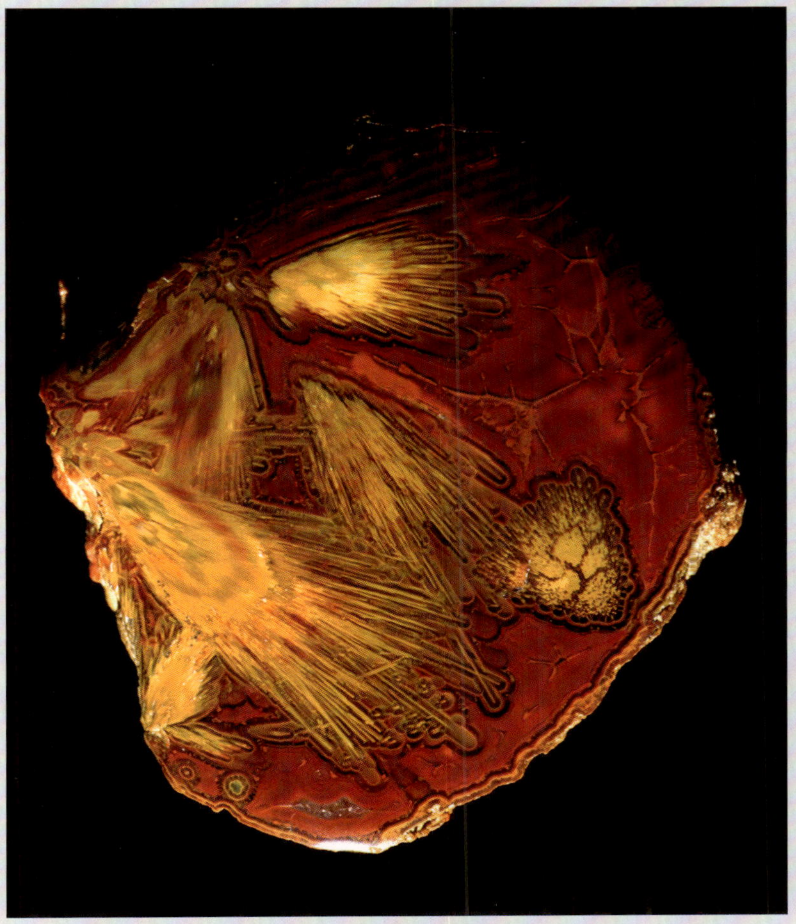

Unusual red wood limb cast showing sprays of an acicular mineral that has been replaced or pseudomorphed by silica. Morrison Formation, Late Jurassic. Moab, Utah, USA. J. Walker specimen, 6 cm.

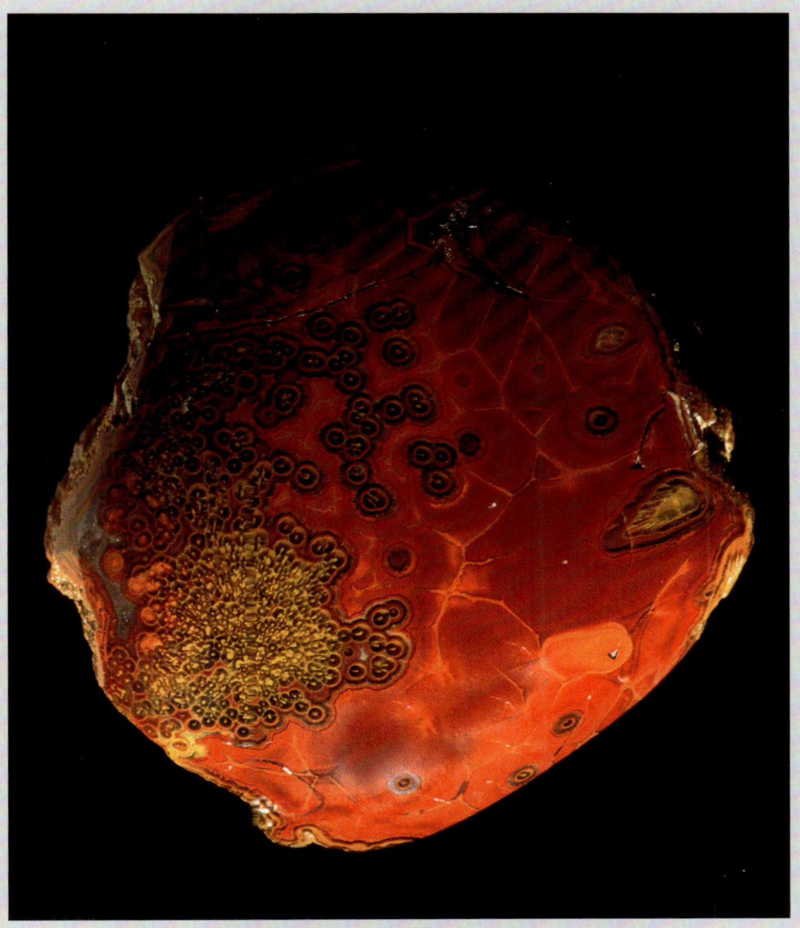

Back side of same specimen shown at left. The sprays of acicular mineral converge and are represented by small tubes. The polygonal pattern in the reddish chalcedony is suggestive of textures found in devitrified rhyolite glass.

TIME REQUIRED TO FORM PETRIFIED WOOD

The time required to produce silicified wood in nature depends on how extensive one defines silicification and what special conditions are considered. The molecular film of polymerized silicic acid that forms hydrogen bonds with polysaccharides of woody tissue probably forms rapidly. As previously noted, it can happen in a matter of years to decades. Wood that falls into a hot spring can be converted into silicified wood fairly rapidly, in a matter of a decade or less, as reported by Akahane and others. However, the percentage of petrified wood that forms in hot spring environments is miniscule compared to the total volume of petrified wood in the world. Wood petrified in hot spring environments is not an accurate representation of the average time required for conversion of wood to stone in a sedimentary basin.

The eventual conversion of polymerized silicic acid to amorphous opal and chalcedony—with associated decreases in permeability and water content and increases in refractive index, hardness, and specific gravity—require a much longer period. This process depends on continuous silica and the pressure and temperature of burial; all these factors are affected by diminishing infiltration rates because of the decreased porosity of the woody material. Probably the analogous path followed by the conversion of marine biogenic opal to chert after burial is more appropriate, which requires several millions of years to complete. The youngest age for sediments containing opalized wood that is sufficiently mineralized to allow cutting and polishing is several million years old. Sometimes wood does not become silicified and instead remains fairly unaltered for millions of years. A famous example is the occurrence of Eocene logs on Axel Heilberg Island in the Canadian Arctic that have remained relatively unaltered. In general, conversion of a once living tree into a lapidary material requires vast amounts of time. Siever estimates that the complete transformation of a limb from initial opal to chalcedony may require 10 to 100 million years.

III

CHAPTER THREE

Fossil Woods of the Northwestern United States

Stinking Water, Cherry Creek, Sweet Home, Hubbard Basin, Yakima Valley, Grassy Mountain, McDermitt, Deschutes, Saddle Mountain—these place names and others evoke visions of lovely and unique fossil wood specimens to collectors worldwide. Fossil woods in the northwestern United States have been studied and collected since the 19th century. In some locales, the plant fossils have been studied, studied more, and studied again, with each new look bringing more expertise to, and shedding more light on, some of the best preserved and extensive accumulations of fossil plant material in existence. As with most great fossil wood locations, extensive volcanism resulting from shifting land masses drove the preservation process. During the Tertiary Period, conditions for perfect fossil preservation ruled the Northwest, especially from the Eocene to the Miocene epochs.

The Clarno Formation, with outcrops common in present-day north-central Oregon, consists of andesitic lava flows and coarse mudflow breccias that yield extensive fossil remains, including a unique assortment of wood, seeds, leaves, and nuts. Consequently, Clarno Formation fossils have been heavily collected and studied. The formation is 1,800 meters (6,000 feet) thick in places and dates from 54 to 39 million years before present (mybp). Continental drift at work some 40 to 42 million years ago shifted volcanism westward to the Cascade Arc where a new mountain range was forming. Tremendous volumes of rhyolitic ash spewed and settled over the faulting, folding, and eroding Clarno deposits, building into what we now call the John Day Formation. The John Day Formation overlies the Clarno except in areas of ancient Clarno-age volcanic highlands where the Clarno protrudes through the John Day into portions of the Middle Miocene Columbia Plateau basalt system. The John Day Formation dates from 39 to 18 mybp, spanning Late Eocene, Oligocene, and Early Miocene epochs. A paleoclimate change at the Eocene–Oligocene boundary shifted the area from a subtropical to a temperate environment, resulting in major flora shifts.

The Crook County courthouse in Prineville, Oregon, USA. Prineville has long been an important center for fossil wood collecting in the state.

One well-known Clarno deposit is the Clarno Nut Beds, dated approximately 44 mybp. Results of an extensive effort to analyze and identify the myriad fossil wood remains found in the Nut Beds culminated in the 2002 paper *Woods of the Middle Eocene Nut Beds Flora, Clarno Formation, Oregon, USA* by E. A. Wheeler and S. R. Manchester. The authors identified 66 genera and 76 species and called the Clarno Nut Beds "the most diverse fossil wood assemblage ever described from a single locality." They determined from the woods present that the climate was warm temperate to subtropical. Present were *Palmoxylon*, numerous angiosperms, and four gymnosperms (*Pinus*, *Sequoia*-like, ginkgo, and an unknown wood with similarities to Taxaceae). It is interesting to note that all the Clarno Formation woods with living members currently survive in eastern Asia and about half now live in eastern North America.

Leaf impression seen in Clarno Formation rocks at the Clarno Unit, John Day Fossil Beds National Monument, Oregon, USA.

Eroding ancient volcanic rocks on the Oregon Coast form sea stacks such as 72 meter (235 feet) tall Haystack Rock seen here in the background offshore from Cannon Beach, Oregon.

Leaf impression seen in Clarno Formation rocks at the Clarno Unit, John Day Fossil Beds National Monument, Oregon, USA.

Leaf impression seen in Clarno Formation rocks at the Clarno Unit, John Day Fossil Beds National Monument, Oregon, USA.

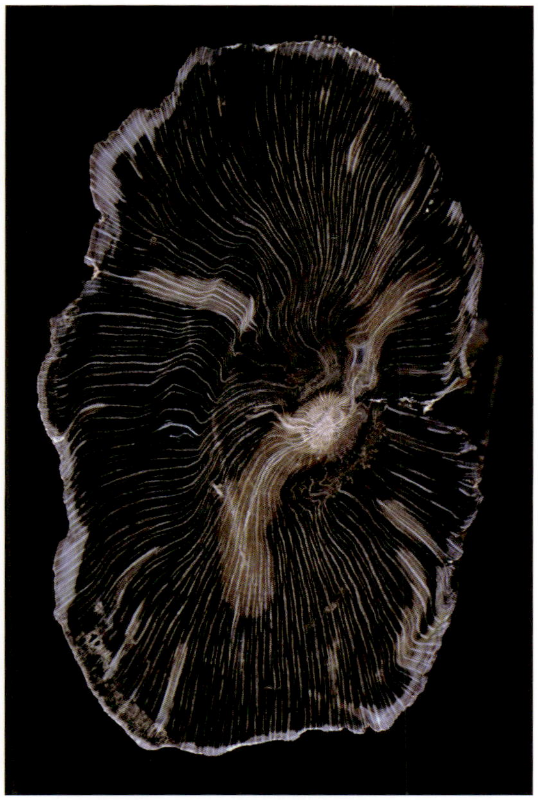

Sycamore-like fossil wood. Clarno Formation, Eocene. Post area, Oregon, USA. Daniels specimen, 7.5 cm.

Leaf impression seen in Clarno Formation rocks at the Clarno Unit, John Day Fossil Beds National Monument, Oregon, USA.

Fossil conifer cone. John Day Formation, Oligocene. Fossil, Oregon. Dayvault specimen, 4.5 cm.

Sycamore-like fossil wood. Clarno Formation, Eocene. Post area, Oregon, USA. Daniels specimen, 4.5 cm.

Wizard Island in Crater Lake, Oregon, USA. Witness to the continuing volcanic activity of the northwestern United States, Crater Lake is a 6-mile-wide caldera formed by the eruption and collapse of Mt. Mazama some 7,000 years ago. The cone that is now Wizard Island formed subsequent to the demise of the mountain. Crater Lake National Park, Oregon, USA.

Lower John Day Formation hills composed of colorful claystone layers. The colors come from various combinations of aluminum, calcium, hydrogen, iron, magnesium, manganese, oxygen, phosphorous, potassium, silicon, sodium, and titanium. Rust-colored layers are rich in iron oxide; yellows are a blend of iron and magnesium oxides; black is manganese oxide. Painted Hills Unit, John Day Fossil Beds National Monument, Oregon, USA.

The colorful clays at the center are an outcrop of the John Day Formation. The white outcrops contain Oligocene Bridge Creek flora. Most of the surrounding terrain is Clarno Formation. East of Post, Crook County, Oregon, USA.

Upper John Day Formation exposure on Logan Butte, Crook County, Oregon, USA, between Post and Hampton Butte.

John Day Formation exposure. Sheep Rock Unit, John Day Fossil Beds National Monument, Oregon, USA.

Photograph from *The Ancient Forests of Oregon* by Ralph W. Chaney, showing early fossil collectors near Sheep Rock, Oregon.

Two famous fossil wood locations within the Clarno Formation are Post, Oregon, known especially for sycamore-like stems, and Hampton Butte, known for beautiful fossil wood in shades of green and carnelian. A survey of Hampton Butte fossil woods identified woods similar to modern-day bald cypress, cypress, laurel, magnolia, oak, pine, *Sequoia*, and sycamore. A few Clarno locations yield occasional osmundaphytes, including *Osmunda dowkerii* and *Osmunda oregonesis*.

One fossil wood collecting area in Hampton Butte is currently home to an old-growth juniper forest. This Hampton Butte juniper measures 4.67 meters (15 feet, 4 inches) in circumference at the base.

Unidentified brecciated fossil wood. Clarno Formation, Eocene. Hampton Butte, Oregon. Daniels specimen, 10.5 cm.

Enough wood structure remains in this brecciated specimen to identify it as a conifer. Clarno Formation, Eocene. Hampton Butte, Oregon. Daniels specimen, 22 cm.

Pine-like double-hearted fossil wood. Grant County, Oregon, USA. Daniels specimen, 13 cm.

Here the Miocene Rattlesnake Formation caps Oligocene John Day Formation sediments, source of pink, green, and blue/gray limb casts, along the South Fork of the Crooked River, Crook County, Oregon, USA.

The climate was temperate in the time of the thriving woods that became fossils in the John Day Formation. One location in the Lower John Day dated to 38.8 mybp yielded leaves, fruits, and seeds of numerous deciduous plants and some ferns but no *Palmoxylon* or conifers. In 1923, Dr. Ralph Chaney, Professor of Paleontology at the University of California, quarried 2.8 cubic meters (98 cubic feet) of shale at the Bridge Creek, Oregon, site, yielding 20,611 specimens of 25 different species of plants. Bridge Creek floras often show a dominance of *Metasequoia*, *Sequoia*, and *Taxodium* with lesser number of deciduous trees, such as alder, elm, oak, and sycamore. The location famous for pink and dendritic limb casts along the South Fork of the Crooked River in Crook County, Oregon, lies within the John Day Formation. This fossil wood material is predominately cast, yet frequently includes segments of poorly preserved permineralized wood.

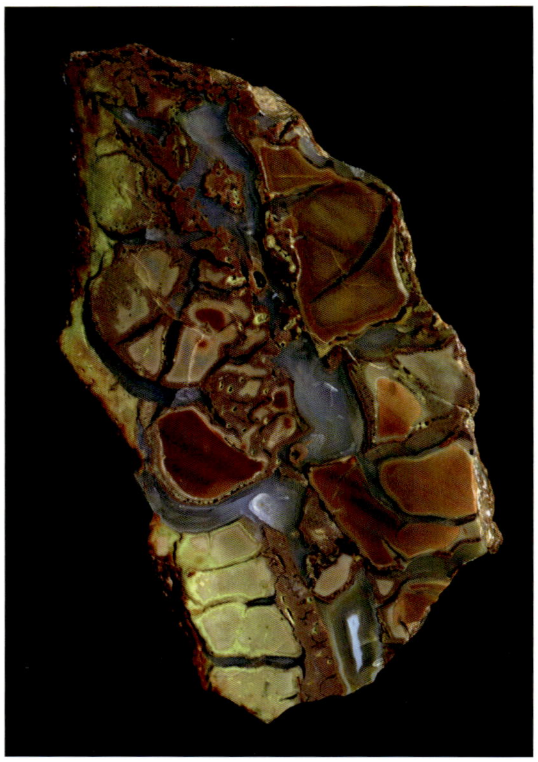

Unidentified brecciated fossil wood. Clarno Formation, Eocene. Hampton Butte, Oregon. Daniels specimen, 14 cm.

Blue/gray chalcedony limb cast with a hollow center and a large amount of green-to-tan permineralized wood. John Day Formation, Oligocene. South Fork of the Crooked River, Crook County, Oregon, USA. Haggerty specimen, 13.5 cm.

For many years this amazing cone cast has been the main attraction at Quant's Rock Shop in Prineville, Oregon. John Day Formation, Oligocene. South Fork of the Crooked River, Crook County, Oregon, USA. Quant specimen, 14.5 cm long.

Set of limb casts from the South Fork of the Crooked River, Crook County, Oregon, USA. John Day Formation, Oligocene. Daniels specimens, largest 12 cm diameter.

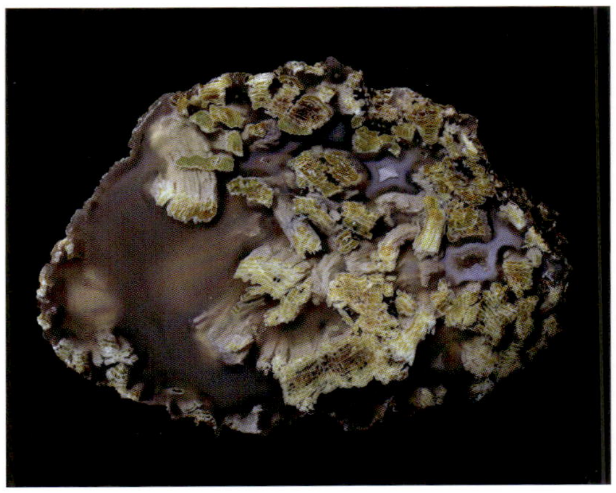

Limb cast with a substantial amount of residual permineralized wood. John Day Formation, Oligocene. South Fork of the Crooked River, Crook County, Oregon, USA. Daniels specimen, 7.5 cm.

Beautifully preserved acorn cast with green exterior. John Day Formation, Oligocene. South Fork of the Crooked River, Crook County, Oregon, USA. Elkins specimen, 3.5 cm.

Limb cast with tightly banded agate in alternating bands of translucent and opaque chalcedony, remarkably like the structure seen in specimens of Laguna agate from Mexico; demonstrates shadow effect. John Day Formation, Oligocene. South Fork of the Crooked River, Crook County, Oregon, USA. Daniels specimen, 7 cm.

A small percentage of South Fork limb casts demonstrate an "iris effect," wherein light refracted through thinly sliced sections produces varied plays of color. John Day Formation, Oligocene. South Fork of the Crooked River, Crook County, Oregon, USA. Daniels specimens, 6 and 6.5 cm.

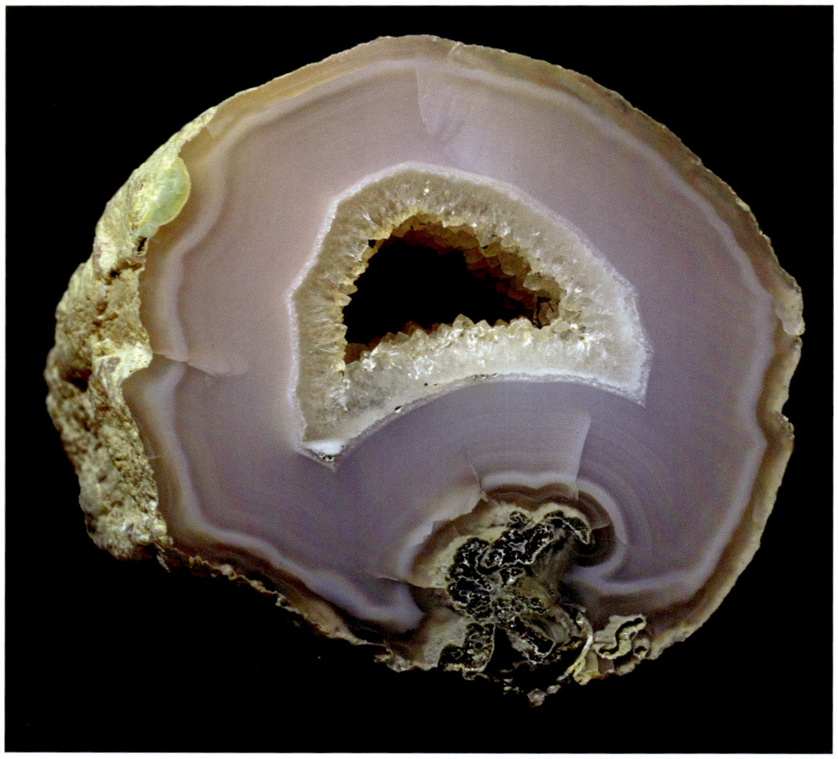

Pink limb cast with quartz-lined center and small section to lower right of poorly preserved petrified wood fragments. John Day Formation, Oligocene. South Fork of the Crooked River, Crook County, Oregon, USA. Daniels specimen, 11 cm.

Blue/gray banded agate limb cast with tiny black inclusions. John Day Formation, Oligocene. South Fork of the Crooked River, Crook County, Oregon, USA. Haggerty specimen, 13 cm.

South Fork limb casts faithfully preserve external structure such as the burl in this specimen. John Day Formation, Oligocene. South Fork of the Crooked River, Crook County, Oregon, USA. Daniels specimen, 17.5 cm wide.

Limb cast. John Day Formation, Oligocene. South Fork of the Crooked River, Crook County, Oregon, USA. Daniels specimen, 7.5 cm.

Limb cast. John Day Formation, Oligocene. South Fork of the Crooked River, Crook County, Oregon, USA. Daniels specimen, 9 cm.

Limb cast. John Day Formation, Oligocene. South Fork of the Crooked River, Crook County, Oregon, USA. Daniels specimen, 7.5 cm.

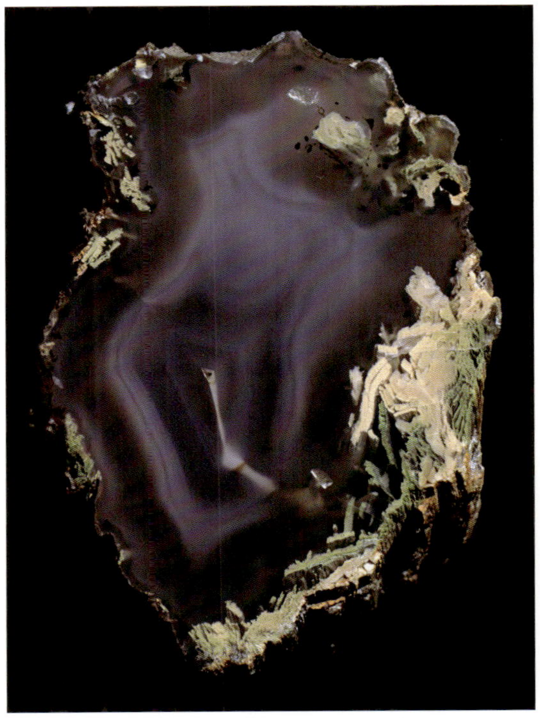

Limb cast. John Day Formation, Oligocene. South Fork of the Crooked River, Crook County, Oregon, USA. Daniels specimen, 8.5 cm.

Limb cast with many tiny tube-like green filaments. John Day Formation, Oligocene. South Fork of the Crooked River, Crook County, Oregon, USA. Daniels specimen, 5.5 cm.

Limb cast. John Day Formation, Oligocene. South Fork of the Crooked River, Crook County, Oregon, USA. Daniels specimen, 10 cm.

Limb cast with encapsulated green permineralized wood. John Day Formation, Oligocene. South Fork of the Crooked River, Crook County, Oregon, USA. Daniels specimen, 9 cm.

The Middle to Late Miocene Succor Creek Formation, composed of lakebed sediments, spans the Oregon–Idaho border. It has been extensively studied, and specimens have been collected at the site since 1900. Identified fossil woods are similar to modern-day alder, ash, birch, chinquapin, cottonwood, elm, hackberry, laurel, madrone, maple, pine, redwood, sycamore, and wild cherry. Some of the most beautifully preserved fossil woods in the northwestern United States hail from Grassy Mountain, Hoodoo Basin, and the Owyhee Mountains in this vicinity.

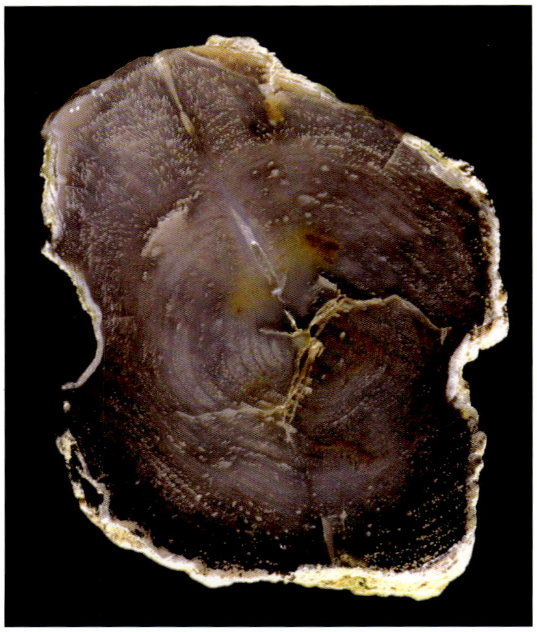

Fossil hardwood preserved as opal with little wood structure remaining. Tertiary. Brogan, Malheur County, Oregon, USA. Daniels specimen, 8 cm.

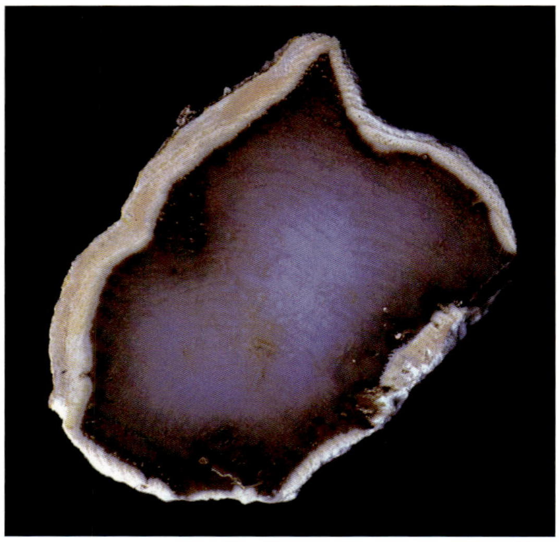

Fossil hardwood preserved as opal with little wood structure remaining. Tertiary. Brogan, Malheur County, Oregon, USA. Daniels specimen, 5 cm.

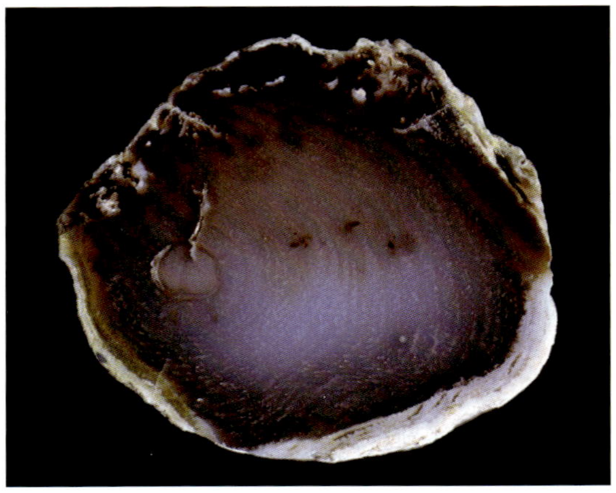

Fossil hardwood preserved as opal with little wood structure remaining. Tertiary. Brogan, Malheur County, Oregon, USA. Daniels specimen, 4 cm.

Ring porous fossil hardwood. Succor Creek Formation, Miocene. Grassy Mountain, Malheur County, Oregon, USA. Daniels specimen, 8.5 cm.

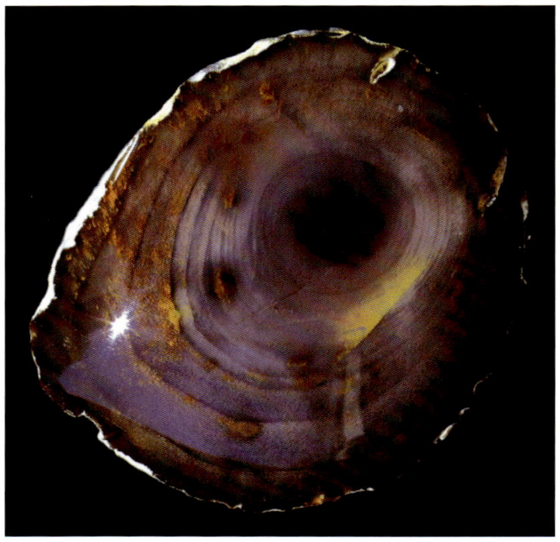

Fossil hardwood. Succor Creek Formation, Miocene. Grassy Mountain, Malheur County, Oregon, USA. Daniels specimen, 8.5 cm.

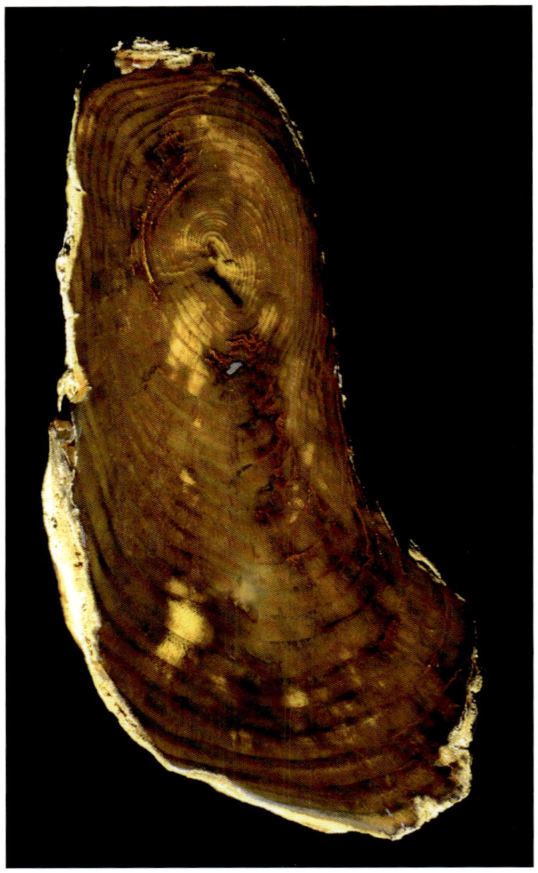

Elm-like fossil hardwood. Succor Creek Formation, Miocene. Hoodoo Basin, Malheur County, Oregon, USA. Daniels specimen, 10.5 cm.

Fossil conifer specimens from this locale are rarely more than colorful chunks of much larger trees, yet they are glassy, translucent, and colorful specimens. Tertiary. Owyhee Mountains, Malheur County, Oregon, USA. Daniels specimen, 7.5 cm.

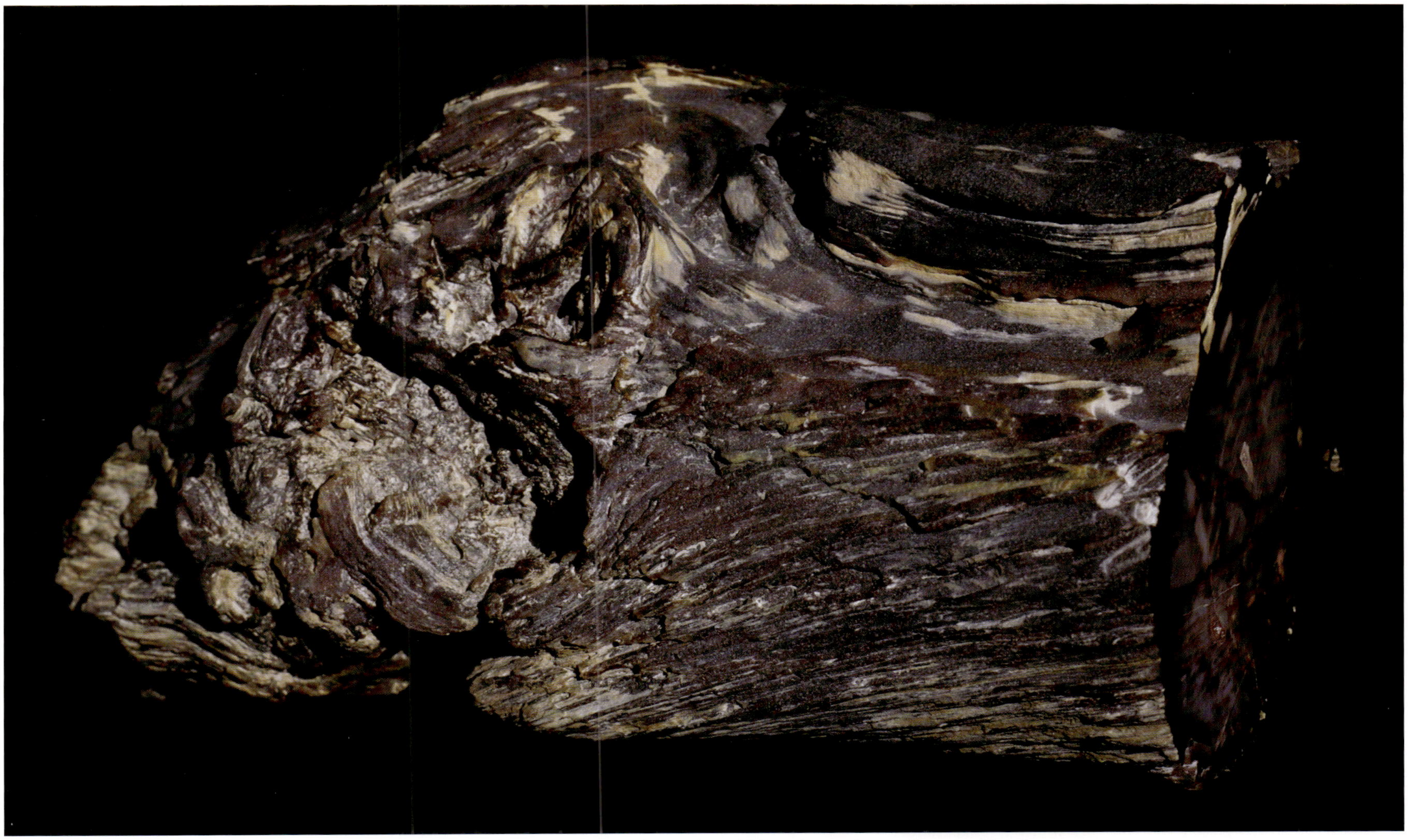

This large gnarly log has a smooth exterior, driftwood appearance, and is red and blue throughout. Succor Creek Formation, Miocene. Grassy Mountain, Malheur County, Oregon, USA. Daniels specimen, 32.5 cm long.

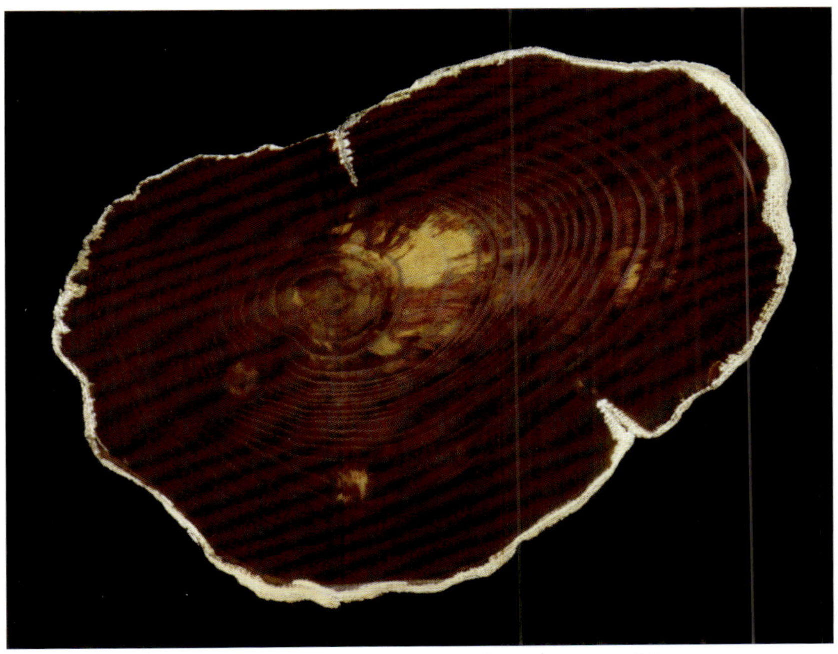

Ring porous fossil hardwood. Succor Creek Formation, Miocene. Succor Creek, Idaho/Oregon, USA. Daniels specimen, 9 cm.

Ring porous fossil hardwood. Succor Creek Formation, Miocene. Hoodoo Basin, Malheur County, Oregon, USA. Daniels specimen, 10.5 cm.

Glassy with typical Grassy Mountain colors of gold, red, and blue and discernable ring porous hardwood structure. Succor Creek Formation, Miocene. Grassy Mountain, Malheur County, Oregon, USA. Daniels specimen, 11 cm.

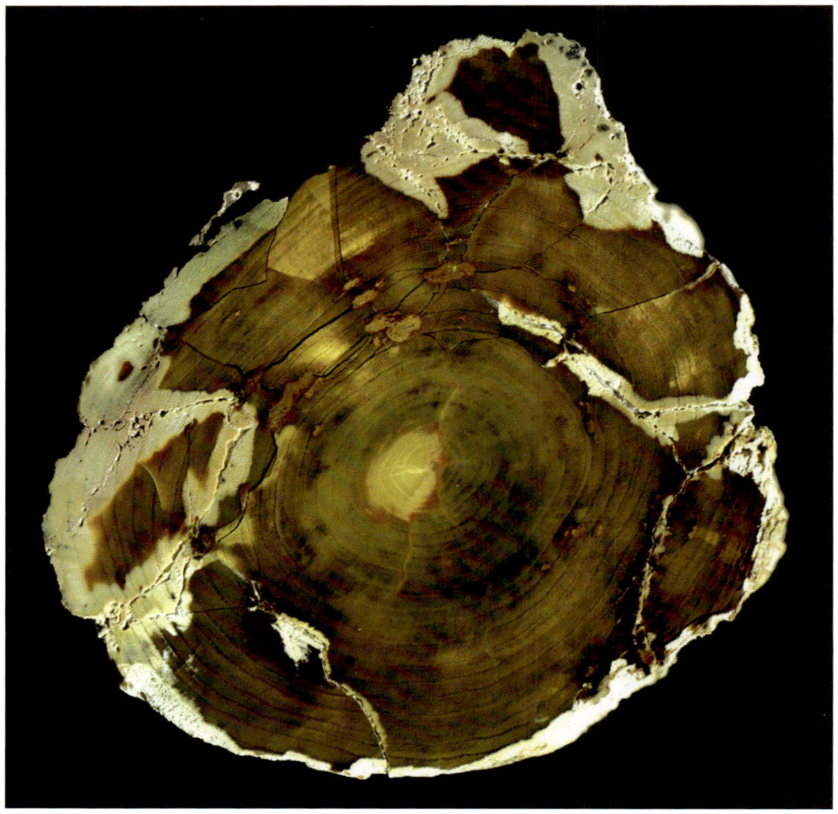

Hoodoo Basin elm-like fossil hardwood with typical coloring and mineralization. Succor Creek Formation, Miocene. Hoodoo Basin, Malheur County, Oregon, USA. Daniels specimen, 10 cm.

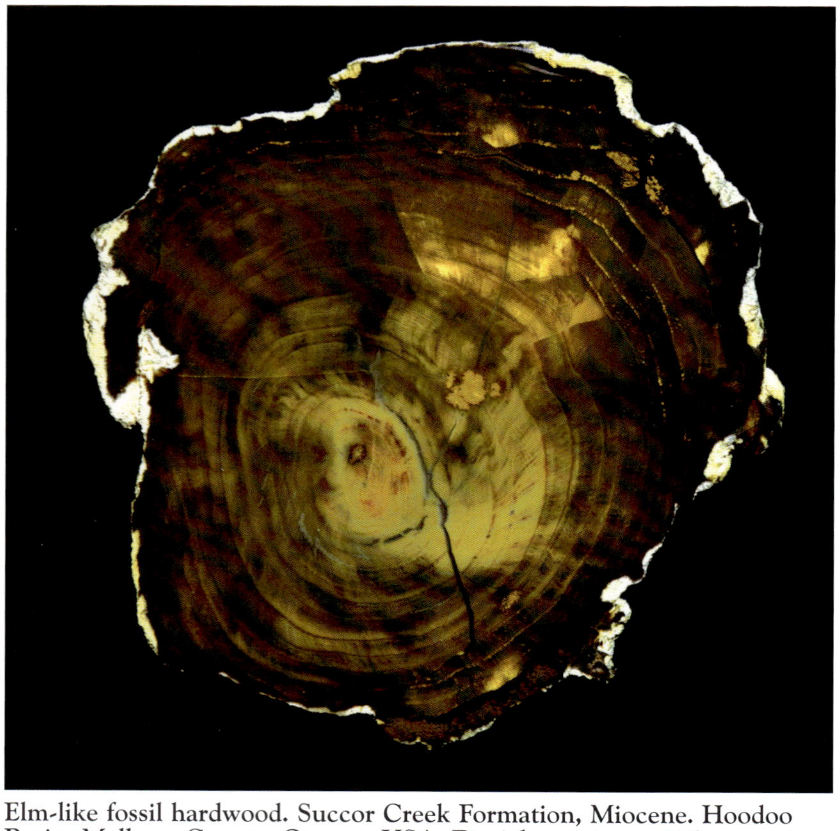

Elm-like fossil hardwood. Succor Creek Formation, Miocene. Hoodoo Basin, Malheur County, Oregon, USA. Daniels specimen, 7.5 cm.

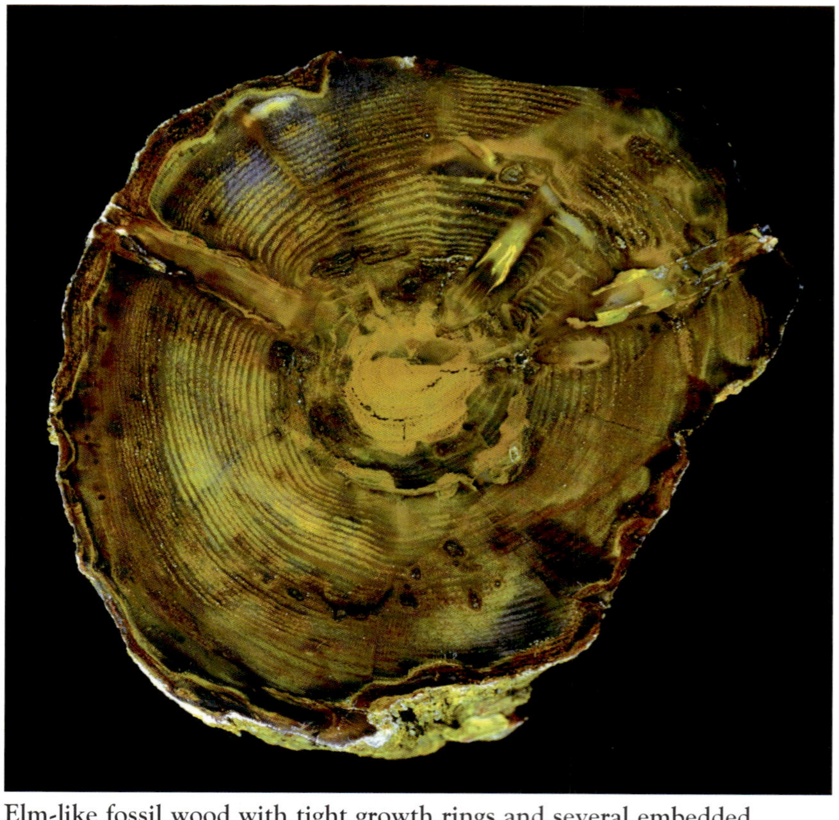

Elm-like fossil wood with tight growth rings and several embedded branches. Succor Creek Formation, Miocene. Grassy Mountain, Malheur County, Oregon, USA. Daniels specimen, 8.5 cm.

Elm-like fossil hardwood. Succor Creek Formation, Miocene. Hoodoo Basin, Malheur County, Oregon, USA. Daniels specimen, 10.5 cm.

The Sweet Home collecting location in Linn County, west of the Cascade Mountains, resides in the Oligocene to Early Miocene Little Butte Volcanic Series. These horizons yield a large variety of hardwood and softwood fossils. Reports of the vesselless angiosperm *Trochodendron* from Sweet Home date as early as the 1950s. Fossil woods from this area are noteworthy because of their faithful preservation of wood structure detail. Unusual Sweet Home fossil wood specimens include possible zeolite crystals replaced by quartz (quartz after zeolite pseudomorphs). Other Linn County collecting areas include Rogers Mountain, Scio, Crabtree Creek, and Brush Creek. Fossil wood from Brush Creek is similar in appearance to the Sweet Home wood. Rogers Mountain, Scio, and Crabtree Creek yield solid and colorful specimens including *Sequoia* and sycamore-like wood. The origin of these specimens is undetermined as they evidence considerable external abrasion as though transported by water for some distance; full round specimens are rare.

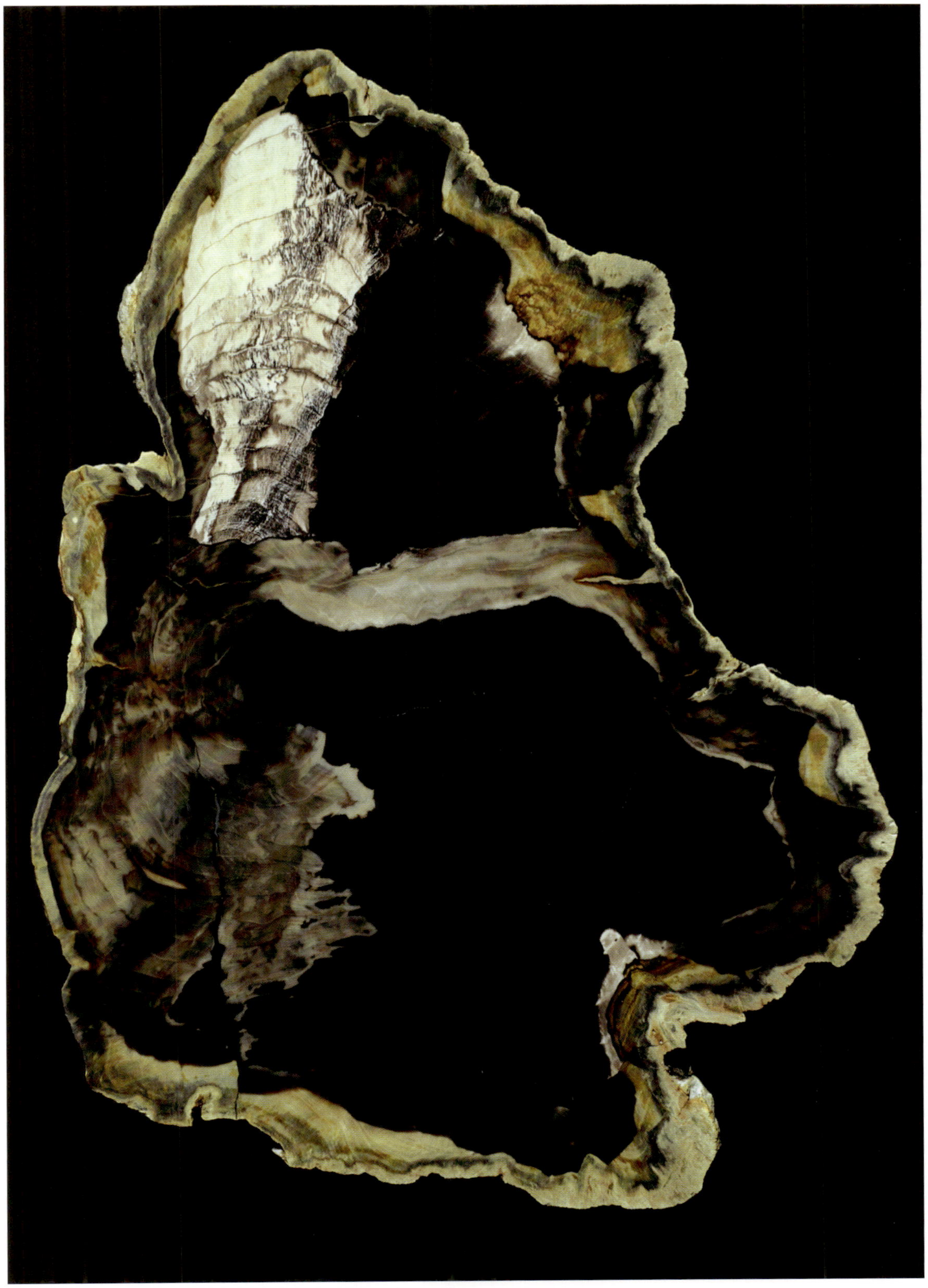

Fossil conifer. Tertiary. Rogers Mountain, Linn County, Oregon, USA. Daniels specimen, 17 cm.

Fossil wood. Tertiary. Rogers Mountain, Linn County, Oregon, USA. Daniels specimen, 15.5 cm.

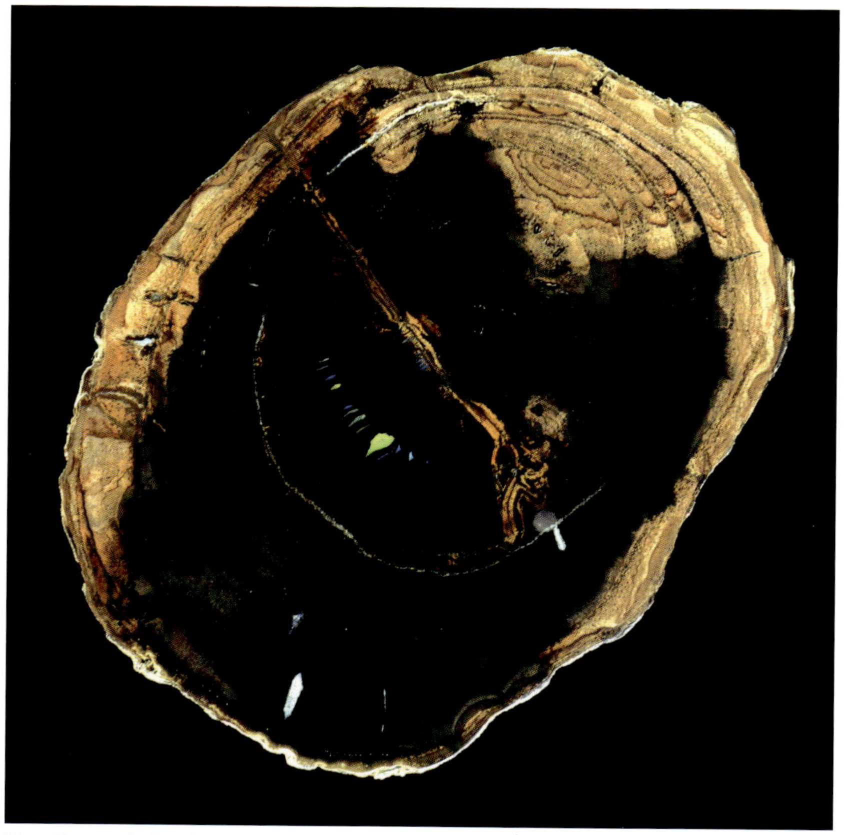

Fossil wood. Little Butte Volcanic Series, Oligocene to Early Miocene. Sweet Home area, Linn County, Oregon, USA. Daniels specimen, 17 cm.

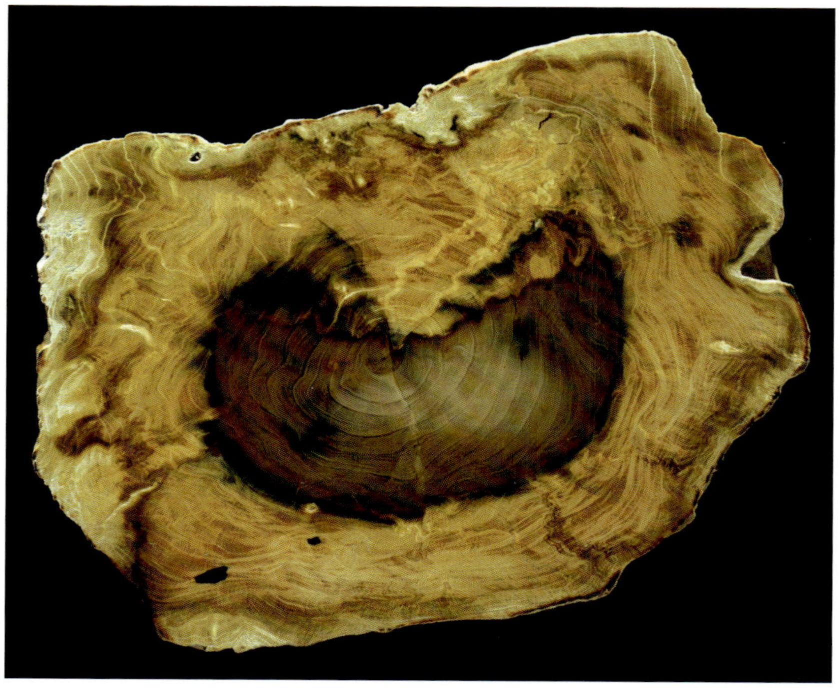

Fossil conifer. Tertiary. Rogers Mountain, Linn County, Oregon, USA. Daniels specimen, 17 cm.

Fossil wood showing minute desiccation cracks or insect borings that have received opportunistic infilled agate. Silica solutions follow openings in the wood and readily permineralize these areas. Little Butte Volcanic Series, Oligocene to Early Miocene. Sweet Home area, Linn County, Oregon, USA. Daniels specimen, 26 cm.

Set of fossil wood logs from the Sweet Home area. Little Butte Volcanic Series, Oligocene to Early Miocene. Sweet Home area, Linn County, Oregon, USA. Daniels specimens, 28 cm longest.

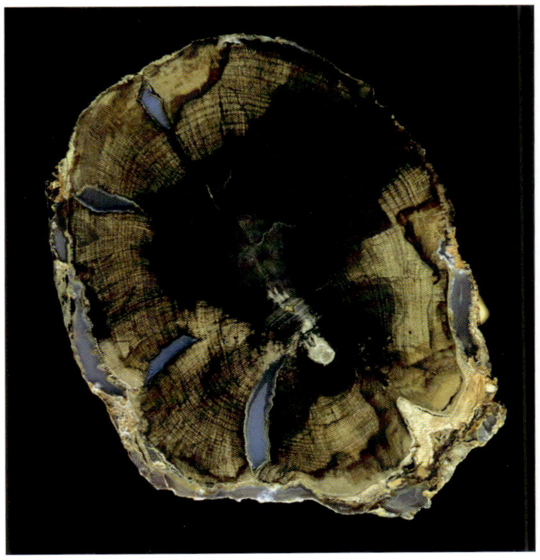

Fossil wood. Little Butte Volcanic Series, Oligocene to Early Miocene. Sweet Home area, Linn County, Oregon, USA. Daniels specimen, 9.5 cm.

Fossil wood. Little Butte Volcanic Series, Oligocene to Early Miocene. Sweet Home area, Linn County, Oregon, USA. Daniels specimen, 4.5 cm.

Fossil wood. Little Butte Volcanic Series, Oligocene to Early Miocene. Sweet Home area, Linn County, Oregon, USA. Daniels specimen, 8 cm.

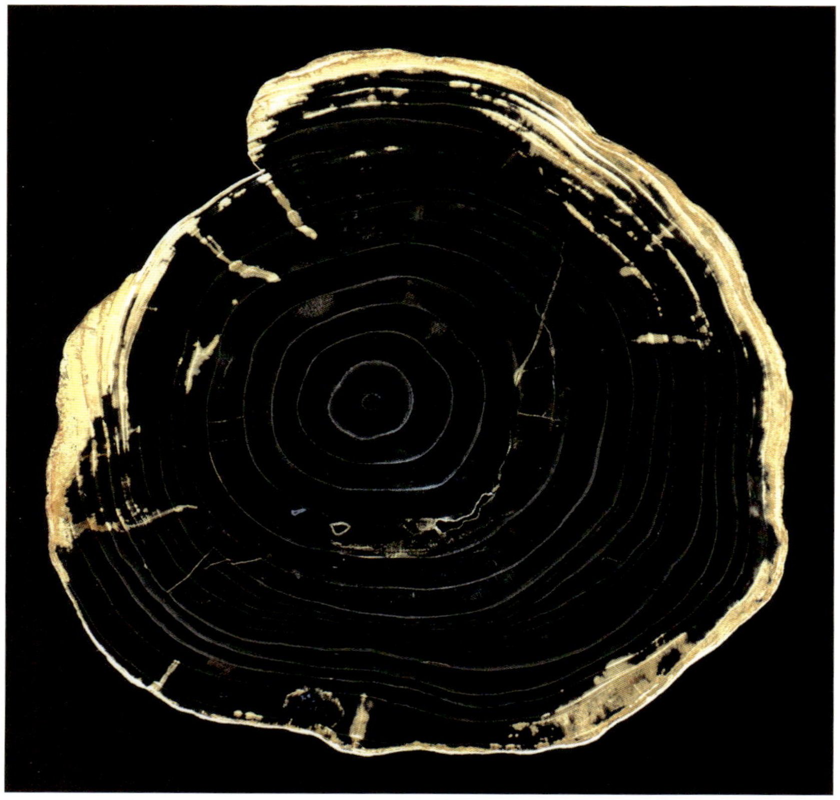

Fossil conifer. Little Butte Volcanic Series, Oligocene to Early Miocene. Brush Creek, Linn County, Oregon, USA. Daniels specimen, 13 cm.

Fossil wood with possible zeolite crystals replaced by quartz (quartz after zeolite pseudomorphs). Little Butte Volcanic Series, Oligocene to Early Miocene. Sweet Home area, Linn County, Oregon, USA. Haggerty specimen, 15 cm.

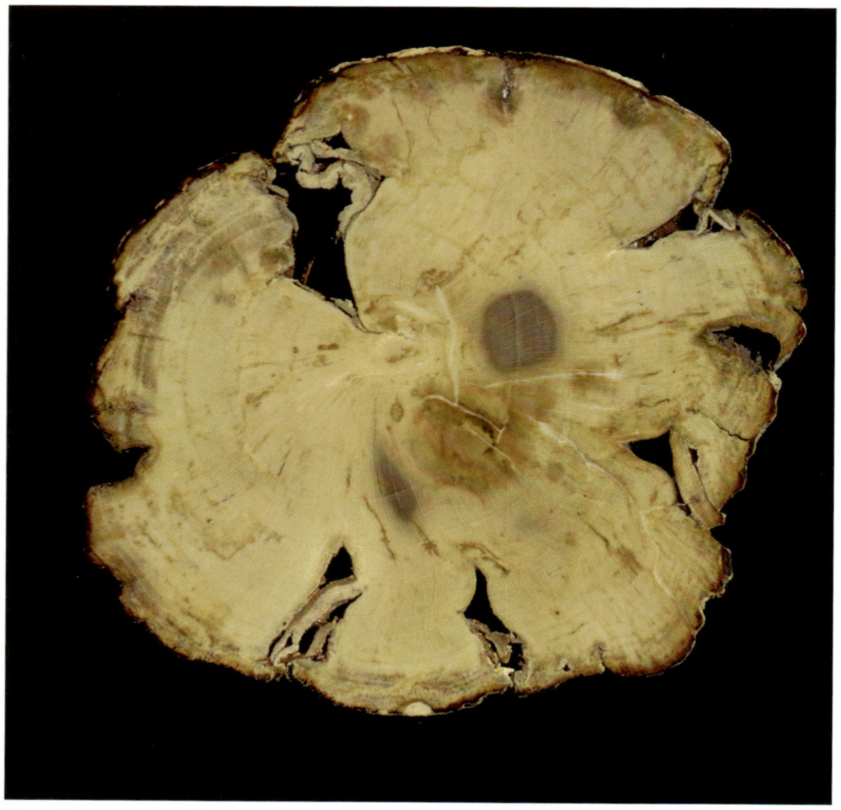

Fossil sycamore-like hardwood. Little Butte Volcanic Series, Oligocene to Early Miocene. Sweet Home area, Linn County, Oregon, USA. Daniels specimen, 11 cm.

Fossil sycamore-like hardwood. Little Butte Volcanic Series, Oligocene to Early Miocene. Sweet Home area, Linn County, Oregon, USA. Daniels specimen, 9.5 cm.

Fossil hardwood. Little Butte Volcanic Series, Oligocene to Early Miocene. Sweet Home area, Linn County, Oregon, USA. Daniels specimen, 8 cm.

The McDermitt area, which spans the Oregon–Nevada border, produces abundant fossil wood specimens from the Trout Creek Formation (13.8 mybp). The formation consists of diatomite with numerous lenses of volcanic ash. The Trout Creek Formation is slightly younger than Succor Creek flora. Trees in the McDermitt area grew in cooler conditions than at the northern Succor Creek sites, and conifers are more diverse and abundant. McDermitt specimens tend to be highly silicified with well-preserved wood structure. Fossil wood from a location known as Zimmerman Ranch in the McDermitt area tends to be especially well preserved and often has an attractive reddish patina.

Fossil fruitwood-like hardwood. Trout Creek Formation, Miocene. McDermitt area, Oregon/Nevada, USA. Daniels specimen, 16 cm.

Double-hearted fossil wood. Trout Creek Formation, Miocene. McDermitt area, Oregon/Nevada, USA. Daniels specimen, 4 cm.

Macro photograph of borer tunnels in fossil conifer. Trout Creek Formation, Miocene. McDermitt area, Oregon/Nevada, USA. Daniels specimen.

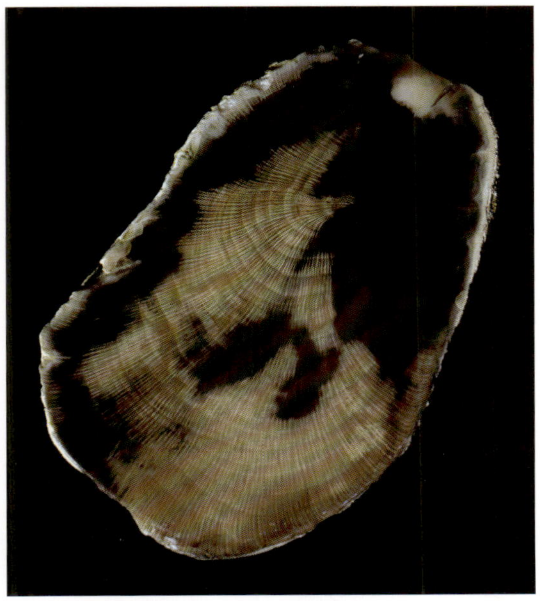

Fossil *Trochodendron*-like wood. Trout Creek Formation, Miocene. McDermitt area, Oregon/Nevada, USA. Daniels specimen, 5.5 cm.

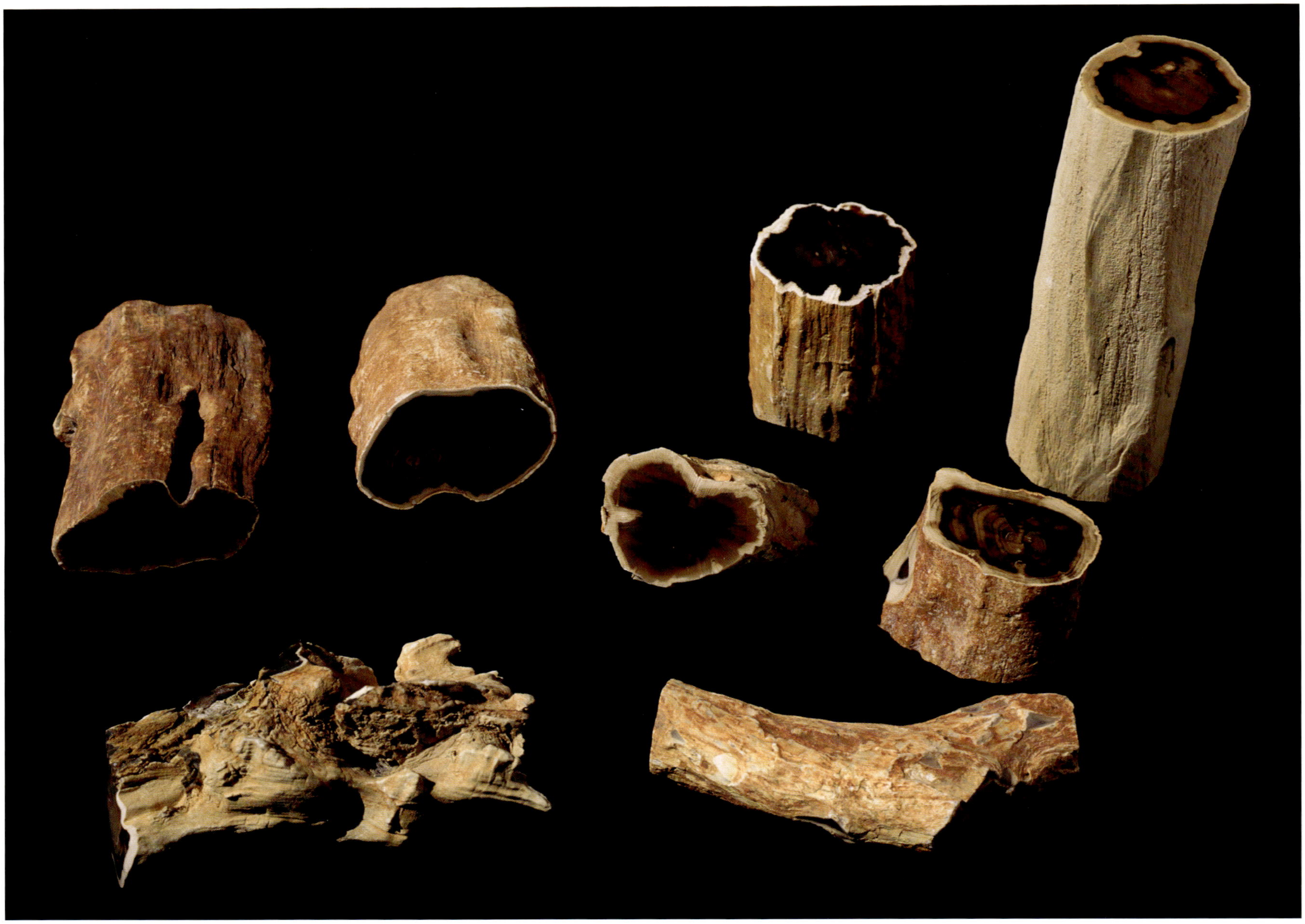

Set showing some of the variety seen in fossil woods from the McDermitt area. Trout Creek Formation, Miocene. McDermitt area, Oregon/Nevada, USA. Daniels specimens, 16.5 tallest.

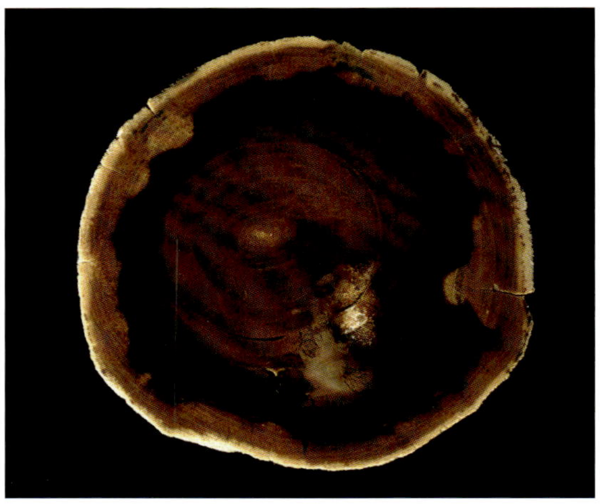

Fossil hardwood. Trout Creek Formation, Miocene. McDermitt area, Oregon/Nevada, USA. Daniels specimen, 5.5 cm.

Fossil *Trochodendron*-like wood. Trout Creek Formation, Miocene. McDermitt area, Oregon/Nevada, USA. Daniels specimen, 6 cm.

Oregon's youngest Tertiary fossil woods reside in the Deschutes flora, an Early to Middle Pliocene deposit near the Columbia River that dates 5.3 to 4.3 mybp. Cherry, elm, hickory, maple, oak, and willow dominate. The oldest well-silicified plant fossils in Oregon are the Lower Cretaceous *Tempskya* fern stems from the Blue Mountains. Cretaceous floras are rare in Oregon. Sea levels during the Cretaceous were about 300 meters (975 feet) higher than now, and much of the state was under a huge inland sea. The Blue Mountain *Tempskya* are found close to Cretaceous ammonite fossils and apparently grew on an island. Top quality specimens are golden in color, glassy, and have well-preserved structures.

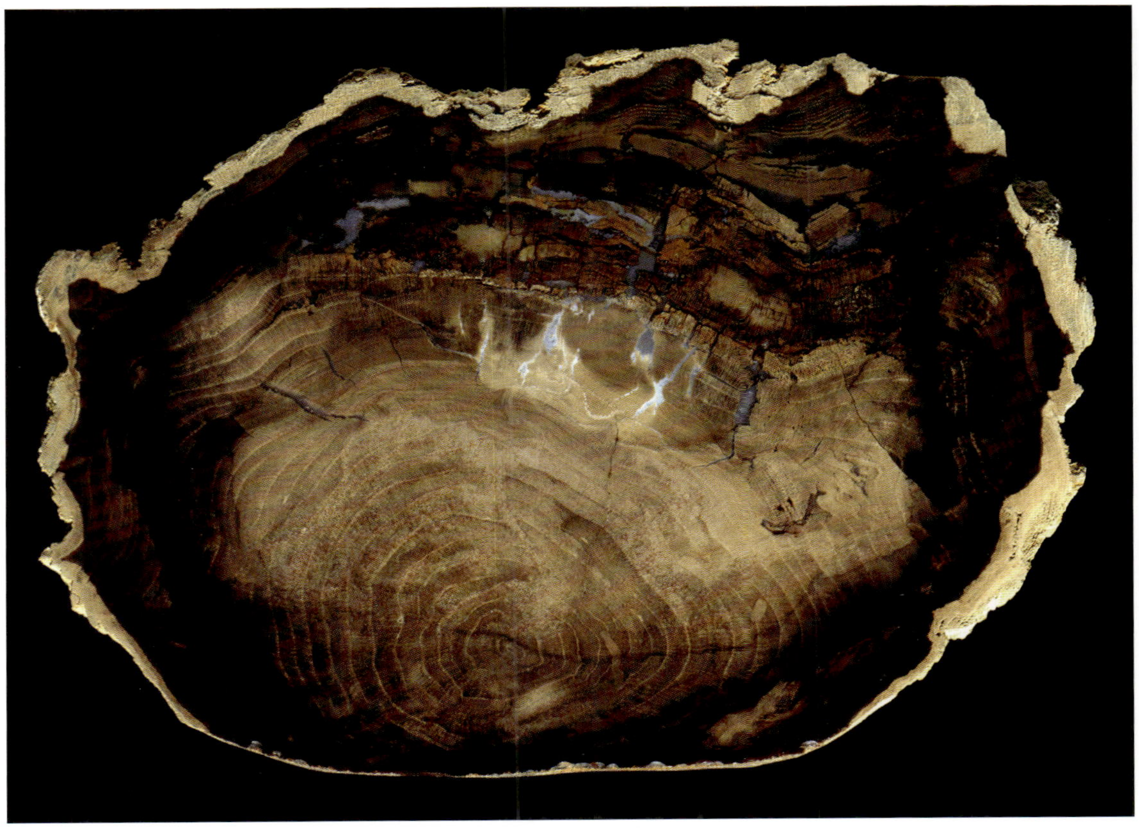

Fossil hardwood. Trout Creek Formation, Miocene. McDermitt area, Oregon/Nevada, USA. Daniels specimen, 16 cm.

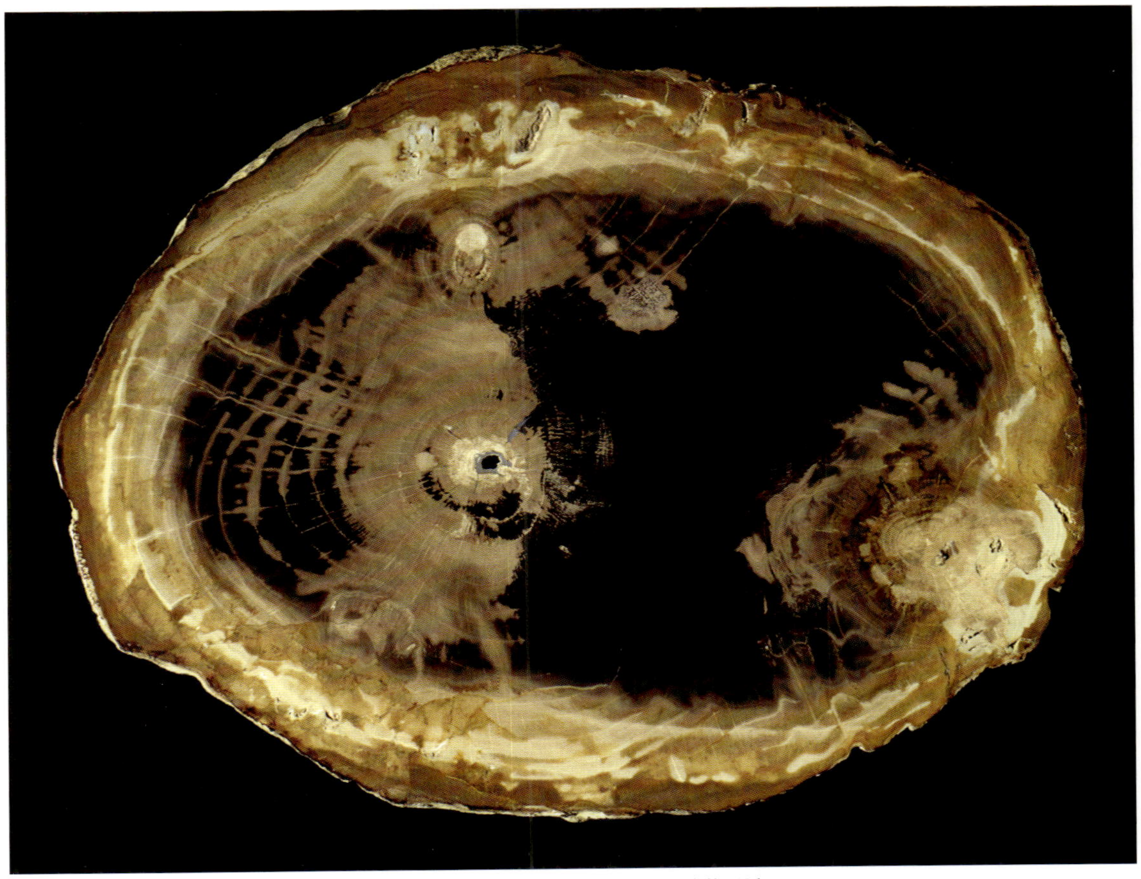

Triple-hearted fossil hardwood. Deschutes flora, Early to Middle Pliocene. Deschutes River area, Oregon, USA. Daniels specimen, 22 cm.

Ginkgo Petrified Forest is a state park located near Vantage, Washington. The fossil wood-bearing strata are within the Columbia Plateau Basalt Group dated as Middle Miocene and estimated 15.4 mybp. It was here that the first known fossil ginkgo log in the world was discovered. The Columbia Plateau Basalt Group is a formation of staggering proportions. Individual lava flows cover thousands of square miles and are as much as 4,600 meters (15,000 feet) thick! Walls of lava 15 meters (50 feet) high flowed across the terrain, incinerating the vast majority of trees. Two categories of wood survived the conflagration as fossils. Some of the trees grew in a swampy environment with portions under water. Fire consumed most of the trees, while the trunks, roots, and whatever else managed to crash into the swamp escaped the inferno and went on to fossilize. This explains why much of the wood from locations such as Saddle Mountain is cypress root.

The other wood that escaped incineration was stripped from its habitat by thundering mudflows that preceded the lava. The forest at Vantage, Washington, appears to be of the latter category. The majority of the identified fossil woods at Vantage are similar to present day Douglas fir, fir, and spruce. Others are similar to bald cypress, beech, birch, elm, ginkgo, maple, oak, sweetgum, sycamore, water tupelo, and yew. Although named Ginkgo Petrified Forest, perhaps just 1 percent of the wood or less is ginkgo. A separate accumulation of logs nearby at Squaw Creek, stratigraphically a few hundred feet higher than the Ginkgo Flow at Vantage, was studied by Professor George Beck of Central Washington College in nearby Ellensburg in the 1940s. His estimate of fossil woods for that accumulation was 15 percent each of bald cypress, Douglas fir or spruce, sweetgum, and sycamore, 5 percent elm, and 1 percent ginkgo. Columbia Plateau Basalt flows were more widespread than just present-day Washington State, flowing across much of Idaho and Oregon as well. Fossil wood specimens have been collected for generations from Saddle Mountain, Yakima Valley, and nearby locales. The famous fossil wood location known as Stinking Water Creek in Oregon is within a Columbia Plateau flow dated to Middle Miocene (12.1 mybp). The associated Stinking Water flora (12.5 to 13.4 mybp) produces leaf impressions of oak, along with alder, cypress, elm, poplar, and sycamore. The picture presented is a swamp habitat where conifers were rare.

Double-hearted oak-like fossil hardwood. Deschutes flora, Early to Middle Pliocene. Deschutes River area, Oregon, USA. Daniels specimen, 19.5 cm.

Set of *Tempskya* stems. Cretaceous. Blue Mountains, Oregon. Daniels specimens, 23.5 longest.

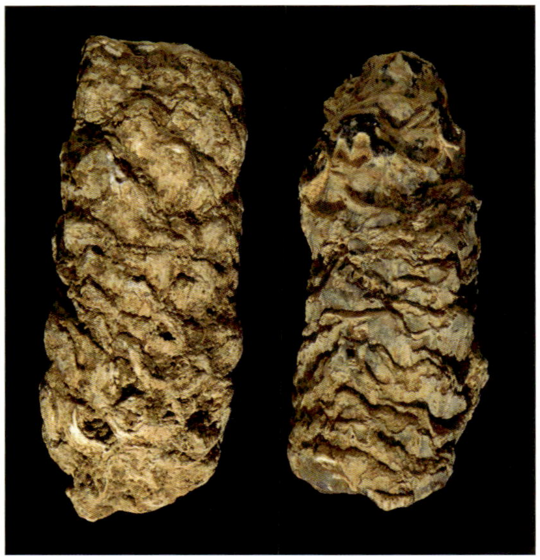

Fossil cones. Columbia Plateau Basalt flow, Miocene. Yakima Canyon, Washington, USA. Daniels specimens, 6 cm.

Taxodioxylon. Columbia Plateau Basalt flow, Miocene. Yakima, Washington, USA. Daniels specimen, 6 cm.

Macro photograph with borer evidence in fossil wood. Columbia Plateau Basalt flow, Miocene. Saddle Mountain, Washington, USA. Daniels specimen.

Yakima Valley, Washington, USA.

Cupressinoxylon. Columbia Plateau Basalt flow, Miocene. Saddle Mountain, Washington, USA. Daniels specimen, 15 cm.

Diffuse porous (maple-like) fossil hardwood. Columbia Plateau Basalt flow, Miocene. Yakima, Washington, USA. Daniels specimen, 24.5 cm.

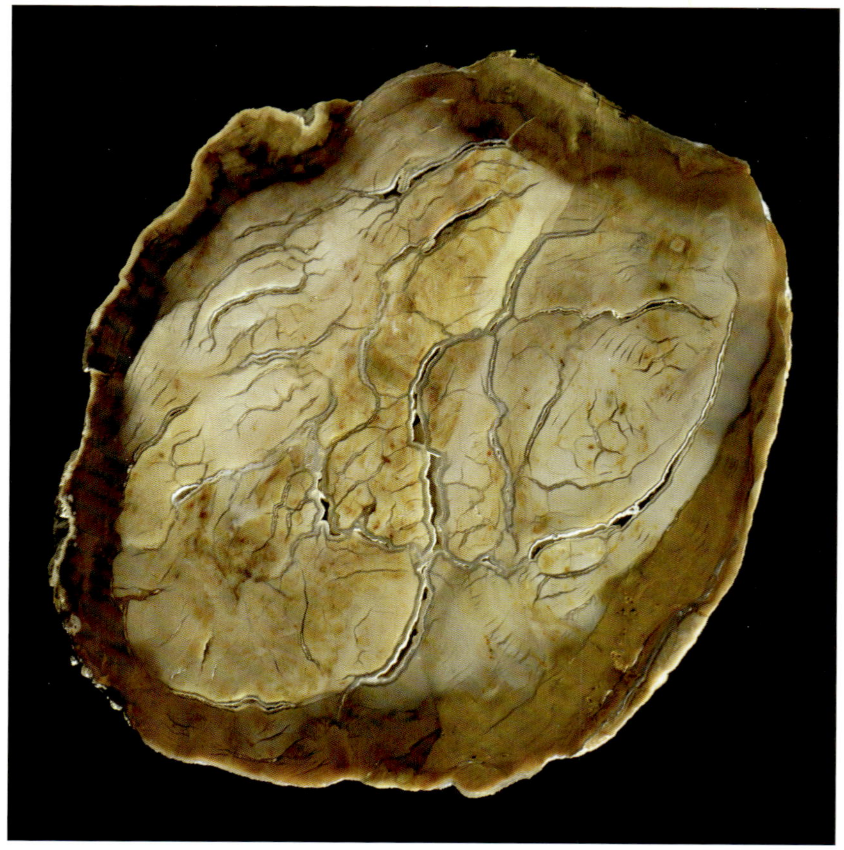

Fossil wood with little residual wood structure. Columbia Plateau Basalt flow, Miocene. Saddle Mountain, Washington, USA. Daniels specimen, 15.5 cm.

Variety of *Cupressinoxylon* known as "Saddle Mountain picture wood" because of the dendritic patterns and coloration. Columbia Plateau Basalt flow, Miocene. Saddle Mountain, Washington, USA. Daniels specimen, 38 cm.

View of Columbia River and basalt cliffs of the Miocene age Columbia Plateau Basalt flow. Ginkgo Petrified Forest, Vantage, Washington, USA.

Accumulation of fossil logs with Columbia River in the background. Ginkgo Petrified Forest, Vantage, Washington, USA.

Oak-like fossil wood. Columbia Plateau Basalt flow, Miocene. Stinking Water Creek, Oregon, USA. Daniels specimen, 29 cm.

Unusual *Taxodioxylon* found in flattened sections with nearly parallel growth rings and uniform, diamond-shaped agate inclusions. The local collectors called it "ribbon wood" because thin polished slices resemble decorative ribbons. It is likely that the buried wood underwent considerable compression, causing the tracheid alignment to buckle into diamond-shaped voids that later infilled with chalcedony and formed agates. Columbia Plateau Basalt flow, Miocene. Yakima area, Washington, USA. Daniels specimen, 17 cm.

Macro photographs of ribbon wood. Largest growth rings approximately 1.3 mm.

Micro images of ribbon wood. Largest growth rings approximately 1.3 mm.

Double-hearted elm-like fossil hardwood. Columbia Plateau Basalt flow, Miocene. Vantage, Washington, USA. Daniels specimen, 24.5 cm.

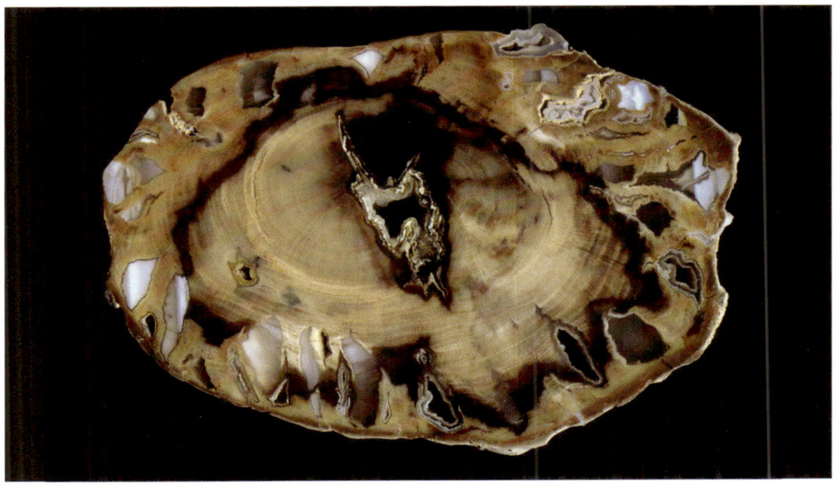

Cupressinoxylon with areas of chalcedony that infilled desiccation voids. Columbia Plateau Basalt flow, Miocene. Saddle Mountain, Washington, USA. Daniels specimen, 11.5 cm.

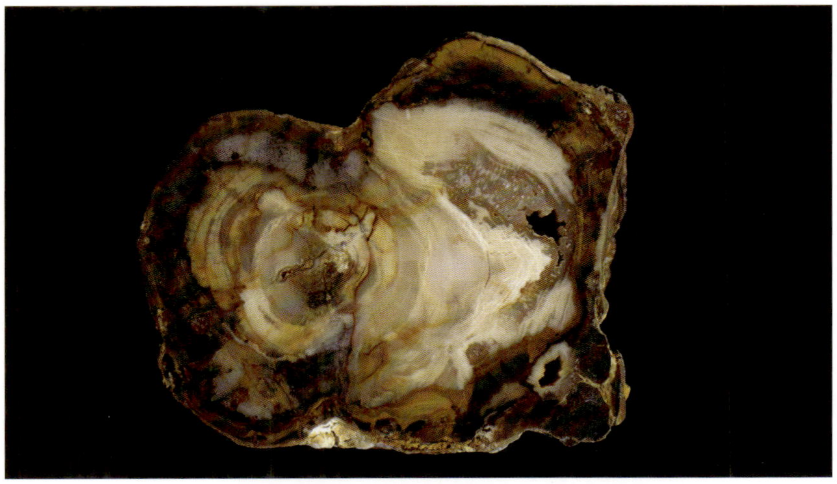

Cupressinoxylon. Columbia Plateau Basalt flow, Miocene. Saddle Mountain, Washington, USA. Daniels specimen, 9 cm.

Taxodioxylon. Columbia Plateau Basalt flow, Miocene. Yakima Valley, Washington, USA. Daniels specimen, 20 cm.

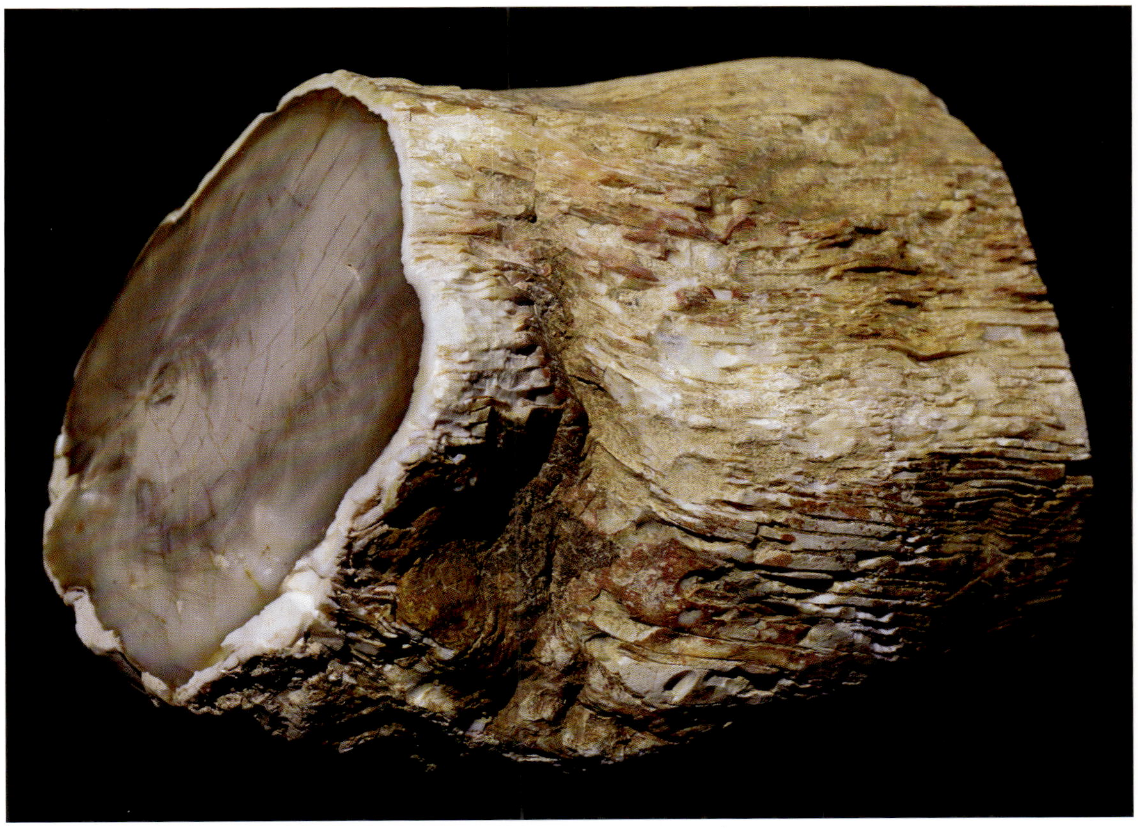

Cupressinoxylon log. Columbia Plateau Basalt flow, Miocene. Yakima, Washington, USA. Daniels specimen, 38 cm long.

Walnut-like fossil hardwood. Columbia Plateau Basalt flow, Miocene. Vantage, Washington, USA. Daniels specimen, 26 cm.

Just west of Ginkgo Petrified Forest is the Ellensburg flora of south-central Washington, dated 10.5 to 13.5 mybp, depending on the horizon. Here we find fossil plant remnants of three habitats: cypress swamp, deciduous hardwood forest (ash, beech, birch, elm, maple, oak, sassafras, and sycamore), and a younger semi-arid oak forest (buckthorn, elm, holly, and oak).

Around the time of the Eocene–Oligocene boundary (34 mybp), severe volcanism and mountain building commenced in the present-day Sierra Nevada region—sea levels fell and precipitation began to become more seasonal. During the Miocene Epoch, the Western Great Basin of Nevada saw habitats of shrublands, evergreen-oak-cypress-pine woodlands, and mixed hardwood forests. A plant fossil survey of the Early to Middle Miocene Buffalo Canyon flora of western Nevada (15.6 mybp), representing a mixed conifer–deciduous forest, established the largest families as Pinacea (10 species), Rosaceae (11 species), Salicaceae (10 species), Aceraceae (5 species), and Betulacaea (5 species). The largest genera were *Salix* (willow; 7), *Acer* (maple; 5), *Betula* (birch; 4), *Populus* (poplar; 4), *Ribes* (currant; 4), *Picea* (spruce; 3), and *Prunus* (cherry; 3). The Pyramid flora in western Nevada (15.6 mybp) is dominated by deciduous hardwoods.

By Late Miocene, most trees we now consider Asian exotics were gone from North American flora, largely because of the decline of summer rainfall. The state of Nevada has long been a favorite with collectors of fossil wood. Hubbard Basin is known for beautifully colored conifers in shades of translucent blues and yellows; Cherry Creek and nearby locations are known for pink, red, and white woods with well-preserved, although compressed, wood structure; Texas Springs is known for lovely, translucent pink limb casts; Virgin Valley is known for fossil wood infilled by precious opal. Across the Sierra Nevada, redwood forests grew during the Pliocene Epoch. The Petrified Forest at Calistoga, California, preserves trees entombed by an eruption of nearby Mount St. Helena dated 3.4 mybp. All the trees identified in this petrified forest are redwood except for one lonely pine tree. These fossil trees are among the largest in the world, with trunks as thick as 2.4 meters (8 feet) in diameter and more than 32 meters (105 feet) in length. The oldest trees in this forest grew for 2,000 years before being entombed and turned to stone.

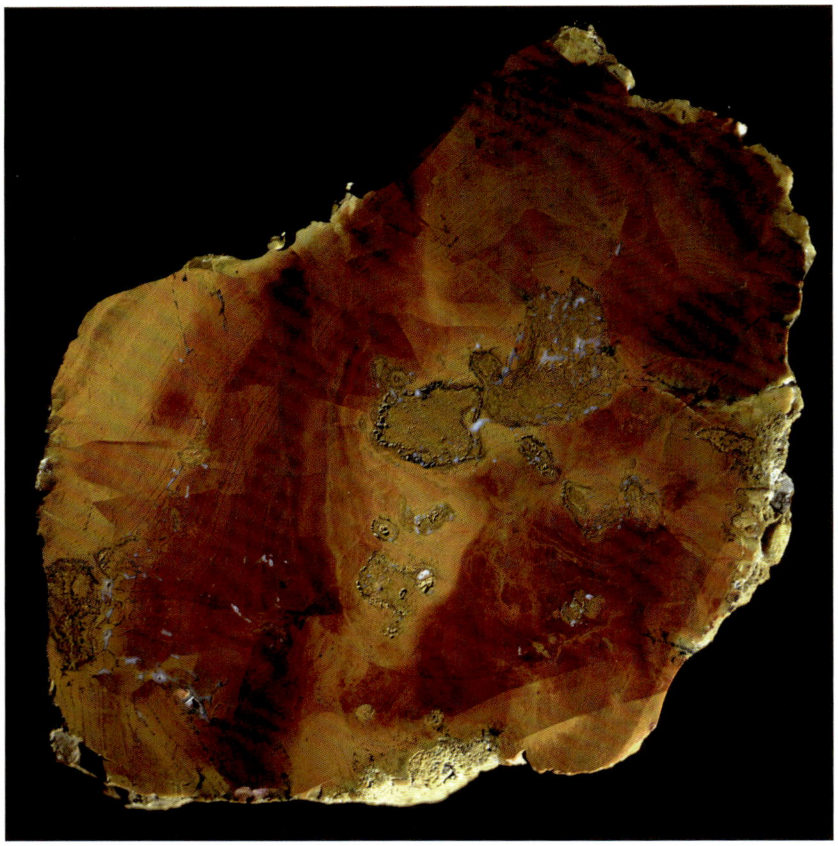

Fossil ring porous hardwood. Tertiary. Basalt, Mineral County, Nevada. Daniels specimen, 18 cm.

Taxodioxylon. Virgin Valley Formation, Miocene. Virgin Valley, Humboldt County, Nevada, USA. Daniels specimen, 14.5 cm.

Taxodioxylon. Tertiary. Ely, White Pine County, Nevada. Daniels specimen, 16.5 cm.

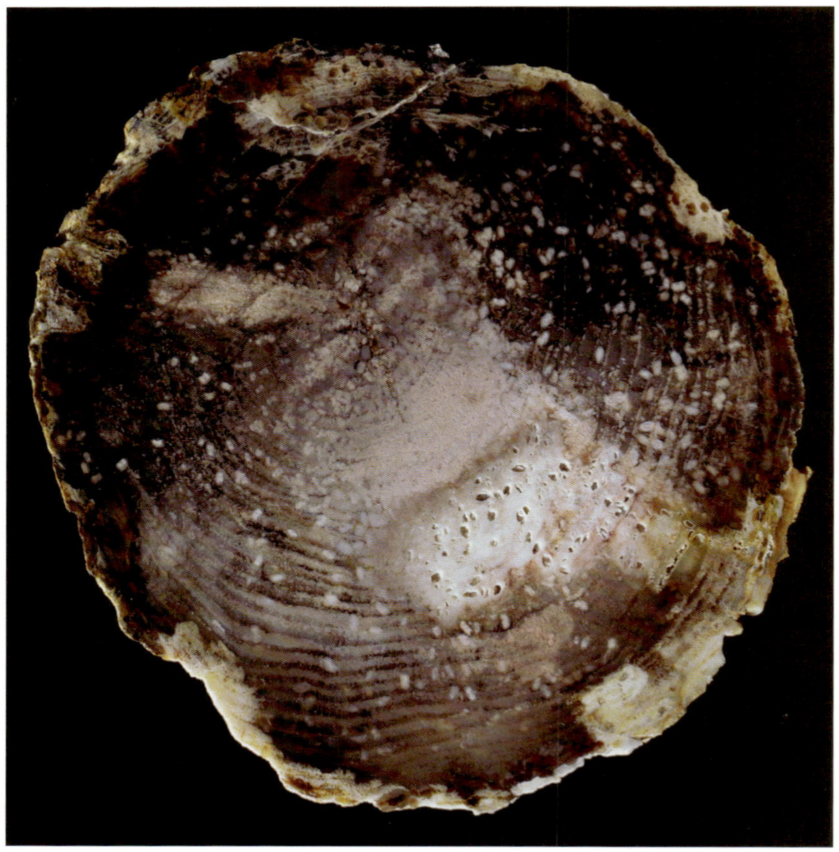

Taxodioxylon included by crystals of carbonate(?) replaced by silica. Tertiary. Gabbs, Nye County, Nevada, USA. Daniels specimen, 16.5 cm.

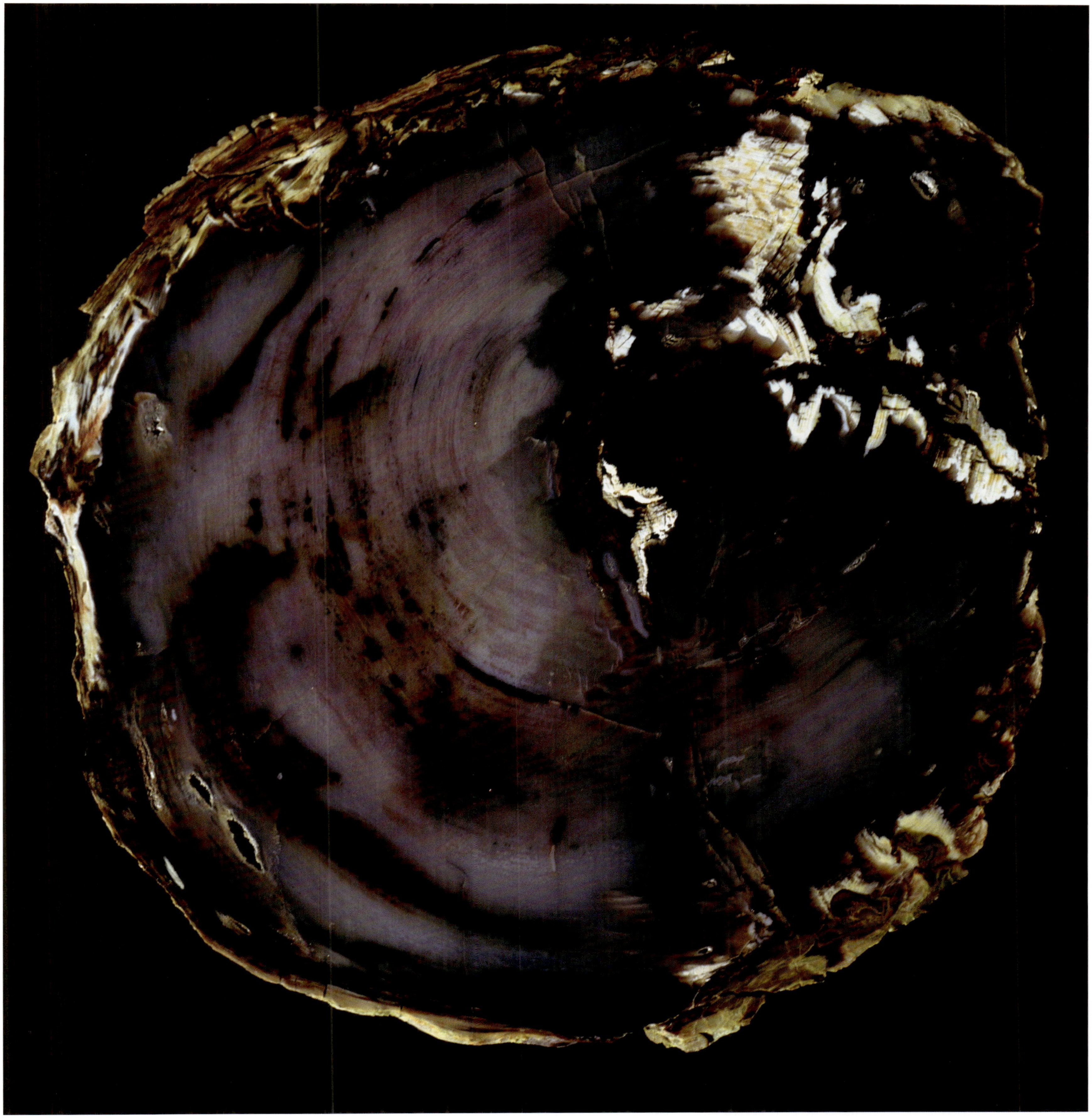

Taxodioxylon knot. Tertiary. Hubbard Basin, Elko County, Nevada, USA. Daniels specimen, 22 cm.

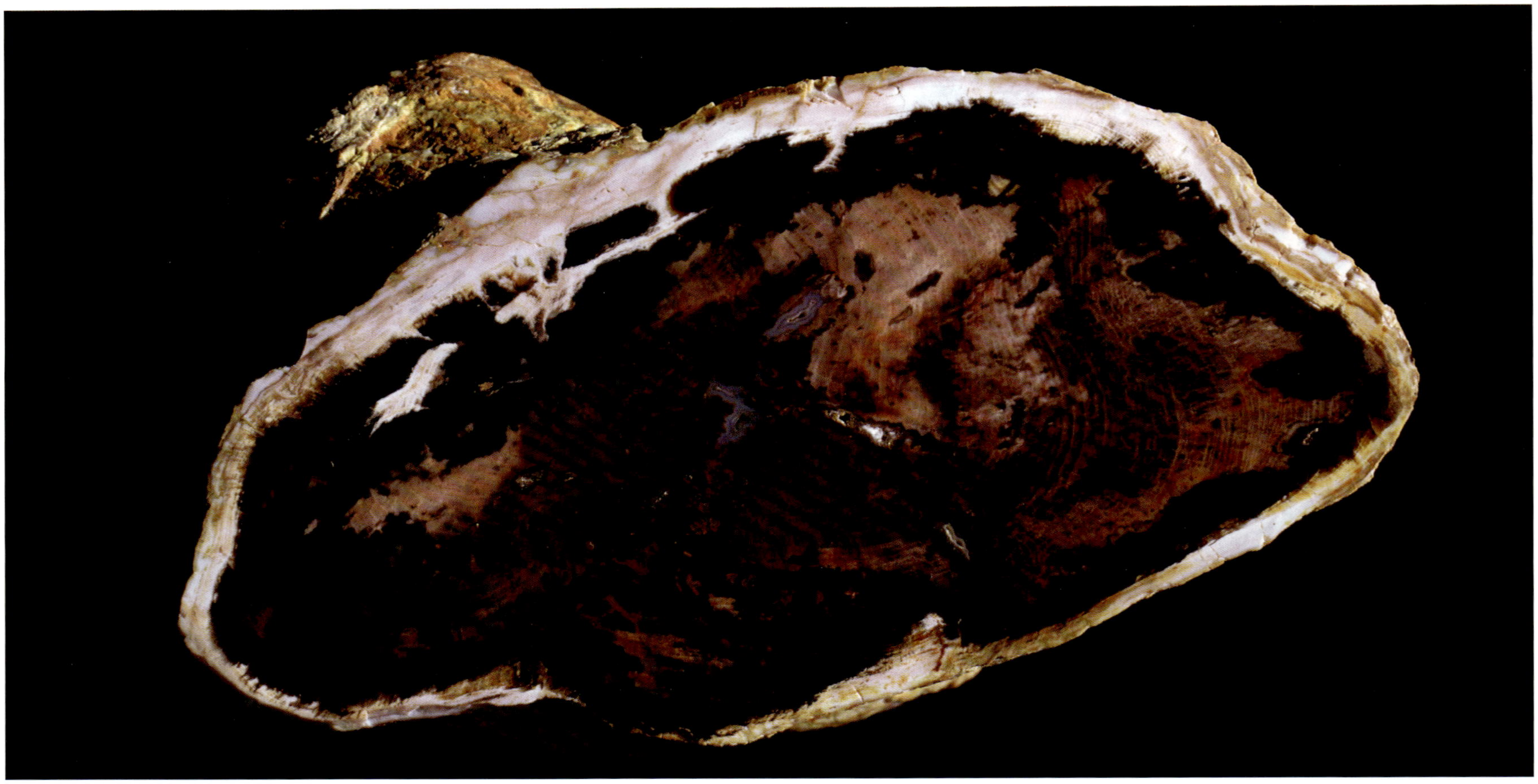

Cupressinoxylon. Kalamazoo Creek Tuff Formation, Oligocene. Cherry Creek, Nevada, USA. Daniels specimen, 15 cm.

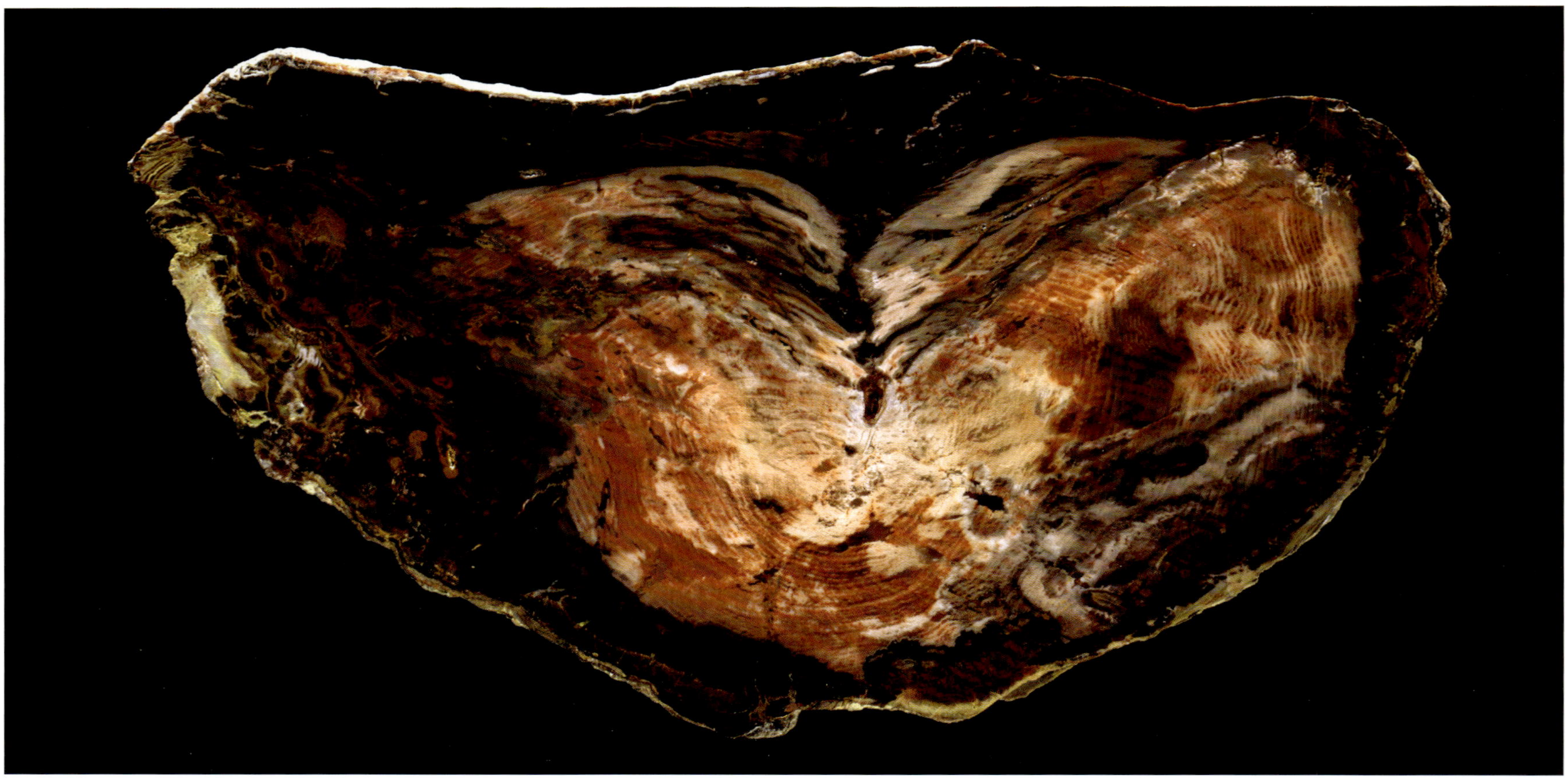

Cupressinoxylon. Kalamazoo Creek Tuff Formation, Oligocene. Cherry Creek, Nevada, USA. Daniels specimen, 15 cm.

Texas Springs limb cast set. Rhyolite Flows, Miocene. Texas Springs, Elko County, Nevada, USA. Daniels specimens, 9.5 cm longest.

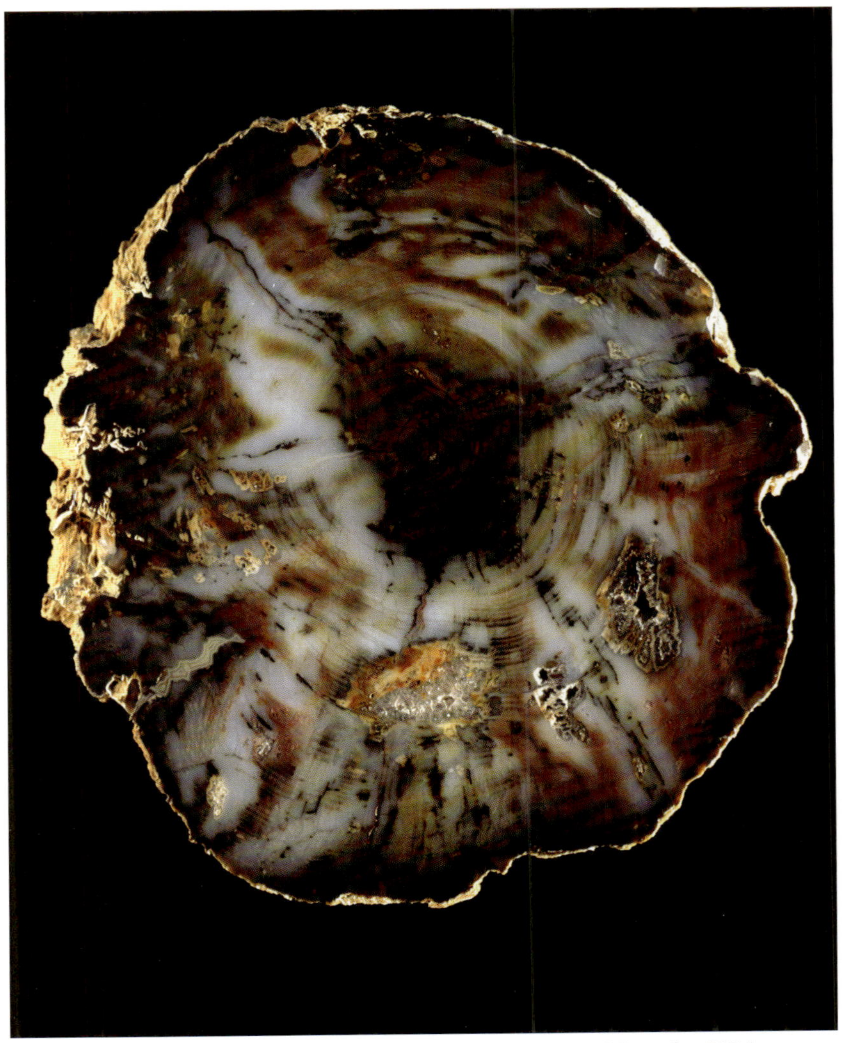

Taxodioxylon. Tertiary. Hubbard Basin, Elko County, Nevada, USA. Daniels specimen, 13 cm.

Taxodioxylon. Tertiary. Hubbard Basin, Elko County, Nevada, USA. Daniels specimen, 16 cm.

Texas Springs limb cast. Rhyolite flows, Miocene. Texas Springs, Elko County, Nevada, USA. Daniels specimen, 4.5 cm.

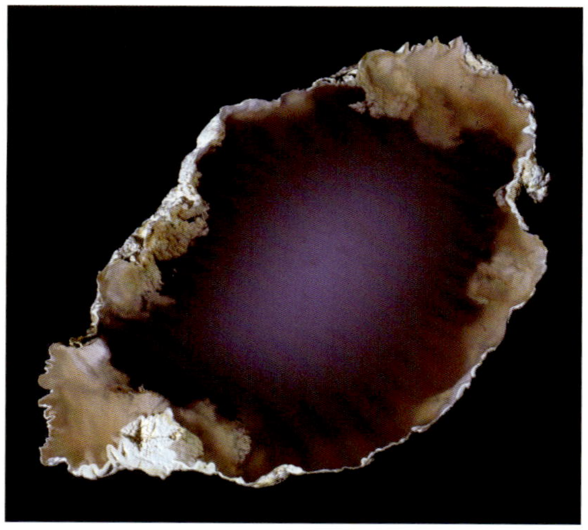

Texas Springs limb cast. Rhyolite flows, Miocene. Texas Springs, Elko County, Nevada, USA. Daniels specimen, 6 cm.

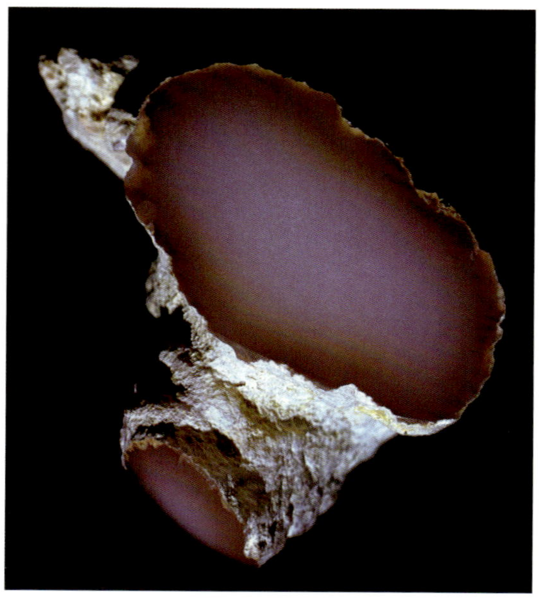

Texas Springs limb cast. Rhyolite flows, Miocene. Texas Springs, Elko County, Nevada, USA. Daniels specimen, 7.5 cm.

Fossil wood. Tertiary. Little Humboldt River, Humboldt County, Nevada, USA. Daniels specimen, 5.5 cm.

Fossil wood. Tertiary. Little Humboldt River, Humboldt County, Nevada, USA. Daniels specimen, 4.5 cm.

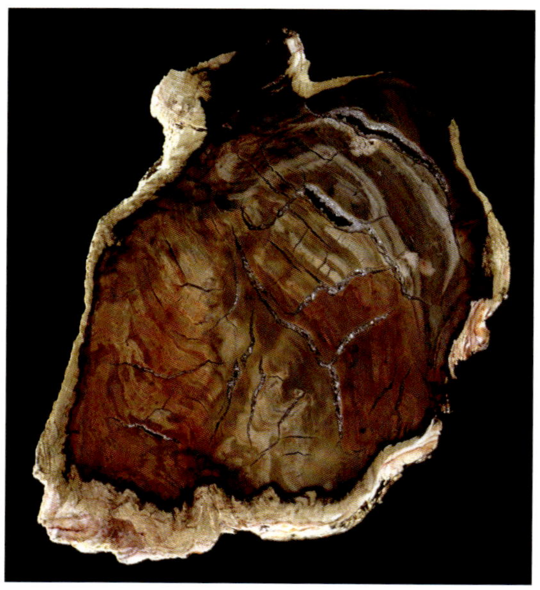

Taxodioxylon. Tertiary. Stonewall Pass near Goldfield, Nevada. Daniels specimen, 8.5 cm.

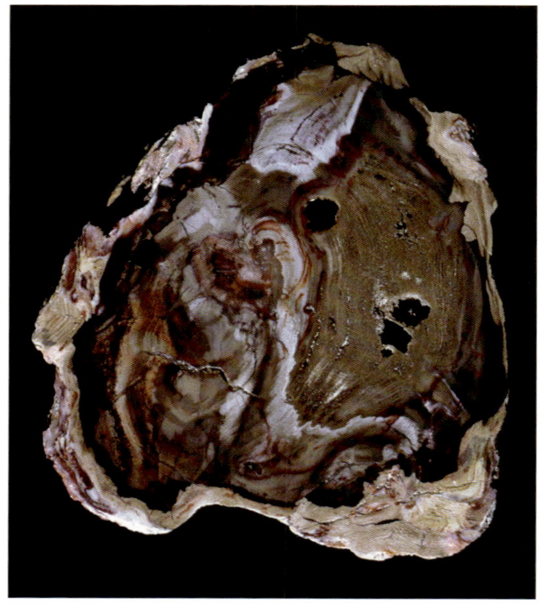

Taxodioxylon. Tertiary. Stonewall Pass near Goldfield, Nevada. Daniels specimen, 9 cm.

Fossil wood. Tertiary. Little Humboldt River, Humboldt County, Nevada, USA. Daniels specimen, 8 cm.

Live oak-like fossil hardwood. Tertiary. Washoe County, Nevada. Daniels specimen, 7 cm.

Fossil wood. Tertiary. Little Humboldt River, Humboldt County, Nevada, USA. Daniels specimen, 8.5 cm.

Fossil wood. Tertiary. Little Humboldt River, Humboldt County, Nevada, USA. Daniels specimen, 7.5 cm.

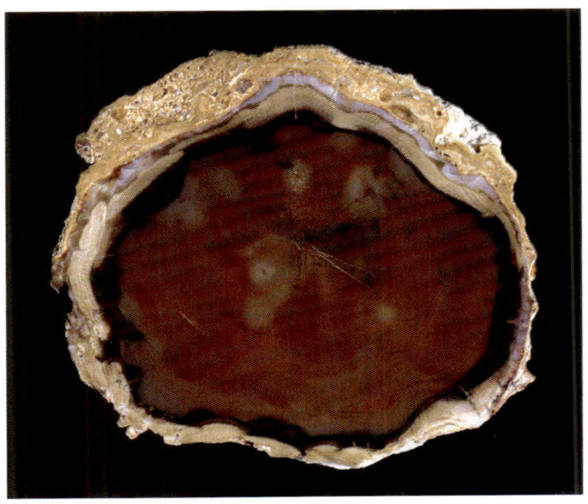

Taxodioxylon. Tertiary. Stonewall Pass near Goldfield, Nevada. Daniels specimen, 5 cm.

Opalized fossil wood. Virgin Valley Formation, Miocene. Virgin Valley, Nevada, USA. Hodson specimen, 8 cm.

Opalized fossil wood. Virgin Valley Formation, Miocene. Virgin Valley, Nevada, USA. Hodson specimen, 6.5 cm.

Opalized fossil wood. Virgin Valley Formation, Miocene. Virgin Valley, Nevada, USA. Daniels specimen, 5 cm.

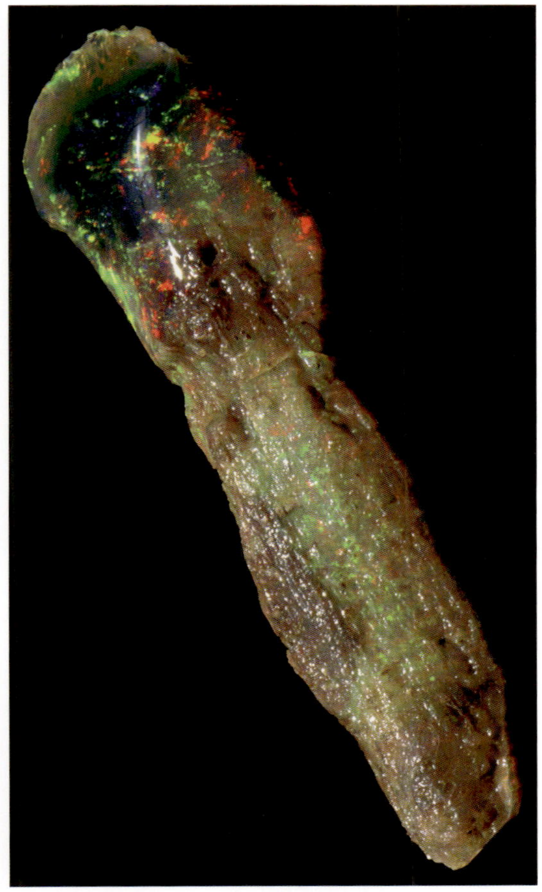

Opalized fossil wood. Virgin Valley Formation, Miocene. Virgin Valley, Nevada, USA. Hodson specimen, 19 cm.

Opalized fossil wood. Virgin Valley Formation, Miocene. Virgin Valley, Nevada, USA. Hodson specimen, 13.5 cm.

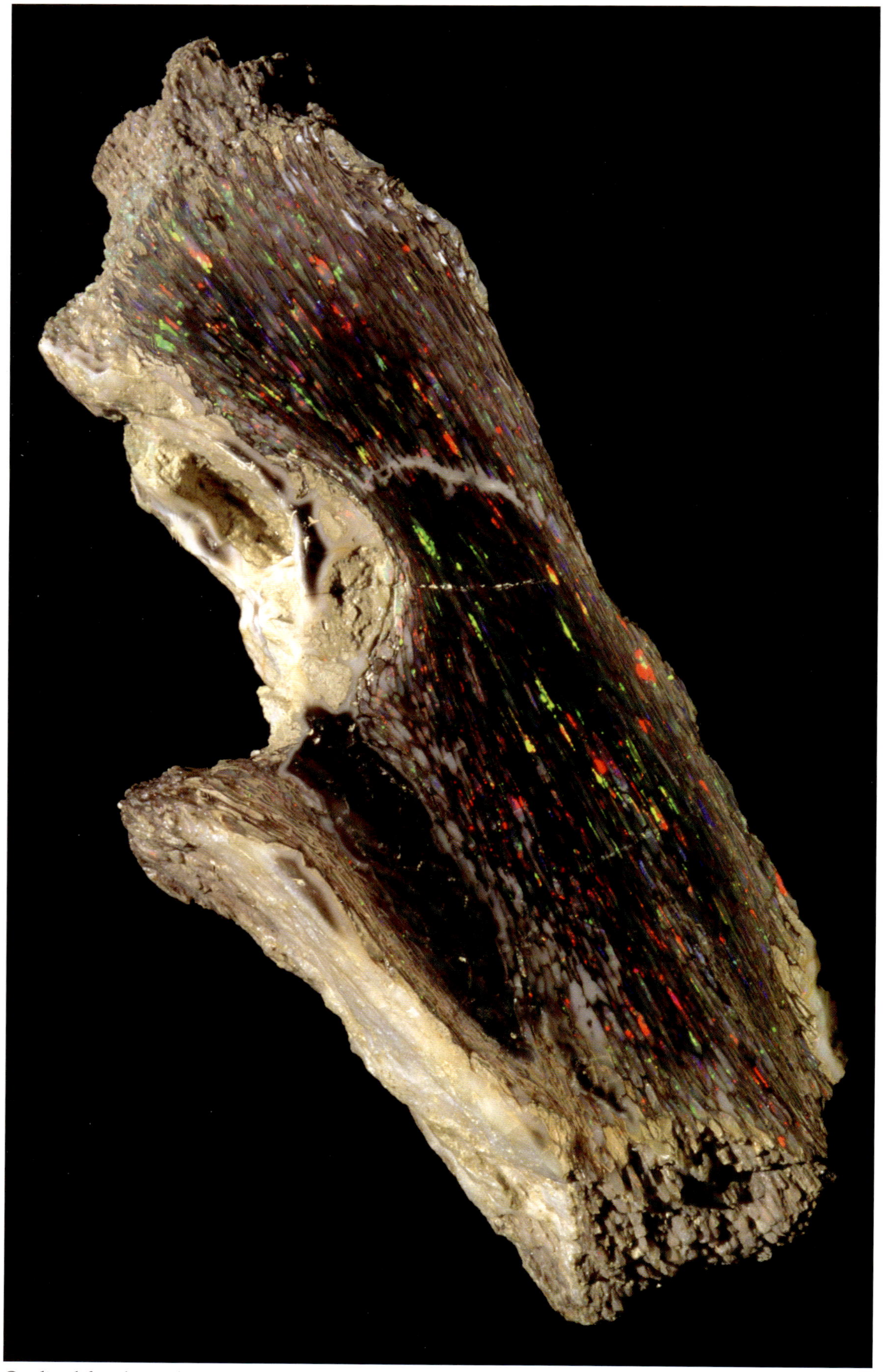

Opalized fossil wood. Virgin Valley Formation, Miocene. Virgin Valley, Nevada, USA. Hodson specimen, 21 cm.

Pityoxylon. Middle Miocene. Cedar Pass, California. Daniels specimen, 10 cm.

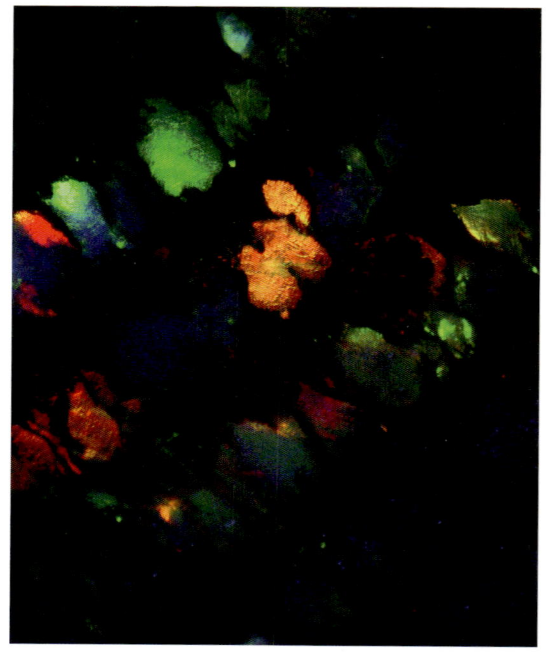

Micro image showing borer tunnel with fecal pellets in fossil conifer. Virgin Valley Formation, Miocene. Virgin Valley, Nevada, USA. Hodson specimen.

Longitudinal cuts of woods from the northwestern United States often display attractive growth ring patterns, accentuated by varied earlywood to latewood transitions, and sometimes display ray fleck in hardwoods.

Old postcard of the petrified forest in Calistoga, California, showing a huge *Taxodioxylon* called "The Giant."

Old postcard of the petrified forest in Calistoga, California, showing a huge *Taxodioxylon* called "The Tunnel Tree."

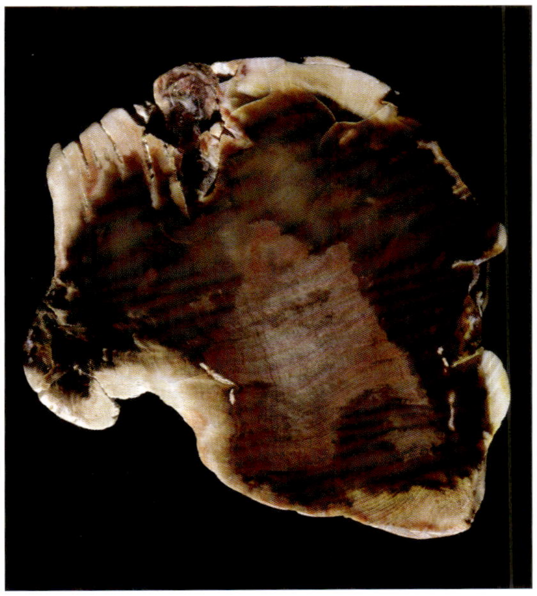

Longitudinally cut fossil softwood. Fossil conifer specimens from this locale are rarely more than colorful chunks of much larger trees, yet they are glassy, translucent, and colorful specimens. Tertiary. Owyhee Mountains, Malheur County, Oregon, USA. Daniels specimen, 8.5 cm.

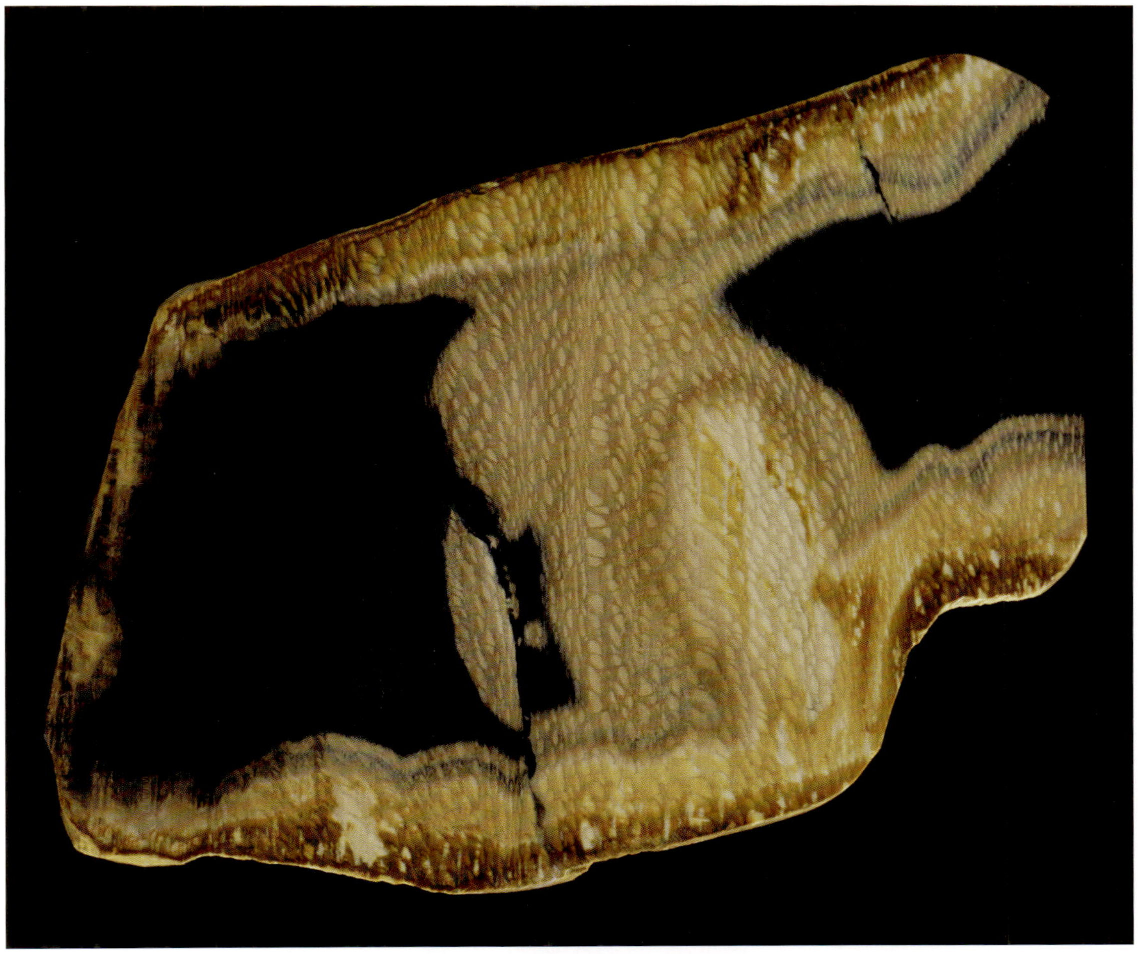

Longitudinally cut sycamore-like fossil hardwood with ray fleck pattern. Tertiary. Rogers Mountain, Linn County, Oregon, USA. Daniels specimen, 16.5 cm.

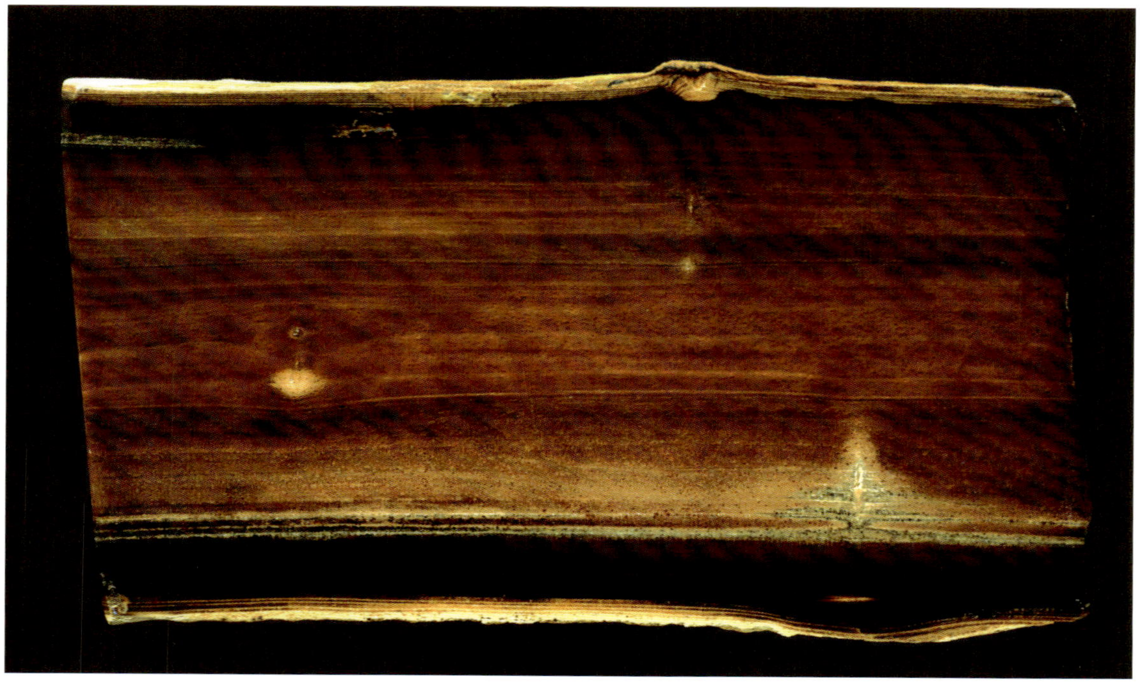

Longitudinally cut fossil hardwood. Trout Creek Formation, Miocene. McDermitt area, Oregon/Nevada, USA. Daniels specimen, 16 cm.

Longitudinally cut sycamore-like fossil hardwood with ray fleck pattern. Columbia Plateau Basalt flow, Miocene. Saddle Mountain, Washington, USA. Daniels specimen, 19.5 cm.

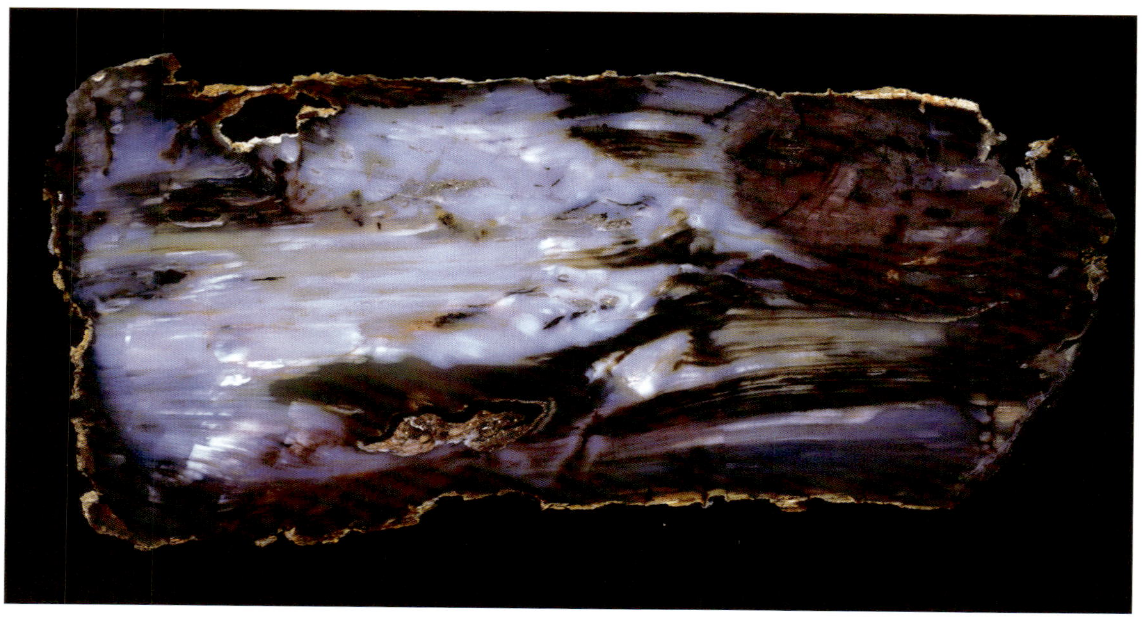

Longitudinally cut fossil conifer. Tertiary. Hubbard Basin, Elko County, Nevada, USA. Daniels specimen, 23 cm.

Borer evidence in longitudinally cut fossil wood. Columbia Plateau Basalt flow, Miocene. Saddle Mountain, Washington, USA. Daniels specimen, 23.5 cm.

Macro photograph of borer evidence in longitudinally cut fossil wood. Columbia Plateau Basalt flow, Miocene. Saddle Mountain, Washington, USA. Daniels specimen, 23.5 cm.

Longitudinally cut fossil conifer. Tertiary. Antelope, Wasco County, Oregon. Daniels specimen, 24 cm.

Longitudinally cut fossil hardwood. Succor Creek Formation, Miocene. Grassy Mountain, Malheur County, Oregon, USA. Daniels specimen, 10.5 cm.

IV

CHAPTER FOUR

Fossil Woods of the Southwestern United States

ARIZONA, UTAH, AND SOUTHWESTERN COLORADO

Mesozoic rocks of eastern Utah, southwestern Colorado, and northern Arizona produce some of the most colorful and unusual petrified wood specimens from anywhere in the world. Specifically, the Late Triassic Chinle Formation, the Late Jurassic Morrison Formation, and the Early Cretaceous Cedar Mountain Formation contain most of the collectable fossil woods in this region. A fortunate set of stratigraphic, tectonic, and geochemical circumstances fostered the preservation and exposure of this resource. This part of the Colorado Plateau contains an unusually complete stratigraphic sequence of Mesozoic rocks deposited by rivers, streams, and lakes.

Forests growing in these environments were uprooted and buried by periodic floods. Much of the sediment contained volcanic ash or volcaniclastic muds that subsequently devitrified to produce copious amounts of mobile silica, which replaced organic matter. Finally, large-scale tectonic activity uplifted the entire Colorado Plateau and exposed the region to erosion for the past 5 million years. Other smaller scale structural features such as salt anticlines and igneous intrusions have warped rock sequences and enhanced localized erosion of sediments to expose fossil-bearing sequences.

This region contains the well-known silicified woods of Arizona but also yields carnelian limb casts historically termed "red wood" from Utah. Other unusual or noteworthy fossil woods are the spectacularly colored cycadeoid trunks of Utah, a variety of small and strangely shaped fossils called short shoots, a diverse array of conifer cones, and some unusual and colorful gymnosperm trunks. The general stratigraphic and structural controls for this area and descriptions of some of the most colorful and interesting fossil woods are the subjects of this chapter. Special attention is paid to some of these lesser known fossil wood treasures, especially in Utah.

Late afternoon drama in Colorado National Monument, Grand Junction, Colorado. View to the northwest of Independence Rock, a 450-foot remnant formed by lowermost red siltstone and shales of the Late Triassic Chinle Formation with vertical cliffs of Jurassic Wingate Sandstone and a small cap of Kayenta Formation sandstone. Dark-colored low hills behind and below Independence Rock are composed of Precambrian metamorphic and igneous rocks. This means a gap in time of about 1.5 billion years exists between the Chinle Formation and Precambrian units.

ARIZONA

The famous Arizona petrified wood forests were growing about 213 million years ago during the Late Triassic. At that time, the area now home to the Chinle Formation, was an enclosed continental basin at the western edge of Pangea, a supercontinent that existed before Triassic time. Over millions of years, this basin saw a variety of shifting river channels, fresh-water lakes, floodplains, deltas, basins, and marshes. These sedimentary environments, along with a tropical climate, were favorable to a varied array of plants including huge conifer trees, cycads, and seed ferns. The deeply rooted *Araucaria*-like trees became the beautiful fossil logs at Petrified Forest National Park in Arizona. Reconstructions of these trees suggest that branches had no systematic arrangement and apparently persisted from low on the trunk to the crown. The lack of growth rings probably indicates a tropical climate with little change in temperature. Volcanoes to the west and southwest contributed massive volumes of ash, providing the raw material for the geologic formation as well as a plentiful source of silica for fossilization. Extensive flooding swept trees from their stands, burying them in sand, ash, and gravel. Fossil woods include trees carried long distances from their growth sites, as well as trees and smaller plants at or near their original sites.

This hillside of fossil logs, known as the Log Jam, eroded from Triassic Chinle clays in northern Arizona. Petrified Forest National Park, Arizona, USA.

Group of green fossil wood specimens. The color in these rare specimens derives from chromium. *Woodworthia* and *Araucarioxylon*. Chinle Formation, Triassic. Northern Arizona, USA. Daniels and Rolater specimens, longest 24 cm.

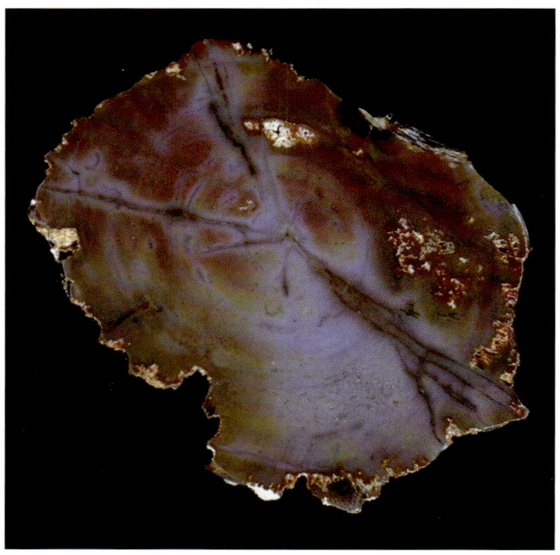

Araucarioxylon. Chinle Formation,Triassic. Northern Arizona, USA. Daniels specimen, 7.5 cm.

Araucarioxylon. Chinle Formation, Triassic. Northern Arizona, USA. Daniels specimen, 8.5 cm.

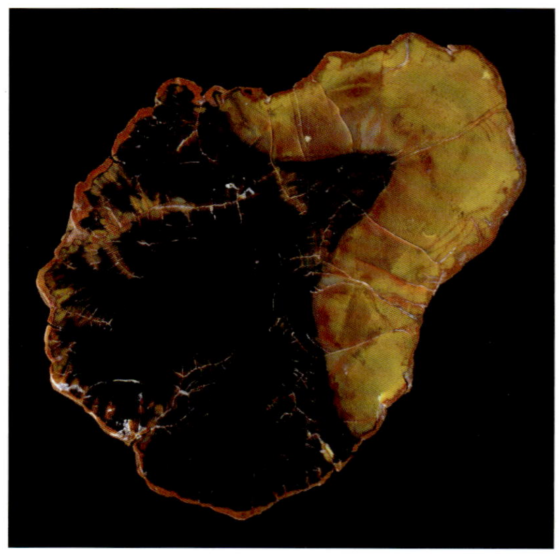

Araucarioxylon. Chinle Formation, Triassic. Northern Arizona, USA. Daniels specimen, 11 cm.

Group of mustard yellow/gold fossil wood specimens. The color in these rare specimens derives from goethite. Star dadoxylon and unidentified specimens. Chinle Formation, Triassic. Northern Arizona, USA. Daniels specimens, longest 16.5 cm.

Araucarioxylon. Chinle Formation, Triassic. Northern Arizona, USA. Daniels specimen, 10.5 cm.

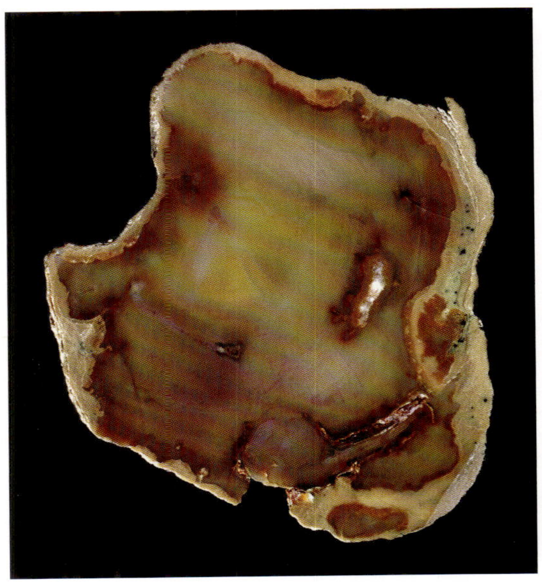

Araucarioxylon. Chinle Formation, Triassic. Northern Arizona, USA. Daniels specimen, 7 cm.

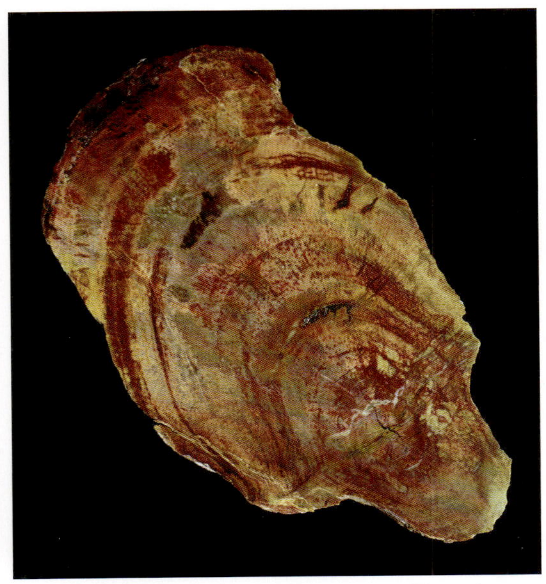

Araucarioxylon. Chinle Formation, Triassic. Northern Arizona, USA. Daniels specimen, 13.5 cm.

Araucarioxylon. Chinle Formation, Triassic. Northern Arizona, USA. Daniels specimen, 19 cm.

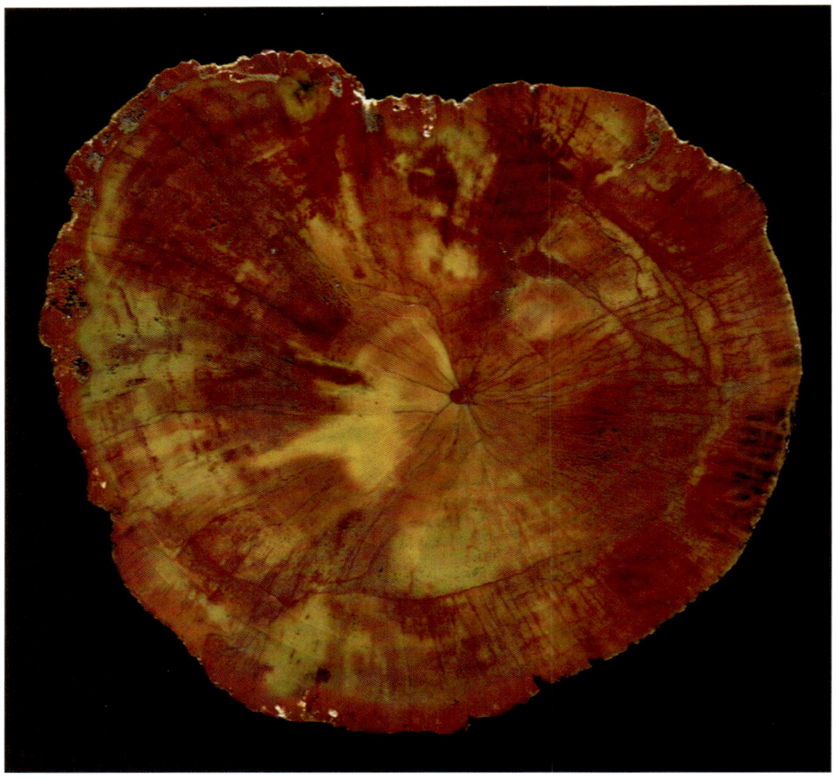

Araucarioxylon. Chinle Formation, Triassic. Northern Arizona, USA. Daniels specimen, 12.5 cm.

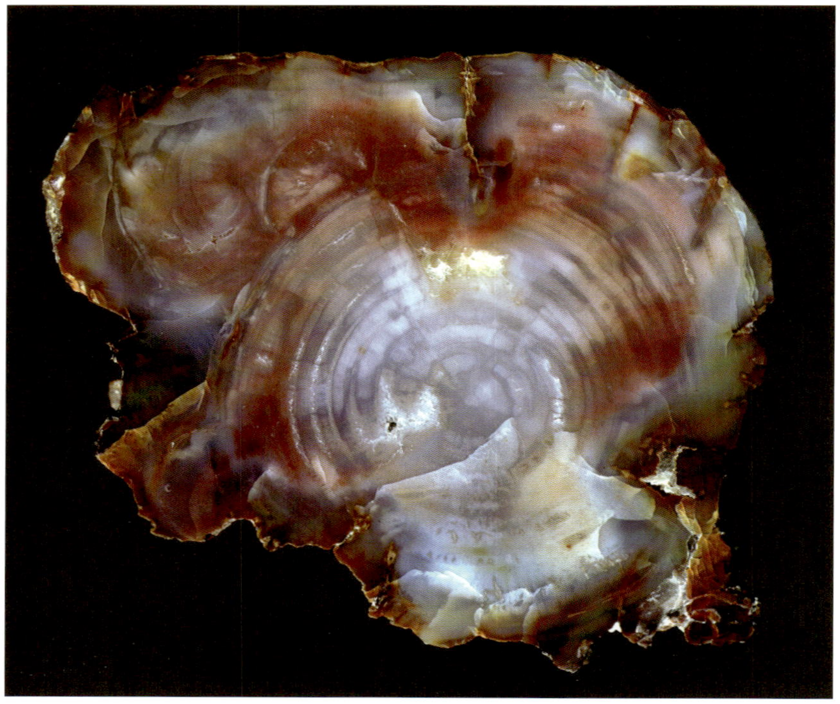

Araucarioxylon. Chinle Formation, Triassic. Northern Arizona, USA. Daniels specimen, 12 cm.

Araucarioxylon. Chinle Formation, Triassic. Northern Arizona, USA. Daniels specimen, 15 cm.

Natural erosion such as that which unearthed millions of pounds of fossil wood in northern Arizona continues today, as in these falls on a tributary to the Colorado River in the Grand Canyon. Havasu Falls, near Supai, Arizona, USA.

Araucarioxylon. Chinle Formation, Triassic. Northern Arizona, USA. Daniels specimen, 18.5 cm.

Araucarioxylon. Chinle Formation, Triassic. Northern Arizona, USA. Daniels specimen, 8.5 cm.

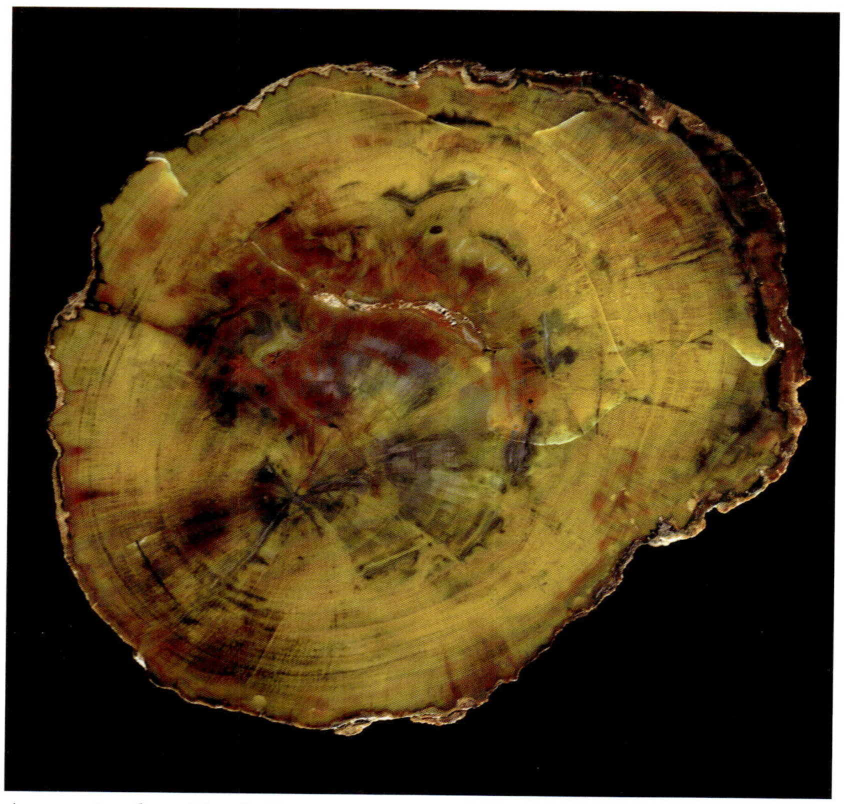

Araucarioxylon. Chinle Formation, Triassic. Northern Arizona, USA. Daniels specimen, 16 cm.

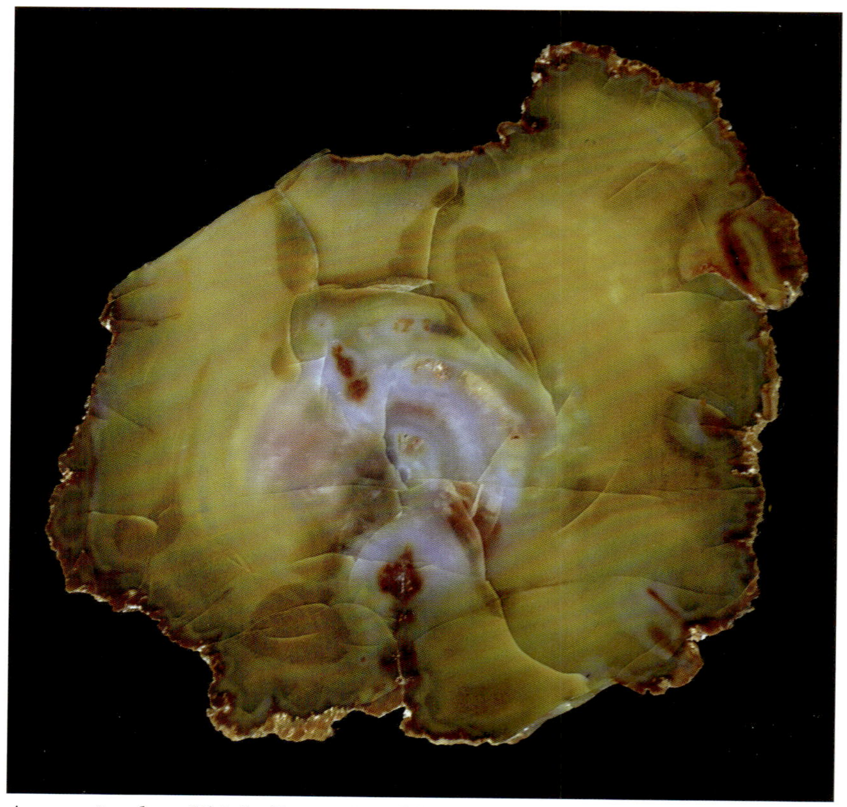

Araucarioxylon. Chinle Formation, Triassic. Northern Arizona, USA. Daniels specimen, 15.5 cm.

The Chinle Formation now extends, with considerable outcrops, from northern Utah to central Arizona and southern Nevada to western Colorado and central New Mexico. But the spectacularly colored fossil wood from the Chinle Formation derives mainly from the Petrified Forest Member in northern Arizona. This member contains greenish to grayish shales and mudstones interlayered with reddish shales and mudstones, often weathering to rounded badlands devoid of vegetation. This colorful badland topography is similar to the landscape produced by the Late Jurassic Brushy Basin Member of the Morrison Formation. The Chinle Formation in Utah produces copious amounts of petrified wood too, but most of it is dark brown to almost black and generally not sought after by collectors. However, colorful Triassic woods are occasionally found in southeastern Utah. The accompanying photographs depict some the incredible varieties of silicified wood from Arizona.

A foggy day in Petrified Forest National Park with a log eroding from the Triassic Period Chinle Formation on Blue Mesa. Note the interspersed layers of gravel among the fine clays. Petrified Forest National Park, Arizona, USA.

Araucarioxylon. Chinle Formation, Triassic. Northern Arizona, USA. Daniels specimen, 7.5 cm.

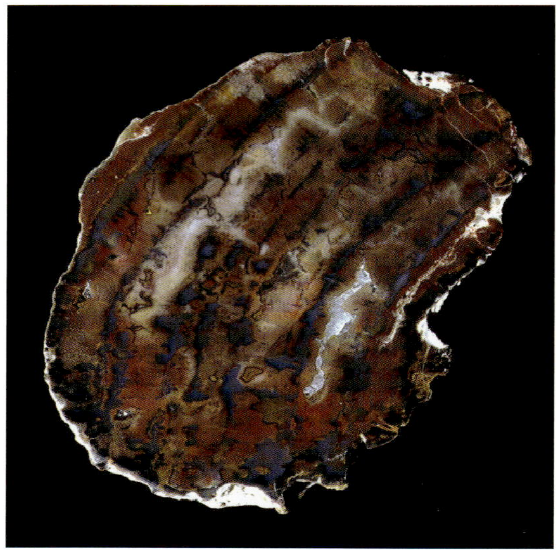

Araucarioxylon. Chinle Formation, Triassic. Northern Arizona, USA. Daniels specimen, 11 cm.

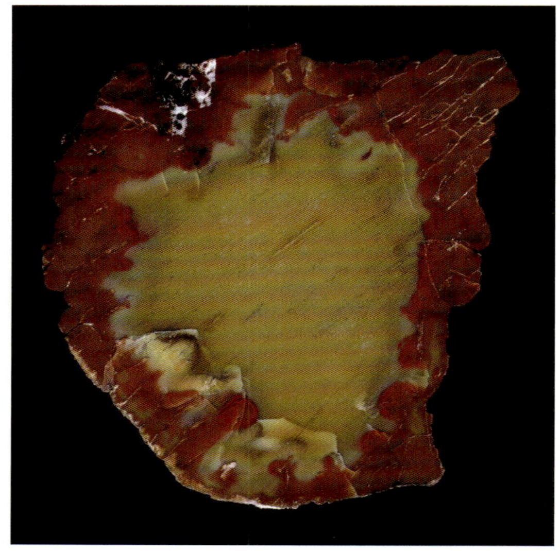

Araucarioxylon. Chinle Formation, Triassic. Northern Arizona, USA. Daniels specimen, 9 cm.

Araucarioxylon. Chinle Formation, Triassic. Northern Arizona, USA. Daniels specimen, 6.5 cm.

Araucarioxylon. Chinle Formation, Triassic. Northern Arizona, USA. Daniels specimen, 6 cm.

Araucarioxylon. Chinle Formation, Triassic. Northern Arizona, USA. Daniels specimen, 10.5 cm.

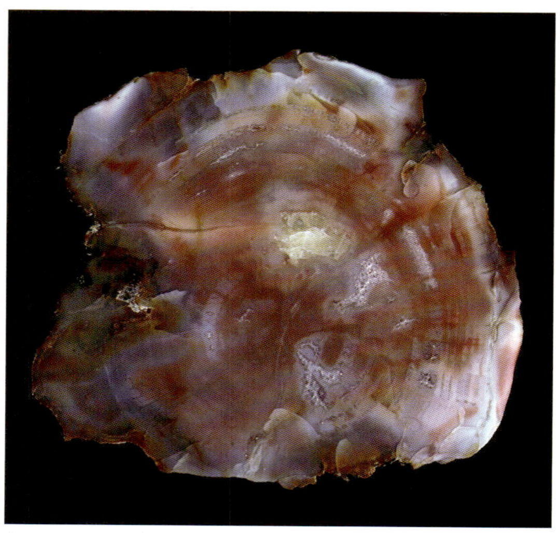

Araucarioxylon. Chinle Formation, Triassic. Northern Arizona, USA. Daniels specimen, 10 cm.

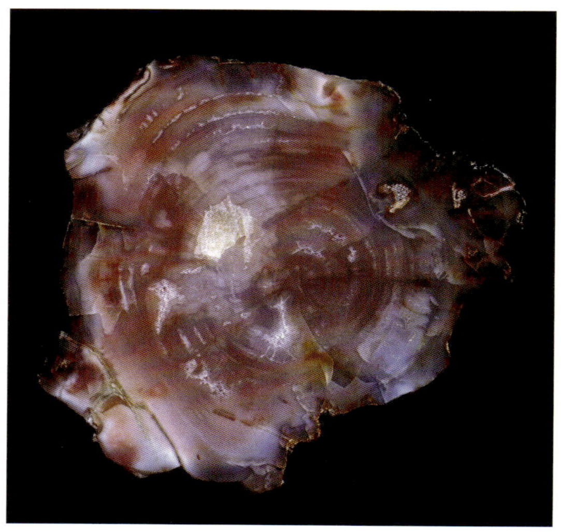

Araucarioxylon. Chinle Formation, Triassic. Northern Arizona, USA. Daniels specimen, 10.5 cm.

Araucarioxylon. Chinle Formation, Triassic. Northern Arizona, USA. Daniels specimen, 8.5 cm.

Eroded Chinle Formation clay hills near Nazlini, Navajo Nation, Arizona, USA.

Araucarioxylon. Chinle Formation, Triassic. Northern Arizona, USA. Daniels specimen, 13 cm.

Araucarioxylon. Chinle Formation, Triassic. Northern Arizona, USA. Daniels specimen, 11.5 cm.

Araucarioxylon. Chinle Formation, Triassic. Northern Arizona, USA. Daniels specimen, 12.5 cm.

Araucarioxylon. Chinle Formation, Triassic. Northern Arizona, USA. Daniels specimen, 9.5 cm.

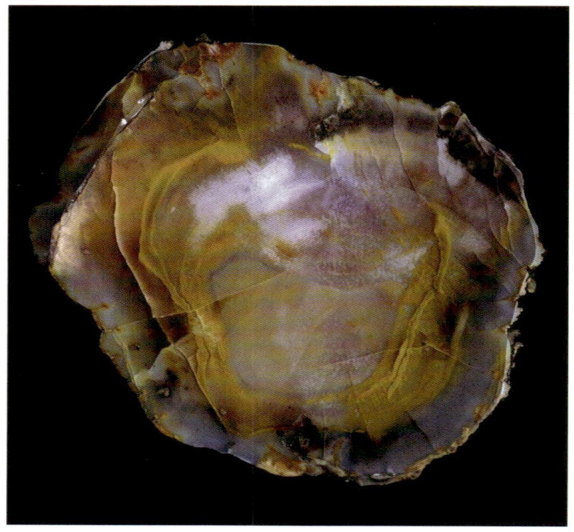

Araucarioxylon. Chinle Formation, Triassic. Northern Arizona, USA. Haggerty specimen, 10 cm.

Araucarioxylon. Chinle Formation, Triassic. Northern Arizona, USA. Daniels specimen, 8 cm.

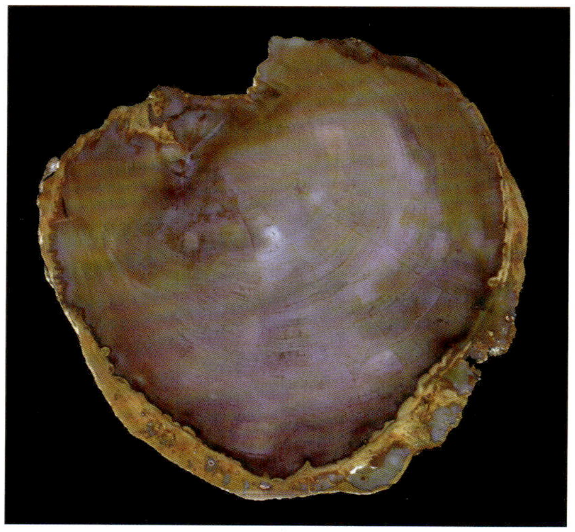

Araucarioxylon. Chinle Formation, Triassic. Northern Arizona, USA. Daniels specimen, 11 cm.

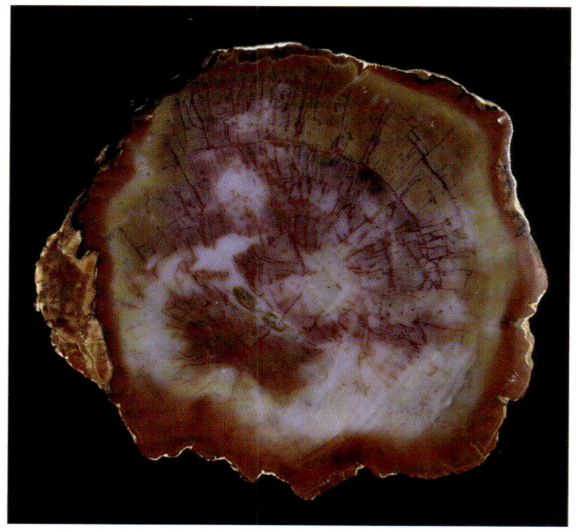

Araucarioxylon. Chinle Formation, Triassic. Northern Arizona, USA. Daniels specimen, 9.5 cm.

Araucarioxylon. Chinle Formation, Triassic. Northern Arizona, USA. Daniels specimen, 7 cm.

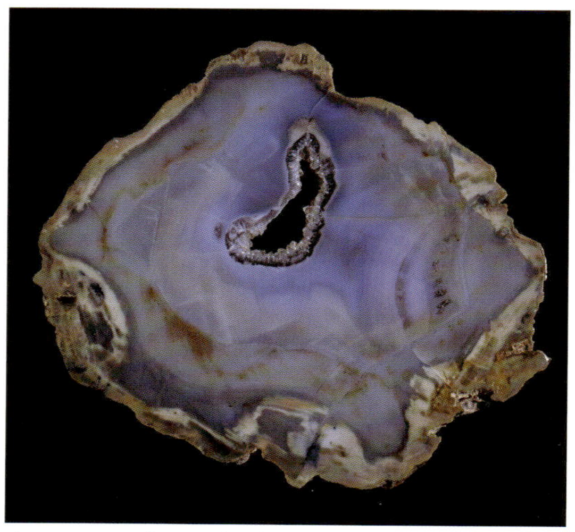

Araucarioxylon. Chinle Formation, Triassic. Northern Arizona, USA. Daniels specimen, 9 cm.

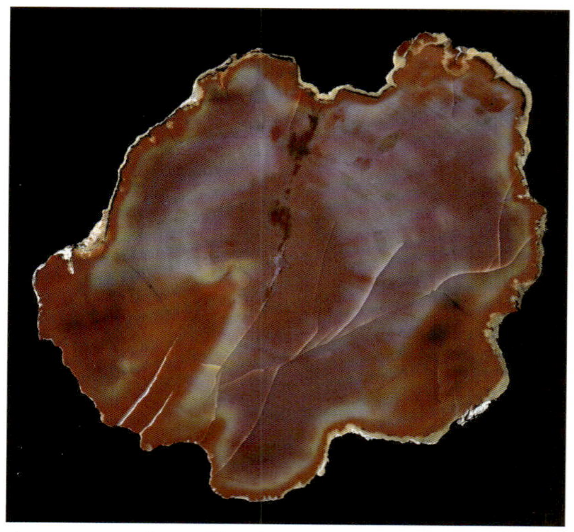

Araucarioxylon. Chinle Formation, Triassic. Northern Arizona, USA. Daniels specimen, 10.5 cm.

These cultivated Norfolk Island pine trees (*Araucaria heterophylla*) on the flank of dormant volcano Haleakala in Maui, Hawaii, USA, offer a hint of how ancient Mesozoic forests may have appeared.

Norfolk Island pine (*Araucaria heterophylla*), native only on Norfolk Island in the South Pacific, flourish together with palms and cycads in tropical and subtropical environments. Maui, Hawaii, USA.

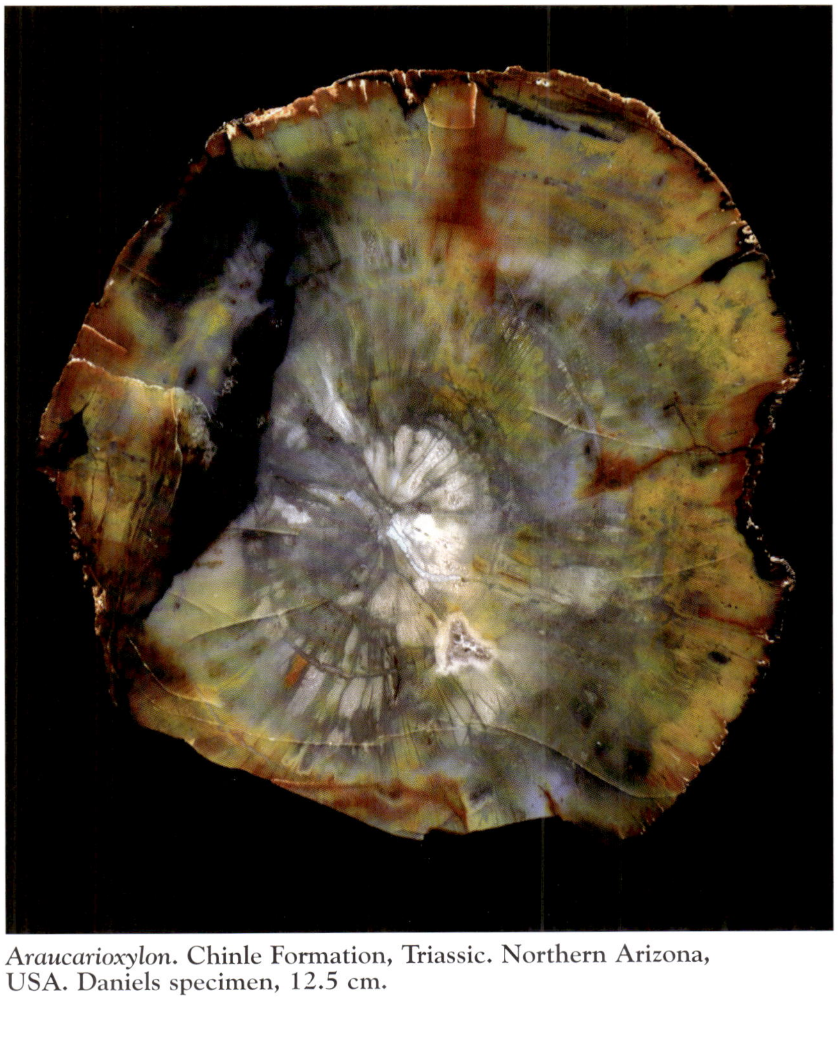

Araucarioxylon. Chinle Formation, Triassic. Northern Arizona, USA. Daniels specimen, 12.5 cm.

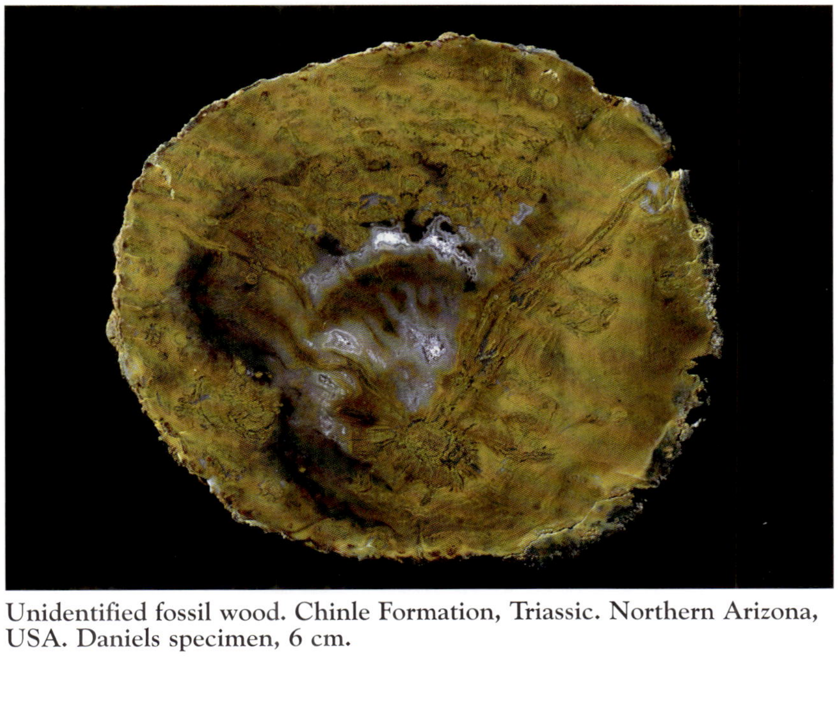

Unidentified fossil wood. Chinle Formation, Triassic. Northern Arizona, USA. Daniels specimen, 6 cm.

80-foot-long *Araucarioxylon* log being excavated on private land in northern Arizona. Navajo County, Arizona, USA.

Macro photograph of termite fecal pellets in fossil wood. Age unknown. Northern Arizona, USA. Daniels specimen.

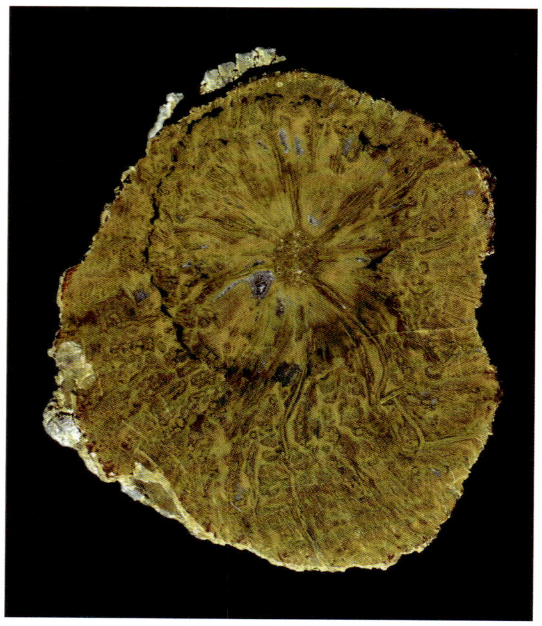

Unidentified fossil wood. Chinle Formation, Triassic. Northern Arizona, USA. Daniels specimen, 6 cm.

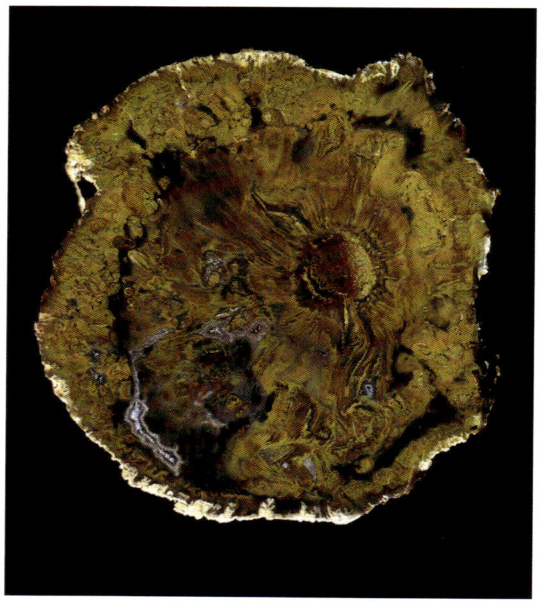

Unidentified fossil wood. Chinle Formation, Triassic. Northern Arizona, USA. Daniels specimen, 6.5 cm.

Unidentified fossil wood. Chinle Formation, Triassic. Northern Arizona, USA. Daniels specimen, 8.5 cm.

View to the north at Mt. Hillers in the southern Henry Mountains, a mid-Tertiary laccolith composed of porphyritic diorite. Hogbacks of sedimentary rocks dip steeply away from or have been incorporated into the intrusive igneous rocks, some slightly baked by the heat of the intrusion. They are sandstones and shales of Permian through Jurassic age. The orange-to-pinkish units are Jurassic Period Entrada Sandstone and the white low-lying sandstone outcrops farther away from the mountain are Morrison Formation sandstones. The low rolling hills in the foreground are composed of Late Cretaceous marine Mancos Shale that also dip away from the laccolith. Farther away and surrounding the mountains are vast expanses of Morrison Formation sandstones and shales that contain conifer and cycad woods.

UTAH

By Late Jurassic time, the supercontinent of Pangea had split into northern and southern sections, with North America in the north. However, the southwestern United States was closer to the equator than today. Several specific tectonic events from that time to the present contributed to the fortuitous circumstances that allowed the the fossil wood contained in several formations to be eroded and uncovered for collectors. The first was the formation of salt anticlines in this region beginning during the Pennsylvanian Period some 300 million years before present (mybp). This was a low region covered by a shallow ocean that periodically evaporated, which produced vast amounts of layered evaporite minerals, including halite (salt), in what came to be called the Paradox Basin. Because salt is plastic during long intervals of time, it tends to flow along lines of least resistance, such as faults, and forms upward-piercing diapirs that dome overlying sediments. These elongated domed structures are called salt anticlines, and they occur throughout this region.

A second tectonic event was the intrusion of mid-Tertiary granitic rocks into structures called laccoliths. The La Sal Mountains near Moab and the Henry Mountains near Hanksville are examples of these distinctively shaped intrusions that resemble, in vertical cross section, the side view of an unfolded umbrella. Magma intrudes the host sediments by feeder dikes or conduits and spreads laterally into the sediments, thickening in the middle and bowing up the overlying sequences.

The third tectonic event was much larger and reshaped the entire western United States. The collision of the East Pacific continental plate with the North American plate, particularly in the past 5 to 10 million years, fractured the western United States into Basin and Range Provenance and uplifted the Colorado Plateau basically as a huge structural block. That uplift resulted in greatly increased regional erosion of near-surface Mesozoic rocks and especially enhanced erosion of tilted sequences surrounding salt anticlines and laccoliths in eastern Utah and southwestern Colorado. For the salt anticlines, erosion allowed ground water to reach the inner core of the anticline and dissolve salt, causing the structure to collapse. This mechanical failure resulted in a topographic valley along the crest of the anticline with rocks dipping away from the axis. Saw-toothed hogbacks of tilted sediments dipping away from these anticlines and laccoliths are common sights in the area. Robert Scott and other geologists calculate that on average, the Colorado River, which flows across this region, is incising 15 centimeters every 1,000 years. While this seems slow in human time, it is rapid in geologic time. In a million years, about 150 meters of incision occurs in this region. Softer rock units such as shales erode at much this same rate, but harder units such as well-cemented sandstones erode somewhat slower. All in all, large quantities of fossil wood continue to erode from Mesozoic age conglomerates, sandstones, siltstones, and bentonitic claystones and mudstones.

View to the south along the Colorado River looking at Fisher Towers and the snow-covered La Sal Mountains in the distance. The Colorado River flows around the bend to the right and across a collapsed salt anticline that produced Castle Valley. The stratigraphic sequence in the middle distance is lower Permian Cutler Formation overlain by the Triassic Moenkopi Formation, which forms the top of the towers. Above the Moenkopi sandstones are slopes of Triassic Chinle Formation, and the highest cliffs on the left are Jurassic Wingate Sandstone. The La Sal Mountains, so named by Spanish explorers because they thought they looked like they were capped with salt, are intrusions of porphyritic diorite emplaced about 25 to 27 mybp. When the La Sal Mountains were formed, sedimentary rocks were tilted up and now erode faster, exposing fossil-bearing units such as the Morrison Formation.

Many fossil wood collectors are familiar with agate limb casts known as "red wood." These casts, found initially and most prolifically in the Yellow Cat area of Grand County, Utah, are among the most beautiful examples of agate fortification wood from anywhere in the world. According to Cannon, the curious appellation "Yellow Cat" is derived from a mountain lion or cougar that was killed near an old campsite located by one of the rare springs. Another version tells of an abandoned, yellow-colored, Caterpillar brand bulldozer as the source of the name. Regardless, the site became known as the Yellow Cat Campsite. Later the name was used to refer to the mining district, and even later to describe the general area. Today the name Yellow Cat Flats appears on topographic maps.

Although collectors have used the name red wood for many years, it is purely a color description and a colloquialism. Prize specimens are entirely agate casts that do not preserve original woody textures, thus making identification impossible. However, the original wood was most likely conifer because most Jurassic age woods were conifers. The fortification agate is typically banded carnelian with accent bands of white, gray, black, yellow, orange, blue, and violet chalcedony. Red wood is found in the Brushy Basin Member of the Late Jurassic Morrison Formation, and areas other than Yellow Cat produce exceptional specimens. Localities are known from northeast of Moab, west to the Green River, south to the Dolores River, and even farther southwest to areas around the Henry Mountains. Although Yellow Cat red wood was not difficult to find 70 years ago, modern collectors must be diligent or lucky to find good specimens.

The Morrison Formation, which contains all red wood casts, has been studied and described by many investigators during the past 100 years. Investigators include Eldridge in the 1800s, Baker and others in the 1930, Stokes in the 1940s, Craig and others in the 1950s, Dare in the 1960s, and, more recently, Peterson, Turner and Fishman, and Anderson and Lucas in the 1980s and 1990s. A two-volume set of *Modern Geology* from 1998 is devoted to describing the Morrison Formation. This formation has traditionally divided into two members, the lower sandstones of the Salt Wash Member containing radium, uranium, and vanadium mineralization that is about 73 meters thick, and shales of the upper Brushy Basin Member containing the red wood, which is about 91 meters thick. The Salt Wash Member was produced by a series of rivers and streams that originated from the west and formed a regional delta system, filling broad plains left by retreating Jurassic seas.

The Brushy Basin Member shales, mudstones, and sandstones represent a change in depositional environment. Deposition in this member is generally considered to occur in a lower energy environment with rivers flowing east to southeast across a broad floodplain, sometimes choked with large volumes of volcanic ash. According to Anderson and Lucas, "Brushy Basin Member deposition took place on a vast muddy floodplain characterized by a mosaic of meandering channels, small lakes or playas and stable pedogenically-modified interfluves (areas between channels that develop soil horizons) with an extensive, long-term infall and/or reworking of volcanic ash." They also suggest numerous shallow lakes or playas in the Brushy Basin that were 5 to 10 miles across. During times of desiccation, these lakes or playas could have become alkaline and saline, producing authigenic minerals like salt, gypsum, and silica. Deposits of nodular carbonates, limestone or micrite beds, silcretes, and smetetic clays exist in the Brushy Basin sediments. Outcrops of the Brushy Basin usually consist of rounded hills showing strikingly banded green, gray, and red shales and mudstones that produce a layer-cake effect, inspiring names such as Pinto Hills along the eastern flank of the Henry Mountains.

All red wood is found in the Brushy Basin Member of the Morrison Formation. Ironically, the authors know of no red wood from red shales and mudstones; all specimens are found in the greenish gray shales and underlying conglomerates. Typically, the favorable horizon contains copious amounts of dark carbonaceous plant debris, often with distinguishable twigs or root fragments. Most of the red wood and other fossil wood are present just above a gravelly layer containing intermittent small pebbles mixed with sand. This placement suggests that the wood was deposited by a pulse of high-energy like a flood, but the eventual depositional environment was generally a much quieter environment, like backwaters in a river system or a lake. Occasionally, a layer containing calcareous concretions, which may grade laterally into a calcareous bed, lies at the base of the wood horizon. Wood is sometimes included in the siliceous-to-calcareous nodules or beds.

View from Clay Point near Ticaboo, Utah, looking south toward Lake Powell. Salt Wash Member Morrison Formation sandstones produce stair steps in the middle distance and the nearest hill is a thick section of Brushy Basin Member Morrison Formation; all are Upper Jurassic. A portion of the Waterpocket Fold is seen on the right. Lighter colored sandstones in the far distance constitute Jurassic age Entrada Sandstone.

A red wood limb section is exposed in place. It lies above a gravelly unit in a carbonaceous clay sequence. This location represents a high-energy pulse that deposited coarse-grain sediments and left wood above the gravels before it was covered with finer grain sands and muds. Many red wood limbs show considerable rounding such as this specimen. The photograph to the right shows a cut and polished cross section of this specimen.

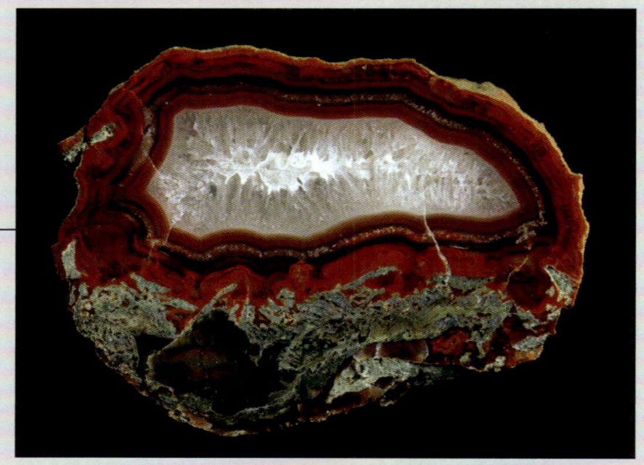

Limb cast with residual silicified wood in the bottom of the cavity mixed with silicified clayey matrix and agate in the top. Brushy Basin Member, Morrison Formation, Late Jurassic. Moab, Utah, USA. Daniels specimen. 7.5 cm.

Another stratigraphic horizon for fossil wood, including red wood limb casts, is at the base of the Brushy Basin directly above and included in a massive basal conglomerate. This conglomerate sometimes contains abundant red chalcedony cement and often contains greenish pebbles, giving it the common name "Christmas conglomerate." Silicified shells of the common elongate Jurassic freshwater clam, *Unio sp.*, are sometimes found with the wood.

As with all limb casts, red wood specimens follow a typical sequence of events to produce agate fortifications. The formation of limb casts resulting from the crystallization of chalcedony from siliceous ground water solutions is described in Chapter Two. Red, violet, orange, and yellow colors in this spectacular agate are the result of trace amounts of iron and manganese oxides; bluish colors are probably due to light dispersion from micron-sized bubbles in colorless chalcedony. One aspect of the limb casts from the classic Yellow Cat locality that distinguishes them from limb casts found in other Brushy Basin localities is the completeness of agatization. Limb casts from Yellow Cat are among the most completely fortified, often having no fragmental or vestigial wood that is partly permineralized. This incomplete fortification is commonly seen in Brushy Basin limb casts found farther south and west.

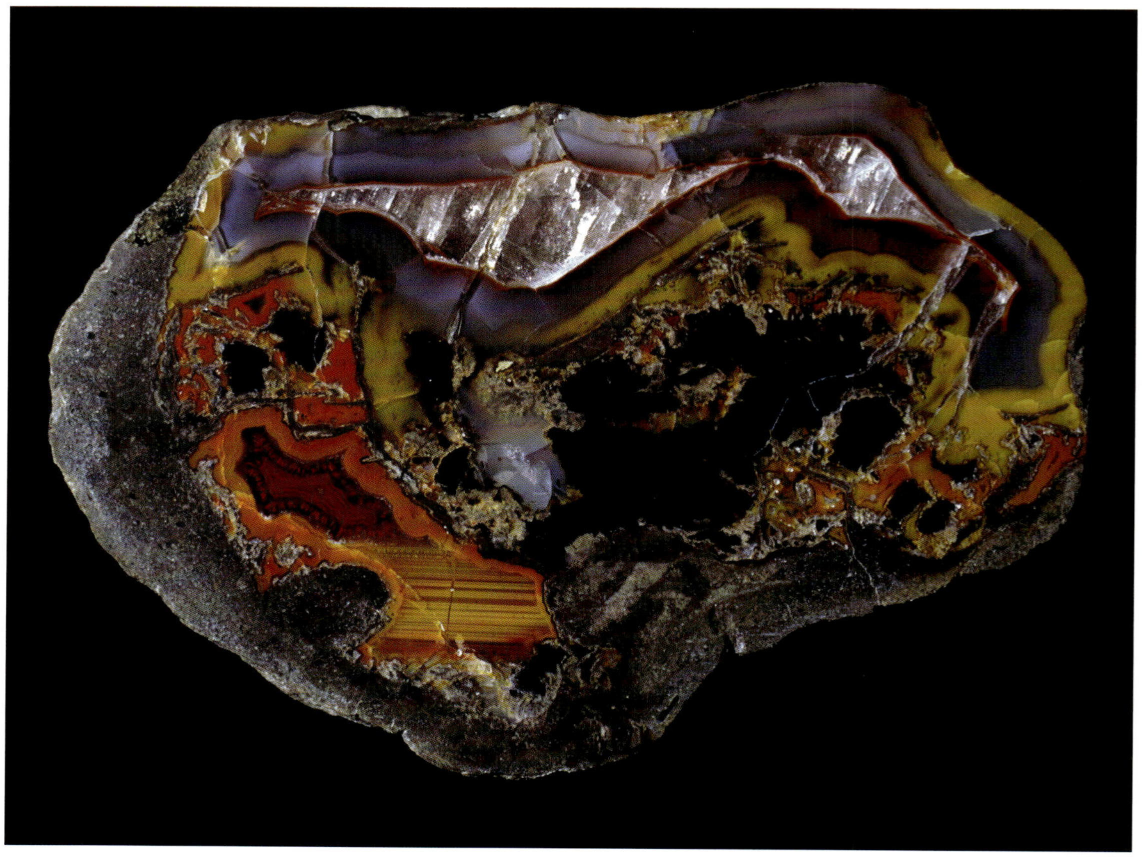

Partial limb cast with a variety of agates infilling open areas. Vestiges of templated and permineralized wood are also present in this specimen, along with a silicified rind that sometimes forms around limb casts from this area. Brushy Basin Member, Morrison Formation, Late Jurassic. Moab, Utah, USA. Dayvault specimen, 7.5 cm.

Partial limb cast with orange and yellow agate infilling open area. Vestiges of templated and permineralized wood are visible in this specimen. Brushy Basin Member, Morrison Formation, Late Jurassic. Moab, Utah, USA. Dayvault specimen, 7.5 cm.

Some incompletely formed limb casts have only a siliceous rind and internal areas that are intermittently silicified or contain calcite or even remain open and pithy. Occasionally, highly silicified templated and permineralized wood is found closely associated with the limb casts. Why one limb completely rots away or disintegrates leaving a void to be filled with agate and an adjacent limb, perhaps less than a meter away, remains intact long enough for templating and permineralization by silica to preserve the original woody texture is a mystery. Partial agate fortification in the otherwise templated and permineralized wood is also fairly common.

The chances of finding additional red wood in the original Yellow Cat area continue to decrease, although new specimens are occasionally uncovered. Collectors should broaden their exploration of similar environments in Brushy Basin sediments elsewhere in the region where fine limb casts also have been discovered. Three examples of fine red wood limb casts that were found many kilometers from Yellow Cat follow.

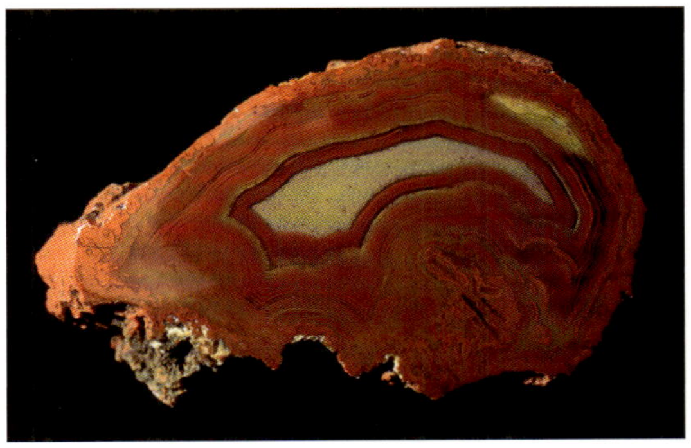

Red-to-orange and yellow agate fills this short shoot stalk section. Brushy Basin Member, Morrison Formation, Late Jurassic. Ticaboo area, southern Henry Mountains, Utah, USA. Dayvault specimen, 3.5 cm.

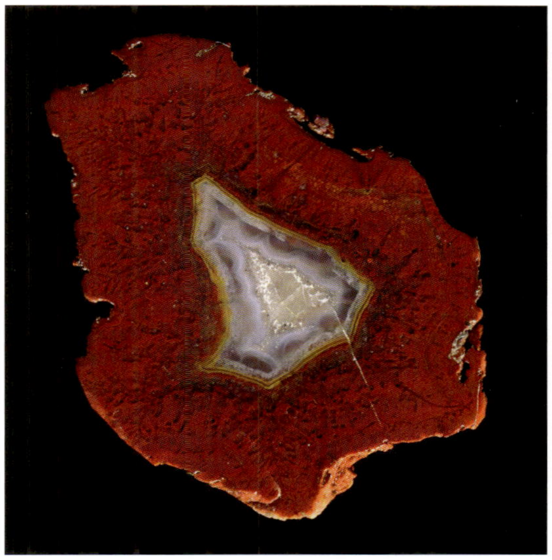

Lacy red agate fills this limb until formation of chalcedony changes with the yellow band. Bluish agate and quartz fill the center. Brushy Basin Member, Morrison Formation, Late Jurassic. Dewey Bridge area, Utah, USA. Dayvault specimen, 7 cm.

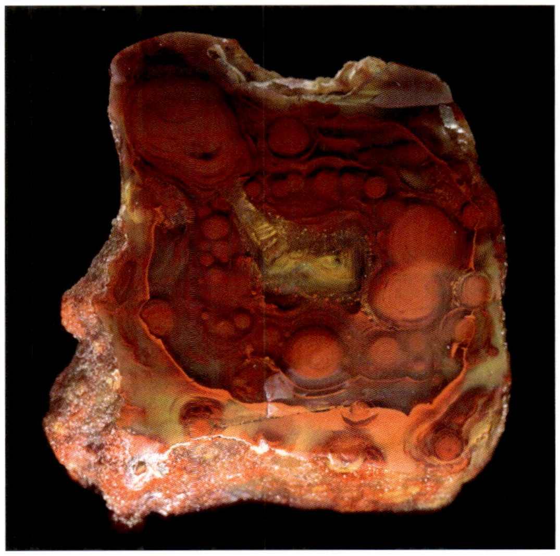

Unusual textures of red spheres (spherulites ?) float in yellowish-clear chalcedony. Brushy Basin Member, Morrison Formation, Late Jurassic. South of Green River, Utah, USA. Dayvault specimen, 3.5 cm.

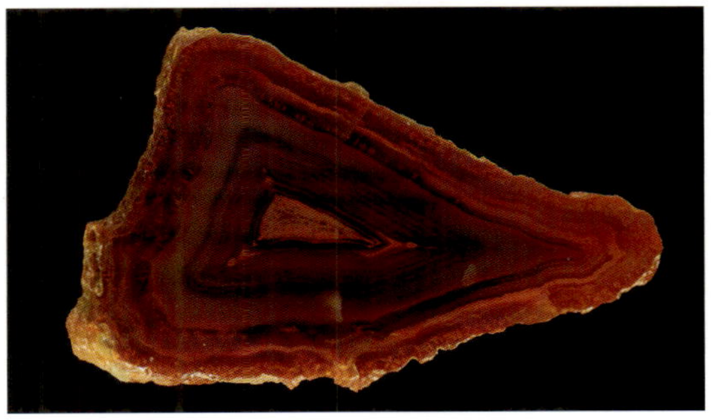

Fossil conifer. Brushy Basin Member, Morrison Formation, Late Jurassic. Moab area, Utah, USA. Daniels specimen, 3 cm.

Fossil conifer. Brushy Basin Member, Morrison Formation, Late Jurassic. Moab area, Utah, USA. Daniels specimen, 2.5 cm.

Fossil conifer. Brushy Basin Member, Morrison Formation, Late Jurassic. Moab area, Utah, USA. Daniels specimen, 2 cm.

Fossil conifer. Brushy Basin Member, Morrison Formation, Late Jurassic. Moab area, Utah, USA. Daniels specimen, 4 cm.

Fossil conifer. Brushy Basin Member, Morrison Formation, Late Jurassic. Moab area, Utah, USA. Daniels specimen, 4 cm.

Fossil conifer. Brushy Basin Member, Morrison Formation, Late Jurassic. Moab area, Utah, USA. Daniels specimen, 4 cm.

Fossil conifer. Brushy Basin Member, Morrison Formation, Late Jurassic. Yellow Cat area, Utah, USA. Daniels specimen, 6 cm.

Fossil conifer. Brushy Basin Member, Morrison Formation, Late Jurassic. Yellow Cat area, Utah, USA. Daniels specimen, 5 cm long.

Fossil conifer. Brushy Basin Member, Morrison Formation, Late Jurassic. Yellow Cat area, Utah, USA. Daniels specimen, 4 cm.

Fossil conifer. Brushy Basin Member, Morrison Formation, Late Jurassic. Yellow Cat area, Utah, USA. Daniels specimen, 3 cm.

Fossil conifer. Brushy Basin Member, Morrison Formation, Late Jurassic. Yellow Cat area, Utah, USA. Daniels specimen, 5 cm.

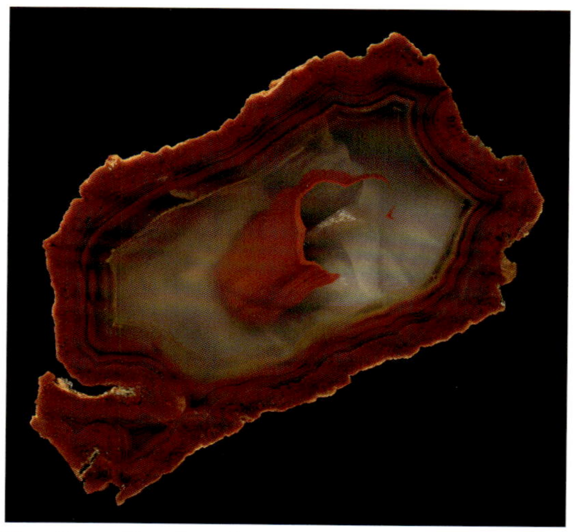

Fossil conifer. Brushy Basin Member, Morrison Formation, Late Jurassic. Yellow Cat area, Utah, USA. Daniels specimen, 5 cm.

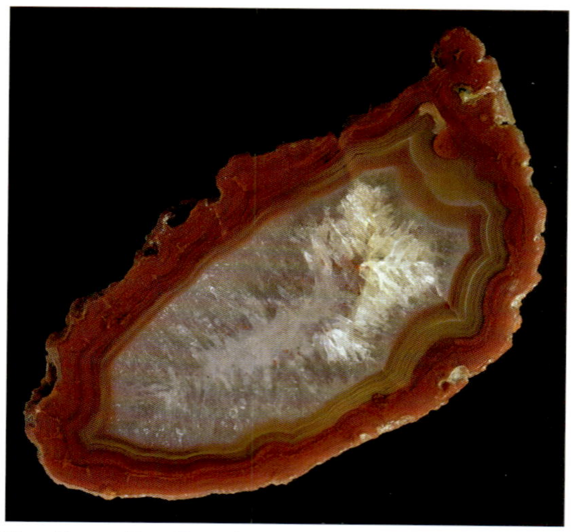

Fossil conifer. Brushy Basin Member, Morrison Formation, Late Jurassic. Yellow Cat area, Utah, USA. Daniels specimen, 4.5 cm.

Fossil conifer. Brushy Basin Member, Morrison Formation, Late Jurassic. Moab area, Utah, USA. Daniels specimen, 5.5 cm.

Fossil conifer specimen from the Yellow Cat area. This specimen retains elements of templated wood and just a small section of red carnelian agate. Brushy Basin Member, Morrison Formation, Late Jurassic. Yellow Cat area, Utah, USA. Daniels specimen, 13.5 cm.

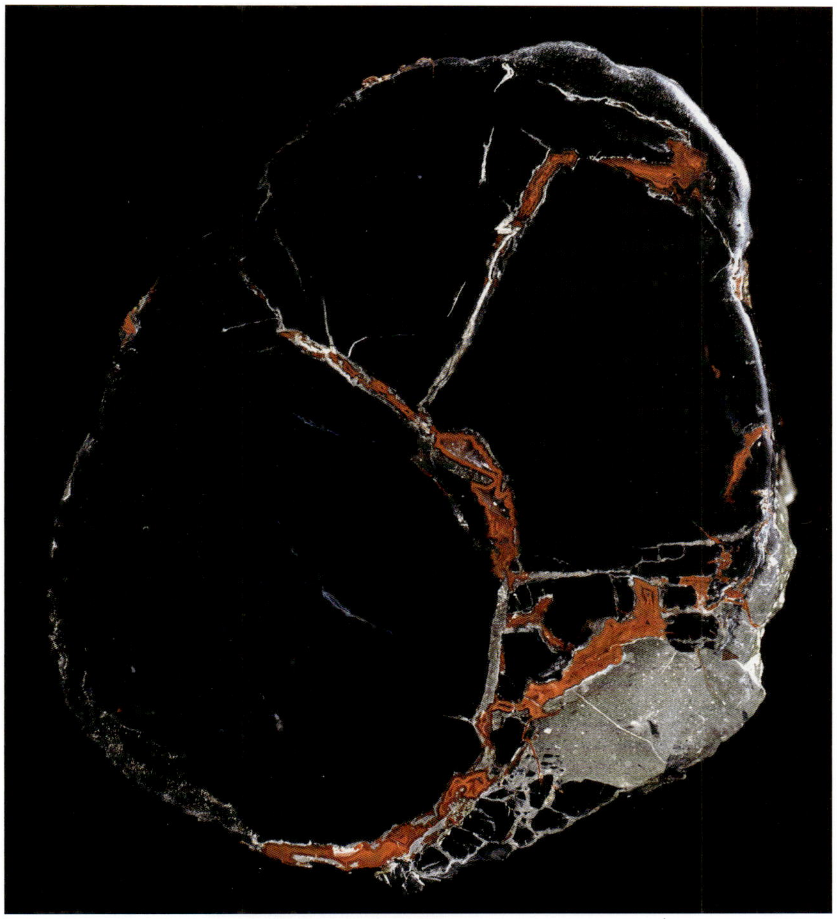

Fossil conifer specimen from the Yellow Cat area. This specimen is mostly black with a few streaks of carnelian agate. Brushy Basin Member, Morrison Formation, Late Jurassic. Yellow Cat area, Utah, USA. Daniels specimen, 12.5 cm.

Portion of one of the larger red wood logs discovered in the Yellow Cat area. Brushy Basin Member, Morrison Formation, Late Jurassic. Yellow Cat area, Utah, USA. Daniels specimen, 9 cm.

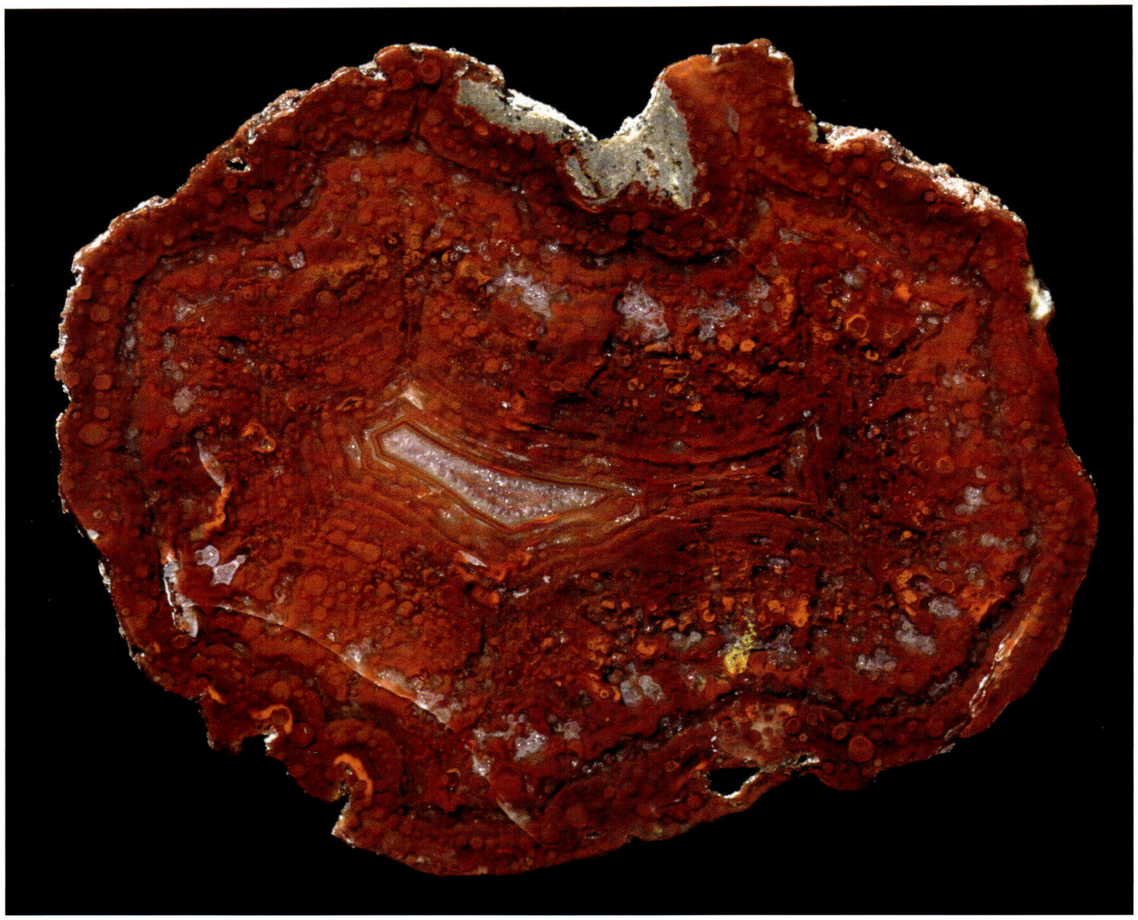

Fossil conifer. Brushy Basin Member, Morrison Formation, Late Jurassic. Yellow Cat area, Utah, USA. Daniels specimen, 6 cm.

Fossil conifer. Brushy Basin Member, Morrison Formation, Late Jurassic. Yellow Cat area, Utah, USA. Daniels specimen, 7 cm.

The Brushy Basin Member of the Morrison Formation in Utah also produces large volumes of colorful silicified wood, particularly in southeastern Utah. The Henry Mountains area is known for colorful wood that weathers from the Petrified Forest Member of the Chinle Formation.

Fossil conifer. Brushy Basin Member, Morrison Formation, Late Jurassic. Yellow Cat area, Utah, USA. Daniels specimen, 4 cm polished face; 8 cm long.

Unidentified fossil conifer short shoot. Morrison Formation, Jurassic. Henry Mountains area, Utah, USA. Daniels specimen, 4 cm.

Thompson Canyon near Ticaboo, Utah, Henry Mountains area, looking north. Lower multicolored mudstones and shales are part of the Upper Jurassic age Brushy Basin Member of the Morrison Formation. The middle sandstone ledge is Upper Cretaceous Dakota Sandstone. Above the sandstone ledge is the Upper Cretaceous Mancos Shale, a sequence of dark marine shale containing occasional invertebrate fossils.

Unidentified fossil conifer. Morrison Formation, Jurassic. Henry Mountains area, Utah, USA. Daniels specimen, 6 cm.

Unidentified fossil conifer. Morrison Formation, Jurassic. Henry Mountains area, Utah, USA. Daniels specimen, 7.5 cm.

Araucarioxylon. Chinle Formation, Triassic. Circle Cliffs area, Utah, USA. Daniels specimen, 9.5 cm.

Araucarioxylon. Chinle Formation, Triassic. Circle Cliffs area, Utah, USA. Daniels specimen, 15 cm.

Unidentified fossil conifer. Morrison Formation, Jurassic. Shitemaring Canyon, Henry Mountains area, Utah, USA. Daniels specimen, 20 cm.

Unidentified fossil conifer. Morrison Formation, Jurassic. North Wash, Henry Mountains area, Utah, USA. Daniels specimen, 11 cm.

View of Waterpocket Fold looking west-northwest. In the foreground, rounded cross-bedded eolian-derived Entrada Sandstone dips toward the viewer. Behind it are reddish-to-light-colored sandstone, mudstone, and gypsiferous sandstone of the marine Carmel Formation. The large cliff is composed of Navajo Sandstone that formed from shifting sand dunes. All formations are Jurassic in age. Capitol Reef National Park, Utah, USA.

Fossil wood embedded in conglomerate rock. Morrison Formation, Jurassic. Henry Mountains area, Utah, USA.

Lichen with fossil wood embedded in conglomerate rock. Morrison Formation, Jurassic. Henry Mountains area, Utah, USA.

Assortment of fossil woods from Utah demonstrating the tremendous variety of fossil wood in the state. Triassic Chinle and Jurassic Morrison formations, Utah, USA. Daniels specimens, largest 13 cm.

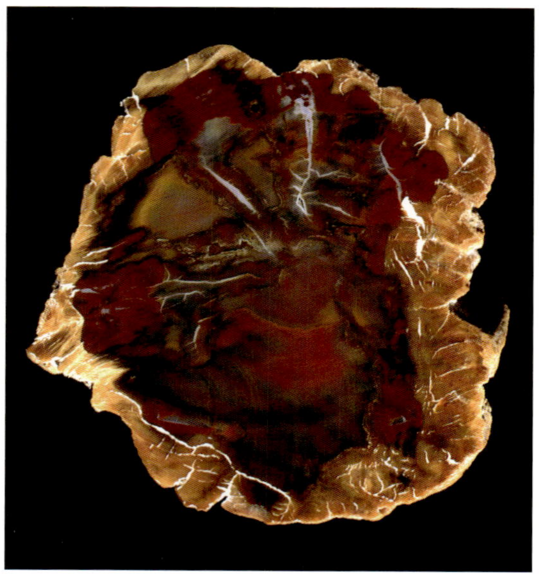

Unidentified fossil conifer. Morrison Formation, Jurassic. Henry Mountains area, Utah, USA. Daniels specimen, 5 cm.

Unidentified fossil conifer. Morrison Formation, Jurassic. Henry Mountains area, Utah, USA. Daniels specimen, 4.5 cm.

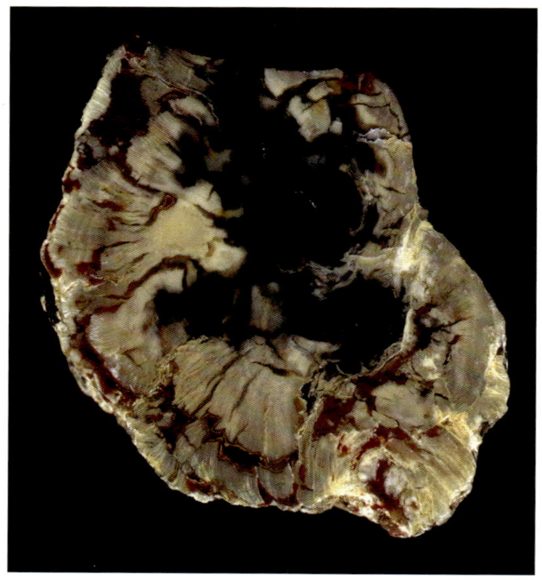

Unidentified fossil conifer. Morrison Formation, Jurassic. Henry Mountains area, Utah, USA. Daniels specimen, 6 cm.

View of Mt. Ellsworth (elevation 8,235 feet) at the southern end of the Henry Mountains looking east-northeast. The mountain is a laccolith formed from intrusions of granitic rocks from about 20 to 30 mybp. Regional erosion produces these majestic isolated peaks in the eastern Utah desert. In the foreground, lowermost rocks are the Jurassic Period Summerville Formation, overlain by light-colored sandstones of the Upper Jurassic Period Salt Wash Member of the Morrison Formation. Flanking the mountain are light-colored sandstones of the Navajo Sandstone surrounded by light-orange-to-pinkish-colored sandstones of the Entrada Sandstone Formation, all of Jurassic age.

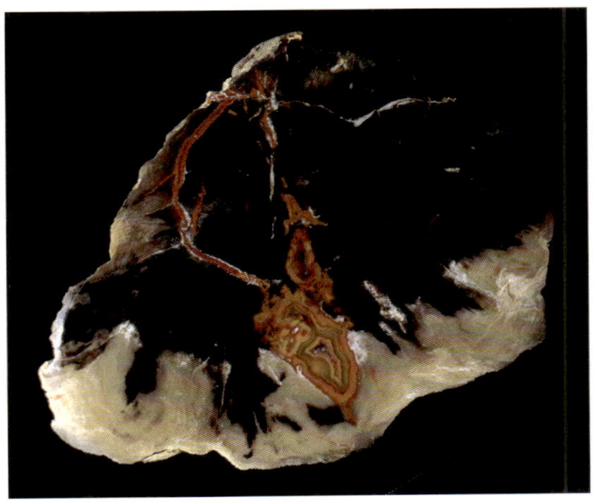

Unidentified fossil conifer. Morrison Formation, Jurassic. Henry Mountains area, Utah, USA. Daniels specimen, 6 cm.

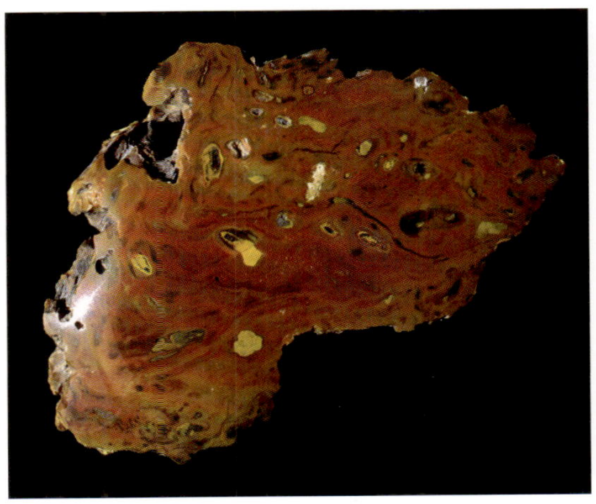

Unidentified fossil conifer. Morrison Formation, Jurassic. Henry Mountains area, Utah, USA. Daniels specimen, 6.5 cm.

Araucarioxylon. Chinle Formation, Triassic. Circle Cliffs area, Utah, USA. Daniels specimen, 6 cm.

Unidentified fossil conifer. Morrison Formation, Jurassic. Henry Mountains area, Utah, USA. Daniels specimen, 8 cm.

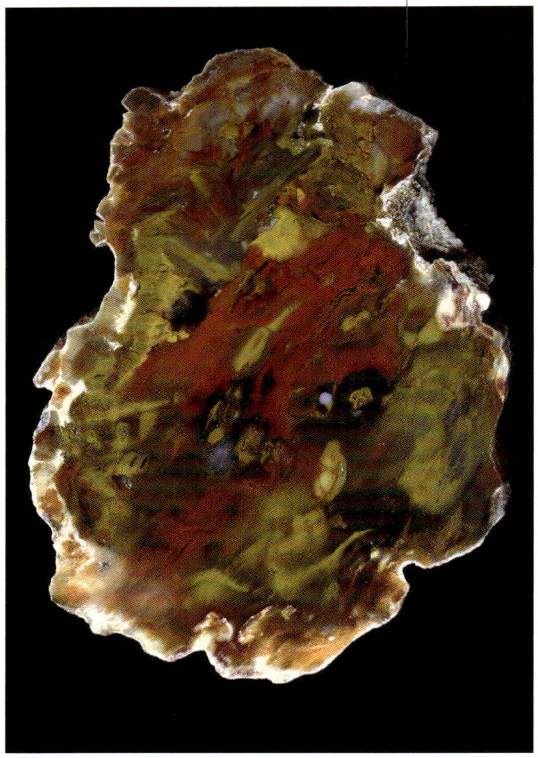

Unidentified fossil conifer. Morrison Formation, Jurassic. Henry Mountains area, Utah, USA. Daniels specimen, 8.5 cm.

Unidentified fossil conifer. Morrison Formation, Jurassic. Henry Mountains area, Utah, USA. Daniels specimen, 6.5 cm.

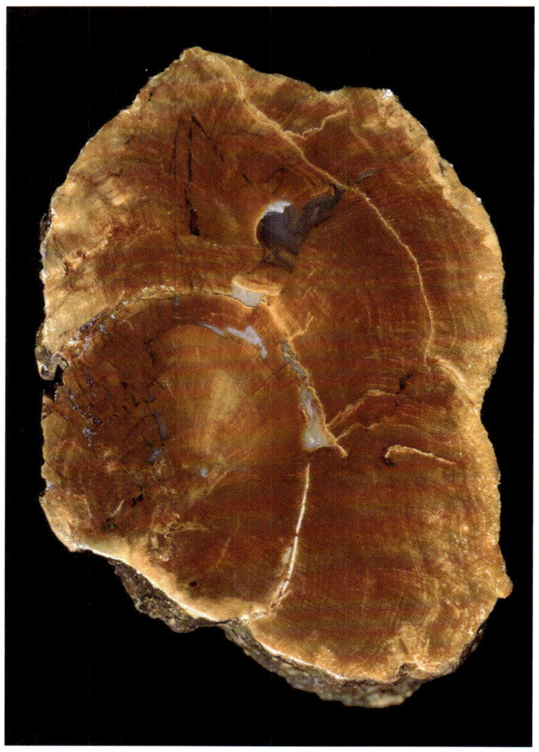

Unidentified conifer. Morrison Formation, Jurassic. Henry Mountains area, Utah, USA. Daniels specimen, 8.5 cm.

Unidentified conifer. Morrison Formation, Jurassic. Henry Mountains area, Utah, USA. Daniels specimen, 6.5 cm.

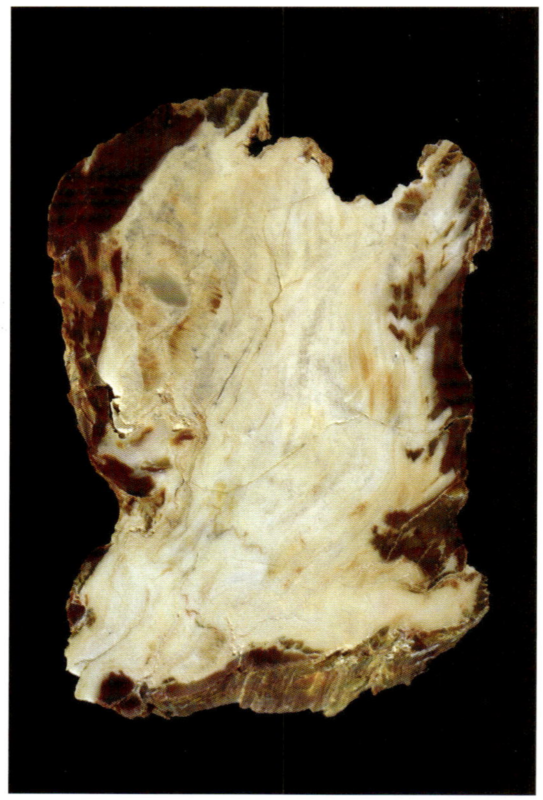

Unidentified conifer. Morrison Formation, Jurassic. Henry Mountains area, Utah, USA. Daniels specimen, 10 cm.

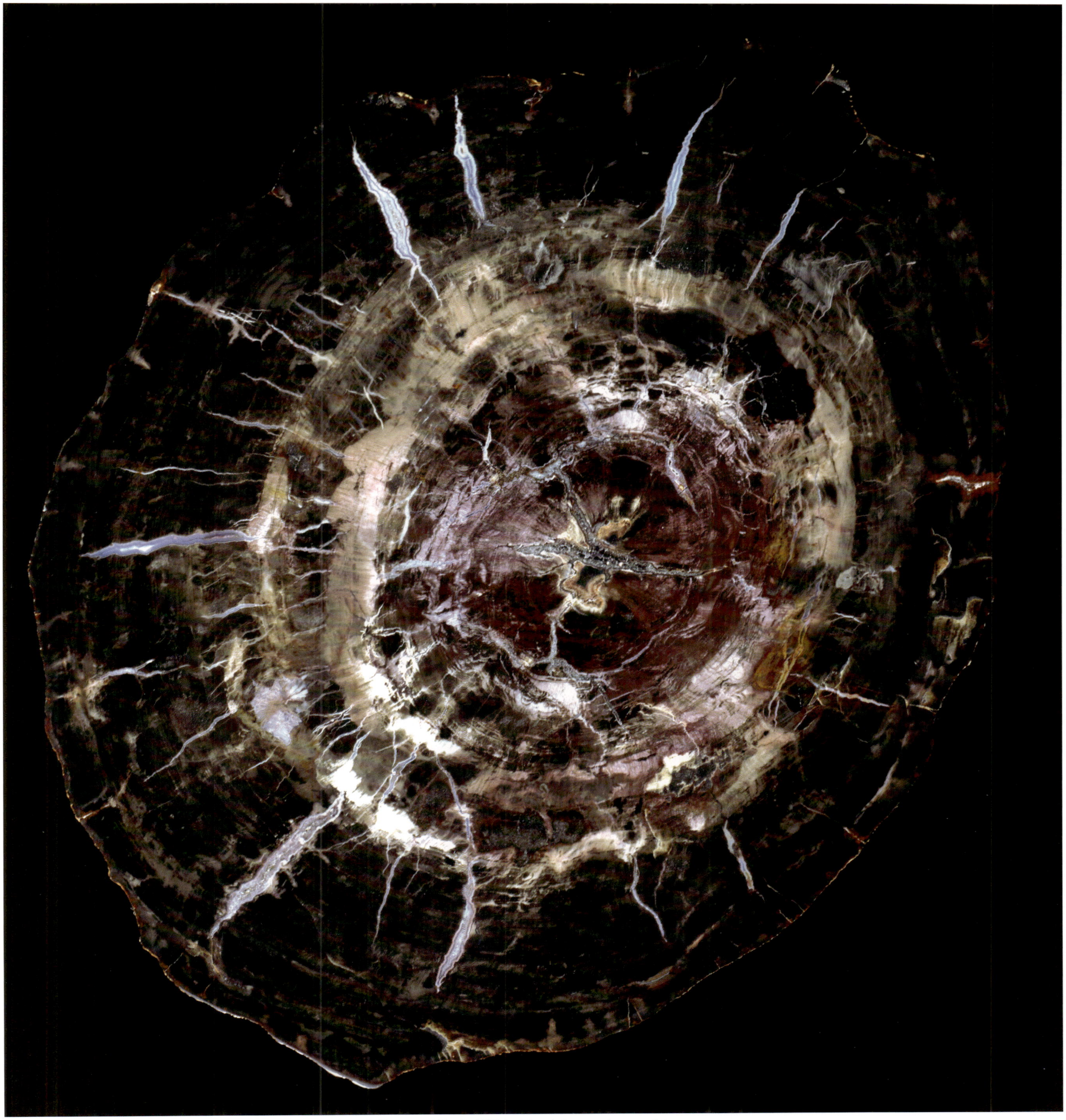

Araucarioxylon. Chinle Formation, Triassic. Circle Cliffs area, Utah, USA. Daniels specimen, 31 cm.

Large, colorful fossil logs weighing thousands of pounds. Morrison Formation, Jurassic. Petrified Forest State Park, Escalante, Utah, USA.

Colorful lichen on a colorful fossil conifer. Morrison Formation, Jurassic. Petrified Forest State Park, Escalante, Utah, USA.

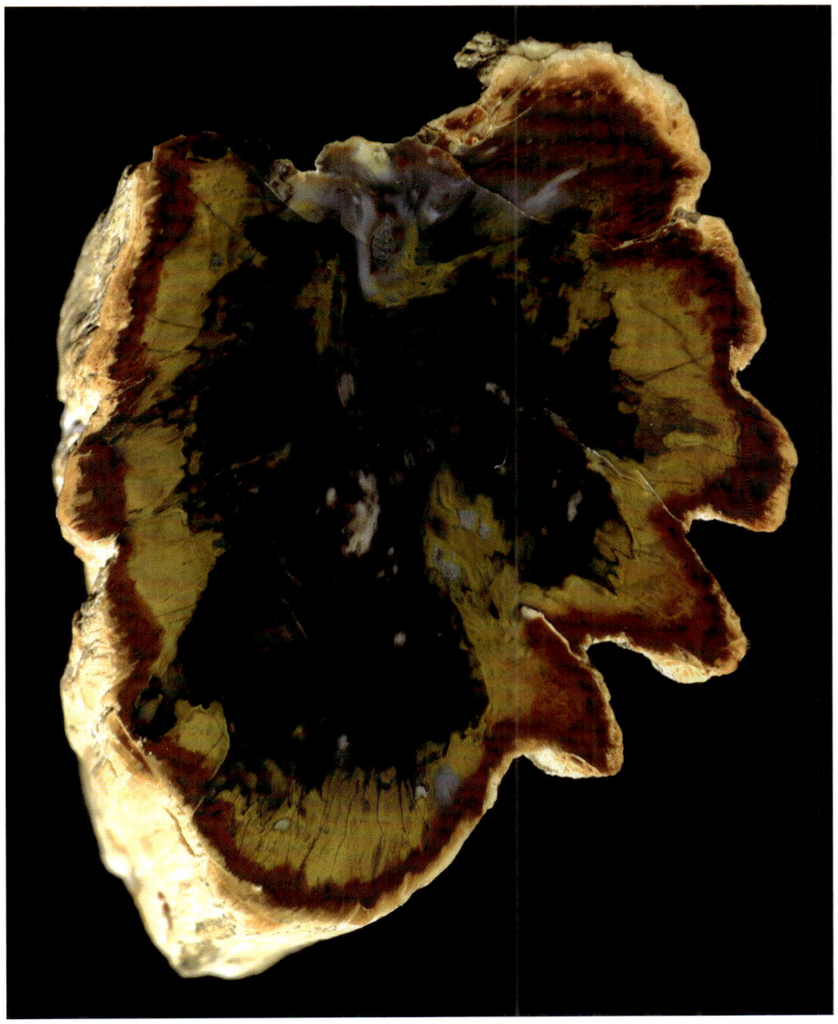

Unidentified fossil conifer. Morrison Formation, Jurassic. Henry Mountains area, Utah, USA. Daniels specimen, 13 cm.

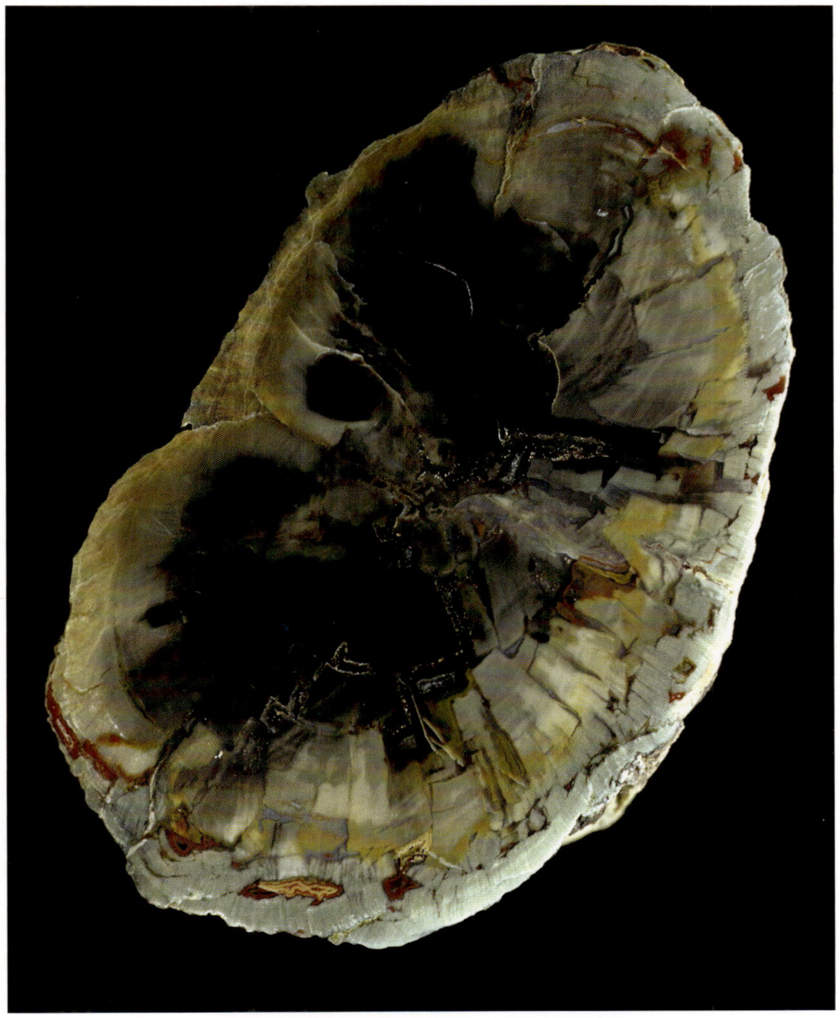

Unidentified fossil conifer. Morrison Formation, Jurassic. Henry Mountains area, Utah, USA. Daniels specimen, 13 cm.

Unidentified fossil conifer. Morrison Formation, Jurassic. Henry Mountains area, Utah, USA. Daniels specimen, 6 cm.

Unidentified fossil conifer. Morrison Formation, Jurassic. Henry Mountains area, Utah, USA. Daniels specimen, 6.5 cm.

Unidentified fossil conifer. Morrison Formation, Jurassic. Henry Mountains area, Utah, USA. Daniels specimen, 6 cm.

Large fossil log weighing thousands of pounds exposed in a gully with tons of other large logs. Morrison Formation, Jurassic. Petrified Forest State Park, Escalante, Utah, USA.

Unidentified fossil conifer. Morrison Formation, Jurassic. Henry Mountains area, Utah, USA. Daniels specimen, 10.5 cm.

Unidentified fossil conifer. Morrison Formation, Jurassic. Henry Mountains area, Utah, USA. Daniels specimen, 8.5 cm.

Unidentified fossil conifer. Morrison Formation, Jurassic. Henry Mountains area, Utah, USA. Daniels specimen, 6.5 cm.

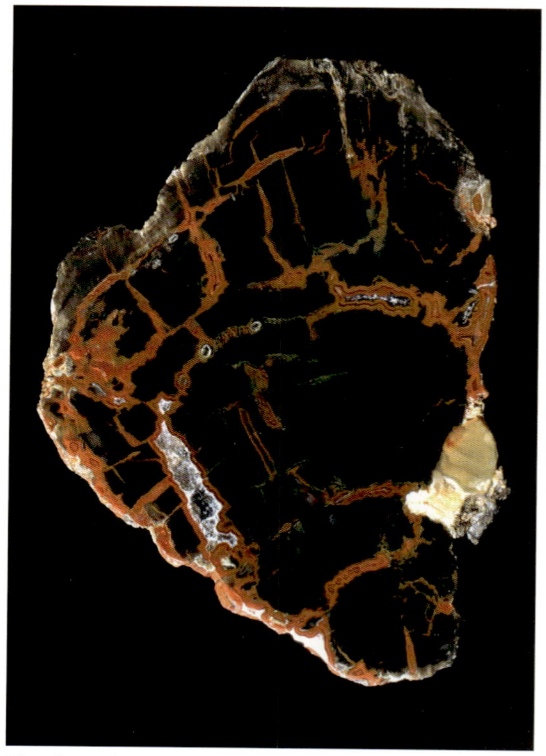

Unidentified fossil conifer. Morrison Formation, Jurassic. Henry Mountains area, Utah, USA. Daniels specimen, 7 cm.

Unidentified fossil conifer. Morrison Formation, Jurassic. Henry Mountains area, Utah, USA. Daniels specimen, 8 cm.

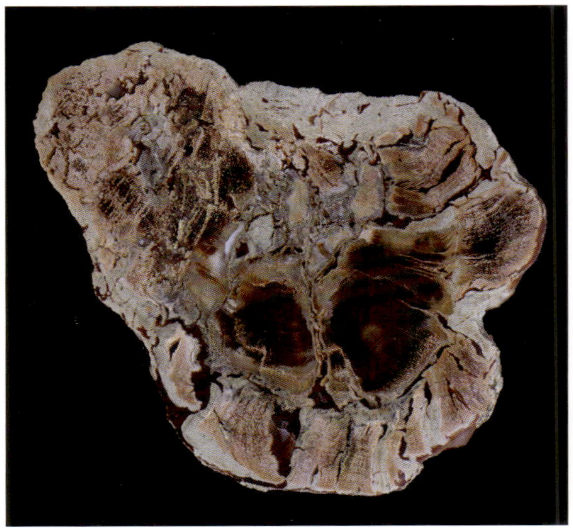
Unidentified fossil conifer. Morrison Formation, Jurassic. Henry Mountains area, Utah, USA. Daniels specimen, 9 cm.

Unidentified fossil conifer. Morrison Formation, Jurassic. Henry Mountains area, Utah, USA. Haggerty specimen, 8.5 cm.

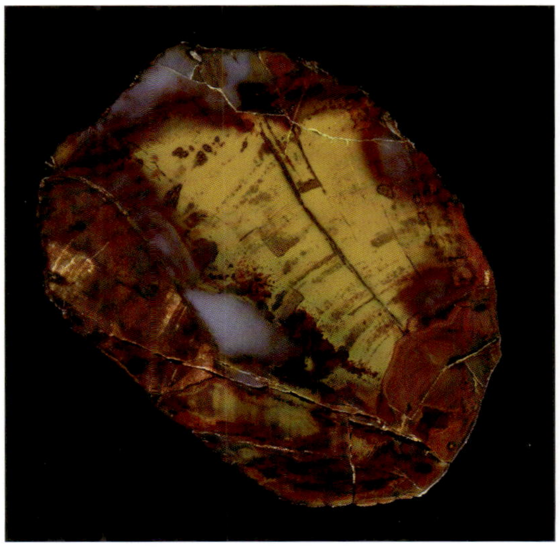
Unidentified fossil conifer. Morrison Formation, Jurassic. Henry Mountains area, Utah, USA. Daniels specimen, 7.5 cm.

Unidentified fossil conifer. Morrison Formation, Jurassic. Henry Mountains area, Utah, USA. Daniels specimen, 8.5 cm.

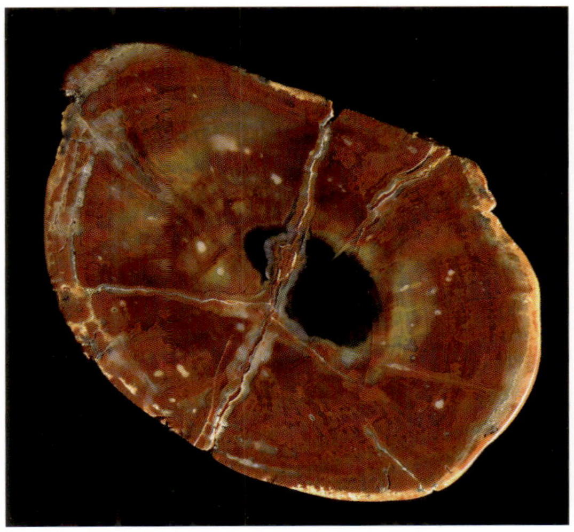
Unidentified fossil conifer. Morrison Formation, Jurassic. Henry Mountains area, Utah, USA. Dayvault specimen, 9 cm.

Unidentified fossil conifer. Morrison Formation, Jurassic. Henry Mountains area, Utah, USA. Daniels specimen, 8.5 cm.

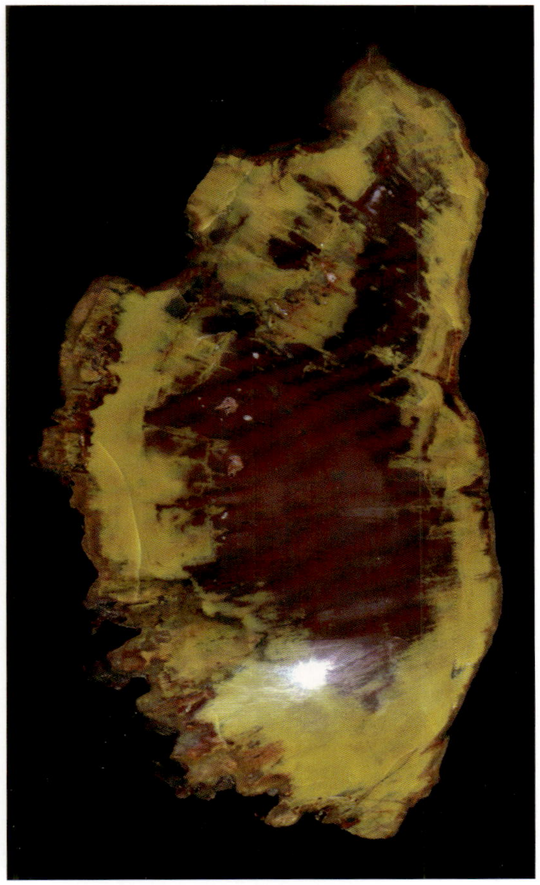
Araucarioxylon. Chinle Formation, Triassic. Henry Mountains area, Utah, USA. Mower specimen, 15 cm.

SHORT SHOOTS

Shorts shoots represent new or budding growth and are known from modern ginkgos and some conifers. They can be short branches producing leaves or seed-bearing structures in contemporary plants and are sometimes called short spurs. Short shoots are fairly common in the Brushy Basin Member of the Morrison Formation, although they have been generally overlooked because of their small size and limited areal extent. For the past 50 years, wood collectors have variously called these unusual structures petrified rose buds, onion sets, figs, or seedlings; other terms include armor plates and little dinosaur bones. They are most commonly observed either as simple rounded woody buds or as elongated woody buds, thicker in the middle with a tapered

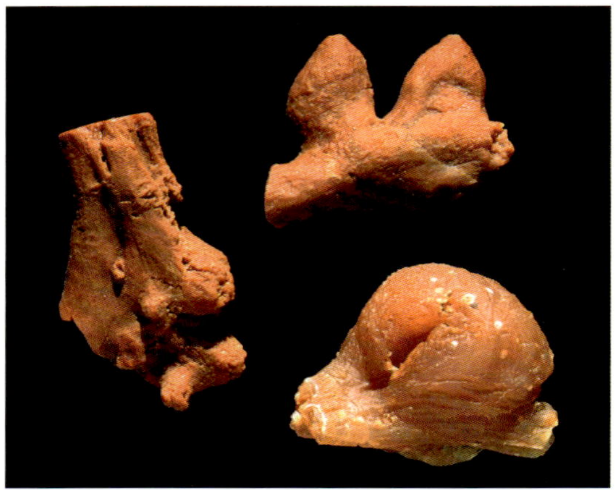

Carnelian casts of short shoot specimens showing different shapes. Brushy Basin Member, Morrison Formation, Late Jurassic. Moab, Utah, USA. L. Walker specimens, maximum length 4 cm.

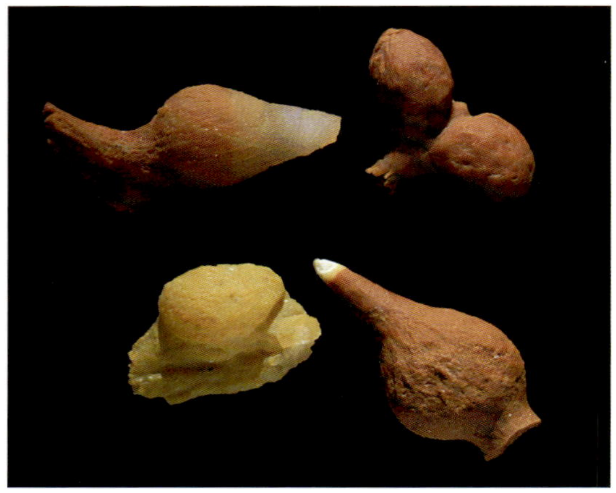

Carnelian casts of short shoot specimens showing different shapes. Top right clockwise: *Behuninia joannei*, *Behuninia provoensis*, unidentified, *Behuninia provoensis*. Brushy Basin Member, Morrison Formation, Late Jurassic. Moab, Utah, USA. L. Walker specimens, maximum length 4 cm.

View near Ticaboo, Utah, looking south-southeast. Variegated mudstones and clays of the Upper Jurassic Brushy Basin Member of the Morrison Formation form mounds in the foreground. Middle distance lower sandstones are formed from Entrada Sandstone overlain by the prominently layered Summerville Formation, which in turn is overlain by the sandstone units of the Salt Wash Member of the Morrison Formation, all of Jurassic age.

tip at one end and a foot or attachment at the other end. Stalks, upon which these structures grew, may have knobby to nodular locations where these buds arose and rarely contain complete short shoot attachments.

At least 30 short shoot localities have been identified in Utah, Colorado, and Wyoming. Fossil woods from the Late Jurassic time that produced short shoot structures seem to be confined to sediments in the lower half of the Brushy Basin Member of the Morrison Formation. One of the most prolific localities for short shoots, off the east flank of Mount Ellen, Utah, produces specimens about 10 meters above the contact with the Salt Wash Member. Specimens from this site are concentrated in a zone less than 1 meter thick that contains a greenish-gray bentonitic claystone overlying a chocolate-to-rust-colored bentonitic claystone. The longest and best-preserved stalks are in a 10 to 15 centimeter thick sequence in the greenish-gray claystone just above this contact. Low areas along the contact seem to contain greater concentrations of material. A location north of the Henry Mountains produces short shoots from higher in the section, and another location south of the Henry Mountains contains short shoots near the middle of a very thick sequence of Brushy Basin sediments. Other short shoot localities near Moab also produce from near the bottom of the Brushy Basin Member. To date, we know of no short shoots from the underlying Salt Wash Member of the Morrison Formation or from the overlying Cedar Mountain Formation (Lower Cretaceous) sediments, even though both contain horizons where numerous small limbs are preserved.

Fossil short shoots from the Morrison Formation of Utah have been described in two papers. In the 1960s, Marjorie Chandler of Great Britain assumed the daunting task of describing a group of new fossils from the Morrison Formation, preserved only as casts, from a site and formation that she never visited. She described three new genera of plants, *Behuninia*, *Jensensispermum*, and *Hillisrobus*, and five new species. An American colleague, Jim Jensen (famous for dinosaur research), who visited the site, suggested a pond or lake environment separated from a stream or river by higher ground. The fossils were considered most interesting because they did not closely resemble any living counterparts. A superficial but misleading resemblance to angiosperms was noted, and a closer resemblance to conifers (especially from the associated cones) and cycads (as seeds) was also noted. *Behuninia joannei* was described on the basis of fragmentary material as seeds arranged in opposite to subopposite pairs on one side of a stalk extending to its terminus. *Jensensispermum redmondi* was also described as a seed, and both genera were thought to have affinities with the group Cycadophyta. As a postscript, Chandler mentioned a new locality described by Charles Bass on the east flank of Mount Ellen in the Henry Mountains. Similar plant fossils and perhaps several other varieties had been recovered from the upper half of the Morrison Formation. The material was collected from the surface where pieces ranged to 3 cm in diameter and 7 cm in length. Bass suggested that the fossils might represent fructifications of angiosperms or gymnosperms or perhaps both, noting that formal identifications had not been made.

In the early 1990s, Tidwell and Medlyn published an article on short shoots from the Morrison Formation in Utah, Wyoming, and Colorado. They revised Chandler's interpretations after studying thousands of specimens from the Mount Ellen locality and other sites in Utah, Wyoming, and Colorado. Specimens from one Wyoming locality retained internal textures that were necessary for proper descriptions and identifications. They emended the description of *Behuninia* as a conifer short shoot (instead of a seed) and added three new species, *B. provoensis*, *B. bassii*, and *B. scottii*. A new short shoot genus, *Steinerocaulis*, was proposed to replace the seed designation, *Carpolithus radiatus*. The short shoots were placed with conifers instead of cycads because some shoots had leaf scars preserved, suggesting needlelike attachments. Several seeds and seed cones were studied and described. *Jensenispermum* was mentioned later by Tidwell in his 1998 book as berry-like seeds that resemble living cycad seeds (referring to the extant order Cycadales instead of the extinct, but then more widespread, order Cycadeoidales). The nomenclature suggested by Tidwell and Medlyn is currently used for specimens shown here.

Most short shoots are calcite or chalcedony casts; however, detailed silicification of internal textures occurs at a few localities. Specimens showing the most detailed internal preservation are found near the uranium-mining town of Ticaboo (said to be Paiute for "friendly place"). Other specimens collected near Moab, Utah, have been converted to carnelian limb casts and are generally known as red wood. As fossils weather out of their matrix, they break into pieces and disperse; therefore, complete specimens are rare. The larger specimens shown here were carefully excavated and reassembled and represent the most complete examples of short shoot stalks ever found. We have identified three types of short shoots from the Henry Mountains area based on external morphology: *Behuninia provoensis*, *Behuninia joannei*, and *Steinerocaulis radiatus*. *Behuninia bassii* is probably represented by a fine carnelian specimen from Moab. By far the most spectacular and probably the most common is *B. provoensis*, followed by *B. joannei*; *S. radiatus* is much less common.

Section of a *Behuninia provoensis* displaying numerous small divots along one layer of preserved wood. These divots may represent leaf scars or may be the outer surface of xylem that can have a similar texture. Brushy Basin Member, Morrison Formation, Late Jurassic. West of Hanksville, Utah, USA. Hatch specimen, maximum length 6.5 cm.

Carnelian cast of *Behuninia cf. B. bassii*. Brushy Basin Member, Morrison Formation, Late Jurassic. North of Moab, Utah, USA. L. Walker specimen, maximum length 4 cm.

Behuninia provoensis

Behuninia provoensis specimens typically occur as fragments that resemble bottle stoppers and buds. The sequence of growth starts with budding structures that continue to elongate and become new stalks, which can produce new buds and start the entire process over again. These types of short shoots are usually fragmentary because the specimens are not dug and reassembled. Some of the small but spectacular red wood casts found at a number of localities in eastern Utah are stalk fragments of short shoots. The next series of photographs show specimens of short shoot stalks and polished sections of stalks collected at two red wood localities near Moab. They are completely agatized and demonstrate some of the color combinations sought by petrified wood collectors. The degree of abrasion or weathering shown by the larger stalks is typical for most short shoot stalks.

Behuninia provoensis showing varying degrees of inflation for the bulb and attachments. Brushy Basin Member, Morrison Formation, Late Jurassic. Mount Ellen locality. Henry Mountains, Utah, USA. Hatch specimens, maximum length 5 cm.

Behuninia provoensis showing tiny divots apparently indicating leaf (needle) scars. Specimens with detailed surface textures have been found in several locations. Brushy Basin Member, Morrison Formation, Late Jurassic. West of Hanksville, Utah. Hatch specimen, 4 cm.

Classic specimen of *Behuninia provoensis* showing elongate bulb and attachment base. Brushy Basin Member, Morrison Formation, Late Jurassic. Mount Ellen locality. Henry Mountains, Utah, USA. Hatch specimen, length 7 cm.

Silicified *Behuninia provoensis* specimens collected in the 1940s. Brushy Basin Member, Morrison Formation, Late Jurassic. Red Canyon, west of Lander, Wyoming, USA. Sims specimens, maximum length 4 cm.

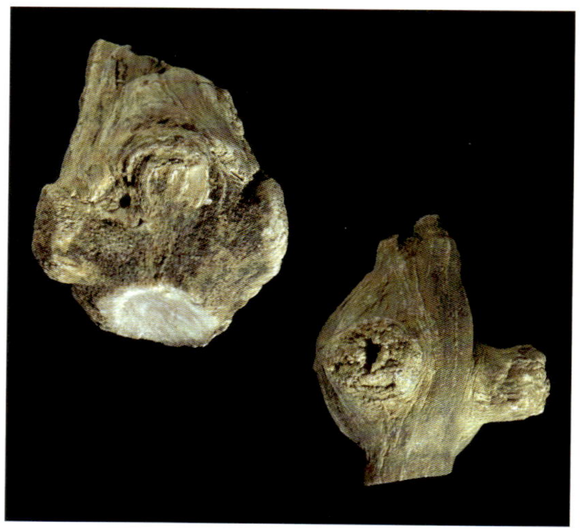

Silicified *Behuninia provoensis* stalk sections collected in the 1940s. Brushy Basin Member, Morrison Formation, Late Jurassic. Red Canyon, west of Lander, Wyoming, USA. Sims specimens, maximum length 4.5 cm.

Carnelian cast of *Behuninia provoensis* with a hole possibly caused by an insect. Brushy Basin Member, Morrison Formation, Late Jurassic. North of Moab, Utah, USA. L. Walker specimen, maximum width 9 cm.

Carnelian cast of *Behuninia provoensis* showing opposing buds aligned along a stalk. Brushy Basin Member, Morrison Formation, Late Jurassic. North of Moab, Utah, USA. L. Walker specimen, maximum width 9 cm.

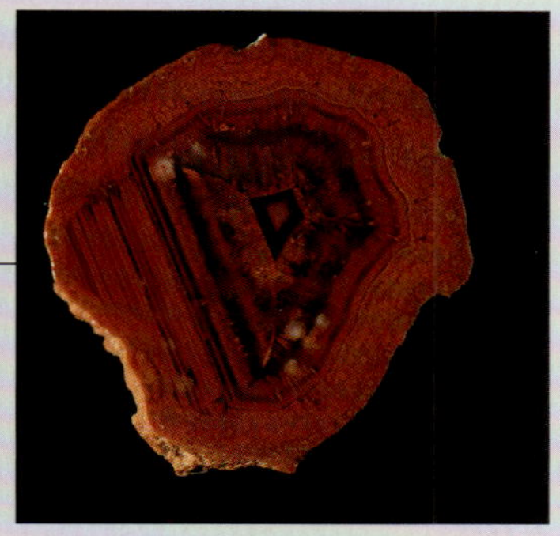

Section of *Behuninia* stalk showing agate banding. Brushy Basin Member, Morrison Formation, Late Jurassic. North of Moab, Utah, USA. L. Walker specimen, 2.5 cm.

Two *Behuninia* carnelian stalk casts from a red wood site. One specimen has been cut and polished. Brushy Basin Member, Morrison Formation, Late Jurassic. North of Moab, Utah, USA. L. Walker specimen, 17 and 18 cm.

Behuninia carnelian stalk showing bud attachments along one side of the stalk, which is not uncommon. Brushy Basin Member, Morrison Formation, Late Jurassic. North of Moab, Utah, USA. Dayvault specimen, length 25.5 cm.

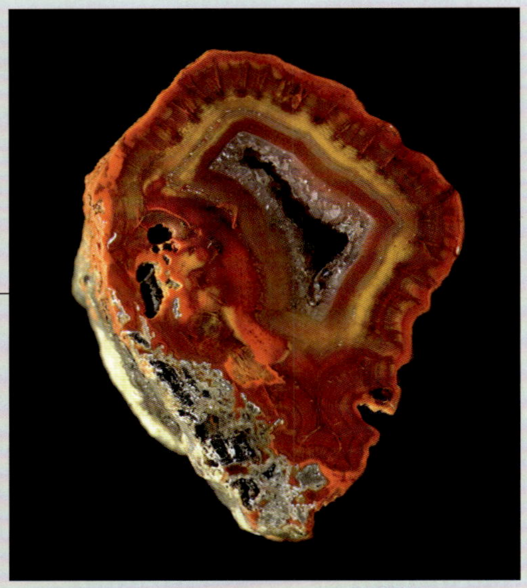

Behuninia carnelian stalk casts from a red wood site. This specimen has been cut and polished to show the agate banding. Photograph at right shows the internal banding. Brushy Basin Member, Morrison Formation, Late Jurassic. North of Moab, Utah, USA. L. Walker specimen, 20.5 cm.

Section of *Behuninia* stalk showing agate banding. North of Moab, Utah, USA. L. Walker specimen, 4 cm.

The internal texture of a short shoot is demonstrated at the Ticaboo locality more clearly than any other site in the Henry Mountains. Some collectors have erroneously named the material *Schilderia* because of its superficial resemblance to this genus of Triassic wood found in Arizona and Utah. Highly silicified stalks to 10 cm in diameter have been found at this locality.

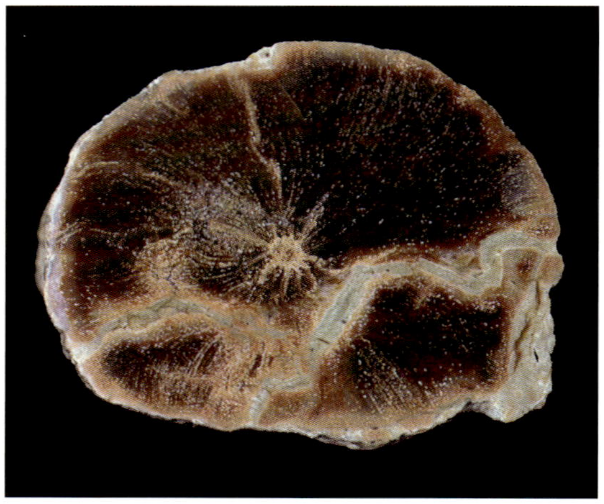

Polished transverse view of short shoot stalk from the Ticaboo area, Utah. The internal texture often shows a strong radial and poorly developed concentric growth pattern. Some specimens from this locality are red agate limb casts. Brushy Basin Member, Morrison Formation, Late Jurassic. Daniels specimen, 3.5 cm.

Behuninia stalk from the Mount Ellen locality. This robust siliceous cast branches twice and displays prominent *B. provoensis* short shoots. Brushy Basin Member, Morrison Formation, Late Jurassic. Mount Ellen site, Henry Mountains, Utah, USA. Dayvault specimen, length 26 cm.

This V-shaped *Behuninia* specimen is the largest short shoot stalk yet found. It shows seven sets of buds along the left main branch side, and it branches four times along the right main branch. The specimen also shows a split at the base. Most importantly, this is the only specimen yet found that shows obvious bark texture, apparent along the lower part of the thicker main branch. Collection and reconstruction of this piece required about 10 hours. Brushy Basin Member, Morrison Formation, Late Jurassic. Mount Ellen site, Henry Mountains, Utah, USA. Dayvault specimen, left side 40 cm long; right side 47 cm.

Desert lag near Ticaboo, Utah, yields fragments of short shoot stalks. At least five obvious stalk sections are visible in the photograph. The internal radiating structure is easily seen in the larger specimen in the lower left. Red agate fortifications and red wood stalk casts are found at this locality; small fragments of red chalcedony are scattered on the surface. Brushy Basin Member, Morrison Formation, Late Jurassic. Largest stalk fragment about 3.5 cm.

Behuninia sp. stalk from the Mount Ellen locality. This cast is unusual because of the numerous weathered short shoot locations. The larger end of the specimen looks like a burl because of the density of growth sites. The specimen was surface collected and has undergone recent weathering that apparently enhanced the textures of the growth areas. Brushy Basin Member, Morrison Formation, Late Jurassic. Mount Ellen site, Henry Mountains, Utah, USA. Dayvault specimen, length 20.5 cm.

Polished transverse view of small short shoot stalk from the Ticaboo area, Utah. The internal texture demonstrates a strong radial structure that is more pronounced in small limbs than in larger diameter specimens. Brushy Basin Member, Morrison Formation, Late Jurassic. Daniels specimen, 5 cm.

Behuninia joannei

Behuninia joannei fossils are common at most of the localities. Colloquial names for these specimens include "rose buds" and "onion sets." These flattened, bud-like structures typically occur as doubles on opposing sides of a stalk or at its termination, although lopsided and single-lobed specimens are also common. Most specimens show small attachment areas where buds were attached to a stalk, which accounts for the lack of articulated specimens. Many of these short shoots are less than 1 cm across and were attached to even smaller diameter stalks. It is probable that many of the numerous small stalk fragments found in these deposits belonged to *B. joannei.*

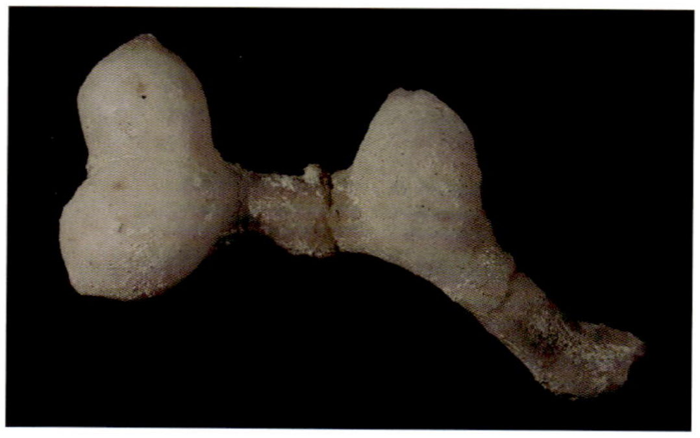

Calcite cast of *Behuninia joannei* terminal attachment with single bud along stem. *B. joannei* attachments are rare and this one shows typical double terminal buds. Brushy Basin Member, Morrison Formation, Late Jurassic. Mount Ellen site, Utah. Dayvault specimen, 8 cm.

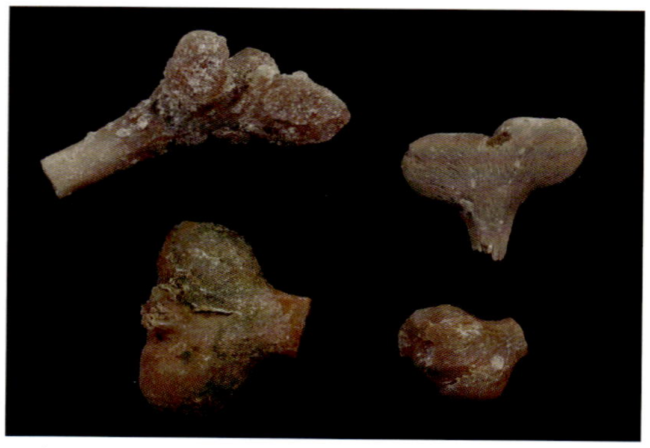

Behuninia joannei specimens show flattened buds and may consist of one, two, or even three nodes; however, the most common morphology is the double-lobed version. Buds are opposite to subopposite and arise at acute angles to the small limbs. Shoots are usually broad at the base and broad toward the top and almost always have a small diameter stem, which accounts for the paucity of articulated specimens. Brushy Basin Member, Morrison Formation, Late Jurassic. Mount Ellen site, Utah. Dayvault specimens, largest 3 cm.

Spectacular *Behuninia joannei* stalk with attachments. Six sets of opposing buds and a terminal bud are connected by a delicate stalk. The specimen, which is composed of red-to-colorless chalcedony, is the only short shoot found so far from this red wood locality. Brushy Basin Member, Morrison Formation, Late Jurassic. Mount Ellen site, Henry Mountains area, Utah, USA. Dayvault specimen, length 29 cm.

Steinerocaulis radiatus

Steinerocaulis radiatus are less common and are mostly found around the Henry Mountains. They may resemble a golf tee or a petrified fig or mushroom. Attachments are rare and the specimens found show little to no stem. The *Steinerocalis* simply sit on the stalk.

Small *Steinerocaulis radiatus* specimens. The five outer specimens demonstrate a bulbous shape and resemble small figs. They are all from the Brushy Basin Member, Morrison Formation, Mount Ellen site, Henry Mountains, Utah, USA. Dayvault specimens, maximum size 15 mm. The central lighter colored specimen is more flattened but has the same general shape. It was found in the Jurassic Portland Stone Formation, southwestern Wiltshire, England. Needham specimen, diameter 9 mm.

Steinerocaulis radiatus specimens may resemble figs, mushrooms, or a golf tee if the attachment is long enough. They are unusual structures and are far less common than varieties of *Behuninia*. The occurrence of these budding structures attached to a stalk is rare. The largest specimen may not be a *Steinerocaulis* but an aerial lignotuber. Brushy Basin Member, Morrison Formation, Late Jurassic. Henry Mountains area, Utah, USA. Dayvault specimens, maximum size 3 cm.

It is unclear whether some specimens that resemble *Steinerocaulis* may in fact be something else, possibly related to the "seedlings" found in the Jurassic Period Cerro Cuadrado Petrified Forest of Patagonia, Argentina. The maximum diameter of the South American specimens is about 10 cm, much larger than similar fossils found in the Henry Mountains region. Investigators since Wehrfeld in the 1930s noted that these "novelties" from Argentina were associated with the araucarian trees and cones of Patagonia. Later, Calder studied them and also decided that they had affinities with araucarians. Stockey and Taylor in the 1970s studied thin sections of many specimens and determined they were woody and could belong to *Araucaria*. Specimens of one or more seedlings reattached to a limb with one or more pararaucarian cones reattached to the same limb have been discovered in Patagonia. Both the seedlings and cones (always pararaucarian) were found together with the stems in the tuffaceous matrix. Specimens of similar structures from the Henry Mountains are co-located with similar-looking pararaucarian cones. More recently, Ruth Stockey redefined the seedlings as aerial lignotubers that are apparently associated with pararaucarian cones. Stockey considers the lignotubers to represent adventitious reproductive structures capable of falling off a limb and germinating. Other so called seedling varieties from Patagonia resemble *Behuninia* varieties.

The lighter specimen on the left with the longer stem is a pararaucarian lignotuber from the Cerro Cuadrado petrified forest of Patagonia, Argentina. The specimen on right may also be a lignotuber but is from the Brushy Basin Member of the Morrison Formation, Henry Mountains, Utah, USA. Both are from Jurassic age sediments. Dayvault specimens, maximum length 7 cm.

Associated Fossils

A number of other plant fossils are associated with the short shoots. Most of them are minute and require nose-to-the-ground searching. These associations aid in establishing the affinity of the short shoots as conifers and help with paleoecological interpretations.

Jensensispermum redmondi is represented by small ovate-to-subovate seeds to 10 millimeters in diameter; most are less than half that size. Some specimens show external textures and some demonstrate two halves that may separate to reveal an internal seed cavity. The outer or pulpy integument of the seed may have a wrinkled appearance. Nodes resembling attachments are formed on the inner integument of the seed. These seeds are common at the Mount Ellen site and are found sparingly at several other short shoot sites. Chandler thought the seeds most closely resembled those of true (modern) cycads.

***Jensensispermum redmondi* seeds. These round or ovate seeds have an outer shell and an apparent attachment. They are abundant in association with the smaller short shoot specimens at the Mount Ellen site and are found sparingly at other sites. *J. redmondi* resemble seeds of some living cycads; however, no cycad fragments have been found at any of the short shoot sites. (The two specimens in the middle left are *B. joannei*). Brushy Basin Member, Morrison Formation, North of Moab Utah, USA. L. Walker specimens, maximum diameter 15 mm.**

"Four-wing" is the name we use for unidentified seeds or seed shells that are uncommon at the Mount Ellen site and are very rare at all other localities. This descriptive name relates to their four-sided ridged appearance. They are elongate, blunt at one end and tapered at the other, and average about 10 mm in length with little variation. Broken specimens show the robustness of the shell. Insect damage may be present on some specimens. A few three-cornered specimens and one five-cornered specimen have also been found.

Undescribed seeds referred to here as four-wings because of their distinctive four-cornered appearance. A few specimens with three corners and one specimen with five corners have also been found. Their affiliation is unknown. They are uncommon at the Mount Ellen site and rare at other short shoot locations. The top left specimen may have been damaged by an insect. Brushy Basin Member, Morrison Formation, Late Jurassic. Mount Ellen site, Henry Mountains, Utah, USA. Dayvault specimens, maximum length 12 mm.

Conifer cones, cone scales, cone seeds, and tiny scaly conifer stems are found at a number of short shoot sites. These fossils may have originated from the same trees that produced the short shoots. Cones are always a prized find for collectors. About 60 mostly complete cones have been recovered from the Mount Ellen site and at least that number from all the other short shoot sites. Most cones are less than 1.2 cm in the longest direction; although one partial cone is 7 cm long and is probably missing another 2 to 3 cm. Loose cone scales are abundant in the sediments at the Mount Ellen site and are found infrequently at other sites. They average about 8 mm in length but range to 15 mm long; shape is uniform with two uneven halves that split to reveal the seed. Tiny round to oblate seeds, 0.5 to 2 mm in diameter, probably the seeds from these seed scales, are also abundant at the Mount Ellen site. Small scaly conifer twigs, usually less than 2 mm in diameter and 10 mm in length, are found sparingly at the Mount Ellen site. Rhombic scales, probably representing leaf scars, appear to be spirally arranged.

Calcite casts of seed scales. These tiny fossils practically litter the surface of certain areas at the Mount Ellen site. Seed scales always show bifold symmetry, and some are broken in half to reveal the oblong seed compartment within. Brushy Basin Member, Morrison Formation, Late Jurassic. Mount Ellen site, Henry Mountains, Utah, USA. Dayvault specimens, maximum length 15 mm.

Paleoecology

Investigators since Lee Stokes in the 1940s have thought the Brushy Basin sediments were deposited in a generally dry climate. More recently, Litwin, Turner, and Peterson used pollen and spore varieties to determine that the climate for the Brushy Basin may have had moderate moisture levels. Tidwell and Medlyn concluded in their short shoot article that the forests in the Brushy Basin floodplains contained ferns and cycads and that many of the conifer trees lacked or had only faint growth rings, suggesting a seasonless and possibly moist, warm climate. Modern trees in seasonless climates, such as those near the equatorial regions, have poorly developed growth rings. In addition, many of the Brushy Basin trees had numerous wood-rotting fungal growths, which supports a climate with sufficient moisture to support fungal activity. The abundance of dinosaur fossils associated with contemporaneous strata suggests that forests had to be well developed to support the diet of large herbivores. Such forests would probably not have grown in an arid environment.

Short shoots are found associated with cycad and fern fossils that probably grew near water, such as areas around lakes and ponds that developed in overbank regions of a river system. The even-layered sediments that enclose short shoot specimens from the Mount Ellen locality correspond with relatively quiet deposition for most periods, with pulses of deposition represented by thicker layers of muds and, infrequently, coarser layers. Snails and clams are found with some of the short shoot deposits, which further indicate a wet environment such as a lake, pond, or stream. Tidwell, Britt, and Ash also suggested a pond or lake setting for the Mygatt/Moore dinosaur quarry near Moab, Utah, that also yields short shoots.

Whatever their origin and paleoecology, short shoots are fascinating and unusual examples of fossil wood. They are often found associated with conifer cones and red wood and usually occur as casts. Nomenclature for these interesting fossils will evolve as specimens are discovered around the world and are compared. So far, all known localities in the United States occur in the Brushy Basin Member of the Jurassic Morrison Formation. The chance of finding new localities is good because specimens are small and sites are of limited areal extent; many sites have undoubtedly been overlooked.

CYCADS

Spectacular silicified cycad trunks occur around the Henry Mountains and Moab areas of southeastern Utah. The most colorful examples are in the Late Jurassic Brushy Basin Member of the Morrison Formation, but larger trunks, more than 2 meters long, have been recovered from the Early Cretaceous Cedar Mountain Formation.

Looking south from Crescent Junction, Utah, past gray marine Mancos Shale of Late Cretaceous age to the La Sal Mountains, are a series of three Tertiary laccolithic intrusions. The middle foreground shows an excellent example of a collapsed salt anticline, in this case properly named Salt Valley, with the axis on the right of the photograph and extending south into Arches National Park near Moab. Upper Jurassic Morrison Formation and overlying Lower Cretaceous Cedar Mountain Formation rocks on either side of the structure dip to the east and west away from the central valley. Rock units once extended across the top of the collapsed structure but are lost in the central graben or eroded away. Salt tectonic features such as this are fairly common in this region and are responsible for exposing large areas of wood-bearing units.

Modern cycads are tropical plants whose ancestors can be traced back in the fossil record to the Upper Paleozoic, perhaps 300 mybp according to Stevenson. They resemble palmettos or palms with long graceful fronds emanating from the top of a squat-to-elongated trunk. The fronds eventually fall off as the plant grows, leaving prominent leaf bases or scars along the trunk that typically display a diamond-shaped pattern. When the plant dies, the trunk may be replaced by silica, calcite, or hematite that sometimes preserves fine details of the plant's structure.

***Cycas revoluta*. A modern cycad. Maui, Hawaii, USA.**

Most of the fossil cycad trunks differ from the cycads of today, even though they look quite similar. True cycads (the living ones) are in the order Cycadales. However, most, but not all, Mesozoic cycads are in the order Cycadeoidales, which became extinct at the end of the Cretaceous Period. The relationship between these two varieties of cycads, both of which are represented in the fossil record, continues to be debated today. We will try to lend some clarity to the often-confusing issue of cycad nomenclature.

From the early to middle 1800s (and even before), cycads were being collected and studied in Europe. By the middle 1800s and into the early 1900s, cycad trunks were being collected in Maryland, South Dakota, Wyoming, Colorado, California, and Idaho. American investigators such as Lester Ward and G. R. Wieland made important contributions to the science of cycads during that period. In the last 70 years, beautiful fossil cycad specimens have been recovered from the Cretaceous of Patagonia, Argentina, and from Jurassic and Cretaceous sediments of southeastern Utah.

The host rocks for cycads from this region are either the Late Jurassic Brushy Basin Member of the Morrison Formation or the Early Cretaceous Cedar Mountain Formation. Cycads from the Brushy Basin are usually associated with coarse-grain sediments, such as coarse sandstones and conglomeratic sandstones. In general, most Jurassic cycad deposits are near the top of the Brushy Basin sequence, near the contact with Cedar Mountain Formation sediments. The climate may have been more conducive for the growth of cycads during this time, or preservational conditions may have been more favorable, with abundant mobile silica to replace the cycad wood. According to Kowallis, the age for the Brushy Basin deposits ranges from about 150 to 145 mybp.

Traditionally, investigators such as Craig and Aubrey divided the Lower Cretaceous into the Burro Canyon Formation, generally located east of the Colorado River, and the intertonguing Cedar Mountain Formation, generally located west of the Colorado River and extending farther west into central Utah. This demarcation is fuzzy but still used. Cycads from the Burro Canyon Formation are found in the extreme southeastern part of Utah; otherwise, most cycads are assigned to the Cedar Mountain Formation. The Burro Canyon and Cedar Mountain formations are also famous among wood collectors for well-silicified specimens of the fossil fern *Tempskya*.

Considerable attention has been given to the Cedar Mountain Formation in the past decade because of the abundance of newly described dinosaurs and the realization that the Colorado Plateau contains perhaps the most complete sequences of terrestrial rocks of this age in the world. As with Brushy Basin, the Cedar Mountain sediments were derived from the west and southwest and were deposited in a broad floodplain of low relief that contained numerous river channels and lake systems. However, the amount of volcaniclastic material was somewhat less and the volume of coarse sands and pebbles was greater in Cedar Mountain time than earlier. The type section is located at Cedar Mountain near Castle Dale, Utah, about 65 kilometers (40 miles) northwest of Green River, Utah. It is a thick sequence of variegated pastel-colored sediments that consist predominately of shales and mudstones with sandstone lenses and calcretes, the latter represented by brown carbonate nodules. A massive conglomerate, termed the Buckhorn Conglomerate, was included later as its basal member in certain areas.

The Cedar Mountain Formation was deposited across eastern Utah over a long period of time, approximately 25 million years, and contains diverse lithology. Deposition was influenced by local geologic factors, such as salt dome tectonics, producing great lateral variations and making division into members problematic. Recently, Kirkland and others subdivided the Cedar Mountain Formation into four additional members along with the basal Buckhorn Conglomerate. These new members are, in ascending order, the Yellow Cat Member, Poison Strip Sandstone, Ruby Ranch Member, and the Mussentuchit Member. Most of the fieldwork to support this revised stratigraphy was performed in the region north and northwest of Moab where cycads are found. Sediment varieties in these members are diverse and include abundant shales, calcretes (representing paleosoil horizons), conglomerates, and sandstones. However, the sandstones and conglomerates of the Poison Strip Sandstone host most of the cycads, as well as large amounts of other fossil wood in the Cedar Mountain Formation. The age of this member is approximately 115 million years, based on work done by Kirkland and others; thus a 30-million-year gap exists between the Late Jurassic and Early Cretaceous cycad-bearing strata in this region.

Cycads, along with conifers, ginkgos, *Hermanophyton*, *Pentoxylon*, and other plants, are part of the large group called gymnosperms that produce bare or exposed seeds, usually in cones, or otherwise not fully enclosed by fruiting organs. Division Cycadophyta, the general taxonomic name for all cycads, is divided into the two orders, Cycadales and Cycadeoidales (or Bennettitales). These orders were once considered closely related, but Norstog and Nicholls indicate that by the early 1900s paleobotanists considered them to be only distantly related.

Cycadales, "true cycads," are still with us and are represented by the common sago palm (*Cycas revoluta*) and cardboard palm (*Zamia maritima*), often seen as decorative indoor plants or in outdoor gardens in warmer climates. However, they are not palm trees. Wild cycads sill still grow along the U.S. Gulf Coast, Mexico, Central America, Australia, South Africa, and other locales.

Cycadeoidales ceased to exist after the end of the Cretaceous. They are the most common fossil cycad and constitute most of the specimens shown in this chapter. To further confuse the issue, the name Bennettitales is frequently used instead of Cycadeoidales. This is because the genus name *Bennettites*, published by Carruthers in 1870, was later determined to be a synonymy for the genus *Cycadeoidea*, published by Buckland in 1828. Crepet suggests that poor preservation probably contributed to the duality of names, and the loss of Buckland's type specimen precluded an early end to this nomenclatural controversy. Prominent American investigators such as Ward and Wieland, in the late 1800s and early 1900s, recognized the rule of priority and elected to use Buckland's name *Cycadeoidea*. Much of the scientific community, which was dominated by European researchers at the time, had settled on the name *Bennettities* and continued its usage. Therefore, order Cycadeoidales is equivalent to order Bennettitales, and, concordantly, genus *Cycadeoidea* is equivalent to genus *Bennettites*. Today, most investigators consider either of the ordinal names to be acceptable, but Bennettitales is generally accepted, especially in Europe, as the ordinal name. The chart at left shows the probable taxonomic organization for the variety of cycads found in the southwestern United States as suggested by Norstog and Nicholls.

If you can see cones embedded in the trunk of a cycad, you can be sure that it is an extinct Cycadeoidale. However, the lack of any cones in the trunk does not necessarily mean that the fossil is a modern cycad (see description of *Bucklandia* below). Other than the cones embedded in the trunks, the general appearances of the two orders are extremely similar to all but the expert eye. Most of the Mesozoic cycads from North America and South America available today are of the extinct variety.

There is some agreement that both living and extinct cycads derived from seed ferns, but paloeobotanists still debate exactly if, when, and how this happened. Arnold placed the origin of both types in the Carboniferous. Mamay suggested that seed-bearing plants, such as those found in the late Pennsylvanian rocks of Kansas and Texas, were ancestors of modern cycads. By the late Triassic and into the Jurassic, cycadeoids were a dominant plant. Late Jurassic deposits of carbonized leaf impressions have yielded higher percentages of modern cycad leaves than extinct cycadeoid leaves, suggesting that the modern cycads were at least as common. However, the paucity of true cycad trunks in the fossil record suggests that the true cycad trunks did not fossilize as readily as the extinct variety.

At least two families of extinct cycadeoid are found in southeastern Utah: the more primitive and far less common Williamsoniaceae and the dominant Cycadeoidaceae. *Bucklandia sp.* is most often attributed to the Williamsoniaceae. Specimens display continuous leaf bases with an absence of cone structures on the trunk and a large pith region with thin armor. As with modern cycadales, cones were probably borne near the top of the plant with the fronds. However, the final taxonomic position of *Bucklandia* is still debated; Watson and Sincock indicate that some paleobotanists relate it to the extinct Cycadoidaceae and others to modern cycadales. Typically, Williamsonian cycads from other parts of the world were slender and sometimes branched.

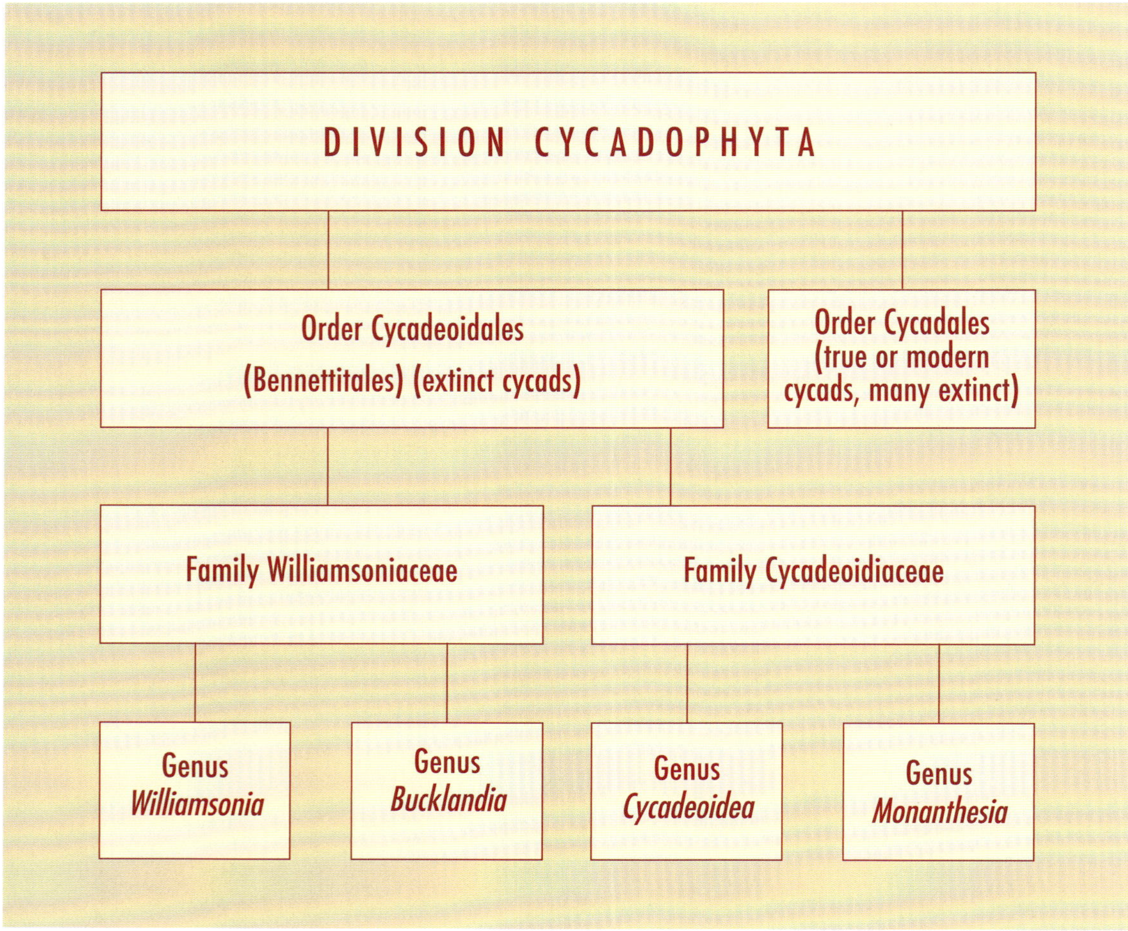

Professor Tidwell lists the general differences between Cycadeoidales and Cycadales that distinguish extinct and extant or true (modern) cycads (both exist as fossils).

Extinct Cycads, Order Cycadeoidales	True (modern) Cycads, Order Cycadales
Most had cones, sometimes called flowers, embedded in the trunk.	Cones on top of trunk, usually on stalks.
Leaf bases originated from vascular cylinder directly below their surface locations.	Leaf bases originate at vascular cylinder and spiral upward before emerging as leaf scars at the surface.

Specimen of probable *Bucklandia sp.* (which is most often attributed to the Williamsoniaceae), a variety of extinct cycad that does not have cones imbedded in the trunk. The regular geometric design of the diamond-shaped leaf bases in the trunk, which are produced when fronds are shed from the top of the plant, is characteristic of most cycads. Cycads were variously considered to be fossil corals, cacti, pineapples, or even beehives before the late 1800s, when little was known of these unusual fossils. Cedar Mountain Formation, Early Cretaceous. North of Moab, Utah, USA. Dayvault specimen, 24 cm.

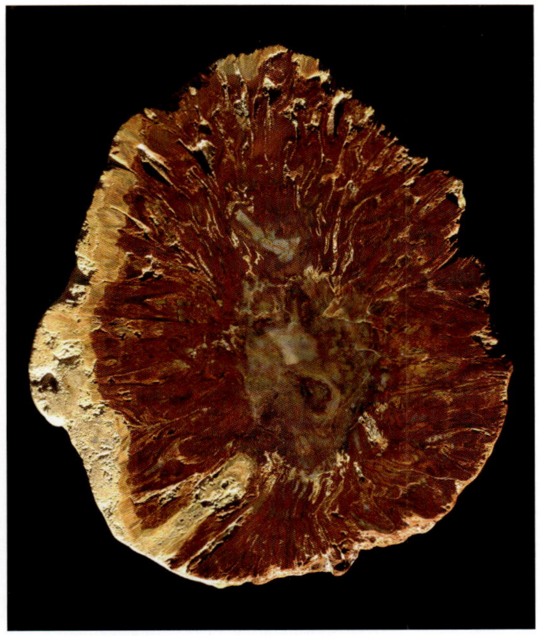

Transverse section of a colorful Brushy Basin cycad. Pith area is small, and the armor of leaf bases is continuous. The pith area was probably hollow and filled with massive chalcedony because little-to-no structure is preserved. Brushy Basin Member, Morrison Formation, Late Jurassic. Henry Mountains, Utah, USA. Dayvault specimen, 16 cm.

Family Cycadeoidaceae consists of two genera, *Cycadeoidea* and the less common *Monanthesia*. Monanthesian cycads might also be represented as a family. Basically, monanthesian cycads have a female reproductive structure (the mature version is a cone) in the axil (the angle between a leaf or branch and the axis from which it arises) of almost every leaf, and all these structures are approximately in the same stage of development. A *Monanthesia* specimen is most spectacular when it displays an almost solid mass of mature cones in its trunk. Cycadeoidean cycads, on the other hand, also develop reproductive structures (cones) in the axil of leaves but do so irregularly, and the degrees of development vary within the same plant. Most of the early *Monanthesia* trunks, studied 80 to 90 years ago, were discovered on Navajo Nation land in northwestern New Mexico and were thought to come from the Late Cretaceous Mesa Verde Formation. Tidwell doubts the historically assigned age of these specimens and thinks they are from Early Cretaceous sediments, as are most monanthesian cycads from the Four Corners states (Colorado, New Mexico, Arizona, and Utah).

Cycad specimens can look completely different depending on whether you see a transverse (across the trunk) or a tangential (parallel or tangential to the trunk) view. When broken or cut transversely, the parts of cycad trunks displayed from the center of the trunk outward are the pith, xylem, phloem, and a lateral view of the armor consisting of leaf bases and cone structures. The diagram on the following page demonstrates these features common to cycadeoids.

Probably the most obvious and striking texture of a cycad trunk is its outer armor consisting of a continuous close-spaced geometric grid of leaf bases interspersed with cones. A spray of leaves or fronds that originated at the apex of the plant was continually displaced downward and old leaves were shed as newer leaves replaced them, leaving diamond-shaped leaf bases of tough leathery tissue. Some earliest cycad specimens were thought to be colonial corals, fossil cacti, or even fossil beehives because of this geometric texture on the surface of the barrel- and spherically-shaped trunks.

Compressed cycad in transverse section. This trunk was flattened and looked much like a donut when found. The pith area completely collapsed and only the leaf bases remain. These textures are sometimes called "flames." Brushy Basin Member, Morrison Formation, Late Jurassic, Henry Mountains, Utah, USA. Dayvault specimen, 16.5 cm.

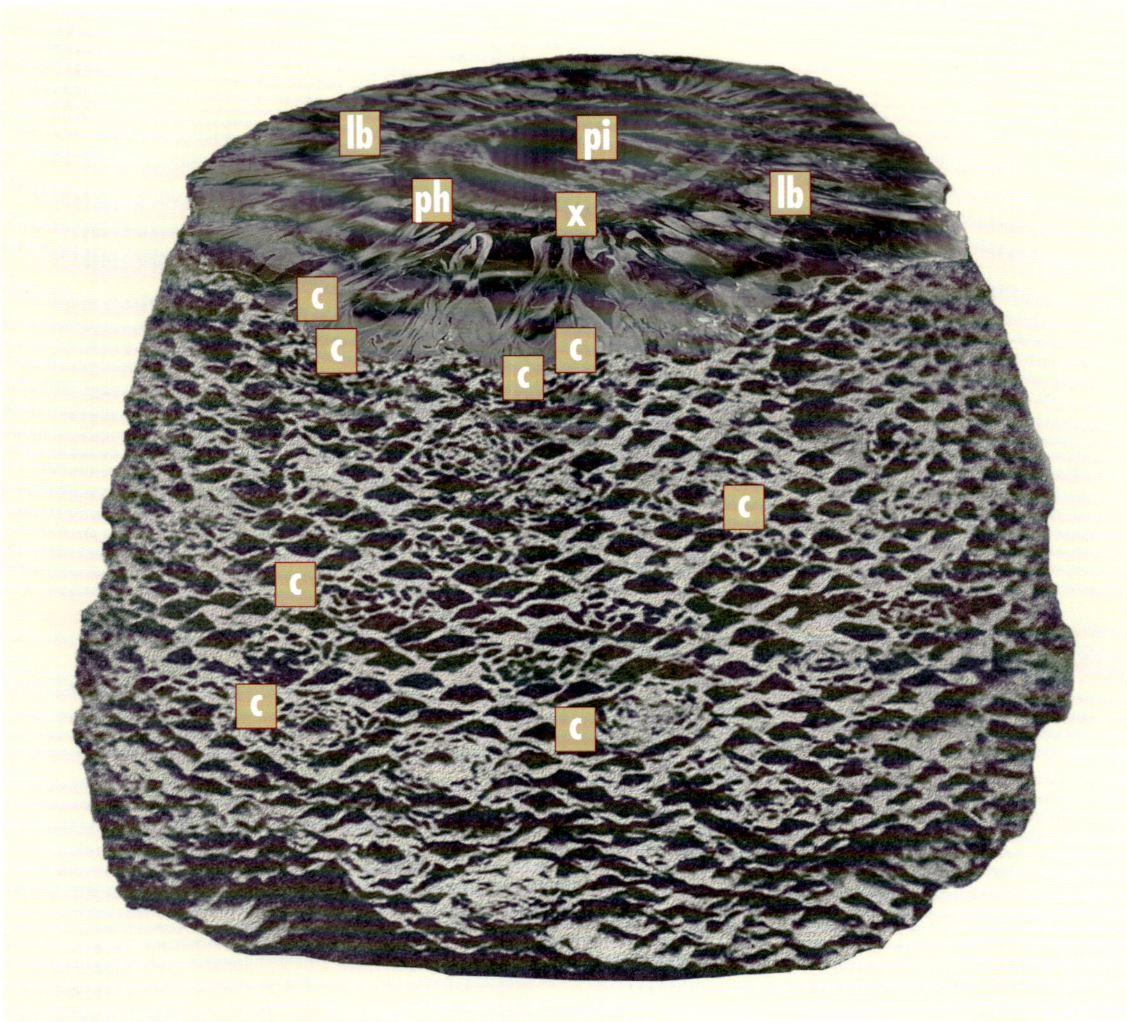

Diagram of a cycadeoid showing important features. Cones (c) in the armor of the trunk. Note two cones are shown in transverse and tangential views. Leaf bases (lb), shown in transverse section, constitute continuous armor around the trunk. Phloem (ph) forms the outer band of tissue, and xylem (x) forms the inner band. Pith (pi) is a pithy-to-spongy material in live cycads that fills the center of the plant. Sometimes the pith is not silicified in fossil trunks. Composite modified from *American Fossil Cycads*, Volume I, by G.R. Wieland and from *Common Fossil Plants of Western North America* by William D. Tidwell.

Transverse cut through a *Monanthesia sp.* specimen showing a continuous battery of cones in the armor. The cone stalks have lengthened, giving this graceful appearance to these structures. Seeds, which would have been in the apex of the cone, were either dispersed or were weathered from the fossil. Cedar Mountain Formation, Early Cretaceous. North of Moab, Utah, USA. Dayvault specimen, 20.5 cm.

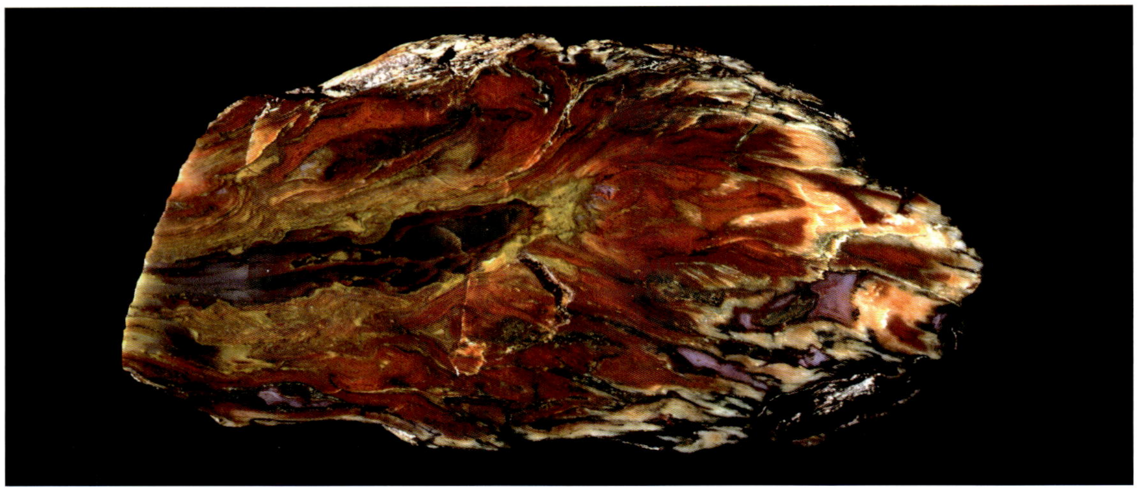

Small compressed, colorful cycad in transverse section. This cycad underwent compression during diagenesis. It is a good example of the flames produced by the leaf bases when viewed in transverse section. Brushy Basin Member, Morrison Formation, Late Jurassic, Henry Mountains, Utah, USA. Dayvault specimen, 5.5 cm.

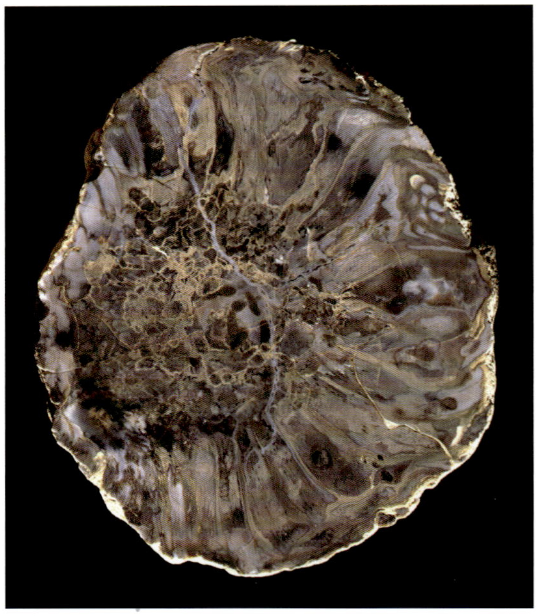

Transverse cut through a small cycad near the crown or apex. Leaf bases and two cones form the armor. Cones contain small crescent-shaped light or dark segments. Cedar Mountain Formation, Early Cretaceous. Yellow Cat area, Utah, USA. Dayvault specimen, 15 cm.

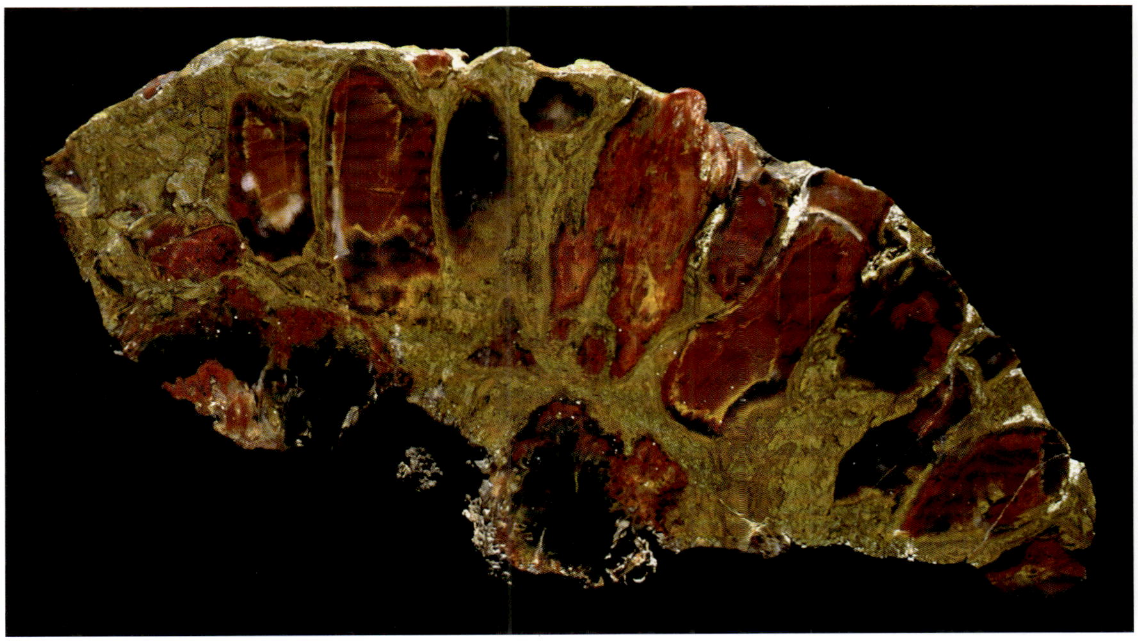

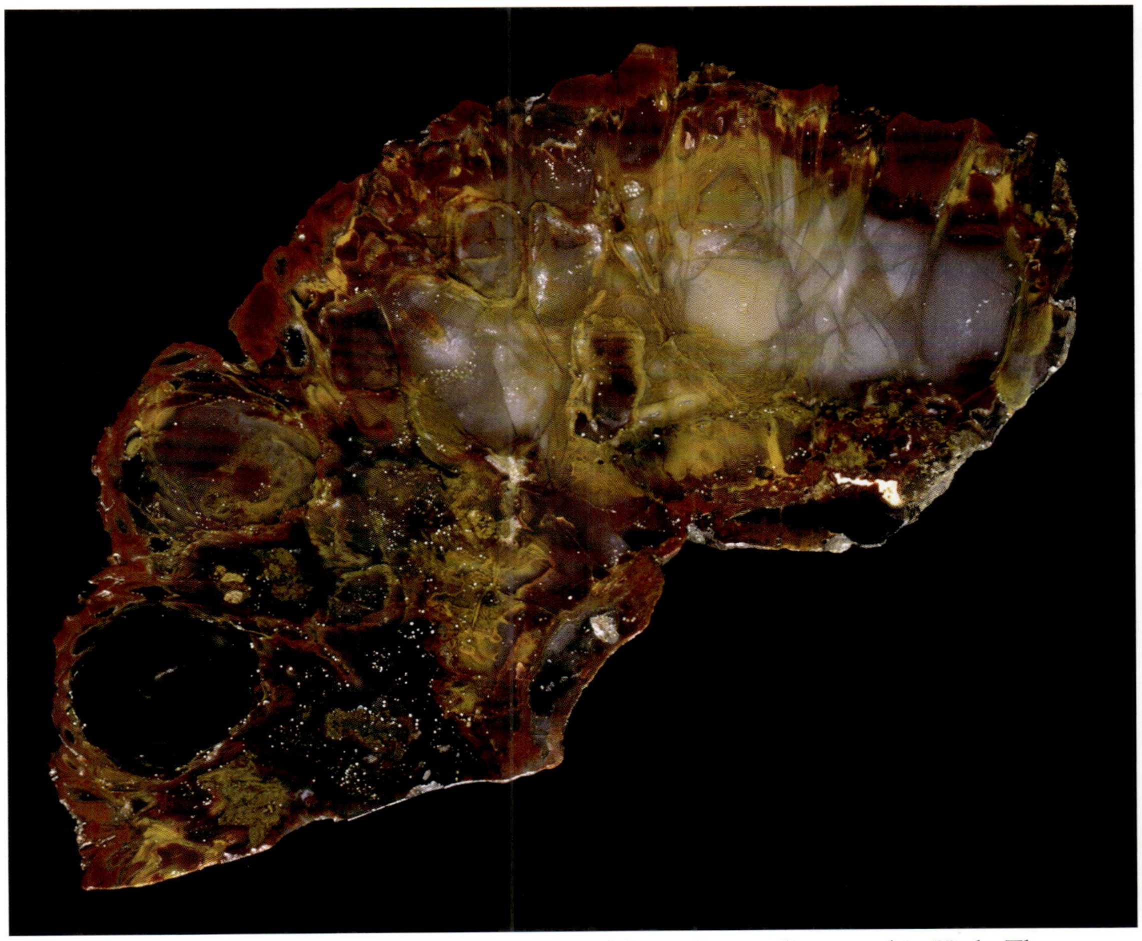

These two specimens are from one of the most colorful cycads ever discovered in Utah. The top specimen shows a section of armor consisting of discrete red-to-black leaf bases separated by greenish-yellow ramenta, a mat of epidermal hairs or scales that serve as protective tissue. The second specimen is a spectacular section of thick armor (approximately one-third of a full round) showing red leaf bases, greenish-yellow ramenta, and two large mature cones on the left side; one is black and the other is red, yellow, and lilac. Silicification on the right side has progressed to the point that structures are somewhat obliterated, but considerable detail remains in the cones. The pith was not preserved. Brushy Basin Member, Morrison Formation, Late Jurassic. Henry Mountains, Utah, USA. Dayvault specimens, upper 32 cm and lower 38 cm.

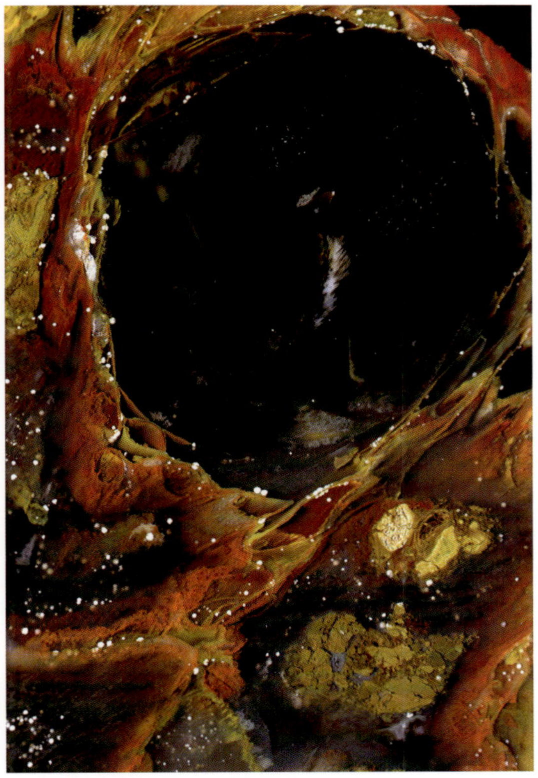

Detail of oxidized cone from previous photograph showing an oblique section through a cone with bracts surrounding the cone, hinting of internal structures in the dark central area. The small white blebs are tiny radiating authigenic calcite(?) rosettes. Oxidation of iron in the permineralizing chalcedony reveals the delicate complexity of structures. Microsporophylls are visible as the yellow tissue in the base of the cone; ramental tissue is the reddish strands that extend toward the distal part of the cone. Brushy Basin Member, Morrison Formation, Late Jurassic, Henry Mountains, Utah, USA. Dayvault specimen, distal diameter of cone 7 cm.

Tangential section of colorful cycad armor. Specimen shows leaf bases slightly compressed. Two cones showing crescent-shaped areas of white chalcedony occur in the upper left center. Brushy Basin Member, Morrison Formation, Late Jurassic, Henry Mountains, Utah, USA. Dayvault specimen, 7 cm.

Tangential section of a cycad fragment that illustrates how uniformly the diamond pattern can be preserved. Small regularly placed dots in the individual leaf bases are the remains of the vascular system that connected the leaf or frond to the trunk. Traces of the vascular system show omega patterns that remain as dark spots in some of the lighter leaf bases. Lightest colored areas in some leaf bases are quartz replacements. Cedar Mountain Formation, Early Cretaceous. North of Moab, Utah, USA. Dayvault specimen, 10 cm.

Section of a large cycad showing oval-shaped leaf bases in armor. Several cones are present. Specimen was collected more than 50 years ago. Cedar Mountain Formation, Early Cretaceous. North of Moab, Utah, USA. Dayvault specimen, width of view 40 cm.

Tangential section of a cycad fragment showing large leaf bases, up to 6 cm in width, suggesting that the original plant was large. An attractive grayish-lilac translucent chalcedony replaces the leaf bases. Vestiges of internal leaf base structures are still visible, but most were lost because of the intense silicification. Brushy Basin Member, Morrison Formation, Late Jurassic. Henry Mountains, Utah, USA. Dayvault specimen, 12 cm.

Macro photo of a spectacularly colored specimen displaying the classic diamond-shaped leaf bases in orange-red to black separated by ramental tissue that was replaced by silica and is now greenish yellow. This detail illustrates the importance of iron oxidation in creating the beautiful colors in this specimen. Rims of individual leaf bases oxidize first, with the oxidation front proceeding inward or across the leaf base. This piece is the outside slice that was cut from a larger specimen about 8 cm thick. The surface of the next specimen inward (the surface that faces this piece) contains more black areas in the same leaf bases, even though it is only separated by the thickness of the saw blade. Oxidation also highlights the internal textures in the leaf bases, except where silicification has advanced too far and obliterated delicate structures. Oxidation of iron to the ferric state produces zoning from surrounding and originally more porous connective tissue to the internal sections of leaf bases. The process produces greenish-yellow to orange-red colors; lack of oxidation produces black leaf bases. Small dark spots in some of the leaf bases represent vestiges of the vascular system that carried water and nutrients to the fronds and sugars back to the trunk. Brushy Basin Member, Morrison Formation, Late Jurassic, Henry Mountains, Utah, USA. Dayvault specimen, field of view 8 cm.

Cycad Cones

Cones embedded in the trunks are one of the cycadeoid's most interesting structures, whether viewed in transverse or longitudinal section. In the early 1900s, Wieland made detailed descriptions of many of the beautifully preserved cycads from the Lower Cretaceous deposits of the Black Hills, South Dakota. He was the first researcher to determine that *Cycadeoidea* cones contained both male and female structures. Crepet and Stewart suggested that pollination was accomplished by self-pollination or through the aid of insects, as in the case of modern cycads. However, modern cycads have separate male and female cones that are not embedded in the trunk.

In transverse section, cones appear as anomalous irregular areas in the trunk surrounded and protected on the trunk's surface by curved scale-like to rhomboid-shaped structures called bracts. In longitudinal section (transverse for the trunk), cones present "Y" shapes from the elongation of ovulate stalks in mature cones as shown in the photograph of the monanthesian specimen shown on the top right of page 296.

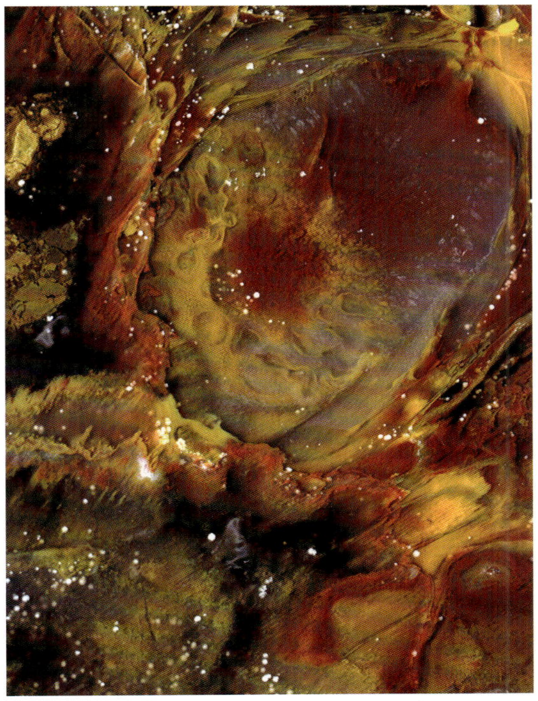

Detail of oxidized cone. Oxidation of iron in the permineralizing chalcedony reveals the delicate complexity of structures in this obliquely sectioned cone. Microsporophylls are visible as the yellow tissue in the base of the cone; ramental tissue is the reddish strands that extend toward the distal part of the cone. Brushy Basin Member, Morrison Formation, Late Jurassic, Henry Mountains, Utah. Dayvault specimen, distal diameter of cone 7 cm.

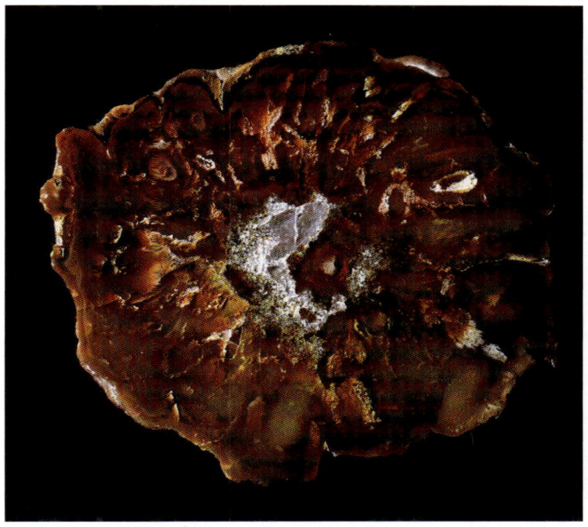

Highly silicified cone(?) possibly of a true cycad. Brushy Basin Member, Morrison Formation, Late Jurassic, Henry Mountains area, Utah, USA. Daniels specimen, 6 cm.

Mature cone of a monanthesian(?) cycad showing small seeds located around the perimeter of the distal structure. The seeds are clearly visible near the outer edge of the cone. Diamond-shaped patterns of leaf-like protective bracts are external to the outer edge of the cone. Brushy Basin Member, Morrison Formation, Late Jurassic, Henry Mountains area, Utah, USA. Dayvault specimen, maximum dimension of cone 4 cm.

Cycads did not branch in complex fashion like modern deciduous trees but usually formed squat-to-elongated solitary trunks. However, they occasionally formed clumps of trunks like barrel cacti, and some cycads had multiple branches similar to the Joshua trees of today. Apex structures were produced at the top of the plant or branch where emerging leaves grew or where a new cycad branch began to depart from the main trunk. These apexes contained the array of tissues that cycads produced to support the production of new leaves and also to protect them. Consequently, this portion of a cycad contains smaller, finer, and more complex features than the rest of the trunk. The enclosing arrangement of protective bracts around apexes and cones can sometimes cause confusion between the two structures in transverse view.

Although extinct cycads are not considered close relatives of modern cycads, they probably occupied similar ecological niches because they have so many similar structures. Modern cycads are found in a variety of environments from tropical rainforests to arid regions, but most species live in the latitudes near the equator from the Tropic of Cancer to the Tropic Capricorn. Therefore, they require relatively warm climates, wet or dry, but grow faster and reproduce more frequently in wetter climates. In South Africa, cycads are common in some remote areas of the arid Transvaal, but they grow faster in wetter areas of that country. A warm, moist climate would have probably been ideal for the growth of the abundant Mesozoic cycads.

The freshwater fossil bivalve *Unio* is often found in coarse-grained sediments containing cycads. These clams were apparently swept up by the same waters that uprooted and carried off the cycads, further suggesting a wet growth environment. Another clue supporting a wet environment for cycads is the small round fossil seed *Jensenispermum*, thought to have been produced by a Cycadophyte. Tidwell and Medlyn noted these seeds often accompany other plant fossils and the freshwater gastropod *Vivaparus reseedii* in quiet water environments, such as river backwater settings or ponds. Cycadeoids probably thrived in similar warm moist environments but may have existed in drier regions, just as their modern counterparts do. Fossil cycads have been found in North America in the greatest abundance, but specimens occur in Mexico, Western and Eastern Europe, India, Argentina, and even Antarctica. Of course, these places were in different latitudes millions of years ago, and the world had a warmer climate during most of the Mesozoic Era than today.

During life, modern cycads are robust plants protected by their thick covering of scaly armor, with resilient pollen and thick tough leaves that are resistant to insect damage and disease. Even forest fires do not always kill modern cycads. These characteristics for survival may have been true 150 million years ago, but their lack of a solid, internal woody tissue prevented many of these plants from becoming fossils. Trunks simply decayed too quickly and were scattered as organic trash. The likelihood of rapid decay would be greater in a moist environment than a dry one.

Fossil cycads are far less common than fossil conifers. This attribute may be a combination of

Cycad trunk showing a preserved crown of new growth tissue. This crown is called the apex structure of a cycad and contains the remains of new leaf growth and associated protective tissues. Leaf base armor is devoid of cones, the pith is large, and the armor is thin, suggesting a *Bucklandia* or perhaps a true cycad. Cedar Mountain Formation, Early Cretaceous. North of Moab, Utah, USA. Dayvault specimen, length of cycad is 32 cm.

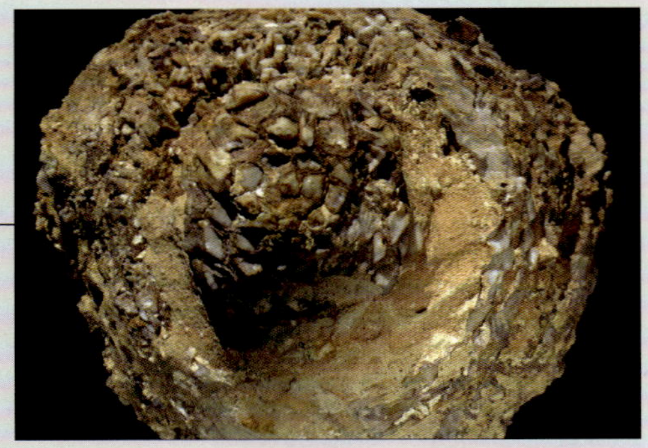

Same specimen, showing detail of apex structure. This view shows the diamond-shaped cross sections of new leaf growth that probably represented a growth spurt for the plant. Modern cycads often produce a new crop of leaves in this manner. The much larger and similar-shaped leaf bases in the main part of the trunk are clearly visible below the apex. Cedar Mountain Formation, Early Cretaceous. North of Moab, Utah, USA. Dayvault specimen, diameter of apex structure is 9 cm.

the greater abundance of conifers compared with cycads and the propensity for cycads to decay before fossilization. Many years of preferential cycad collecting by fossil wood collectors can also bias the apparent abundance of material left on the surface. Whatever the case, modern field collectors in the western United States know that fossil cycads are few and far between.

The demise of the highly successful cycadeoids at the end of the Cretaceous, while the modern cycads persisted, is another paleontological enigma. One possible explanation involves the dependency of cycadeoids on insects as pollinators. These plants had elaborate reproductive structures in the cones embedded in their trunks that may have required specialized insects to help with pollination. Studies of living cycads have shown a similar dependence on specialized beetles for cross-pollination of male and female cones. The plants entice the beetles by lowering natural toxin levels and emitting a chemical signal pleasing to the insect. This pollination sequence requires great coordination between plant and insect. Norstog and Nicholls concluded that if the obligate insect pollinators disappeared at the end of the Cretaceous, then so too might the dependent cycadeoids. Of course, modern cycads did not experience this disaster, but they may have depended on different insects for cross-pollination.

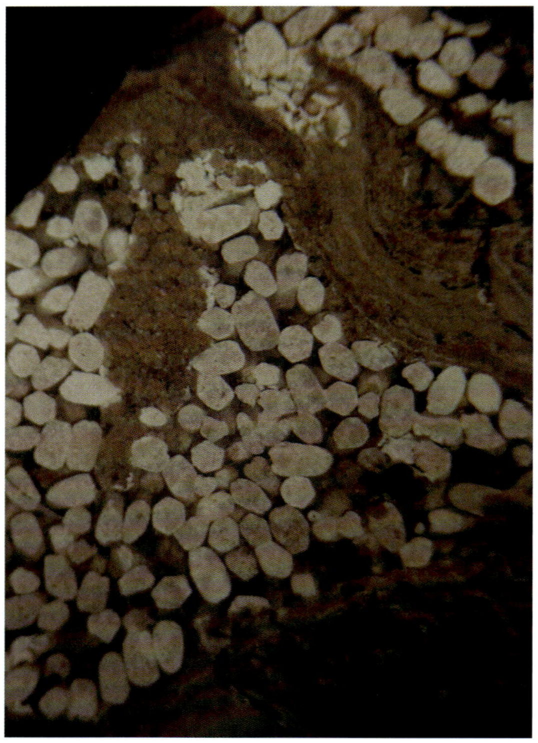

Micro image of cycad specimen with leaf bases containing small white blebs that are hexagonal to subhexagonal in cross section. Conrad Labandeira, Curator of Paleoentomology at the National Museum of Natural History, Smithsonian Institution, considers these blebs to be fecal pellets left by a termite-type insect. Similar hexagonal-to-subhexagonal blebs have also been observed in petrified conifer wood in the same formation and in woods of other ages and regions. Insects probably played an important role in cycadeoid cone fertilization, just as they do in modern cycads. This example shows the remains probably left by a termite that apparently inhabited this particular plant. Pellets are also common on the reverse side of this specimen, where it had been attached to the pith, suggesting that the pith was either missing or accessible when these termites deposited these microcoprolites and possibly indicating that the plant was already dead or in poor health. Cedar Mountain Formation, Early Cretaceous. North of Moab, Utah, USA. Dayvault specimen, fecal pellets average 1 mm.

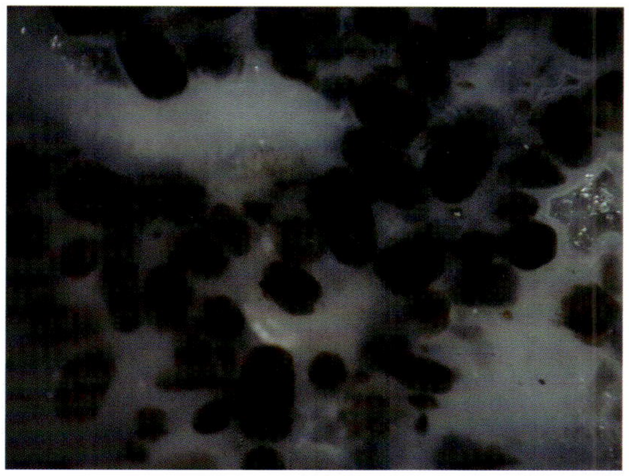

Silicified conifer wood contains termite coprolites in chalcedony. This specimen was found within a mile of the cycadeoid that contains similar coprolites and in the same rock unit. Early Cretaceous Cedar Mountain Formation. North of Moab, Utah, USA. Dayvault specimen, coprolites average about 1 mm.

CONIFER CONES

Bristle cones from the mountains of Colorado, USA.

When most fossil wood enthusiasts think of silicified cones, they think of the well-preserved, colorful cones from Patagonia, Argentina. Utah produces perhaps a greater variety of conifer cones than its South American counterpart, but they are not as large and spectacular and are certainly not as abundant. The great variety of colors in Utah cones reflects the various colors of silica in the Morrison sediments. Formal descriptions of the Utah cones are lacking, so cones cannot be identified to genus and species. Most resemble either a variety of *Araucaria* or *Pararaucaria* and all await study by paleobotanists.

Utah cones are found in four principal sedimentological settings. The vast majority of Utah cones are associated with short shoots, and all of these localities occur in the Late Jurassic Brushy Basin Member of the Morrison Formation. As with the short shoot localities, the vertical and lateral extents of the fruitful sediments are limited, often no thicker than a meter and no larger areally than 1,000 square meters. The general silhouettes of the cones are quite variable although most are small and elongate, no more than 3 to 4 cm in length and 1 to 2 cm in diameter. Preservation is often as a chalcedony or calcite cast, but detailed preservation of internal structures is present in silicified specimens. A large number of calcite cone casts occur with the short shoots in the Mount Ellen locality. In gross morphology, cones from this type of geological setting resemble the pararaucarians, although seed scale structures are dissimilar. The photographs show a number of these specimens from 10 different localities around the Henry Mountains, Moab, and elsewhere in Utah.

Small elongate siliceous cone attached to stem. Brushy Basin Member, Morrison Formation, Late Jurassic. West of Hanksville, Utah. Dayvault specimen, maximum dimension 4.5 cm.

Small equant-to-oval siliceous cones. Whether these cones are immature or a discrete variety is unknown. Brushy Basin Member, Morrison Formation, Late Jurassic. Mount Ellen locality, Henry Mountains, Utah. Hatch specimens, maximum dimension 1.9 cm.

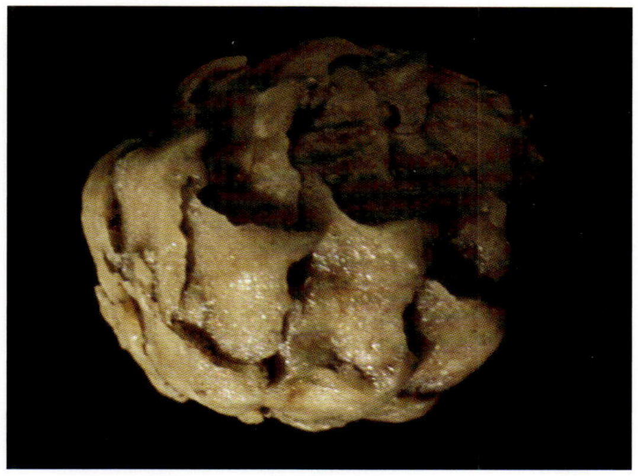

Small equant-to-oval siliceous cone showing prominent seed scales. This cone is a possible example of an immature cone or a discrete variety of small cone. Brushy Basin Member, Morrison Formation, Late Jurassic. Mount Ellen locality, Henry Mountains, Utah. Hatch specimen, maximum dimension 0.8 cm.

Large elongate cones with large seed scales as calcite casts. Brushy Basin Member Morrison Formation, Late Jurassic. Mount Ellen locality, Henry Mountains, Utah. Dayvault specimens, maximum dimension 7 cm.

Complete calcite cone cast with well-formed seed scales. Brushy Basin Member, Morrison Formation, Late Jurassic. Mount Ellen locality, Henry Mountains, Utah. Dayvault specimen, maximum dimension 3 cm.

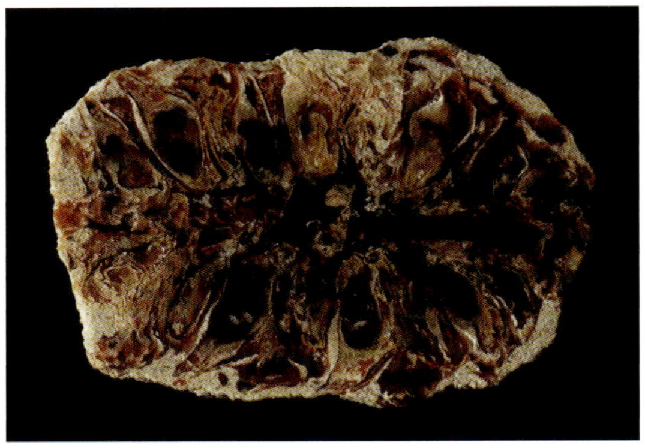

Small siliceous cone with large seeds. Brushy Basin Member, Morrison Formation, Late Jurassic. North of Hanksville, Utah. Dayvault specimen, maximum dimension 3 cm.

Siliceous cone with very regular seed scale pattern. Brushy Basin Member, Morrison Formation, Late Jurassic. Northwest of Hanksville, Utah. Hatch specimen, maximum dimension 4 cm.

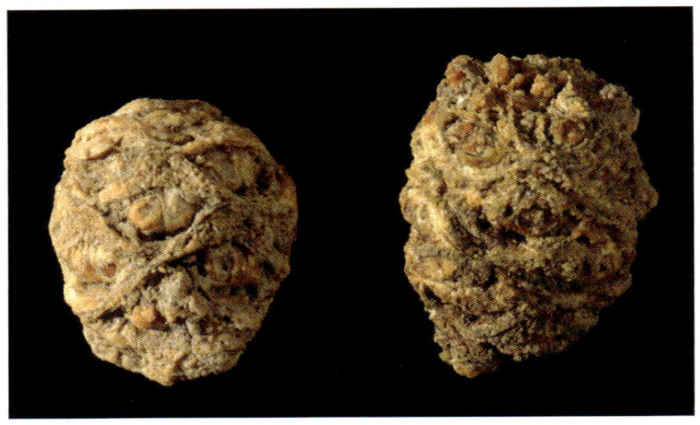

Small equant siliceous cones showing exposed seeds. Brushy Basin Member, Morrison Formation, Late Jurassic. West of Hanksville, Utah. Hatch specimens, maximum dimension 2 cm.

Section of siliceous cone. Brushy Basin Member, Morrison Formation, Late Jurassic. Northwest of Hanksville, Utah. Hatch specimen, maximum dimension 2.5 cm.

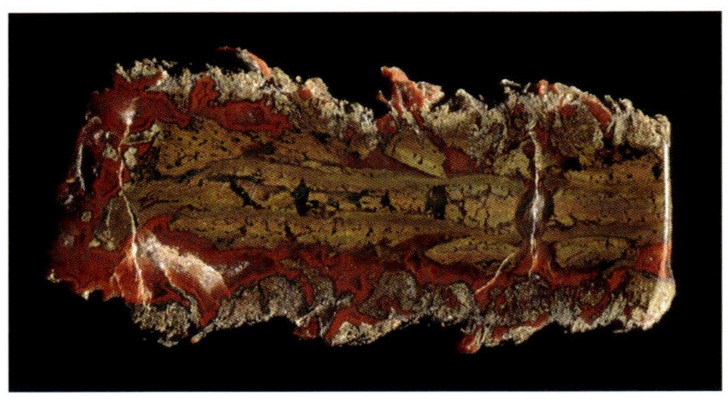

Colorful elongate siliceous cone cut in half. Brushy Basin Member, Morrison Formation, Late Jurassic. West side of Henry Mountains, Utah. Dayvault specimen, maximum dimension 4 cm.

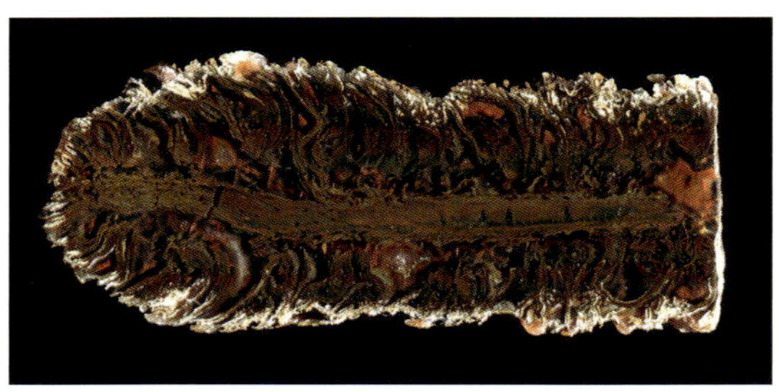

Colorful elongate siliceous cone cut in half. Brushy Basin Member, Morrison Formation, Late Jurassic. West of Hanksville, Utah. Dayvault specimen, maximum dimension 4.5 cm.

Araucaria-like siliceous cone cast. Brushy Basin Member, Morrison Formation, Late Jurassic. North of Hanksville, Utah. Dayvault specimen, maximum dimension 5 cm.

Siliceous cone in matrix. Brushy Basin Member, Morrison Formation, Late Jurassic. North of Moab, Utah. Dayvault specimen, maximum dimension of cone 2.5 cm.

Siliceous cone probably showing compression along long axis. Brushy Basin Member, Morrison Formation, Late Jurassic. Mount Ellen locality, Henry Mountains, Utah. Stoufer specimen, maximum dimension 3 cm.

Araucaria-like siliceous cone cast. Brushy Basin Member, Morrison Formation, Late Jurassic. North of Hanksville, Utah. Hatch specimen, maximum dimension 3 cm.

Carnelian cone cast from a red wood locality. Brushy Basin Member, Morrison Formation, Late Jurassic. North of Moab, Utah. Dayvault specimen, maximum dimension 2.5 cm.

Siliceous cone cut in half. Brushy Basin Member, Morrison Formation, Late Jurassic. Uncompahgre Plateau, Grand County, Utah. Dayvault specimen, maximum dimension 3.5 cm.

Siliceous cone cut in half. Brushy Basin Member, Morrison Formation, Late Jurassic. North of Moab, Utah. Dayvault specimen, maximum dimension 4.5 cm.

The second most important sedimentological setting hosts cones on the erosional surface between the Salt Wash and Brushy Basin Members of the Morrison Formation. In a number of locations, especially around the Henry Mountains where the strata dip only slightly, the softer Brushy Basin Member sediments erode faster than the more resistant sandstones and conglomerates of the Salt Wash Member. This erosional surface can cover many square kilometers. Black and white petrified wood is common on this surface that sometimes produces cones. *Araucarian*-like silicified cones with detailed internal preservation have been collected for at least 60 years from a limited number of locations with this setting. The maximum size of these cones is about 9 cm in length and 4 cm in width. Another locality for Salt Wash Member cones is in McElmo Canyon in Montezuma County, Colorado. Here, elongate cones to 4 cm in length are associated with unusual fossil woods, including *Hermanophyton*, *Cycadeoidea*, and an undescribed wood that resembles *Pentoxylon* or *Rhexoxylon*.

Hermanophyton taylorii. Morrison Formation, Upper Jurassic. Cortez area, Colorado, USA. Daniels specimen, 9 cm.

Hermanophyton. Morrison Formation, Upper Jurassic. Cortez area, Colorado, USA. Daniels specimen, 11.5 cm.

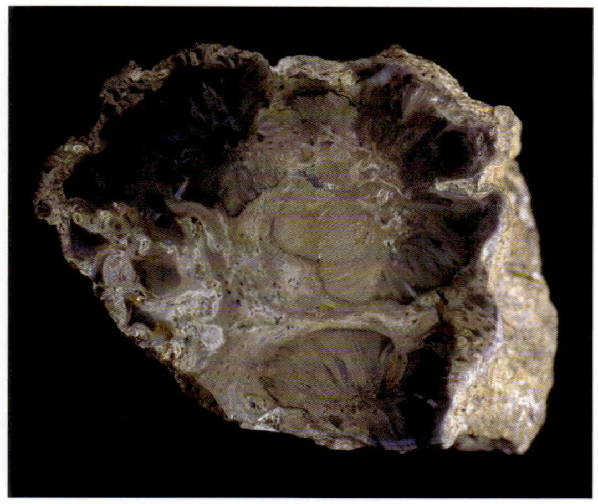

Unidentified polyvascular wood. Morrison Formation, Upper Jurassic. Cortez area, Colorado, USA. Dayvault specimen, 8.5 cm.

Araucaria-like siliceous cone. Polished half of a cone showing seed structure in black and white chalcedony. Salt Wash Member, Morrison Formation, Late Jurassic. Northwest of Hanksville, Utah. Dayvault specimen, maximum dimension 5 cm.

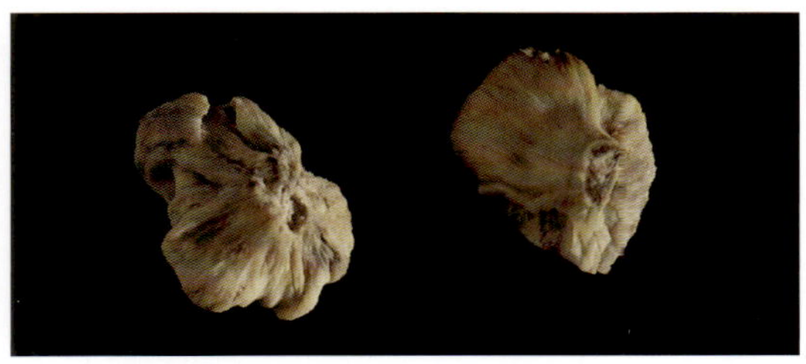

Unusual siliceous wood structures that resemble *Steinerocaulis sp.* from cone locality. Brushy Basin Member, Morrison Formation, Late Jurassic. Vernal area, Utah. Dayvault specimens, 2 cm.

Silicified cones showing detailed internal preservation have been recovered from Brushy Basin Member sediments near Vernal, Utah. This is not an apparent short shoot location, but unusual burl-like structures associated with the cones resemble *Steinerocaulis sp*. The cones mostly occur in a 2-meter-thick sequence of grayish shale where one or more trees have been preserved. Numerous small twigs occur with the cone-bearing units, rarely with well-preserved surface textures. The cones are to 6 cm in length, although the longest specimen is not complete.

Araucaria-like siliceous cone with small seeds. Salt Wash Member, Morrison Formation, Late Jurassic. Northwest of Hanksville, Utah. Dayvault specimen, 9 cm.

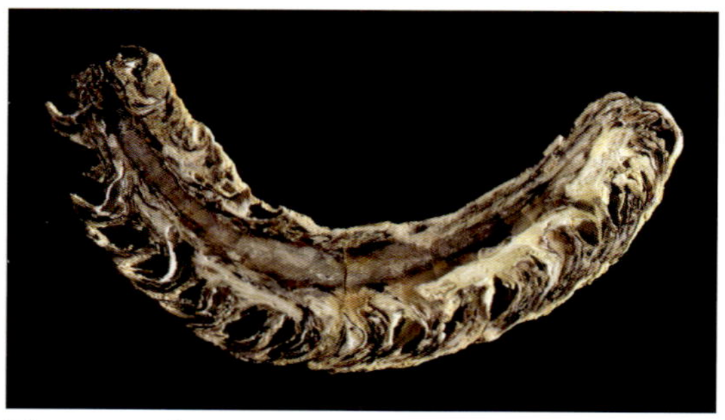

Elongate siliceous cone with distinctive curve. Salt Wash Member, Morrison Formation, Late Jurassic. Northwest of Hanksville, Utah. Dayvault specimen, maximum dimension 6 cm.

Elongate cone showing detailed weathering of small seed scales. Brushy Basin Member, Morrison Formation, Late Jurassic. Vernal area, Utah. Dayvault specimen, 5 cm.

Araucaria-like siliceous cone. Salt Wash Member, Morrison Formation, Late Jurassic. Northwest of Hanksville, Utah. Dayvault specimen, 6 cm.

Siliceous cones with detailed internal structure. Brushy Basin Member, Morrison Formation, Late Jurassic. Vernal area, Utah. Dayvault specimens, maximum dimension 5 cm.

Small siliceous twig with detailed surface textures from Vernal cone locality. Individual attachments for leaves are evident in this specimen. Brushy Basin Member Morrison Formation, Late Jurassic. Vernal area, Utah. Dayvault specimen, 2.5 cm.

Siliceous twigs from cone locality. Brushy Basin Member, Morrison Formation, Late Jurassic. Vernal area, Utah. Dayvault specimens, maximum dimension 3.5 cm.

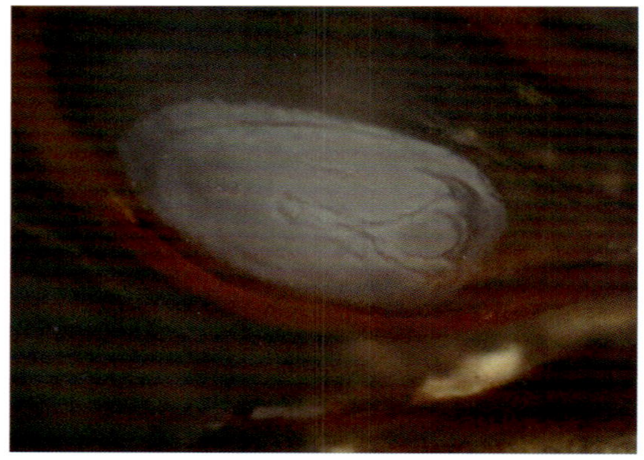

Polished longitudinal section of cone showing vertical section through seed matrix. Nutritive tissue is visible in the seed. Brushy Basin Member, Morrison Formation, Late Jurassic. Vernal area, Utah. Dayvault specimen, view 7 mm.

Cones formed from colorful silica showing detailed internal preservation have been recovered from Lower Cretaceous Cedar Mountain Formation sediments near Castle Dale, Utah. They are found in the first few meters of shales above the Buckhorn Conglomerate. These cones most resemble typical *Araucaria* cones from Argentina, but the colors are pink to red with white highlights. A final type of cone from Utah is a sandstone cast that appears to be part of a sandstone concretion. It was not found in place; therefore, its stratigraphic location is unknown, but its mode of preservation is similar to some fossil wood specimens that occur in the Late Cretaceous Dakota Sandstone. At 14 cm in length and 7 cm in diameter, this is the largest cone that the authors have seen from this region.

Carnelian cone cast showing somewhat weathered seed scale structure. Cones such as this are rarely found with red wood specimens and are almost always associated with small limbs. Brushy Basin Member, Morrison Formation, Late Jurassic. Moab area, Utah. S. Wade specimen, 3.5 cm.

Araucaria sp. showing external texture. Cedar Mountain Formation, Early Cretaceous. Castle Dale, Utah. Smith specimen, 7 cm.

Colorful siliceous *Araucaria sp.* showing internal structure of seed scales. Cedar Mountain Formation, Early Cretaceous. Castle Dale, Utah. Smith specimen, 7 cm. (Same specimen as shown at left.)

Unusual sandstone cone cast as part of a concretion. Possibly from the Dakota Sandstone, Late Cretaceous. West of Hanksville, Utah. Dayvault specimen, 14 cm.

V

CHAPTER FIVE

Fossil Woods of the Upper Interior Western United States

MONTANA, WYOMING, AND NORTHERN COLORADO

At the end of the Cretaceous Period, the Rocky Mountain land mass stood just 1 kilometer (3,281 feet) above sea level, compared with its current elevation of more than 4 kilometers (with numerous peaks reaching above 14,000 feet) above sea level. During the Paleocene Epoch (65 to 56 mybp), the Rocky Mountains rose considerably, in some regions approaching current elevations, altering weather patterns across the region. During the Paleocene Epoch, the number of extant deciduous angiosperm families doubled from the Late Cretaceous for several reasons, including the cooling of temperatures and seasonal instability of waterside habitats. Ancestral beech, birch, elm, sweetgum, and walnut families dominated the angiosperms. A warm-temperate broad-leaved evergreen forest with few deciduous hardwoods grew in the Colorado Front Range near present-day Denver, while a mixed deciduous and evergreen broad-leaved forest inhabited Wyoming. We know little about Early Tertiary Period angiosperms. Paleocene angiosperm woods have been identified in just three areas in the western United States — one is the Denver Basin where Paleocene and Early Eocene woods were examined by Wheeler and Michalski. They found mostly angiosperms, which were all diffuse porous. One principal type was *Paraphyllanthoxylon*, a probable member of Lauraceae. They describe other Denver Basin fossil woods as Betulaceae-like, Lauraceae, *Platanoxylon*, Taxodiaceae-like, Juglandaceae-like, and *Parabombacaceoxlyn*. During the Early Eocene Epoch (56 to 50 mybp), temperatures rose and rainfall increased in the western United States. Numerous lakes formed in the Rocky Mountains as rising and folding terrain created catch basins. Early Eocene flora in the central Rocky Mountains region was tropical to paratropical and included many trees now considered Asian exotics.

Farther north, the forests now preserved at Specimen Ridge (51 to 50 mybp) in Yellowstone National Park thrived between geologically frequent volcanic eruptions. During the Middle Eocene to Early Miocene, the Rocky Mountain uplift continued. Temperatures cooled from Middle to Late Eocene, transitioning flora from subtropical to temperate species, while extensive volcanism in the central Rocky Mountains deposited as much as 1,700 meters (5,500 feet) of ash in basins.

Old postcard of Lamar, Colorado, showing a building made from petrified wood.

In the foreground is a Tertiary age fossil log; the Denver skyline is in the distance. Green Mountain, Jefferson County, Colorado, USA.

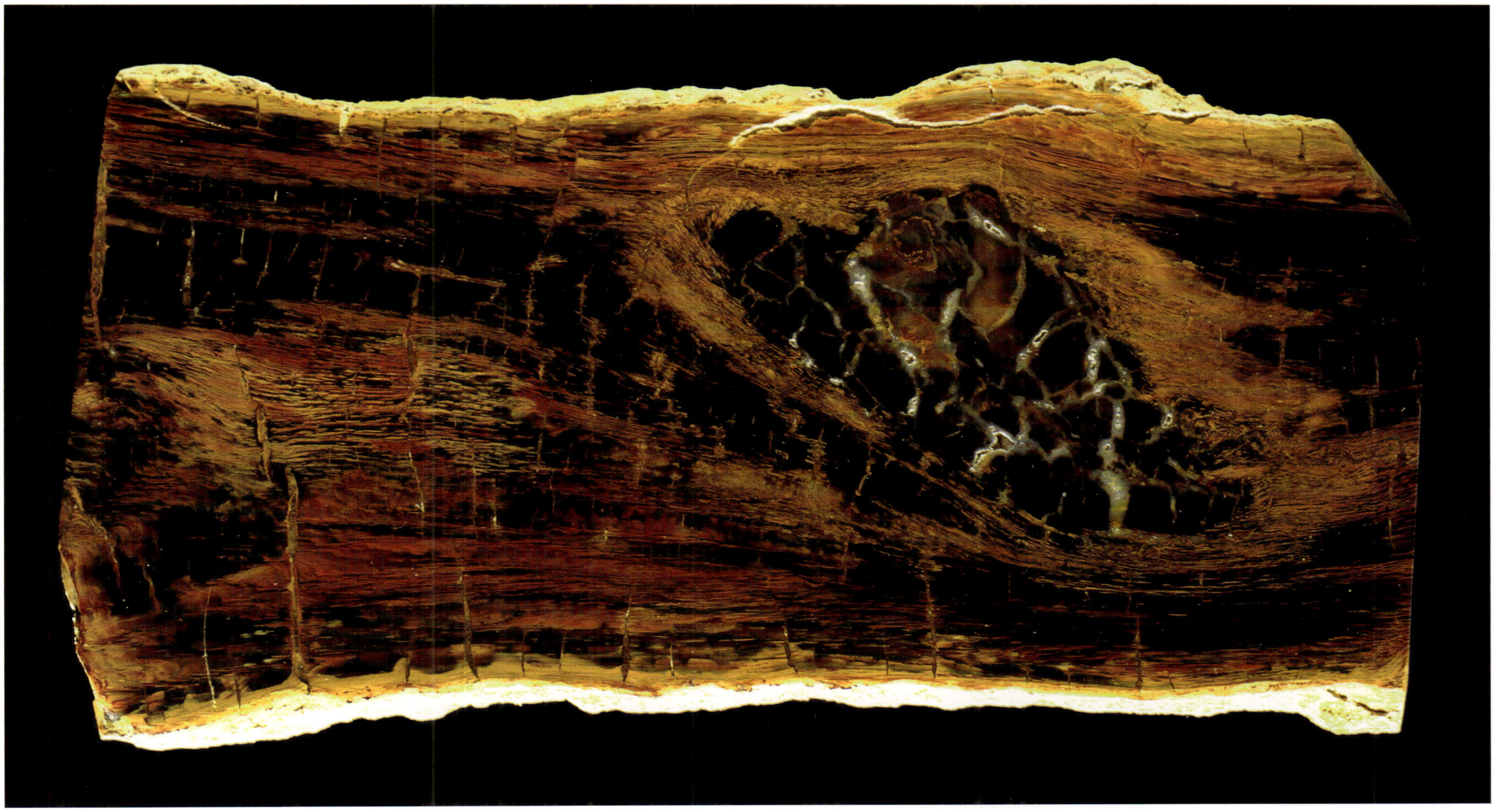

Longitudinal slice of sycamore-like (*Platanoxylon*) fossil hardwood. Denver Basin, Eocene. Douglas County, Colorado, USA. Daniels specimen, 17.5 cm.

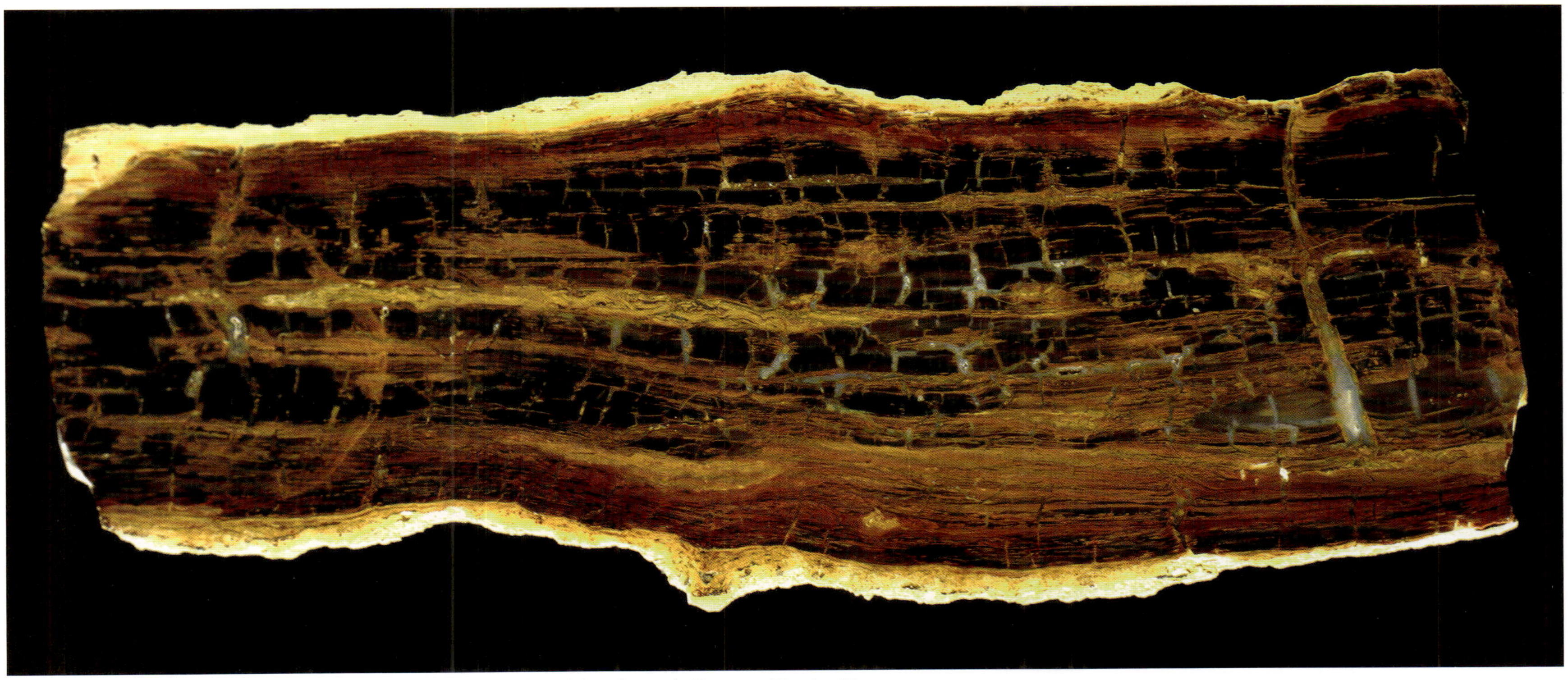

Longitudinal slice of sycamore-like (*Platanoxylon*) fossil hardwood. Denver Basin, Eocene. Douglas County, Colorado, USA. Daniels specimen, 27 cm.

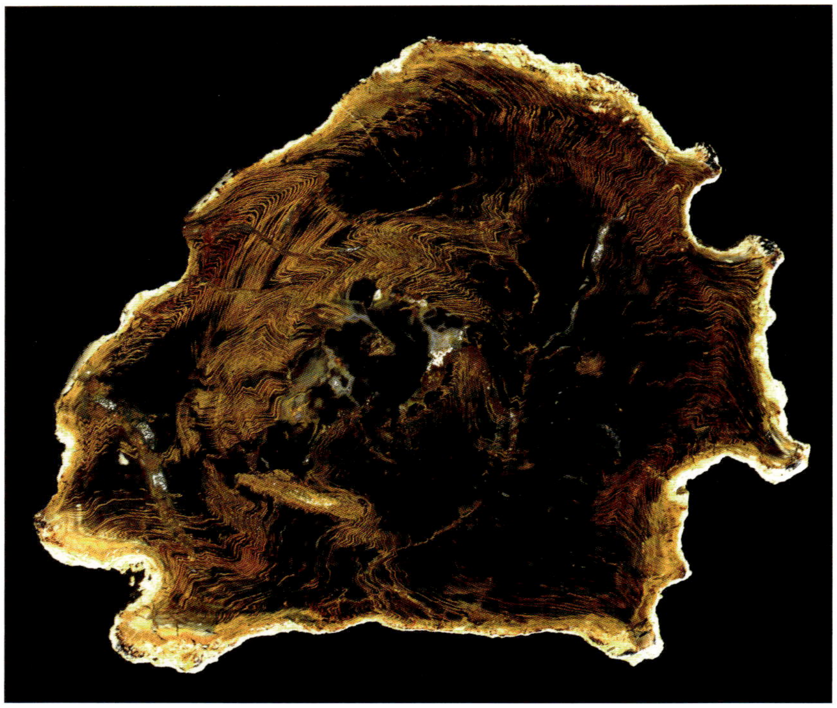

Sycamore-like (*Platanoxylon*) fossil hardwood. Denver Basin, Eocene. Douglas County, Colorado, USA. Daniels specimen, 13 cm.

Longitudinal slice of fossil hardwood. Denver Basin, Tertiary. Arapahoe County, Colorado, USA. Daniels specimen, 18 cm.

Longitudinal slice of sycamore-like (*Platanoxylon*) fossil hardwood. Denver Basin, Eocene. Douglas County, Colorado, USA. Daniels specimen, 18.5 cm.

Sycamore-like (*Platanoxylon*) fossil hardwood. Denver Basin, Eocene. Douglas County, Colorado, USA. Daniels specimen, 13 cm.

In 1914, Frank Knowlton wrote *The Fossil Forests of the Yellowstone National Park* for the United States Geological Survey. Although somewhat primitive relative to modern works (he estimated the trees might be as old as 1 million years old), it was a well-researched and well-written analysis of the forest for the time. Knowlton took special note that the trees in this succession of petrified forests are still upright in their growth positions, a rarity for petrified forests. He wrote: "After the first forest grew and was entombed, there was a time without volcanic outburst—a period long enough to permit a second forest to grow above the first. This in turn was covered by volcanic material and preserved, to be followed again by a period of quiet, and these more or less regular alternations of volcanism and forest growth continued through the time the beds were in process of formation." Knowlton noted no less than 12 levels of forests; later researchers raised this estimate to no less than 27. Knowlton described 10 species of fossilized trees. Ninety-five percent of the trees were conifers related to pine and redwood. The remainder had affinities to bay, buckthorn, laurel, oak, and sycamore. As is normally the situation, a survey of fossil plant impressions in the area revealed far more variety with some 150 types of fossil plants, including conifers (pine, redwood, and yew), ferns, horsetail, and numerous deciduous plants and trees (acacia, aralias, ash, basswood, bay, beech, birch, bittersweet, breadfruit, buckthorn, chestnut, cinnamon, dogwood, elm, fig, grape, hazel nut, hickory, laurel, magnolia, maple, oak, persimmon, poplar, soap berry, sumac, sycamore, walnut, willow, and others). Similar fossil wood preservations occur in related strata in Montana and Wyoming, including the Gallatin Mountains and Tom Miner Basin, where beautifully preserved fossil wood specimens colored green and yellow, some with amethyst-lined hollow centers, were found.

Typical fossil log from the area of Parker, Colorado. Denver Basin, Eocene. Douglas County, Colorado, USA. Daniels specimen, 18 cm tall.

Cover of the 1928 booklet *The Fossil Forests of the Yellowstone National Park* by Frank Knowlton, showing a fossil tree on Specimen Ridge. Yellowstone National Park, Wyoming, USA.

Photograph from *The Fossil Forests of the Yellowstone National Park* by Frank Knowlton, showing fossil tree on Specimen Ridge. Yellowstone National Park, Wyoming, USA.

Diagram from *The Fossil Forests of the Yellowstone National Park* by Frank Knowlton, showing an idealized view of the successive petrified forests on Specimen Ridge.

The same two Specimen Ridge *Sequoia*-like *Taxodioxylons* depicted in *The Fossil Forests of the Yellowstone National Park* by Frank Knowlton, showing the effect of approximately 80 years erosion. Yellowstone National Park, Wyoming, USA.

Upright Eocene age *Taxodioxylon* on Specimen Ridge keep watch over the Yellowstone Valley. Yellowstone National Park, Wyoming, USA

This large *Sequoia*-like *Taxodioxylon* [approximately 5.5 meters (18 feet tall)] seems at home with modern conifers on Specimen Ridge. Yellowstone National Park, Wyoming, USA.

Two Specimen Ridge *Sequoia*-like *Taxodioxylons* with Frank Daniels. Yellowstone National Park, Wyoming, USA.

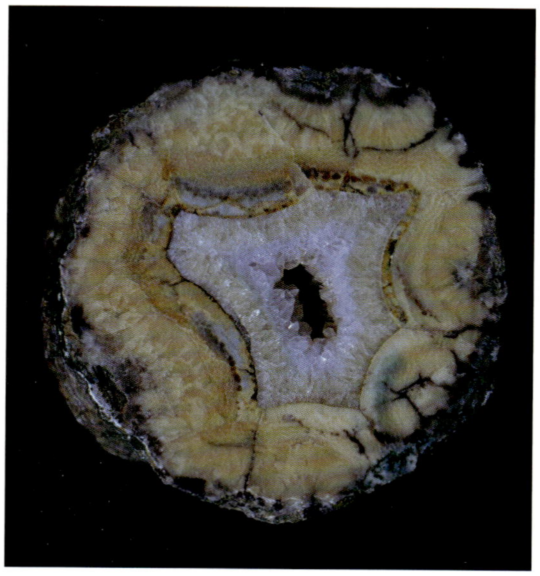

Fossil wood cast. Tertiary. Tom Miner Basin, Montana, USA. Daniels specimen, 5.5 cm.

Fossil wood cast. Tertiary. Tom Miner Basin, Montana, USA. Daniels specimen, 4.5 cm.

Fossil wood cast. Tertiary. Tom Miner Basin, Montana, USA. Daniels specimen, 6 cm.

The Grand Canyon of the Yellowstone lies within a 30 by 45 mile caldera. The most recent major eruptions occurred about 2 million years ago, 1.3 million years ago, and 630,000 years ago. The Yellowstone River flows 305 meters (1,000 feet) below the top of the canyon, having cut its way through the colorful volcanic rock. Yellowstone National Park, Wyoming, USA.

The largest of these Specimen Ridge *Sequoia*-like *Taxodioxylons* measures approximately 1.8 meters (6 feet) in diameter. Yellowstone National Park, Wyoming, USA.

Fossil conifer cone. Jurassic. Red Canyon, Wyoming, USA. Daniels specimen, 4 cm.

Set of fossil wood specimens from Tom Miner Basin, Montana, displaying the variety of types from this area. Tertiary. Tom Miner Basin, Montana, USA. Daniels specimens, 11.5 cm largest face.

Metasequoia-like cones. Cretaceous. Collected from various locations in Montana and North Dakota, USA. Daniels specimens, 2.5 to 3.5 cm.

Fossil wood cast. Tertiary. Tom Miner Basin, Montana, USA. Daniels specimen, 9 cm.

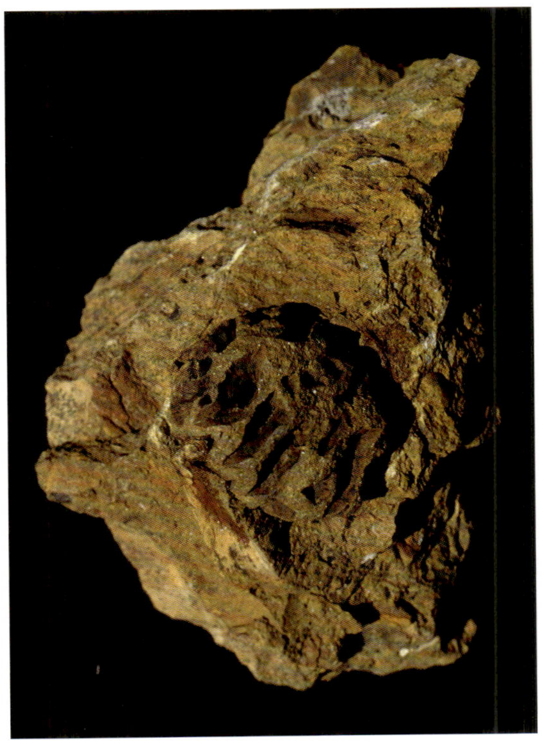

Metasequoia-like cone in matrix. Cretaceous. Collected at various locations in Montana and North Dakota, USA. Daniels specimen, cone 3.5 cm.

Double *Metasequoia*-like cone. Cretaceous. Collected at various locations in Montana and North Dakota, USA. Daniels specimen, 3 cm.

Fossil wood cast. Tertiary. Tom Miner Basin, Montana, USA. Daniels specimen, 7 cm.

Especially rare form of fossil wood with amethyst-lined interior void. Tertiary. Tom Miner Basin, Montana, USA. Daniels specimen, 22 cm long; 12 cm diameter.

Fossil wood cast. Tertiary. Tom Miner Basin, Montana, USA. Daniels specimen, 8.5 cm.

Fossil wood cast. Tertiary. Tom Miner Basin, Montana, USA. Daniels specimen, 8 cm.

Fossil wood cast. Tertiary. Tom Miner Basin, Montana, USA. Daniels specimen, 7 cm.

The steepest red cliffs are Triassic age Chugwater Formation; the tan sandstone at the top right is Jurassic age Nugget Formation. Red Canyon, Wyoming, USA.

In large portions of present-day northern Colorado, northern Utah, and Wyoming, lake sediments set the stage for the fossil-rich Middle Eocene Green River Formation (45 to 48 mybp), extending to 65,000 square kilometers (25,000 square miles). The Green River Formation yields remains of plants from at least four distinct habitats: lake margins, swamps, and floodplains (cottonwood, palm, sycamore, and willow); drier higher slopes (savanna and oak-pine woodland); uplands (broad-leaved deciduous trees); and mixed hardwood conifer forests. We know the Green River Formation for leaf impressions and fish fossils as well as a wide array of silicified stems. Plant fossil surveys of the formation identified 7 species of fern (including *Acrostichum hesperium* and *Equisetum winchesteri*), gymnosperms (including pine and sequoia), 4 species of monocots, and 63 species of dicots. The most well-known collecting locations for fossil wood specimens are the "Blue Forest" and Eden Valley near Farson, Wyoming. Blue Forest specimens receive their name from the coating of botryoidal blue chalcedony that envelopes some of the specimens.

Colorful banded onyx fossil wood cast. Wagon Bed Formation, Eocene. Sand Draw, Wyoming, USA. Albers specimen, 11 cm.

Fossil oak-like hardwood. Wagon Bed Formation, Eocene. Sand Draw, Wyoming, USA. Albers specimen, 9.5 cm.

Fossil hardwood from Eden Valley. As with many of the specimens from this area, the exterior is largely enveloped in blue chalcedony. Green River Formation, Eocene. Eden Valley, Wyoming, USA. Albers specimen, 18 cm.

Green fossil wood cast. Green River Formation, Eocene. Dubois area, Wyoming, USA. Daniels specimen, 11 cm.

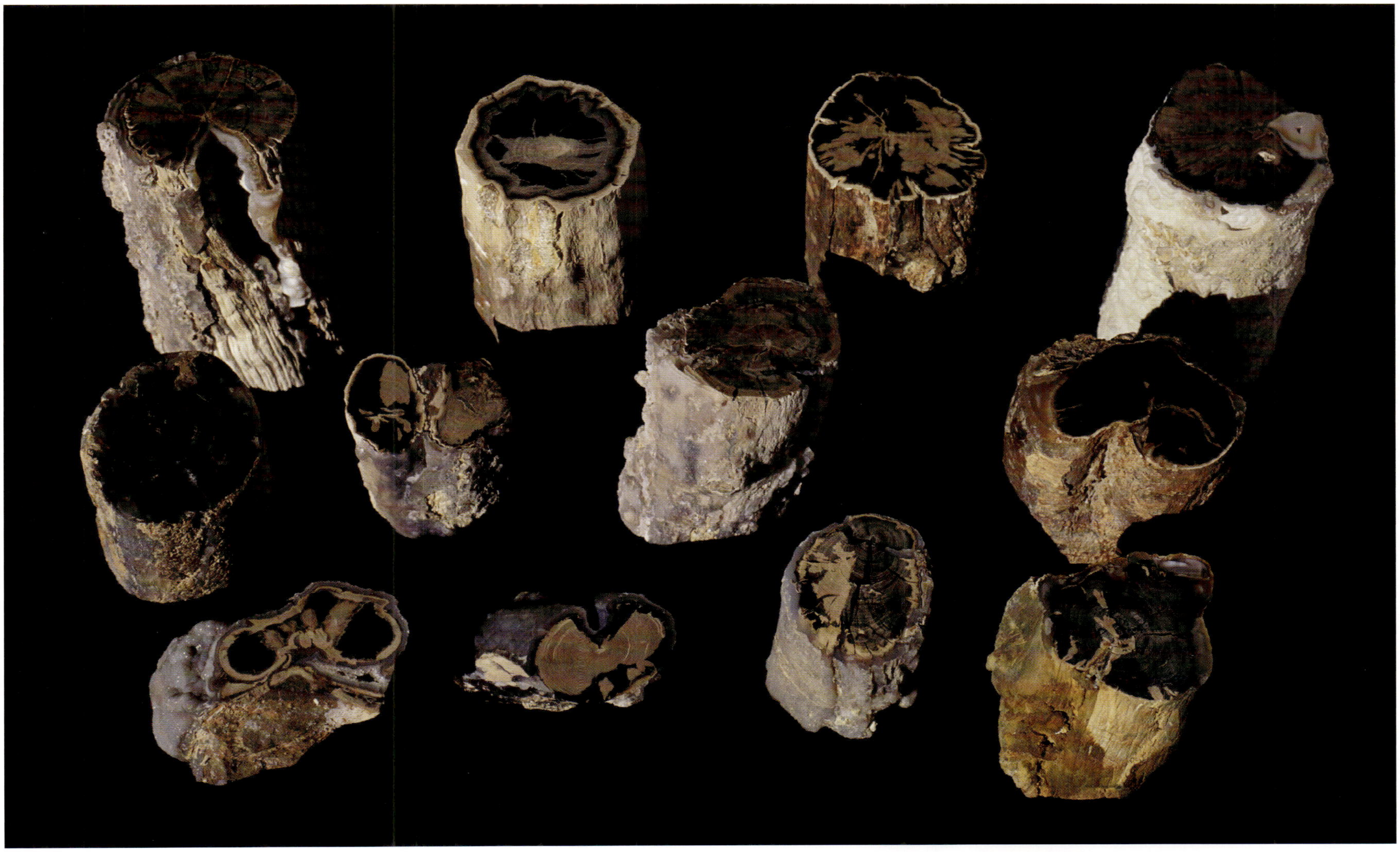

Fossil wood specimens from the area of Farson, Wyoming. Most are from the Blue Forest and Eden Valley. Specimens from these locations frequently exhibit well-preserved cell structure. Green River Formation, Eocene. Farson, Wyoming, USA. Daniels specimens, 18 cm tallest.

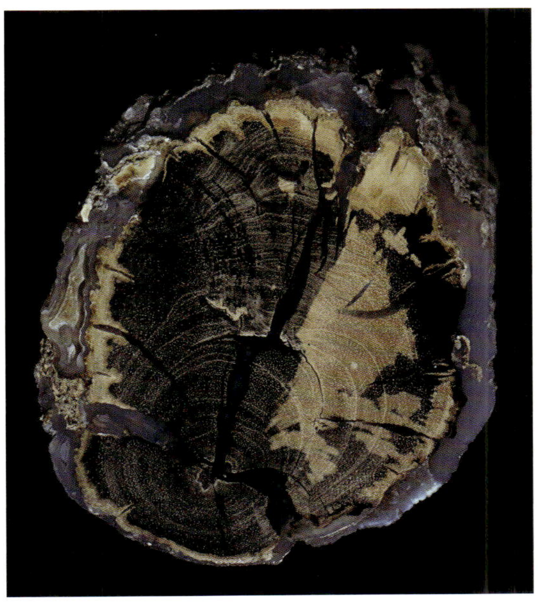

Fossil hardwood. Green River Formation, Eocene. Farson area, Wyoming, USA. Daniels specimen, 5.5 cm.

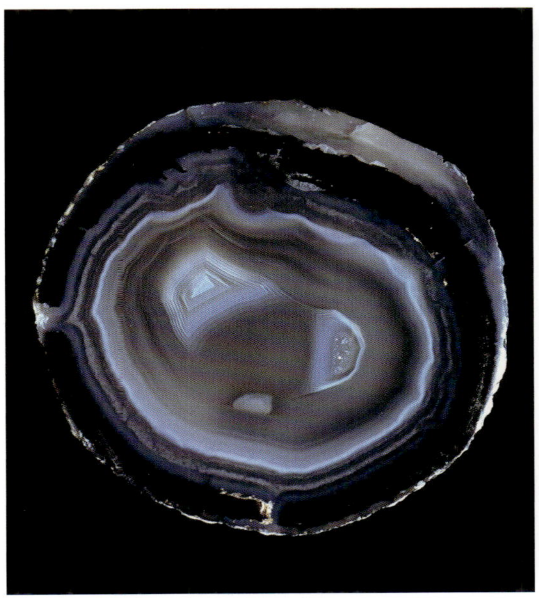

Fossil wood cast. Green River Formation, Eocene. Farson area, Wyoming, USA. Daniels specimen, 3.5 cm.

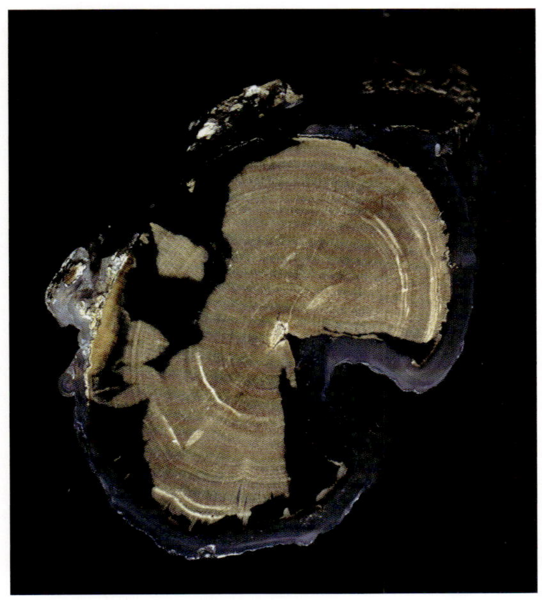

Fossil hardwood. Green River Formation, Eocene. Farson area, Wyoming, USA. Daniels specimen, 5.5 cm.

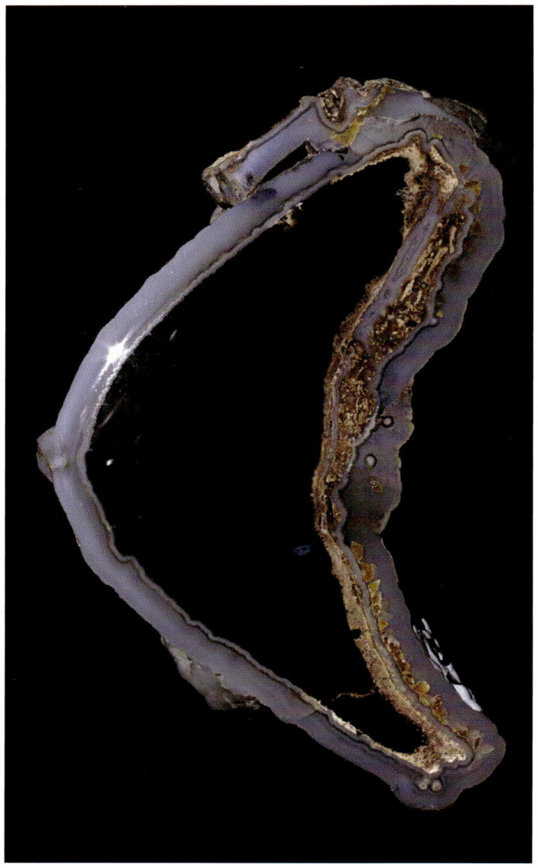

Palmoxylon frond. Green River Formation, Eocene. Big Sandy Reservoir, Wyoming, USA. Albers specimen, 9 cm.

Palmoxylon frond. Green River Formation, Eocene. Big Sandy Reservoir, Wyoming, USA. Albers specimen, 7.5 cm.

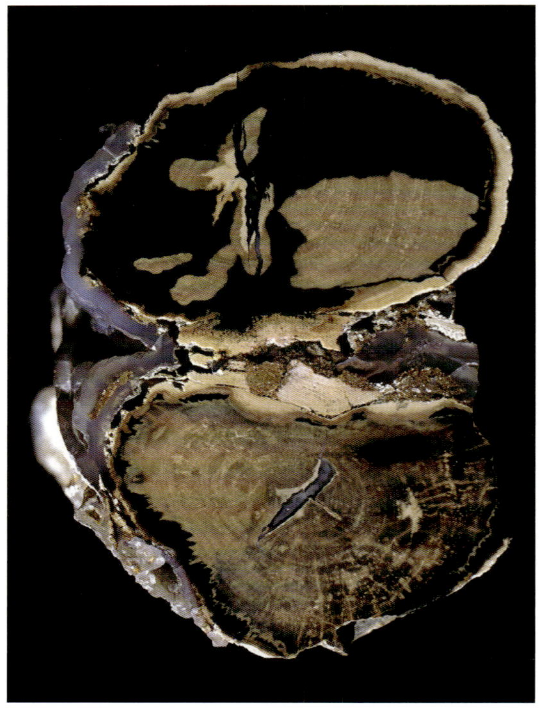

Fossil hardwood fork in branch. Green River Formation, Eocene. Farson area, Wyoming, USA. Daniels specimen, 5.5 cm.

Fossil hardwood. Green River Formation, Eocene. Farson area, Wyoming, USA. Daniels specimen, 6 cm.

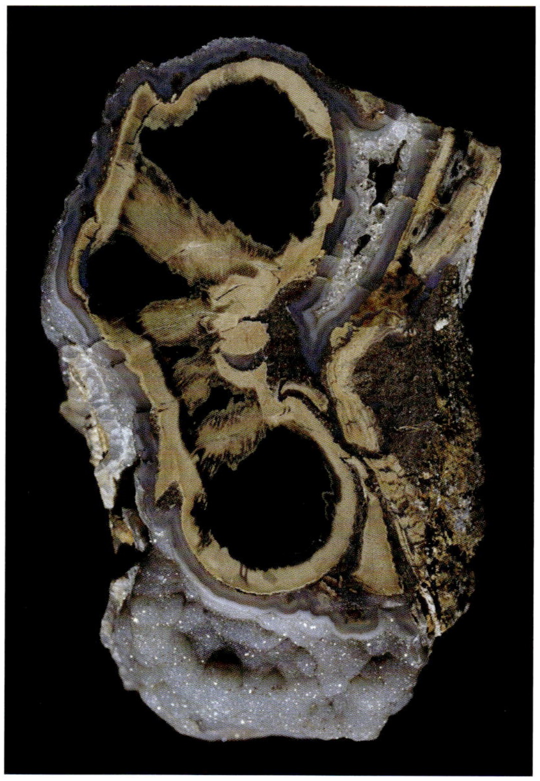

Fossil hardwood fork in branch. Green River Formation, Eocene. Farson area, Wyoming, USA. Daniels specimen, 8 cm.

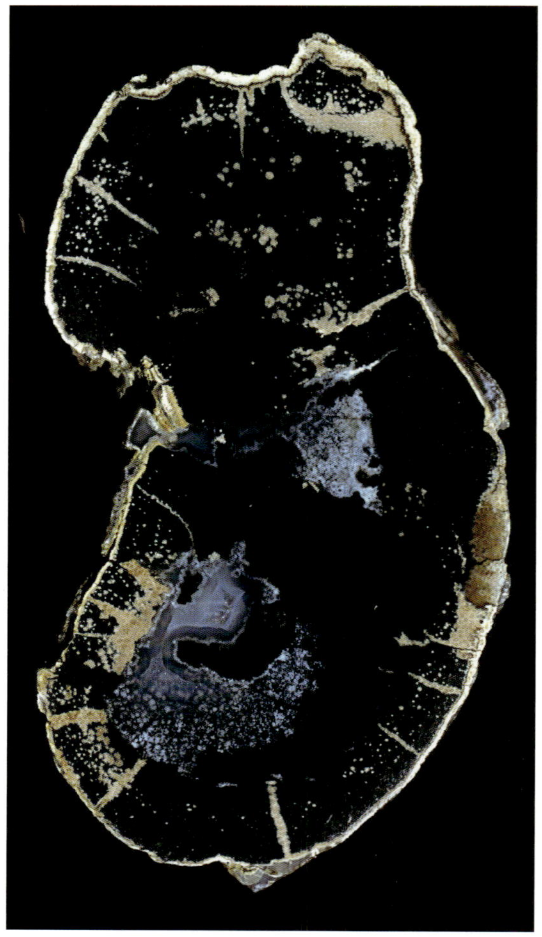

Palmoxylon. Green River Formation, Eocene. Farson area, Wyoming, USA. Daniels specimen, 8.5 cm.

Set of small *Palmoxylon* ("cane") specimens. The original wood evidently dried and shrank within its place of deposition, allowing chalcedony to infill the remainder of the vcid, yet the cell structure is frequently well-preserved. Green River Formation, Eocene. Farson area, Wyoming, USA. Daniels specimens, 4 cm largest diameter.

Fossil wood cast. Green River Formation, Eocene. Dubois area, Wyoming, USA. Daniels specimen, 13 cm long.

Fossil wood cast. Green River Formation, Eocene. Farson area, Wyoming, USA. Daniels specimen, 6.5 cm.

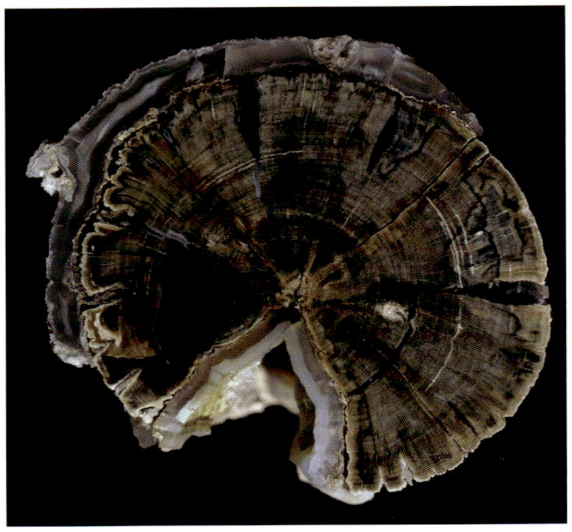

Fossil hardwood. Green River Formation, Eocene. Farson area, Wyoming, USA. Daniels specimen, 7 cm.

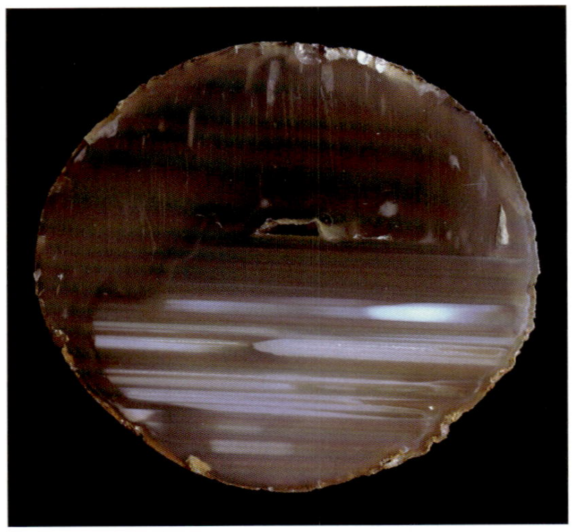

Fossil wood cast. Green River Formation, Eocene. Farson area, Wyoming, USA. Daniels specimen, 5.5 cm.

Fossil wood cast. Wagon Bed Formation, Eocene. Beaver Rim, Wyoming, USA. Albers specimen, 7 cm.

Borer evidence in fossil wood. Green River Formation, Eocene. Farson area, Wyoming, USA. Daniels specimen, 6 cm.

Fossil hardwood. Green River Formation, Eocene. Oregon Buttes, Wyoming, USA. Albers specimen, 5 cm.

Fossil hardwood encased in agate. Green River Formation, Eocene. Eden Valley, Wyoming, USA. Albers specimen, 4.5 cm.

Fossil hardwood. Green River Formation, Eocene. Farson area, Wyoming, USA. Daniels specimen, 2.5 cm.

Fossil hardwood. Green River Formation, Eocene. Farson area, Wyoming, USA. Daniels specimen, 2.5 cm.

Fossil hardwood. Green River Formation, Eocene. Farson area, Wyoming, USA. Daniels specimen, 8 cm.

Fossil hardwood with blue botryoidal chalcedony coating. Green River Formation, Eocene. Blue Forest, Wyoming, USA. Daniels specimen, 7 cm long.

Farther south and east of the Continental Divide, we find the Florissant Formation (35 mybp), representing 5,000 years of lake sediments in present-day Colorado. Fossil remains include ferns, 10 species of gymnosperms, 4 species of monocots, and no less than 95 species of dicots, including 9 oak species. The central Rocky Mountains continued tectonic activity during the Early Miocene Epoch with as much as 2,300 meters (7,000 feet) of volcanic ash deposited in the Jackson Hole, Wyoming, area. In the San Juan Mountains of southern Colorado, an area of 25,000 square kilometers (9,650 square miles) received 20,000 cubic km (4,800 cubic miles) of ash and tuff.

Old postcard showing large *Taxodioxylon* stump that measured 4.3 meters (14 feet) tall and 22.5 meters (74 feet) in circumference. Florissant, Colorado, USA.

Florissant flora offers abundant information on Late Eocene Epoch plant and animal life in the region, representing life from the period preceding significant cooling at the Eocene–Oligocene boundary. A volcanic flow entombed numerous trees, including giant redwood-like trees, to a height of about 5 meters (16 feet). Petrified stumps, predominately trees with affinity to *Sequoia*, remain in their growth positions. The largest is 4.1 meters in diameter at chest height. There are also five types of angiosperms: four are ring porous [with affinities to *Koelreuteria* (golden rain tree), *Robinia* (locust), and *Zelkova* (Caucasian elm)], and one is

The Wheeler Geologic Area near Creede, Colorado, resides encompassed by the southwest part of the La Garita Wilderness Area. This striking landscape of spires, domes, pinnacles, and hoodoos is caused by differential erosion of Tertiary age tuffs. The dark upper tuffs are densely welded and weather more slowly than the lower lighter tuffs that are poorly welded.

diffuse porous (new genus *Chadronoxylon*). One *Sequoiaoxylon* stump complex, called the "Redwood Trio," is a triple trunk of one tree, the only known fossil redwood trio. This habit persists in modern redwoods when several sprouts will grow from the base of a main trunk; eventually the parent tree dies and rots, leaving the sprouts, now grown into large trees. Some of the other fossil wood is similar in structure to *Acer* (maple), *Ailanthus* (tree of heaven), *Cedrelospermum* (an extinct member of the elm family), *Eucommia* (hard rubber tree), and *Hydrangea*. Most of the species information comes from compression and impression fossils where we find evidence of 1,500 species of insects and 150 species of plants. The Florissant Formation has yielded more known species of fossil butterflies than any other site and the only known fossil tsetse fly. Leaves, fruits, seeds, and flowers reveal plants with affinity to conifers (bald cypress, cypress, larch, pines, spruce, and yews), ferns, horsetails, and many angiosperms (including basswood, beech, birch, buckthorn, cashew, citrus trees, cocoa, currant, elder, elm, grape, hard rubber tree, hydrangea, laurel, mahogany, maple, mulberry, oak, palm, rose, spindle tree, sycamore, torchwood, tree of heaven, walnut, and willow). Relatives of many Florissant fossil plants currently live in the southern Rocky Mountains, including currant, mountain mahogany, pine, rose, spruce, and willow.

VI

CHAPTER SIX

Fossil Woods from Worldwide Locations

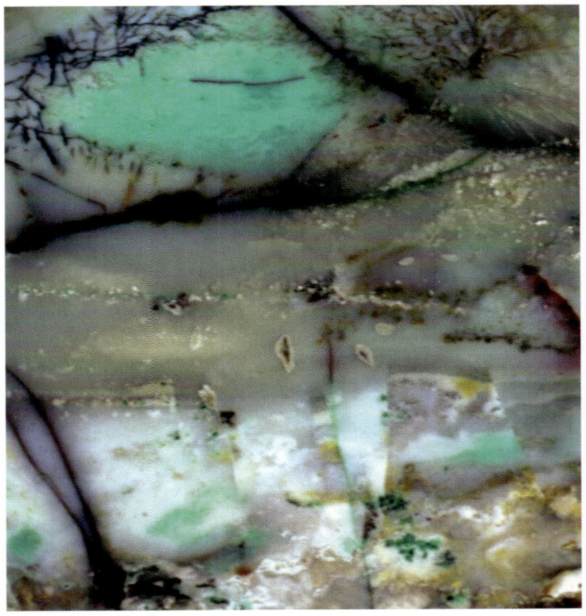

Fossil woods from worldwide locations often become available to collectors around the globe through a network of rock and mineral distributors. Because Europeans have a long history of fossil wood collecting and the pursuit continues, top quality specimens rarely leave that continent. Specimens from some of the more important source countries appear in chapter one. This chapter offers a glimpse of some additional fossil woods from around the world. In recent years, fossil woods from Indonesia have become available to western collectors. In many instances, Indonesian fossil woods have well-preserved cell structure and offer a challenge to those interested in identifying Tertiary age tropical hardwoods by genus. When purchasing imported fossil wood specimens, it is important to seek out reputable dealers who comply with legal requirements for export from the source countries.

Araucarioxylon. Triassic. Wurtzburg, Germany. Daniels specimen, 13 cm.

Cupressinoxylon. Tertiary. Kizilcahaman, Turkey. Daniels specimen, 8.5 cm.

Woodworthia. Triassic. Wurtzburg, Germany. Daniels specimen, 10 cm.

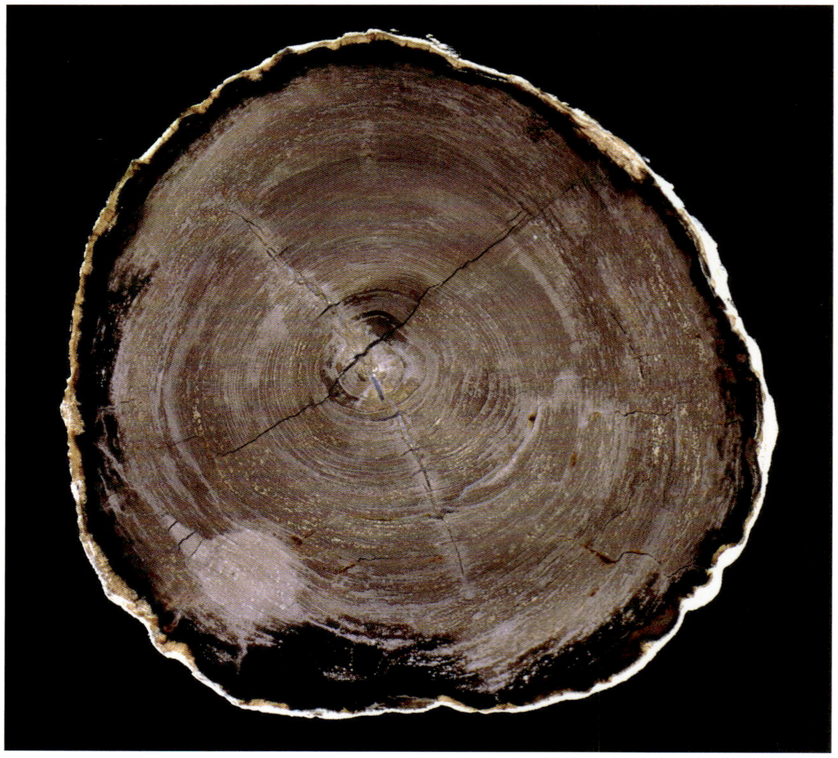

Cupressinoxylon. Tertiary. Kizilcahaman, Turkey. Daniels specimen, 6.5 cm.

Micro image of Wurtzburg *Araucarioxylon*. Daniels specimen.

Micro image of Wurtzburg *Woodworthia*. Daniels specimen.

Taxodioxylon. Miocene. Ratka, Hungary. Daniels specimen, 6 cm.

Green fossil conifer. Tertiary. Northern Turkey. Daniels specimen, 10.5 cm.

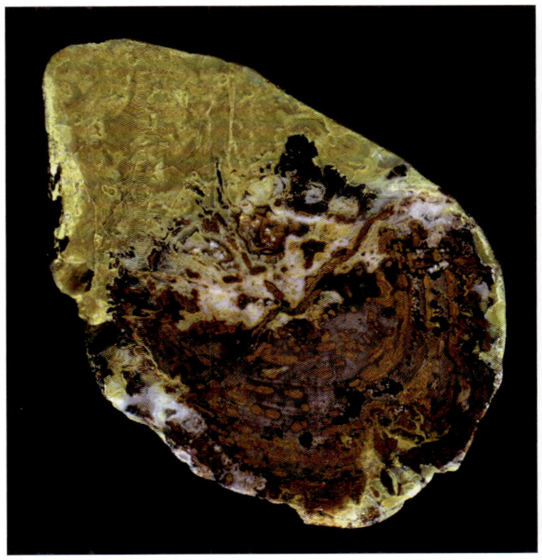

Fossil conifer. Oligocene/Miocene. Lesbos, Greece. Daniels specimen, 8.5 cm.

Cupressinoxylon. Tertiary. Kizilcahaman, Turkey. Daniels specimen, 4.5 cm.

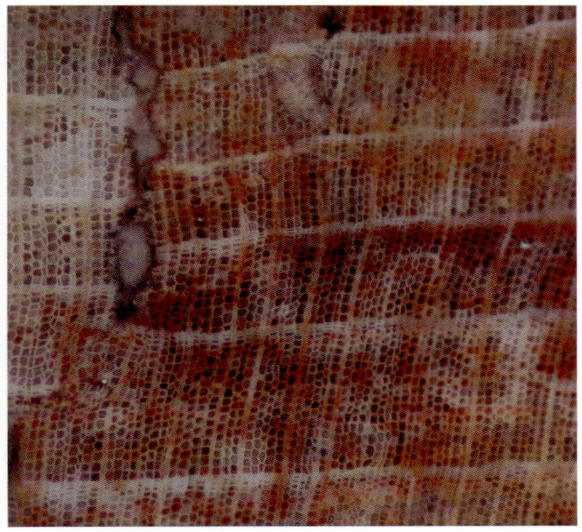

Micro image of Kizilcahaman *Cupressinoxylon*. Daniels specimen.

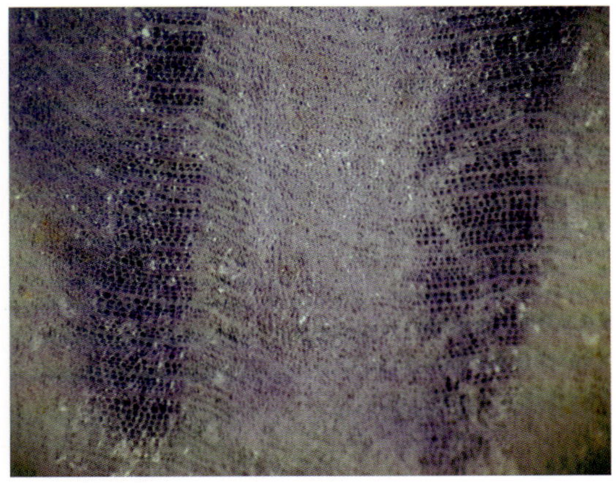

Micro image of Chemnitz fluorite *Dadoxylon*. Daniels specimen. Full specimen appears on page 41.

Taxodioxylon. Miocene. Ratka, Hungary. Daniels specimen, 6.5 cm.

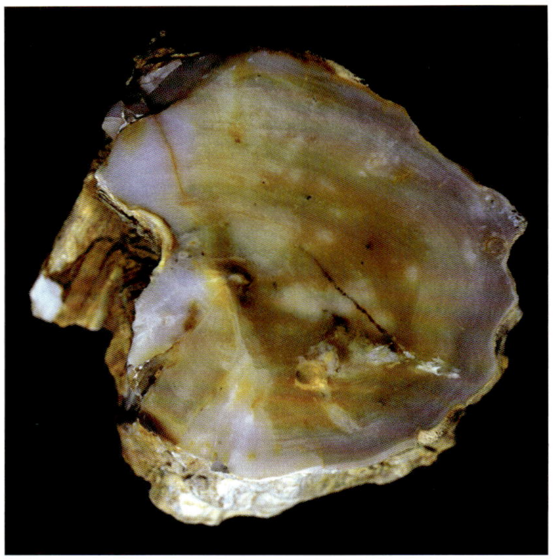

Taxodioxylon. Miocene. Boldogkovaralja, Hungary. Daniels specimen, 6 cm.

Fossil hardwood. Tertiary. Indonesia. Daniels specimen, 9 cm.

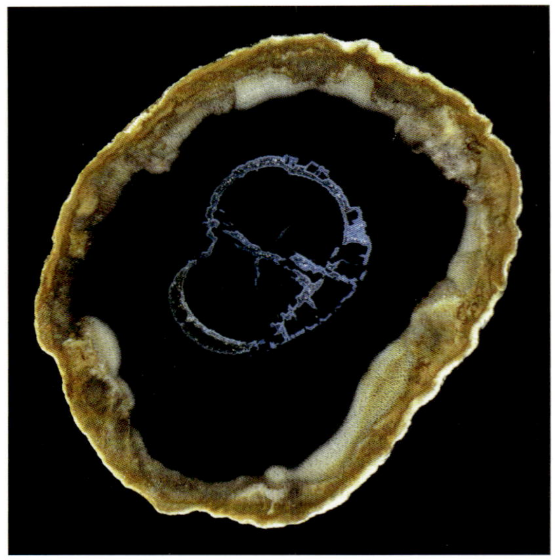

Fossil hardwood. Tertiary. Indonesia. Daniels specimen, 6 cm.

Taxodioxylon. Miocene. Ratka, Hungary. Daniels specimen, 4.5 cm.

Fossil hardwood. Tertiary. Indonesia. Daniels specimen, 7.5 cm.

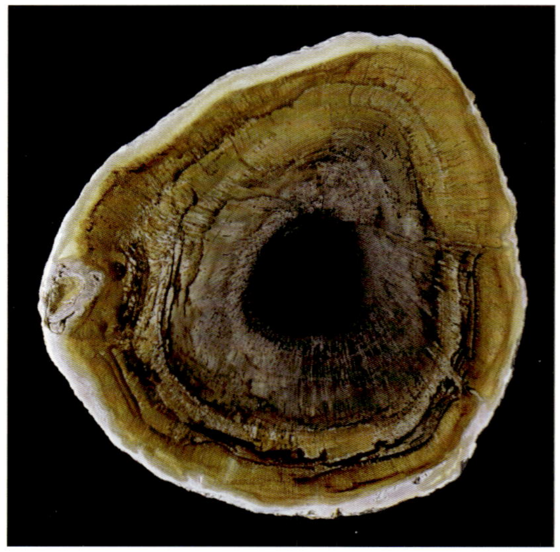

Fossil hardwood. Tertiary. Indonesia. Daniels specimen, 7.5 cm.

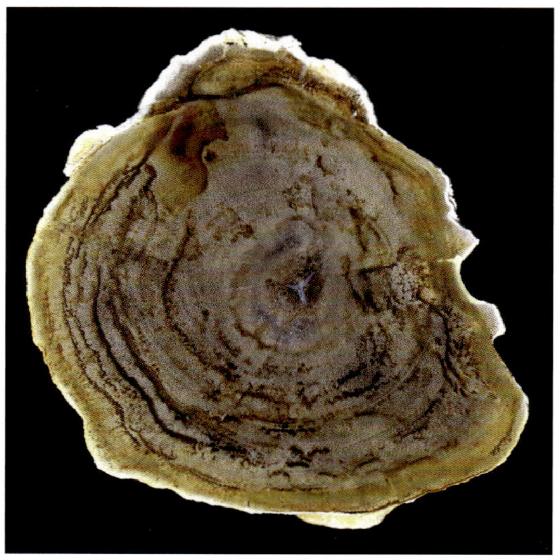

Fossil hardwood. Tertiary. Indonesia. Daniels specimen, 8 cm.

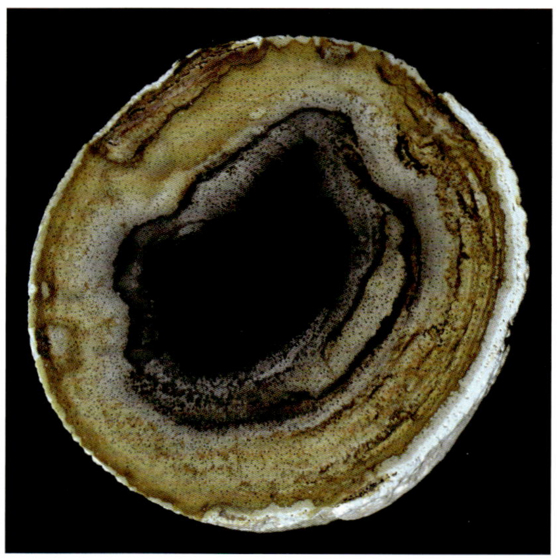

Fossil hardwood. Tertiary. Indonesia. Daniels specimen, 5.5 cm.

Fossil hardwood. Tertiary. Indonesia. Daniels specimen, 8.5 cm.

Fossil wood cast from an area reported to be rich in copper. Tertiary. Indonesia. Daniels specimen, 26.5 cm.

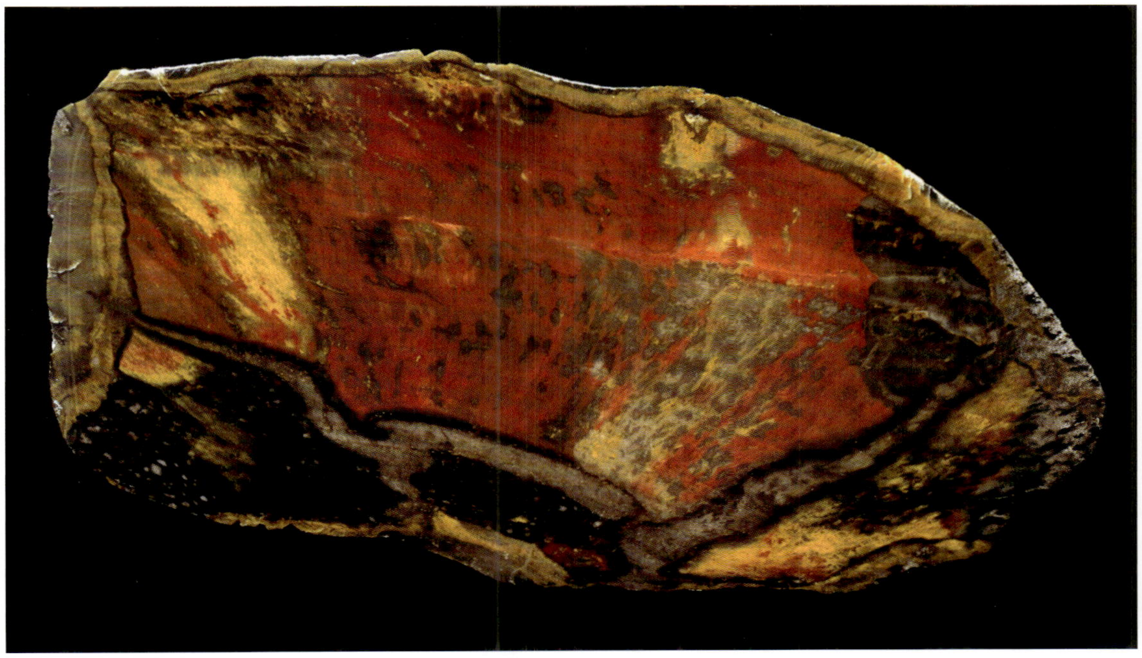

Fossil hardwood. Age unknown. Davis Mountains, Texas. Daniels specimen, 11 cm.

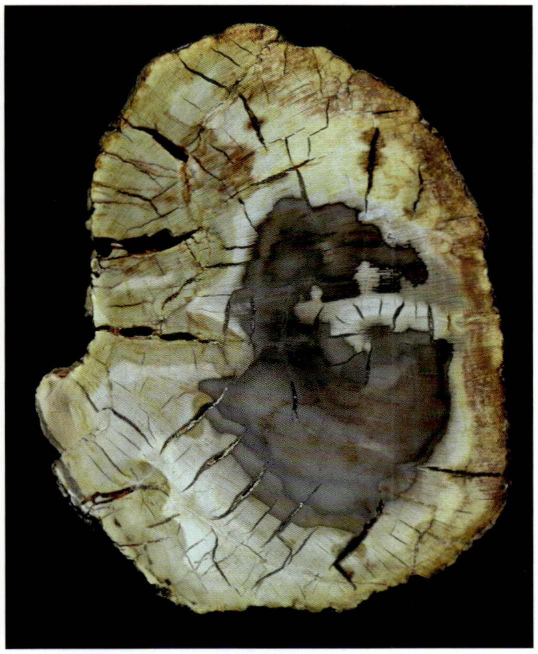

Dadoxylon. Permian. Paraguay. Daniels specimen, 8 cm.

Fossil hardwood from the crotch of a tree. Tertiary. Indonesia. Daniels specimen, 16.5 cm.

Fossil hardwood. Tertiary. Indonesia. Daniels specimen, 12.5 cm.

Fossil hardwood. Tertiary. Indonesia. Daniels specimen, 20.5 cm.

Fossil hardwood. Tertiary. Indonesia. Daniels specimen, 20 cm.

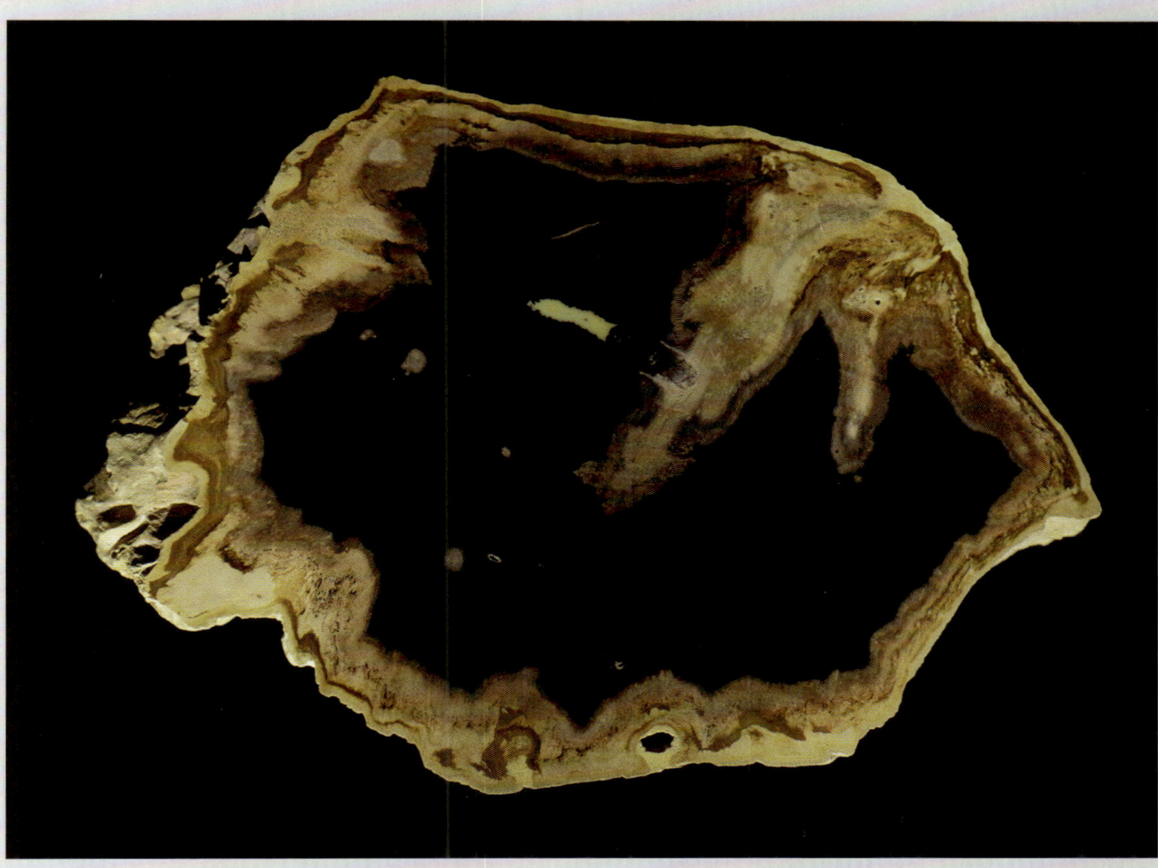

Macro photograph of borer tunnels in fossil hardwood. Tertiary. Indonesia. Daniels specimen, 18 cm.

Borer tunnels in fossil hardwood. Tertiary. Indonesia. Daniels specimen, 18 cm.

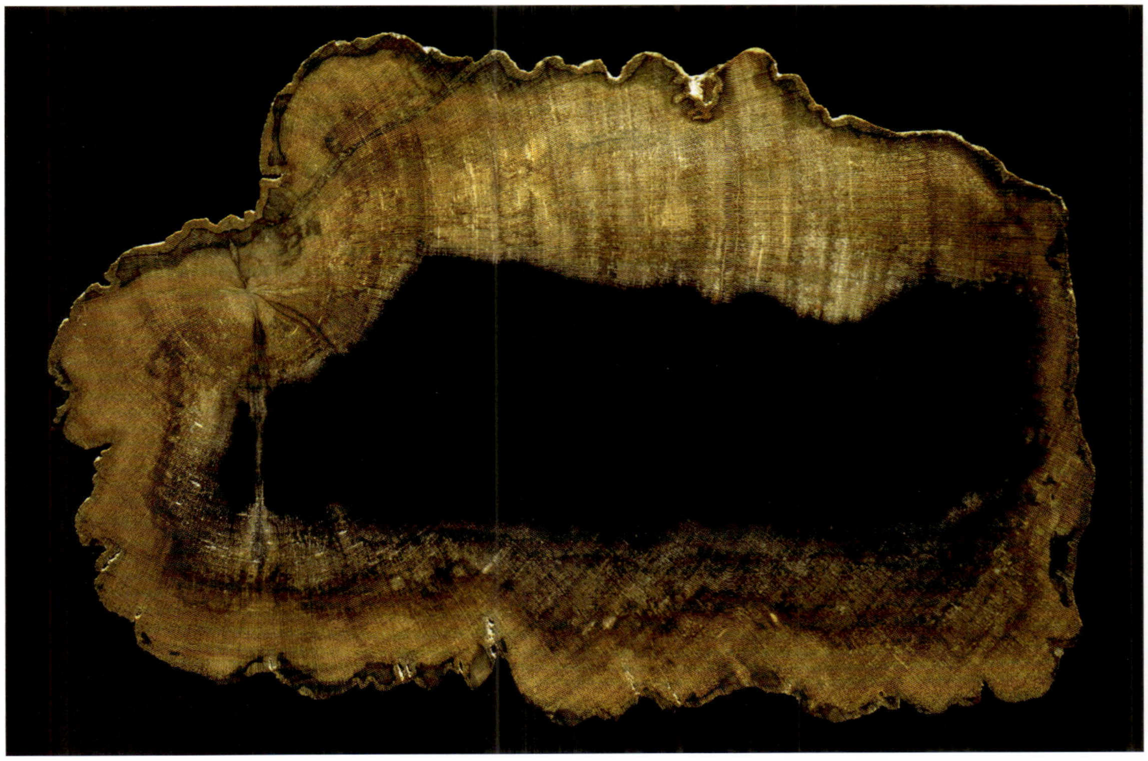

Fossil hardwood. Tertiary. Indonesia. Daniels specimen, 16 cm.

Dadoxylon. Permian. Paraguay. Daniels specimen, 7.5 cm.

Fossil hardwood. Tertiary. Indonesia. Daniels specimen, 15 cm.

Fossil hardwood. Tertiary. Indonesia. Daniels specimen, 13.5 cm.

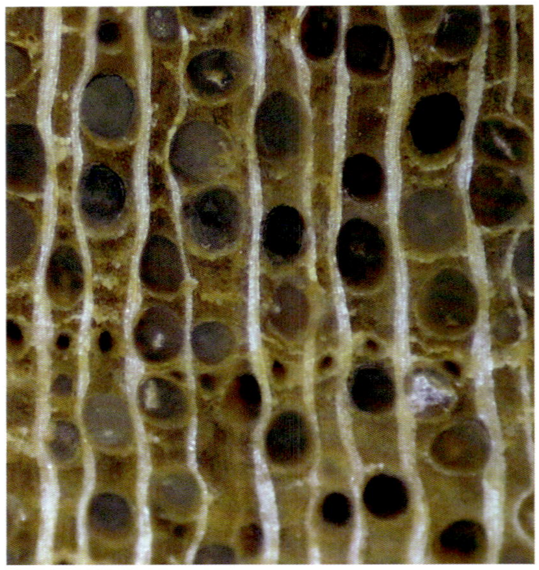

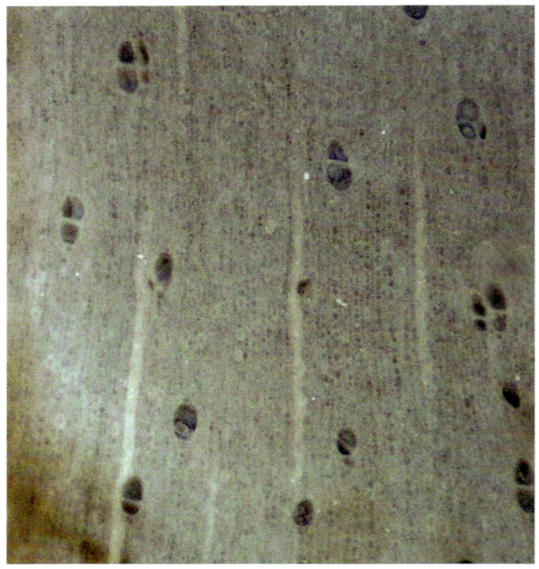

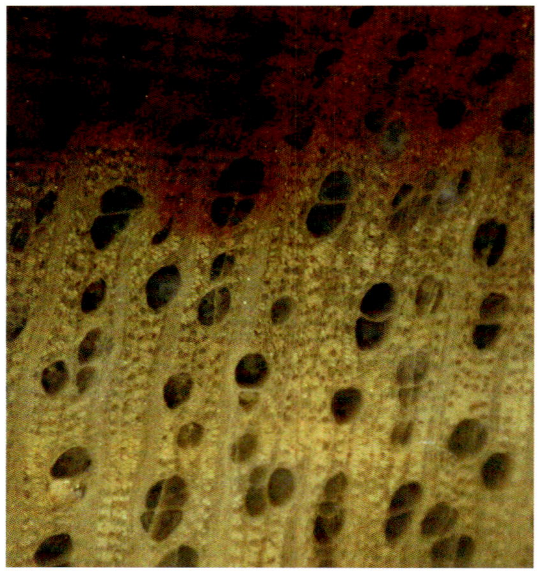

Above: three micro images showing variety in Indonesian Tertiary Period hardwoods. All specimens Daniels.

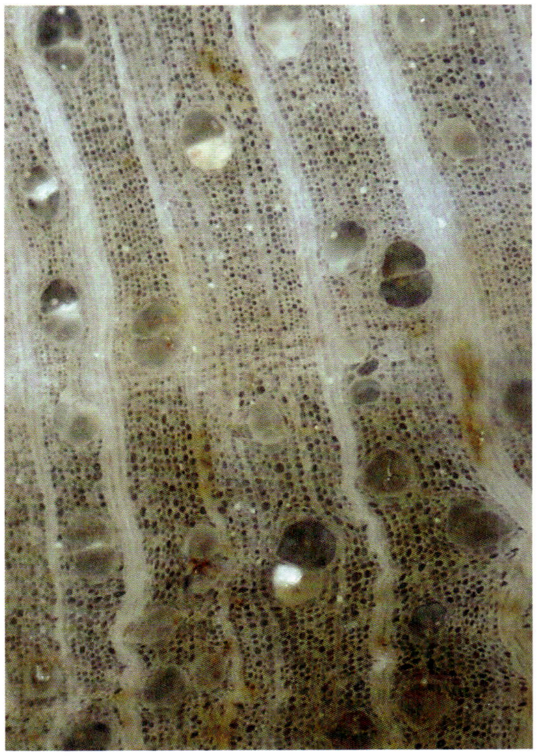

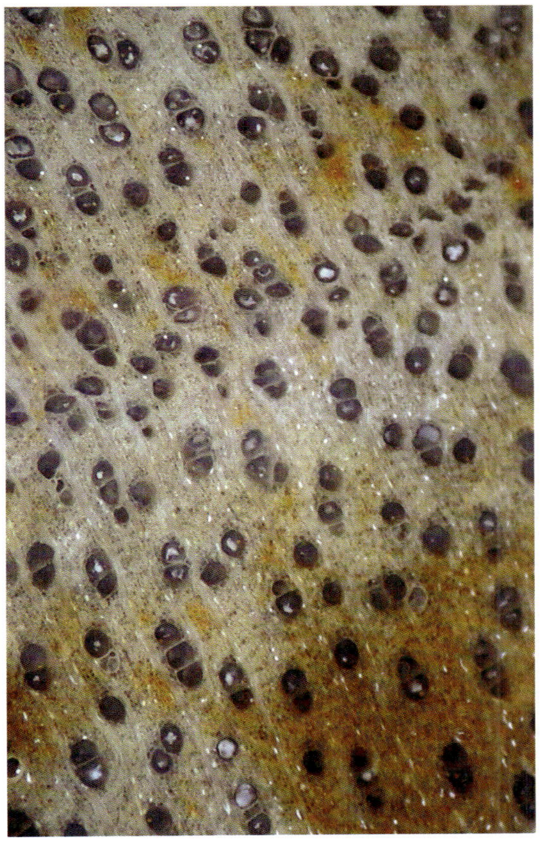

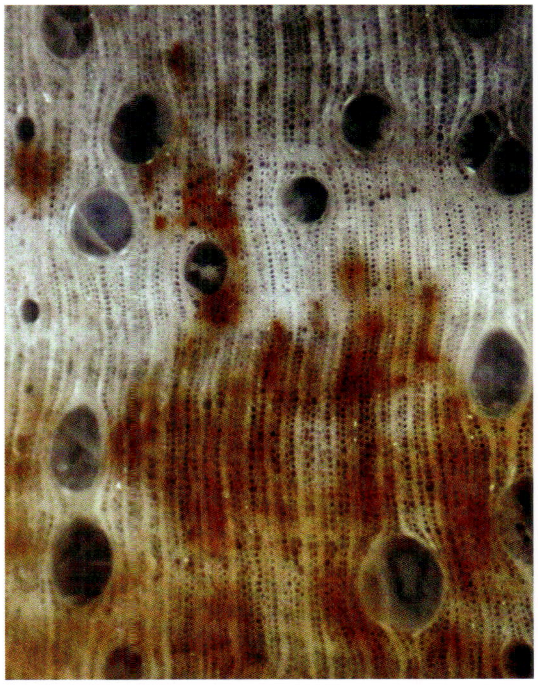

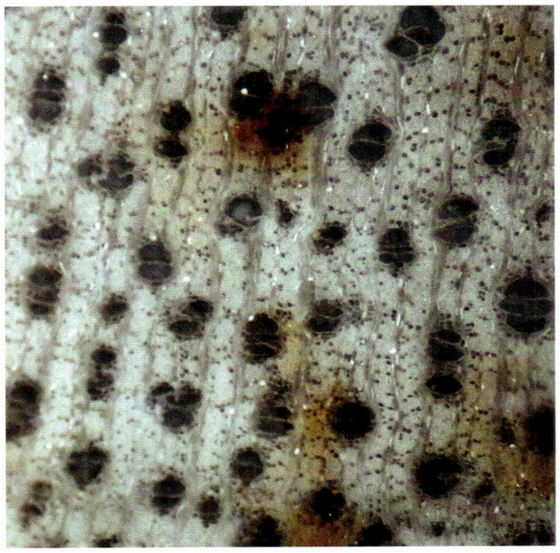

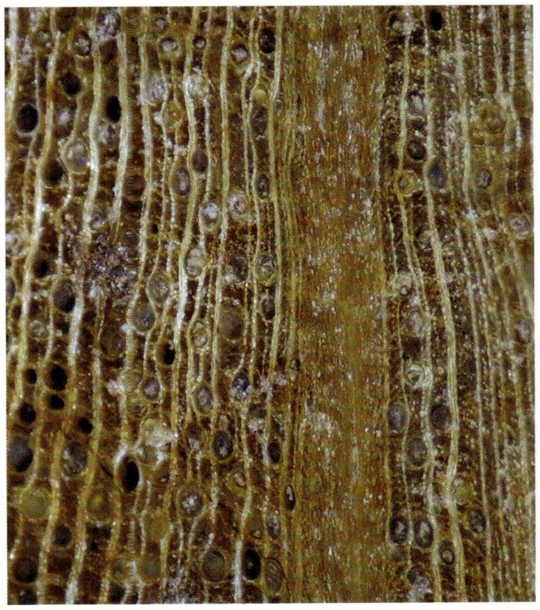

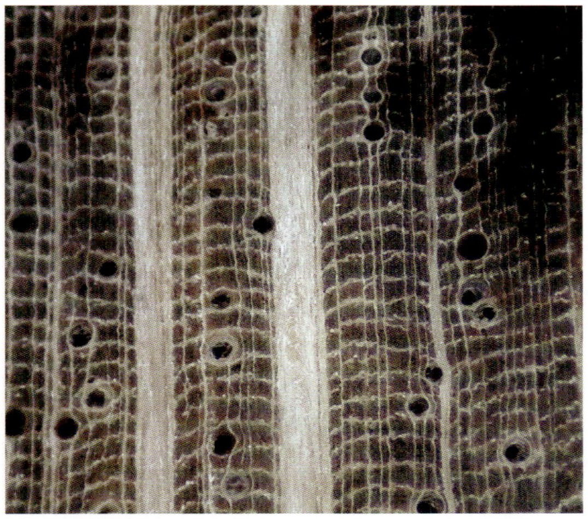

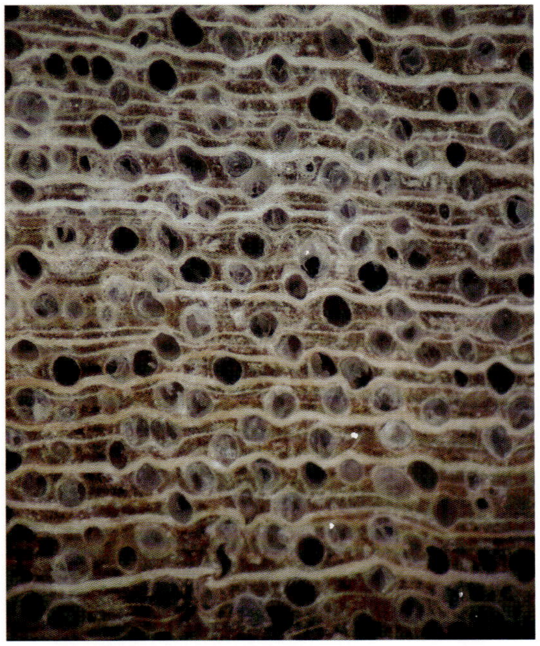

Above: eight micro images showing variety in Indonesian Tertiary Period hardwoods. All specimens Daniels except for the upper right specimen which is *Casuarina*-like, courtesy of Dennis Murphy.

VII

CHAPTER SEVEN

Museum Collections

THE LYLE & LELA JACKLIN COLLECTION

Washington State University; Pullman, Washington.

THE DENNIS & MARY MURPHY COLLECTION

Rice Northwest Museum of Rocks and Minerals; Hillsboro, Oregon.

THE JIM & CATHY GRAY COLLECTION

Jim Gray's Petrified Wood Company Museum; Holbrook, Arizona.

THE LYLE & LELA JACKLIN COLLECTION

Many, but not all, of the speciments are identified to genus. Identifications and locations listed here are those attributed in the collection.

Looking north from the roof of the Webster Physical Sciences Building, which houses the geology department and The Lyle and Lela Jacklin Collection at Washington State University in Pullman, Washington. The older brick buildings are part of the lovely university campus. The rolling green hills in the distance are part of the famous Palouse region, an area of lush grain farms and vast pastoral vistas.

Unidentified fossil wood. Columbia Plateau Basalt flow, Miocene. Saddle Mountain, Washington, USA. The Lyle and Lela Jacklin Collection, 23.5 cm.

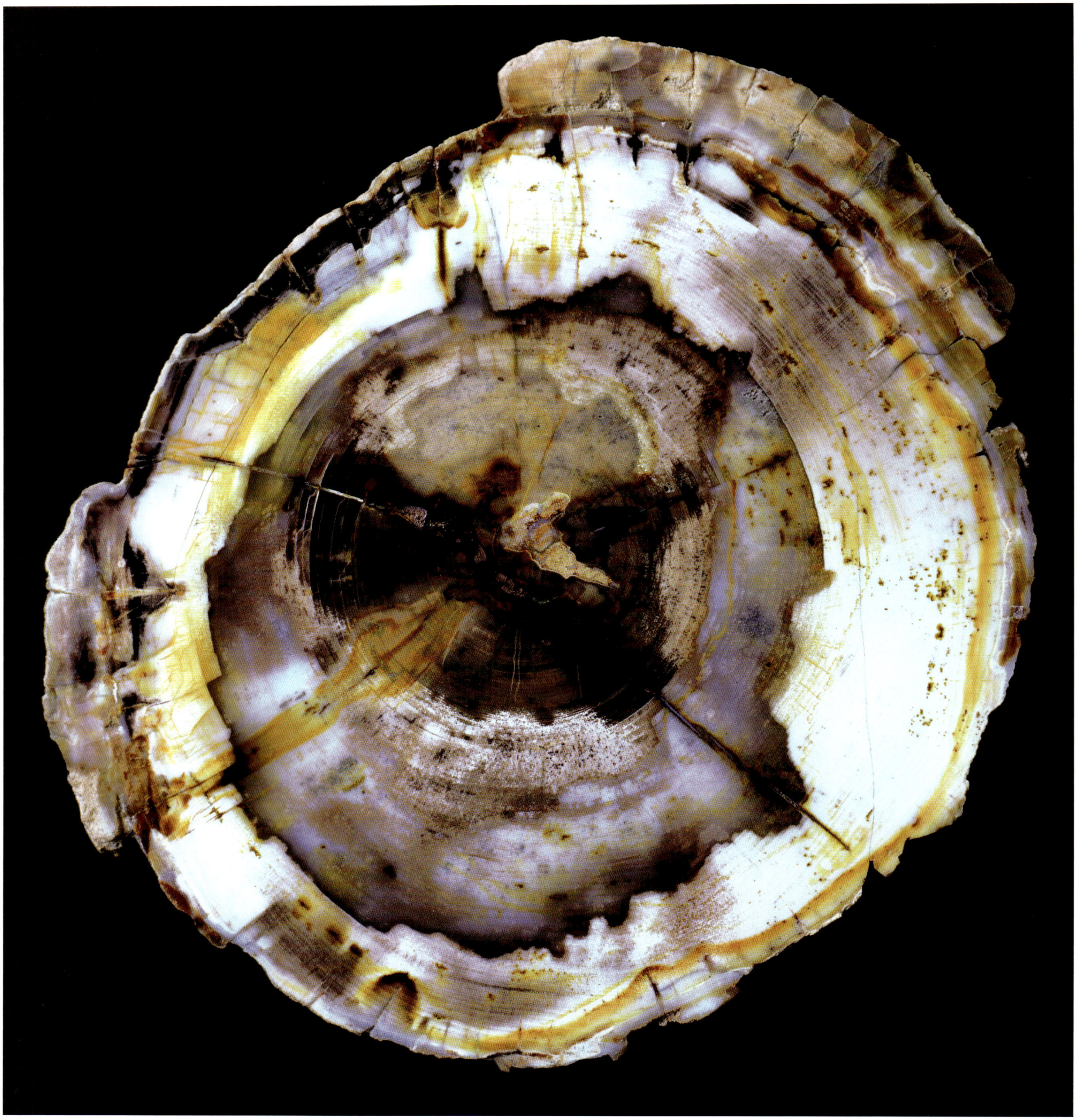

Maple-like fossil hardwood. Columbia Plateau Basalt flow, Miocene. Saddle Mountain, Washington, USA. The Lyle and Lela Jacklin Collection, 23 cm.

Unidentified fossil wood. Columbia Plateau Basalt flow, Miocene. Sunnyside, Washington, USA. The Lyle and Lela Jacklin Collection, 28 cm.

Elm-like fossil hardwood. Columbia Plateau Basalt flow, Miocene. Saddle Mountain, Washington, USA. The Lyle and Lela Jacklin Collection, 30 cm.

Longitudinally cut unidentified "picture wood." Columbia Plateau Basalt flow, Miocene. Saddle Mountain, Washington, USA. The Lyle and Lela Jacklin Collection, 28 cm.

Ginkgo-like fossil wood. Columbia Plateau Basalt flow, Miocene. Vantage, Washington, USA. The Lyle and Lela Jacklin Collection, 27.5 cm.

Fossil conifer. Little Butte Volcanic Series, Oligocene to Early Miocene. Sweet Home area, Linn County, Oregon, USA.
The Lyle and Lela Jacklin Collection, 29 cm.

Oak-like fossil hardwood. Tertiary. Swartz Canyon, Crook County, Oregon, USA. The Lyle and Lela Jacklin Collection, 37.5 cm.

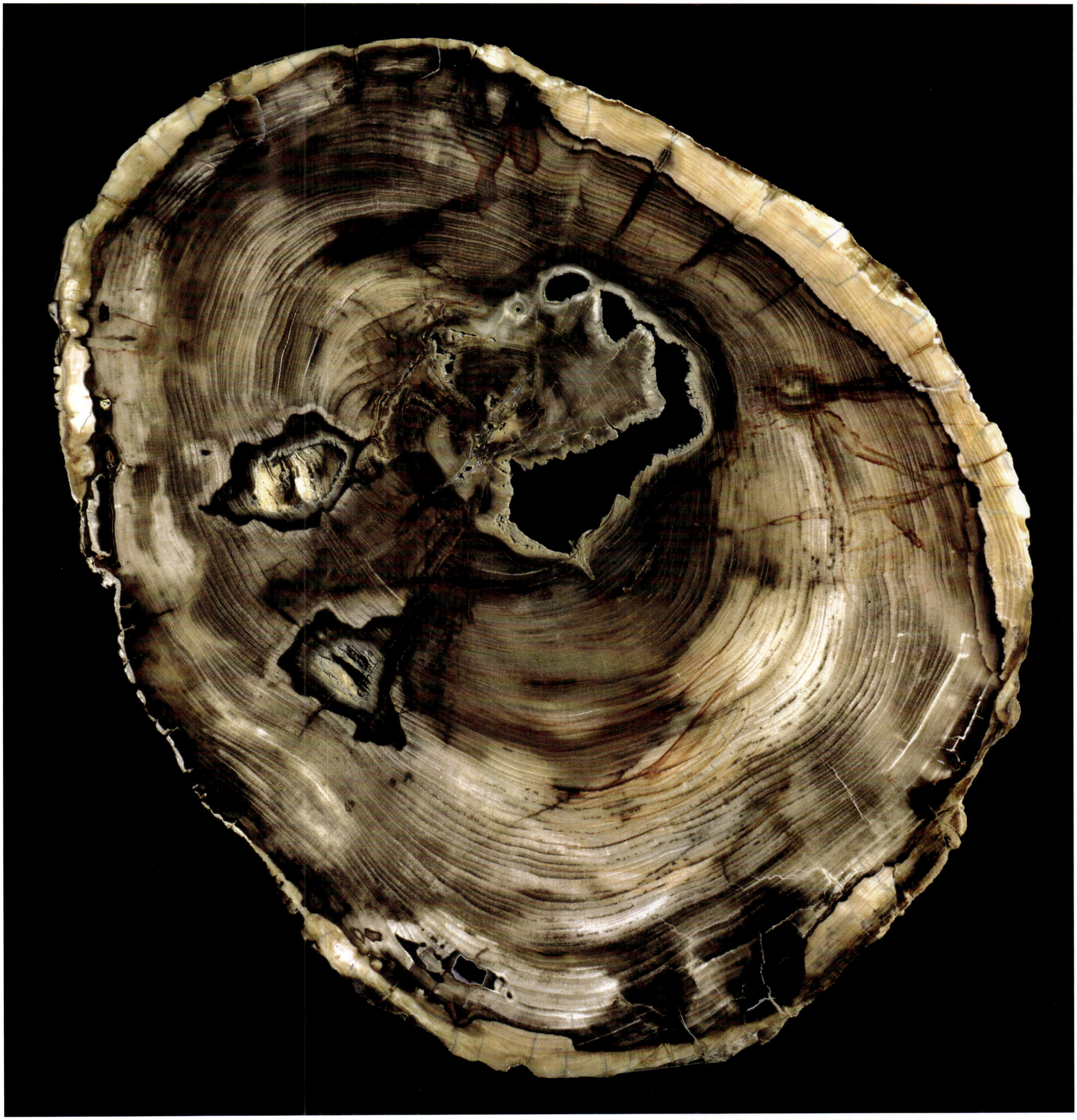

Elm-like fossil hardwood. Columbia Plateau Basalt flow, Miocene. Saddle Mountain, Washington, USA. The Lyle and Lela Jacklin Collection, 40 cm.

Knot in longitudinally cut elm-like fossil hardwood "picture wood." Columbia Plateau Basalt flow, Miocene. Saddle Mountain, Washington, USA. The Lyle and Lela Jacklin Collection, 27 cm.

Unidentified fossil wood. Columbia Plateau Basalt flow, Miocene. Sunnyside, Washington, USA. The Lyle and Lela Jacklin Collection, 31 cm.

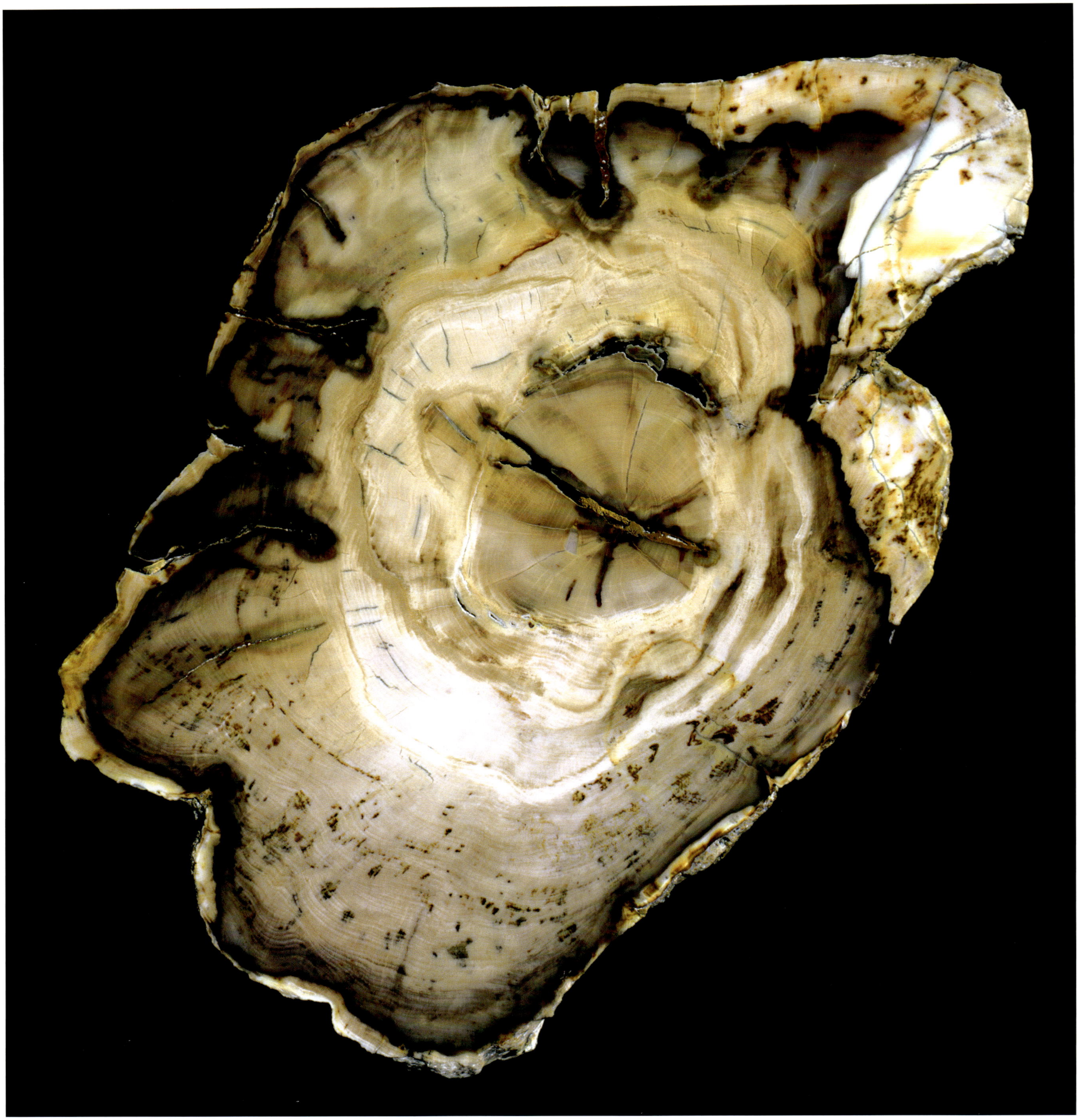

Cupressinoxylon. Columbia Plateau Basalt flow, Miocene. Sunnyside, Washington, USA. The Lyle and Lela Jacklin Collection, 35 cm.

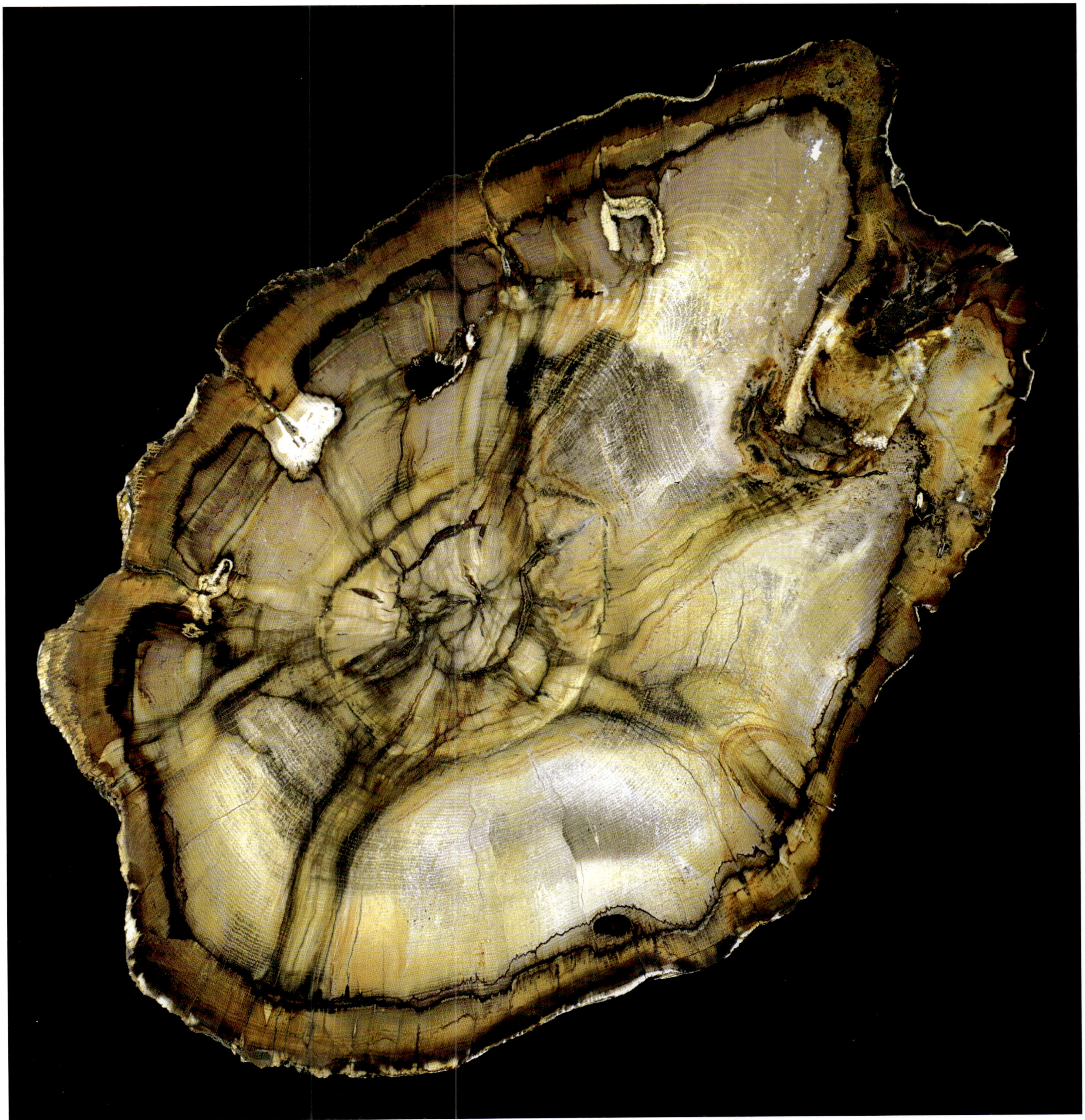

Oak-like fossil hardwood. Columbia Plateau Basalt flow, Miocene. Stinking Water Creek, Oregon, USA. The Lyle and Lela Jacklin Collection, 58 cm.

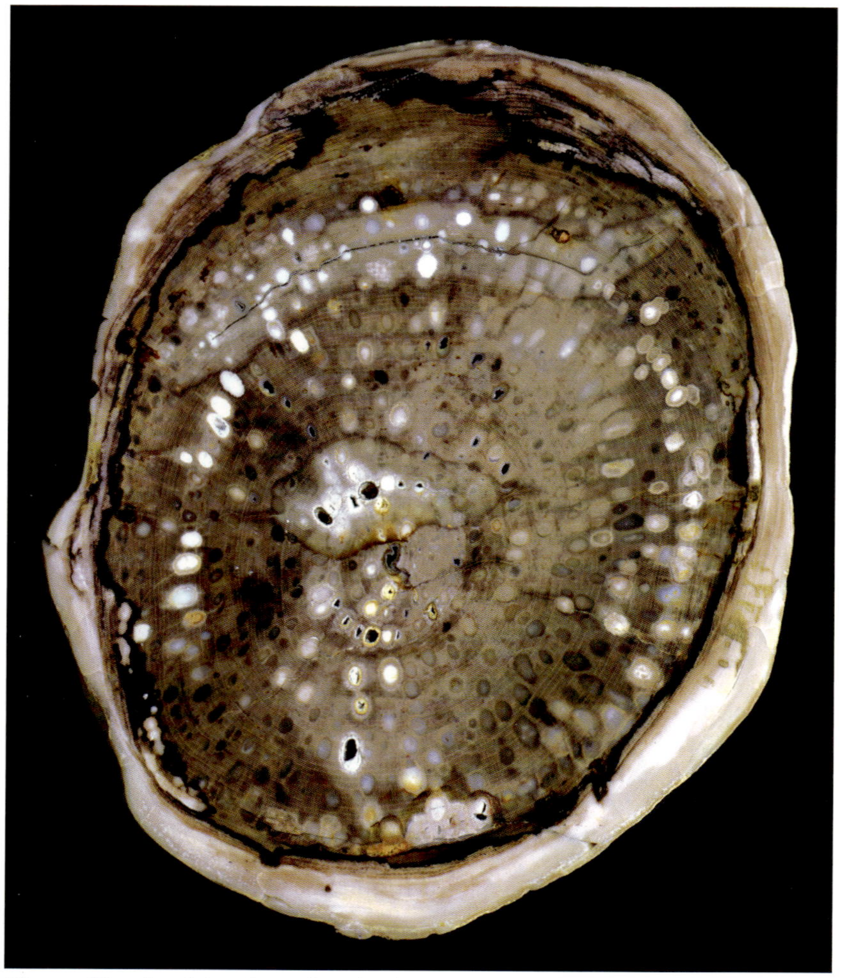

Cherry-like fossil hardwood with abundant insect borings. Columbia Plateau Basalt flow, Miocene. Saddle Mountain, Washington, USA. The Lyle and Lela Jacklin Collection, 17.5 cm.

Double-hearted unidentified fossil hardwood. Columbia Plateau Basalt flow, Miocene. Saddle Mountain, Washington, USA. The Lyle and Lela Jacklin Collection, 21 cm.

Knot in longitudinally cut unidentified fossil wood. Columbia Plateau Basalt flow, Miocene. Saddle Mountain, Washington, USA. The Lyle and Lela Jacklin Collection, 23 cm.

Fossil ring porous hardwood. Succor Creek Formation, Miocene. Grassy Mountain, Malheur County, Oregon, USA. The Lyle and Lela Jacklin Collection, 12 cm.

Knot in longitudinally cut unidentified fossil "picture wood." Columbia Plateau Basalt flow, Miocene. Saddle Mountain, Washington, USA. The Lyle and Lela Jacklin Collection, 24 cm.

Cupressinoxylon. Columbia Plateau Basalt flow, Miocene. Sunnyside, Washington, USA. The Lyle and Lela Jacklin Collection, 19 cm.

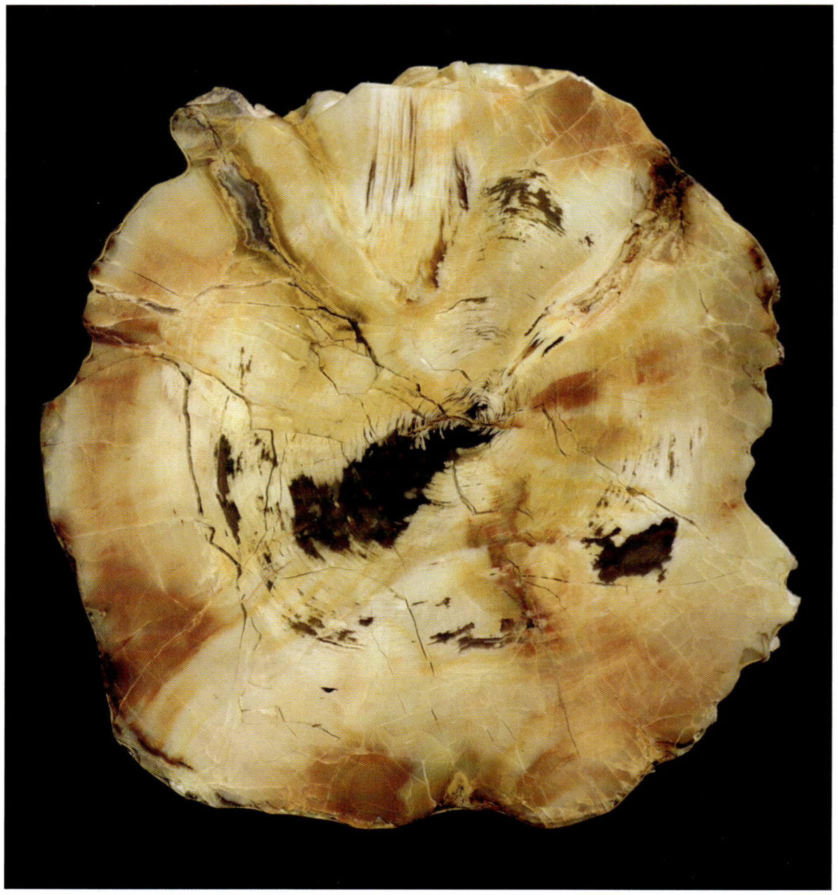

Unidentified fossil wood. Tertiary. Nevada, USA. The Lyle and Lela Jacklin Collection, 18.5 cm.

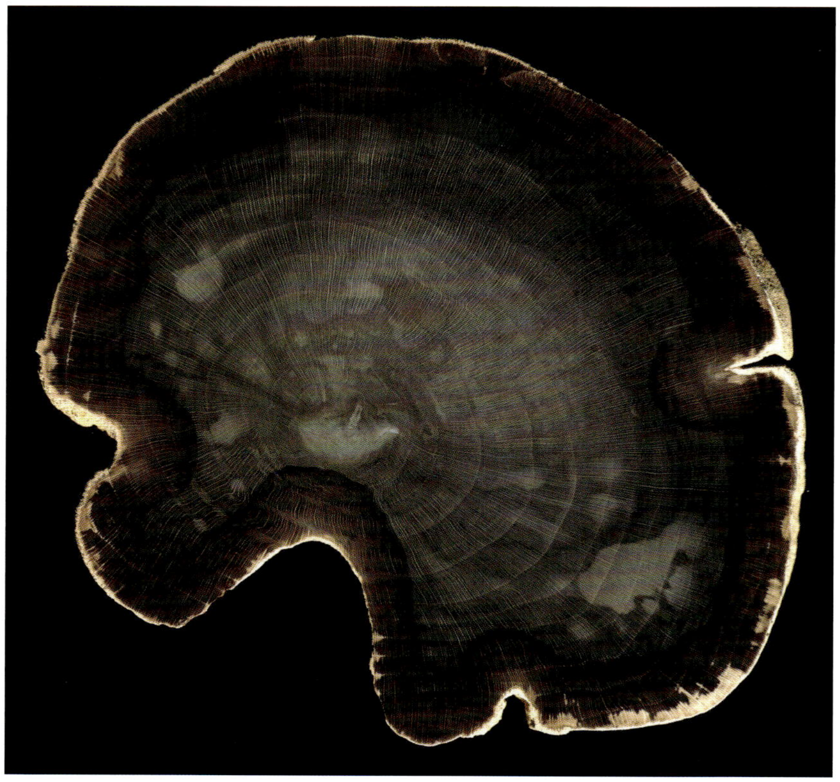

Sycamore-like fossil hardwood. Little Butte Volcanic Series, Oligocene to Early Miocene. Brush Creek area, Linn County, Oregon, USA. The Lyle and Lela Jacklin Collection, 19 cm.

Pityoxylon (spruce-like). Columbia Plateau Basalt flow, Miocene. Saddle Mountain, Washington, USA. The Lyle and Lela Jacklin Collection, 18.5 cm.

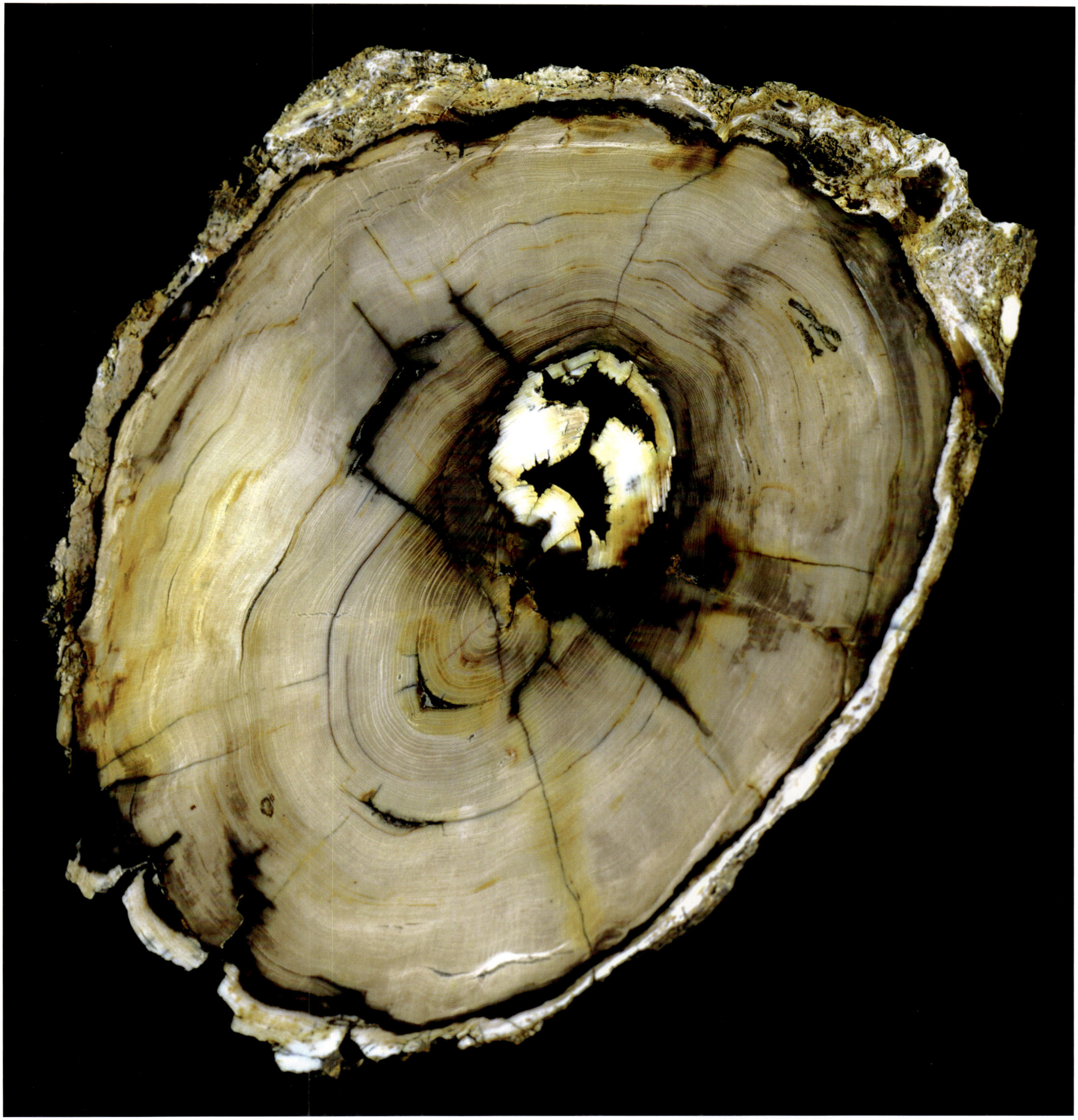

Cupressinoxylon. Columbia Plateau Basalt flow, Miocene. Sunnyside, Washington, USA. The Lyle and Lela Jacklin Collection, 28 cm.

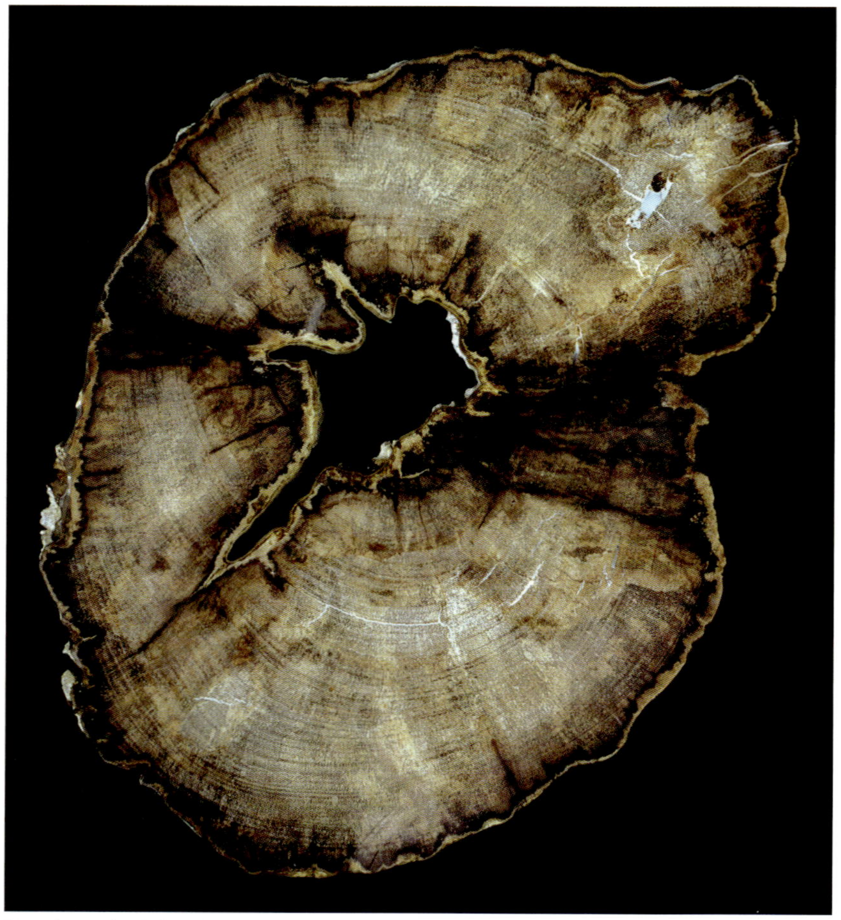

Unidentified fossil wood. Columbia Plateau Basalt flow, Miocene. Pittsburg Landing, Idaho, USA. The Lyle and Lela Jacklin Collection, 24 cm.

Fossil hardwood. Columbia Plateau Basalt flow, Miocene. Asotin, Washington, USA. The Lyle and Lela Jacklin Collection, 10 cm.

Fossil gum-like hardwood. Columbia Plateau Basalt flow, Miocene. Saddle Mountain, Washington, USA. The Lyle and Lela Jacklin Collection, 14 cm.

Ring porous fossil hardwood. Succor Creek Formation, Miocene. Grassy Mountain, Malheur County, Oregon, USA. The Lyle and Lela Jacklin Collection, 13 cm.

Elm-like fossil hardwood. Tertiary. Maury Mountains, Crook County, Oregon. The Lyle and Lela Jacklin Collection, 25 cm.

Fossil hardwood. Columbia Plateau Basalt flow, Miocene. Saddle Mountain, Washington, USA. The Lyle and Lela Jacklin Collection, 10 cm.

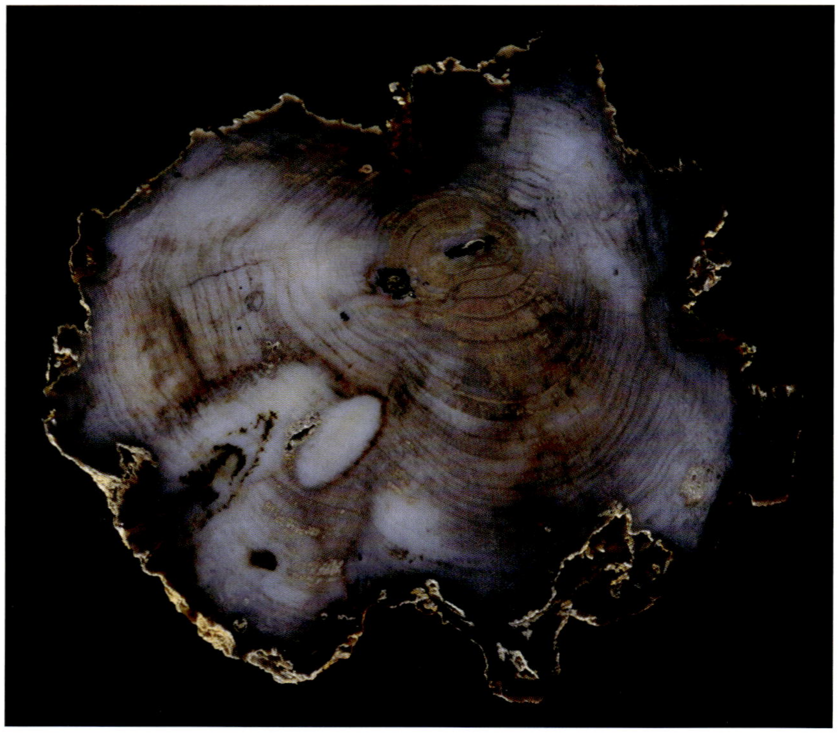

Fossil conifer. Tertiary. Hubbard Basin, Nevada, USA. The Lyle and Lela Jacklin Collection, 16 cm.

Knot in longitudinally cut unidentified fossil "picture wood." Columbia Plateau Basalt flow, Miocene. Saddle Mountain, Washington, USA. The Lyle and Lela Jacklin Collection, 16.5 cm.

DENNIS & MARY MURPHY COLLECTION

Araucarioxylon. Chinle Formation, Triassic. Northern Arizona, USA. The Dennis and Mary Murphy Collection, 38.5 cm.

Double-hearted oak-like fossil wood. Columbia Plateau Basalt flow, Miocene. Stinking Water Creek, Oregon, USA. The Dennis and Mary Murphy Collection, 50 cm.

Araucarioxylon. Chinle Formation, Triassic. Northern Arizona, USA. The Dennis and Mary Murphy Collection, 42 cm.

Taxodioxylon. Tertiary. Hubbard Basin, Nevada, USA. The Dennis and Mary Murphy Collection, 60 cm.

Taxodioxylon. Tertiary. Hubbard Basin, Nevada, USA. The Dennis and Mary Murphy Collection, 55 cm.

Oak-like fossil hardwood. Tertiary. Swartz Canyon, Crook County, Oregon, USA. The Dennis and Mary Murphy Collection, 44 cm.

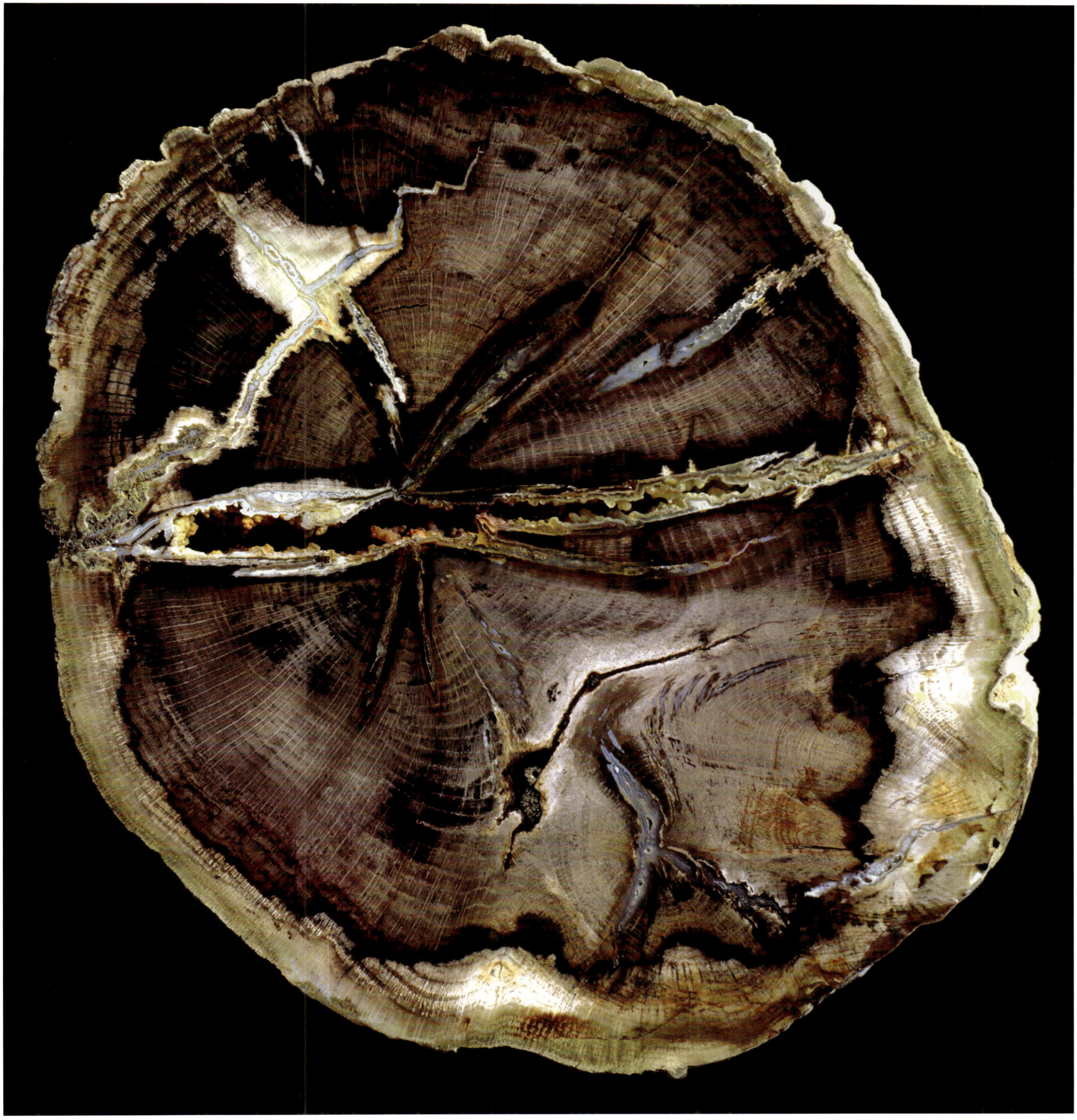

Oak-like fossil hardwood. Deschutes flora, Early to Middle Pliocene. Deschutes River area, Oregon, USA. The Dennis and Mary Murphy Collection, 35 cm.

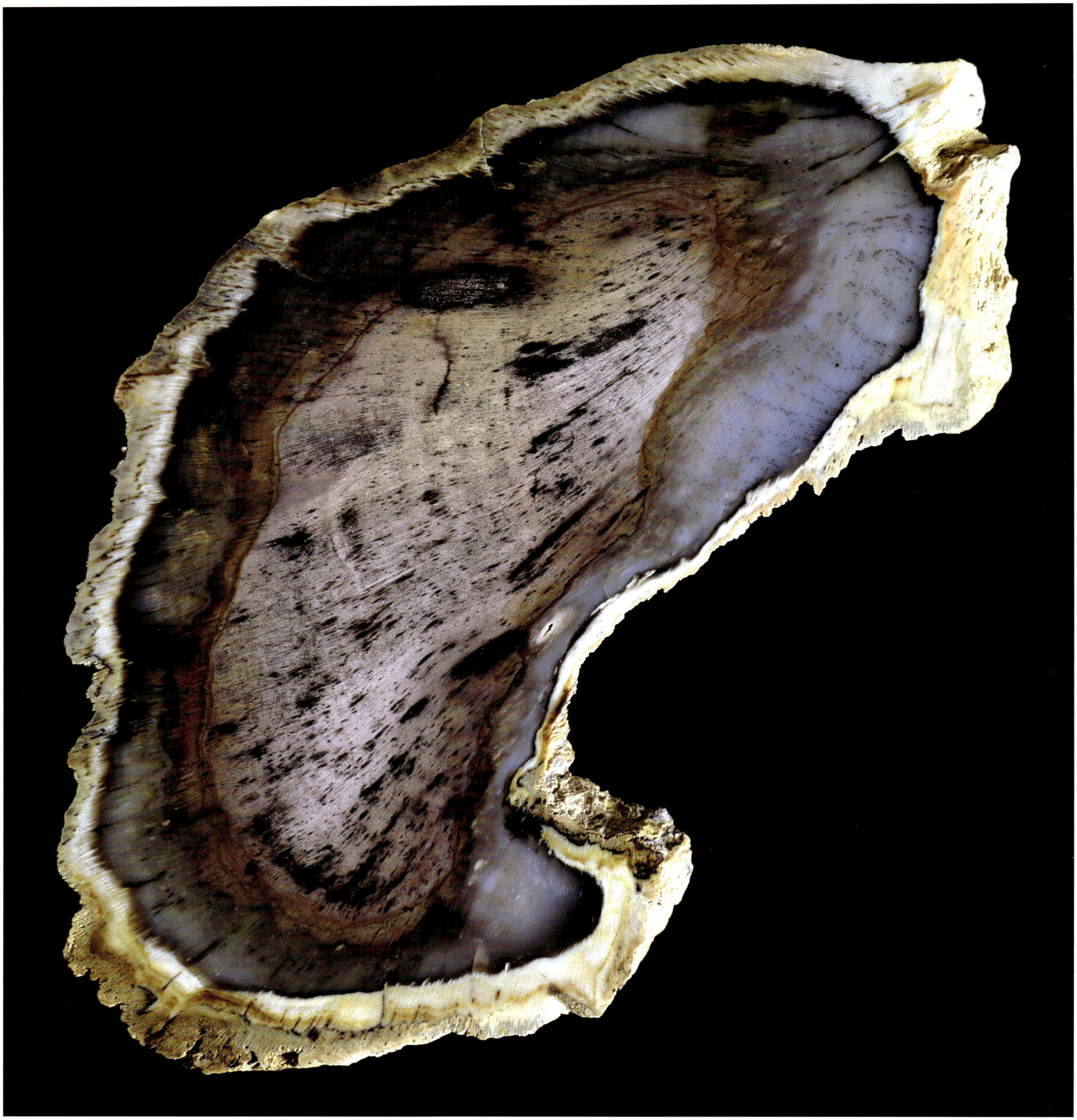

Oak-like fossil hardwood. Tertiary. Swartz Canyon, Crook County, Oregon, USA. The Dennis and Mary Murphy Collection, 39 cm.

Taxodioxylon. Tertiary. Hubbard Basin, Nevada, USA. The Dennis and Mary Murphy Collection, 36 cm.

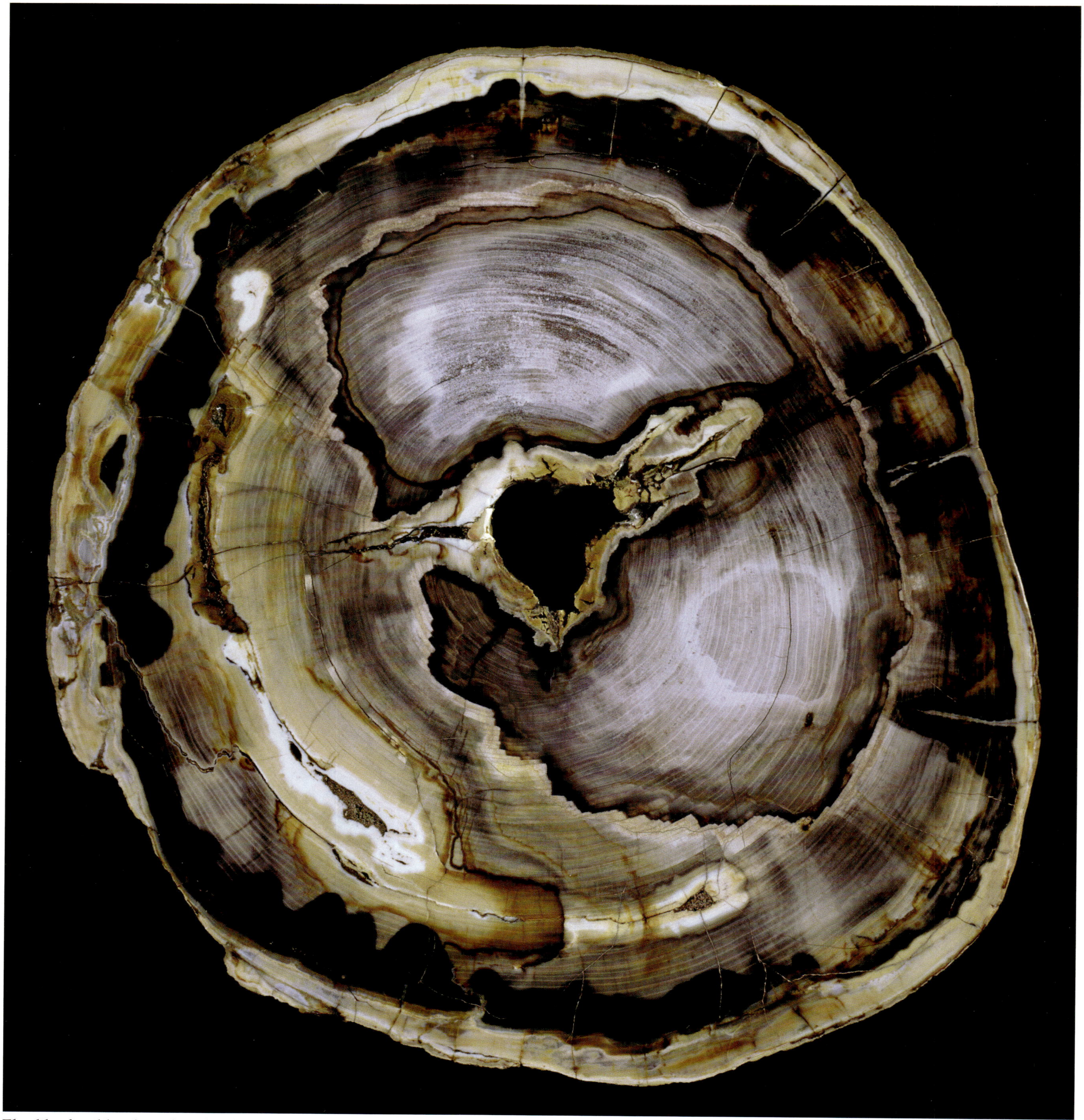

Elm-like fossil hardwood. Columbia Plateau Basalt flow, Miocene. Vantage, Washington, USA. The Dennis and Mary Murphy Collection, 38.5 cm.

Fossil conifer. Morrison Formation, Jurassic. Escalante, Utah, USA. The Dennis and Mary Murphy Collection, 40 cm.

Quadruple-hearted oak-like fossil hardwood. Early to Middle Pliocene. Deschutes River area, Oregon, USA. The Dennis and Mary Murphy Collection, 32.5 cm.

Unidentified fossil wood. Tertiary. Marston Ranch, Ashville, Oregon, USA. The Dennis and Mary Murphy Collection, 35.5 cm.

Brecciated *Metasequoia*. Clarno Formation, Eocene. Crook County, Oregon, USA. The Dennis and Mary Murphy Collection, 33 cm.

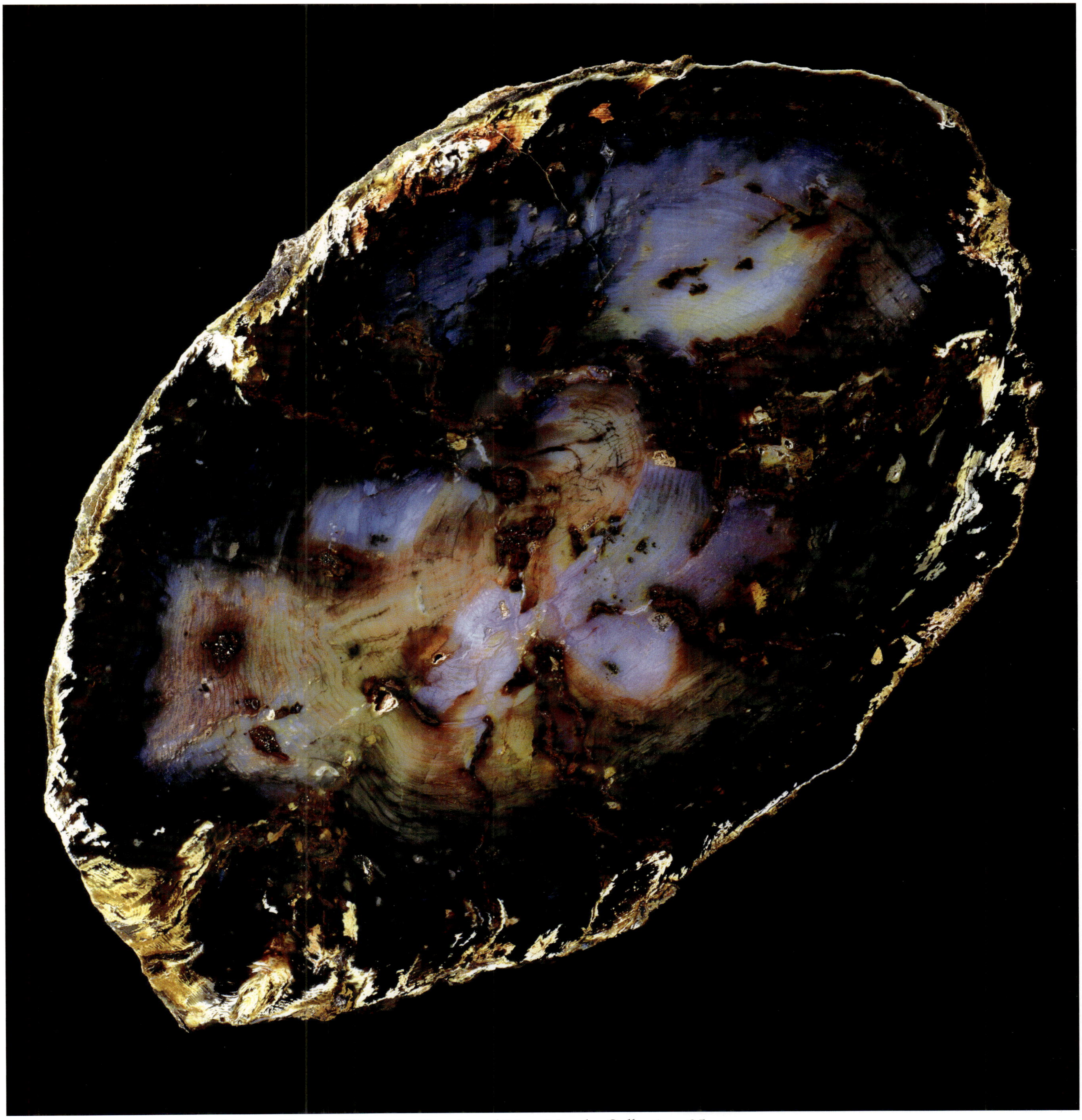

Taxodioxylon. Tertiary. Hubbard Basin, Nevada, USA. The Dennis and Mary Murphy Collection, 35 cm.

Fir-like fossil conifer. Columbia Plateau Basalt flow, Miocene. Vantage, Washington, USA. The Dennis and Mary Murphy Collection, 28 cm.

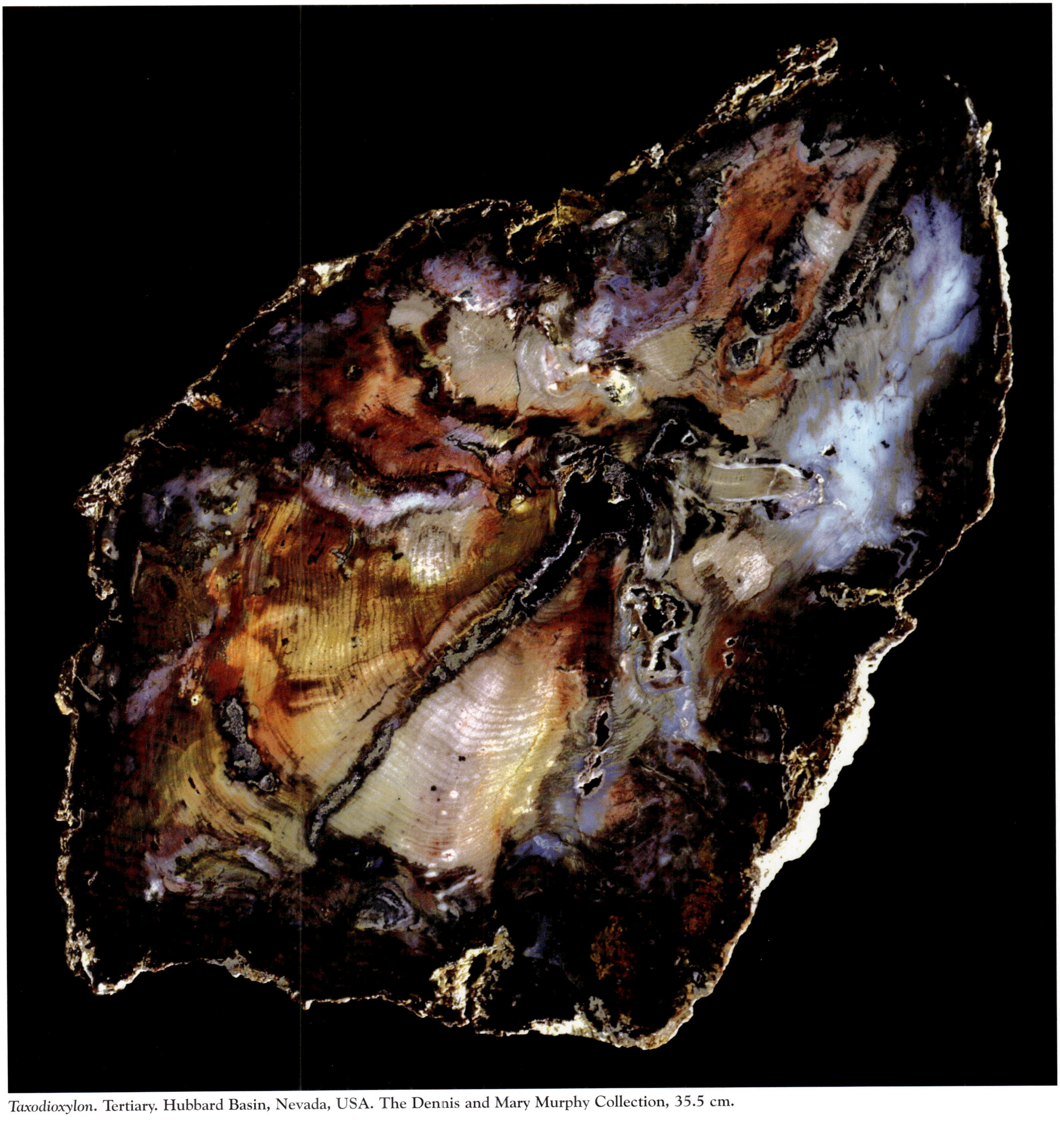

Taxodioxylon. Tertiary. Hubbard Basin, Nevada, USA. The Dennis and Mary Murphy Collection, 35.5 cm.

Oak-like fossil hardwood. Tertiary. Swartz Canyon, Crook County, Oregon, USA. The Dennis and Mary Murphy Collection, 28.5 cm.

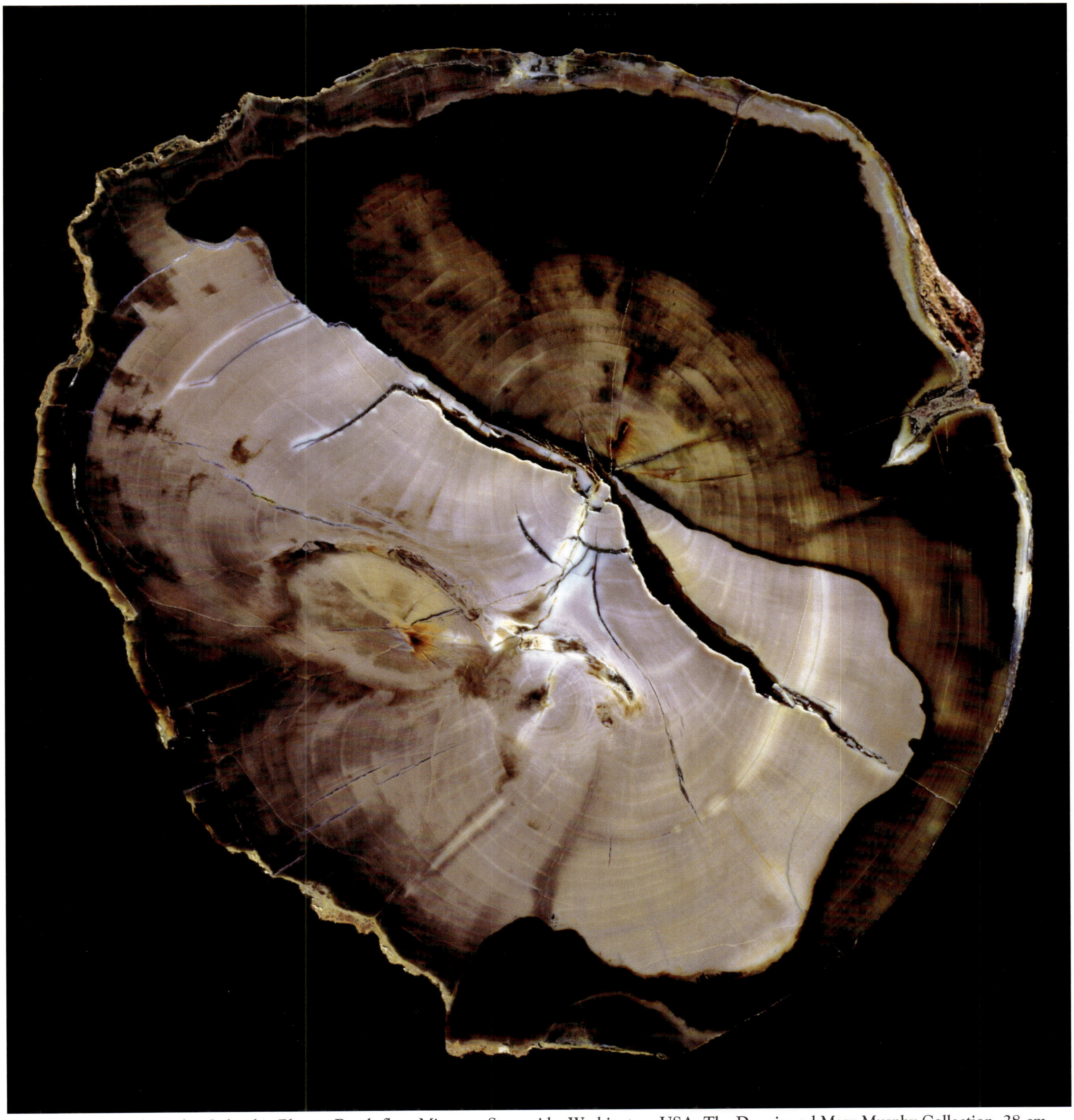

Triple-hearted fossil conifer. Columbia Plateau Basalt flow, Miocene. Sunnyside, Washington, USA. The Dennis and Mary Murphy Collection, 28 cm.

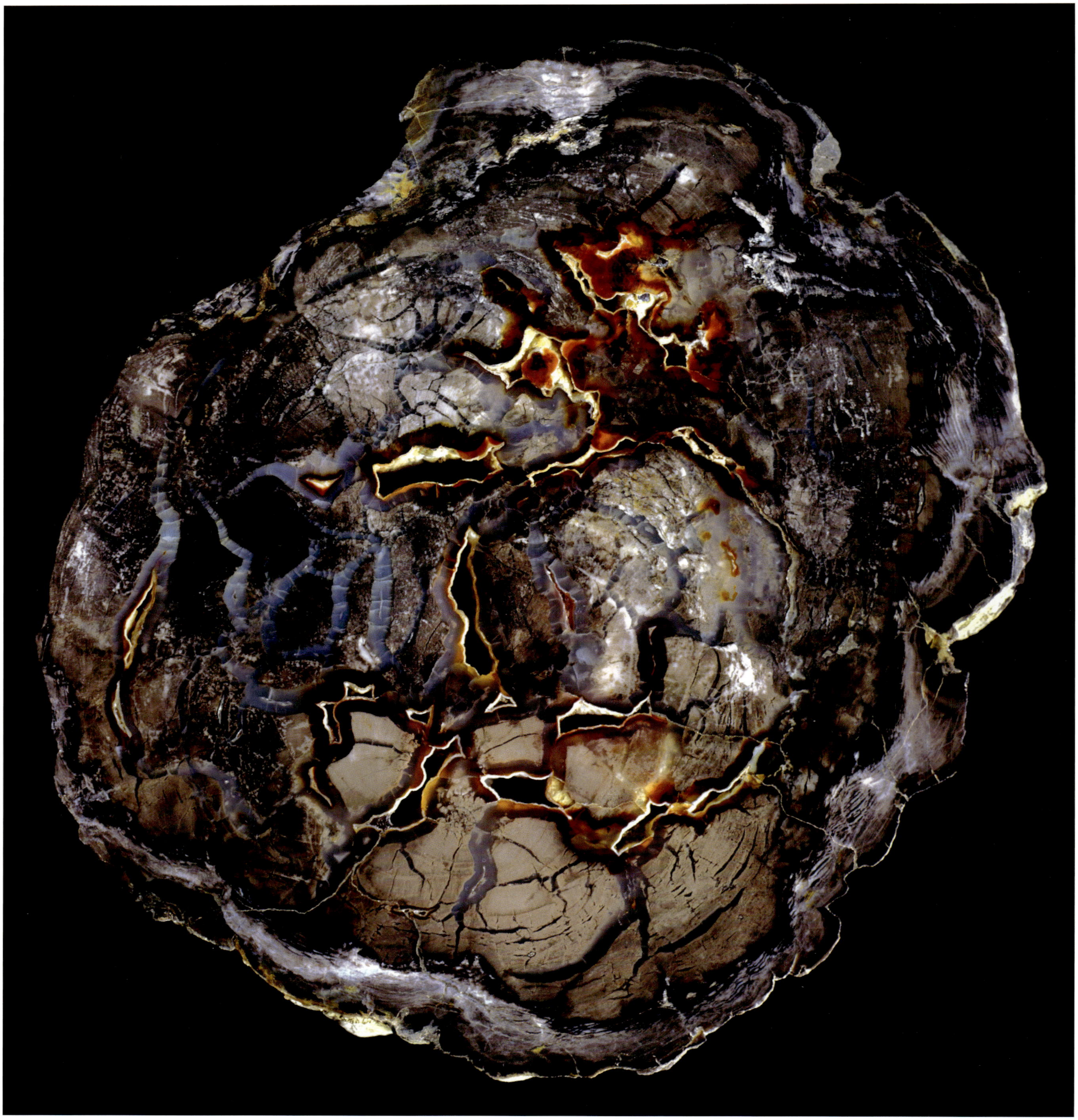

Fossil conifer. Tertiary. Eastern Oregon, USA. The Dennis and Mary Murphy Collection, 22 cm.

Oak-like fossil wood. Columbia Plateau Basalt flow, Miocene. Stinking Water Creek, Oregon, USA. The Dennis and Mary Murphy Collection, 24 cm.

Ginkgo-like fossil wood. Columbia Plateau Basalt flow, Miocene. Vantage, Washington, USA. The Dennis and Mary Murphy Collection, 19.5 cm.

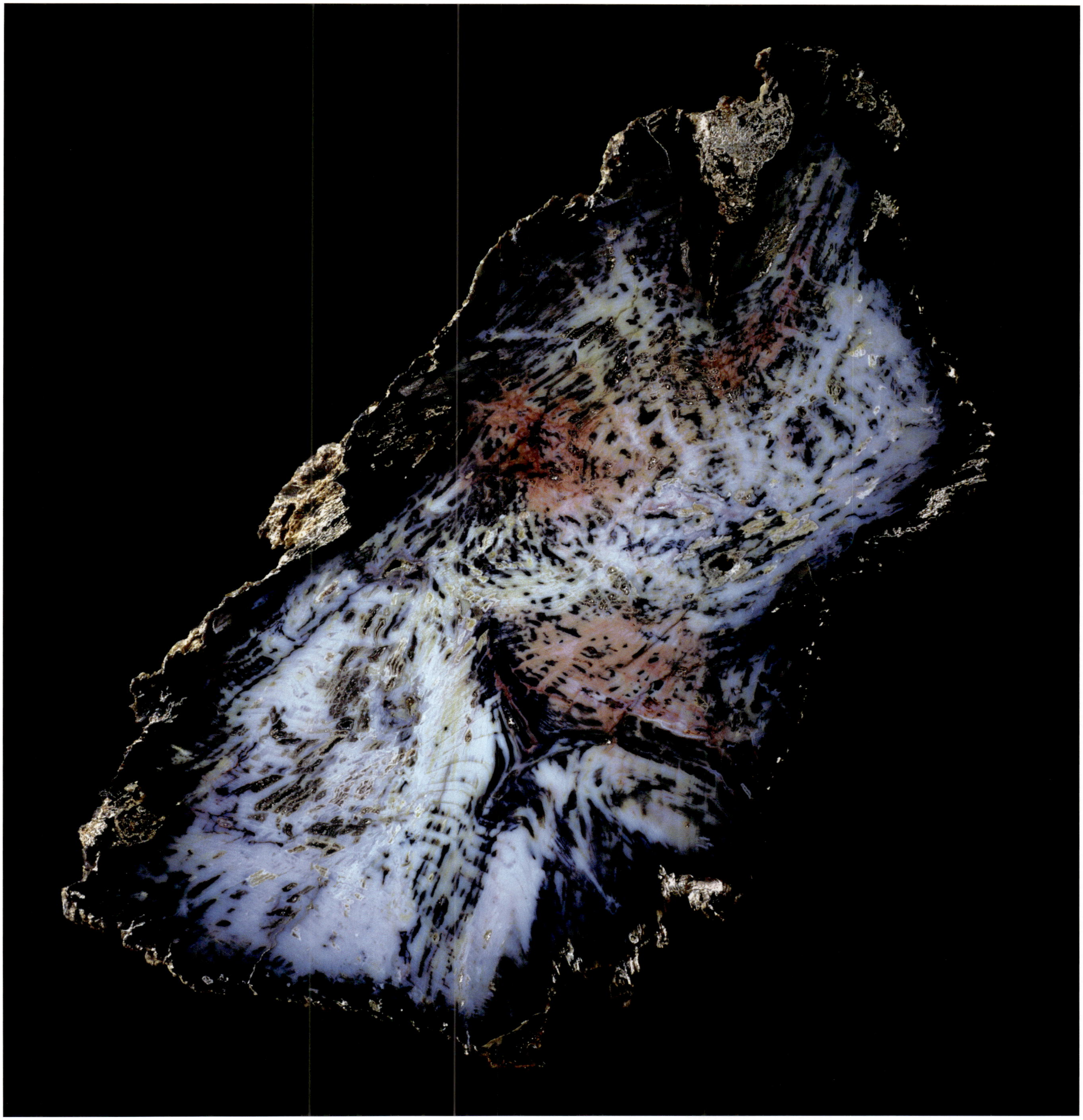

Taxodioxylon. Tertiary. Hubbard Basin, Nevada, USA. The Dennis and Mary Murphy Collection, 23 cm.

JIM & CATHY GRAY COLLECTION

Araucarioxylon. Chinle Formation, Triassic. Winslow area, Northern Arizona, USA. The Jim and Cathy Gray Collection, 21 cm.

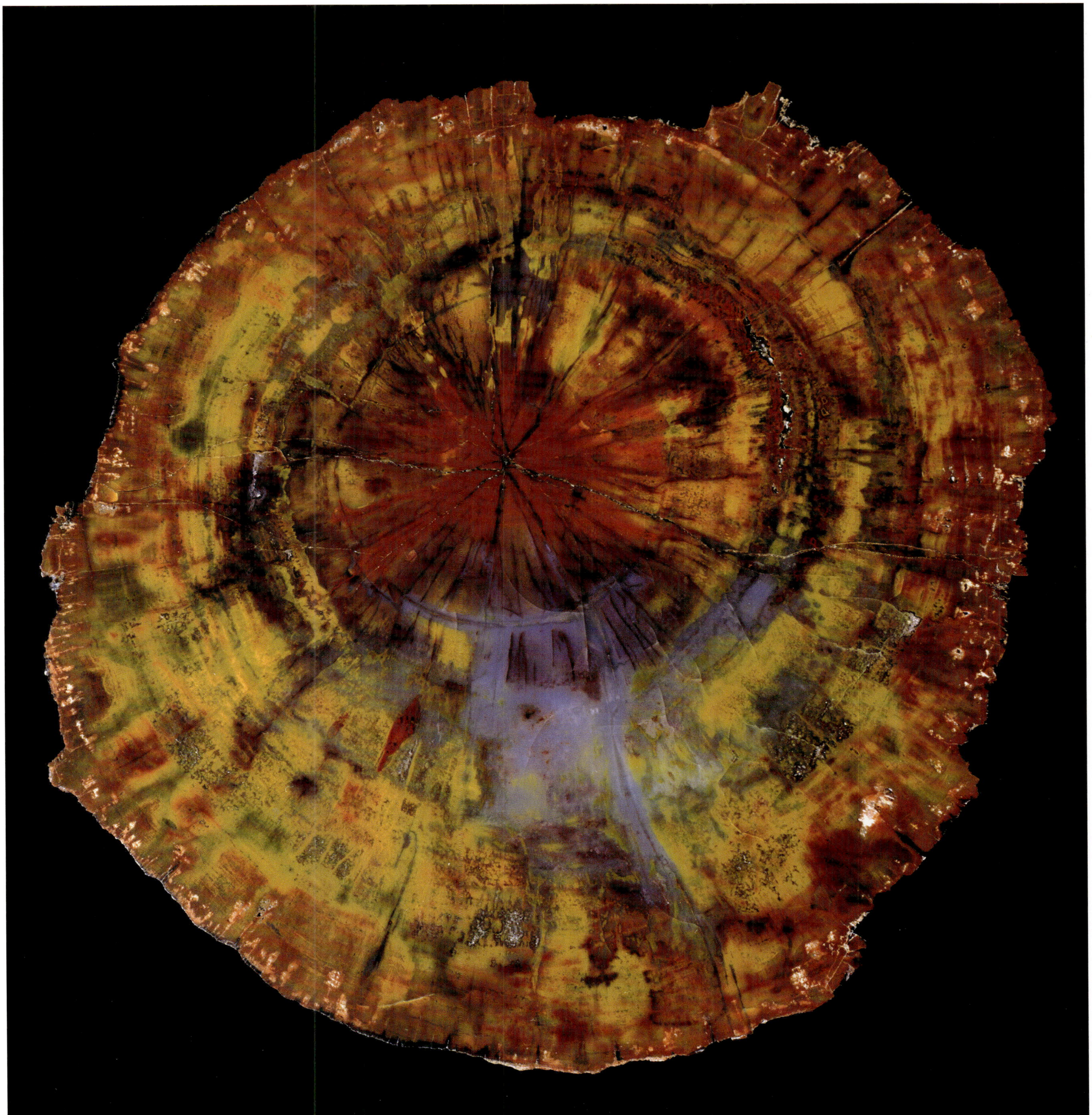

Araucarioxylon. Chinle Formation, Triassic. Northern Arizona, USA. The Jim and Cathy Gray Collection, 19 cm.

Woodworthia. Chinle Formation, Triassic. Northern Arizona, USA. The Jim and Cathy Gray Collection, 24.5 cm.

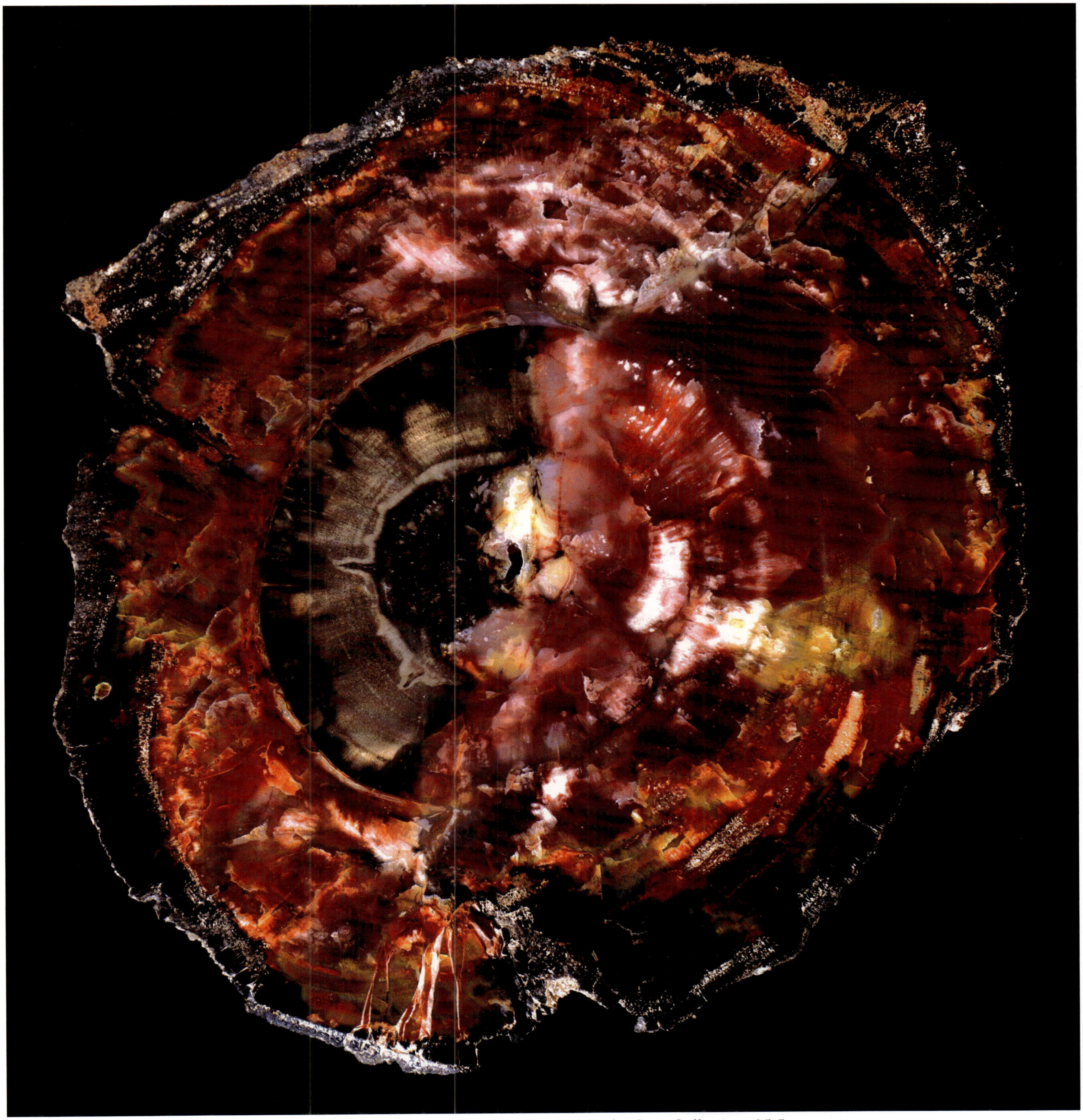

Araucarioxylon. Chinle Formation, Triassic. Northern Arizona, USA. The Jim and Cathy Gray Collection, 25.5 cm.

Araucarioxylon. Chinle Formation, Triassic. Winslow area, Northern Arizona, USA. The Jim and Cathy Gray Collection, 51 cm long.

Araucarioxylon. Chinle Formation, Triassic. Northern Arizona, USA. The Jim and Cathy Gray Collection, 30.5 cm.

Schilderia. Chinle Formation, Triassic. Northern Arizona, USA. The Jim and Cathy Gray Collection, 26 cm.

Araucarioxylon. Chinle Formation, Triassic. St. Johns, Northern Arizona, USA. The Jim and Cathy Gray Collection, 30.5 cm.

Schilderia. Chinle Formation, Triassic. Northern Arizona, USA. The Jim and Cathy Gray Collection, 27 cm.

Schilderia. Chinle Formation, Triassic. Northern Arizona, USA. The Jim and Cathy Gray Collection, 20.5 cm.

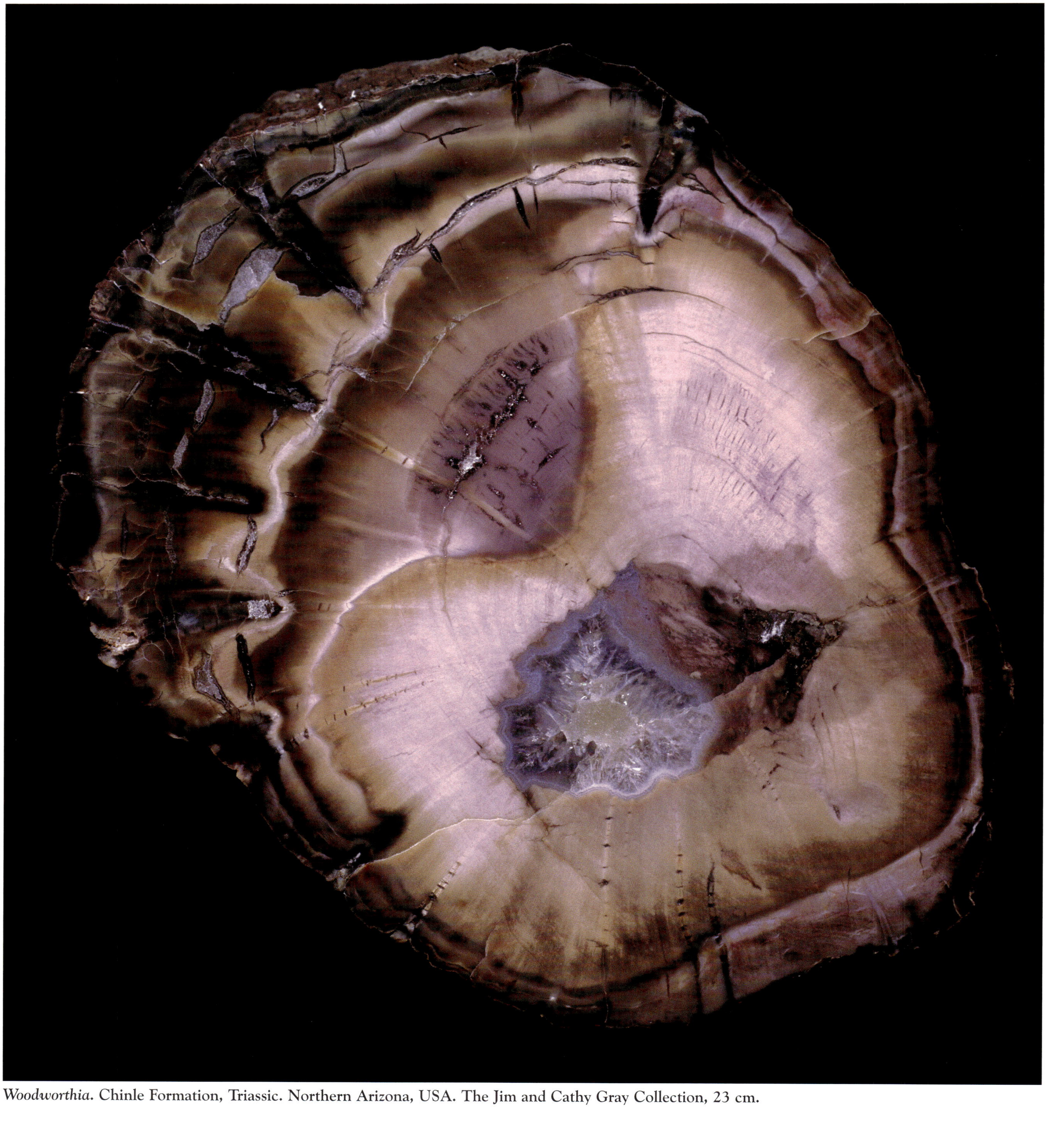

Woodworthia. Chinle Formation, Triassic. Northern Arizona, USA. The Jim and Cathy Gray Collection, 23 cm.

Araucarioxylon. Chinle Formation, Triassic. Northern Arizona, USA. The Jim and Cathy Gray Collection, 15.5 cm.

Cupressinoxylon. Bidahochi Formation, Miocene/Pliocene. Northern Arizona, USA. The Jim and Cathy Gray Collection, 8.5 cm.

Araucarioxylon. Chinle Formation, Triassic. Northern Arizona, USA. The Jim and Cathy Gray Collection, 14.5 cm.

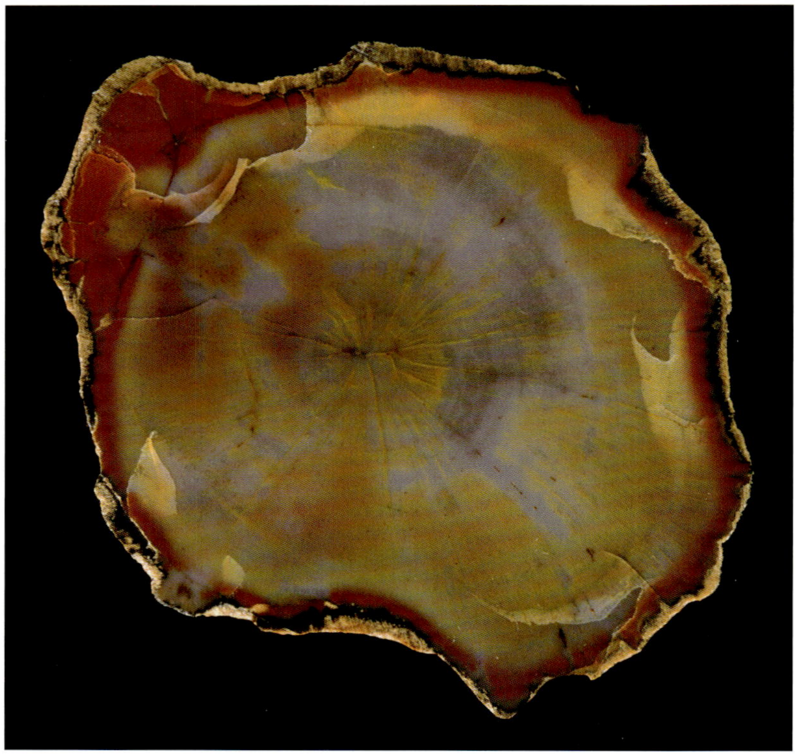

Araucarioxylon. Chinle Formation, Triassic. Northern Arizona, USA. The Jim and Cathy Gray Collection, 8 cm.

Araucarioxylon. Chinle Formation, Triassic. Northern Arizona, USA. The Jim and Cathy Gray Collection, 9 cm.

Araucarioxylon. Chinle Formation, Triassic. Northern Arizona, USA. The Jim and Cathy Gray Collection, 14 cm.

Araucarioxylon. Chinle Formation, Triassic. Northern Arizona, USA. The Jim and Cathy Gray Collection, 14 cm.

Araucarioxylon. Chinle Formation, Triassic. Northern Arizona, USA. The Jim and Cathy Gray Collection, 11 cm.

Araucarioxylon. Chinle Formation, Triassic. Northern Arizona, USA. The Jim and Cathy Gray Collection, 12 cm.

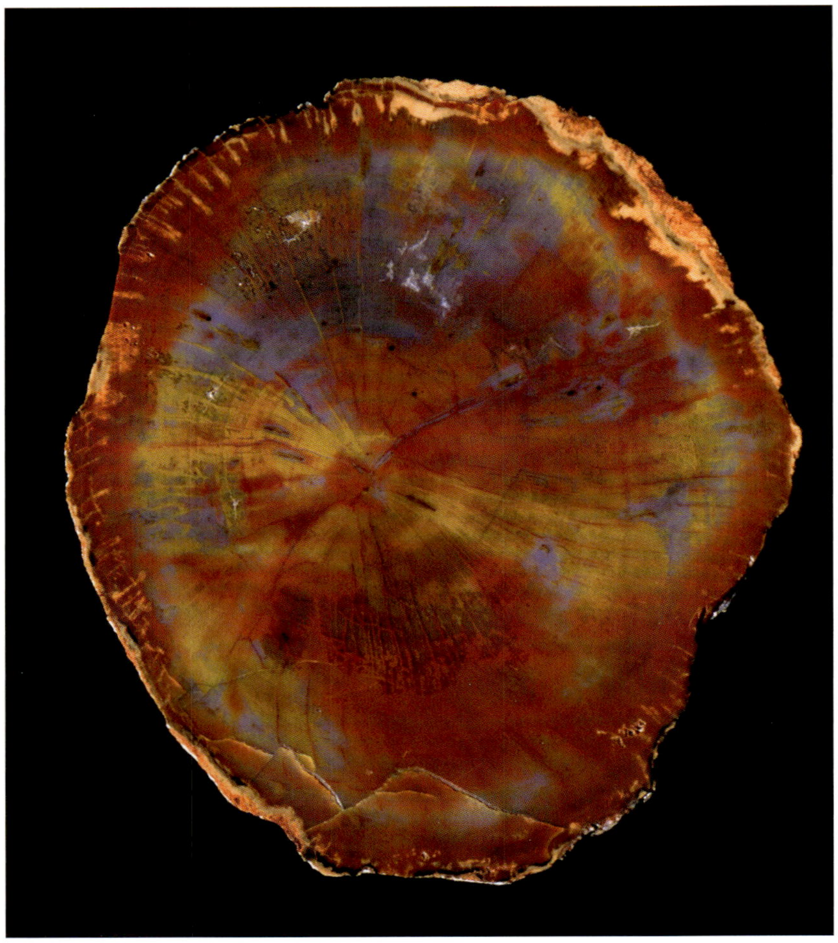

Araucarioxylon. Chinle Formation, Triassic. Northern Arizona, USA. The Jim and Cathy Gray Collection, 9.5 cm.

Araucarioxylon. Chinle Formation, Triassic. Northern Arizona, USA. The Jim and Cathy Gray Collection, 16 cm.

Araucarioxylon. Chinle Formation, Triassic. Northern Arizona, USA. The Jim and Cathy Gray Collection, 13.5 cm.

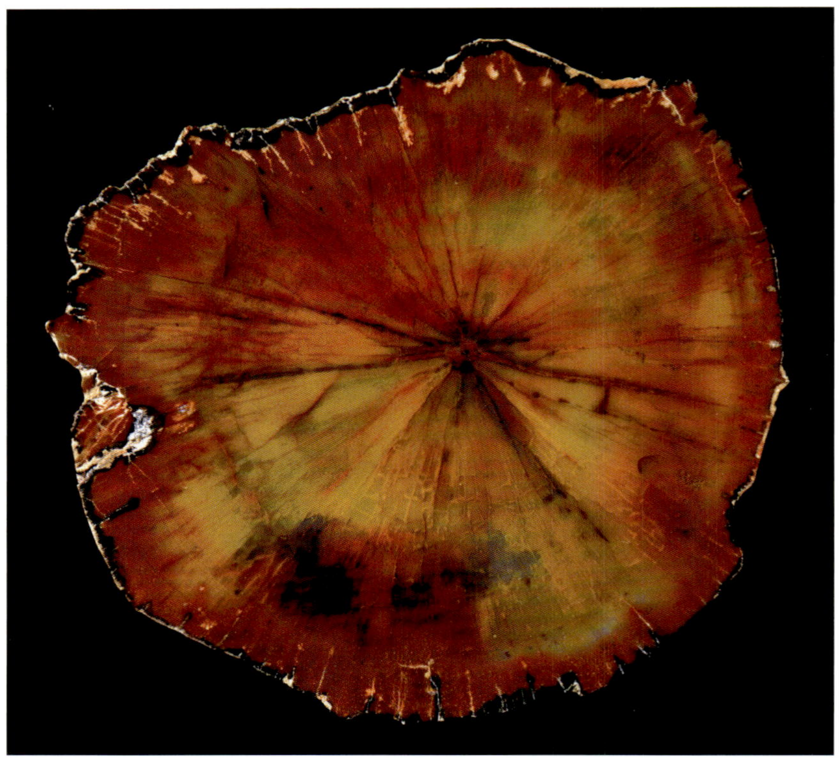

Araucarioxylon. Chinle Formation, Triassic. Northern Arizona, USA. The Jim and Cathy Gray Collection, 9 cm.

VIII

C H A P T E R   E I G H T

Special Collections

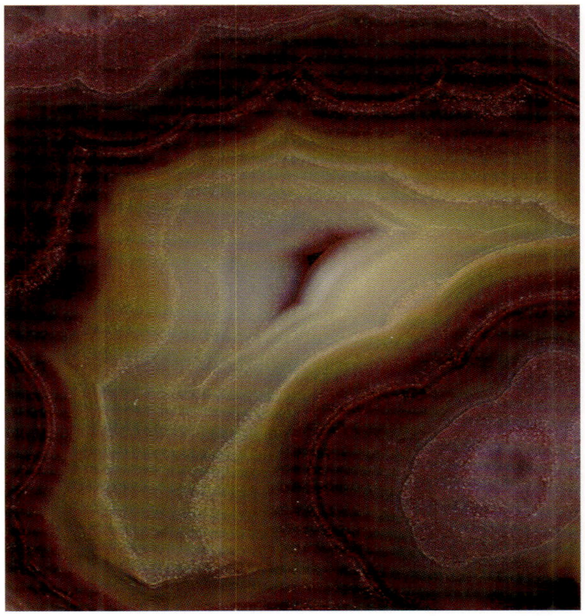

THE SHREVEPORT COLLECTION

THE JOE TAUBR COLLECTION

AGATES

Modern palms as they may have appeared 30 million years ago along what was then a broad coastal plain near present-day central Louisiana.

THE SHREVEPORT COLLECTION

Assembled during the course of half a century by Captain N. A. Foshee of Shreveport, this collection of Louisiana *Palmoxylon* is the finest in the world. *Palmoxylon* from the Oligocene Epoch Catahoula Formation is found in a thin vein across central Louisiana. Now some 200 kilometers from the coast, this area was a broad coastal plain of swamps, beaches, and palm groves during the Oligocene Epoch. The formation has been heavily collected to the point that very few surface specimens remain. Since World War II, Captain Foshee spent countless hours seeking out the finest specimens to cut and polish for this collection. He estimates he handled more than a quarter million pounds of Louisiana palm wood in the course of 50 years, keeping the finest specimens for his collection. A master carpenter by profession, Captain Foshee is also a master rock polisher and achieved the finest polish imaginable on these specimens. Composed of incredibly tough chalcedony, these specimens ring like fine china when tapped.

Present-day central Louisiana in the once *Palmoxylon*-rich outcrop of the Oligocene age Catahoula Formation.

Set of *Palmoxylon* specimens from The Shreveport Collection, largest 28 cm. Because of the nature of palm structure, with the main support composed of longitudinal fibrous tissues, fossilizing wood frequently split longitudinally, exposing an elongated view of vascular bundles. Specimens cut and polished on tangential and radial planes make particularly attractive specimens.

Two *Palmoxylon* "stumps" from The Shreveport Collection, 15 and 17 cm tall.

Two *Palmoxylon* "stumps" from The Shreveport Collection, 8.5 and 13 cm tall. Specimens with bright white exteriors such as the one on the right come from an area the local collectors call "The White Ground," so-named because the ground is bright white.

Macro photograph of termite fecal pellets in *Palmoxylon*. Catahoula Formation, Oligocene. Louisiana, USA. Daniels specimen.

This is one of two *Palmoxylon* specimens from the collection that show evidence of probable termite activity. Termites apparently devoured sections of the tree while still woody, leaving behind pockets of fecal pellets that later fossilized. The Shreveport Collection, 19 cm.

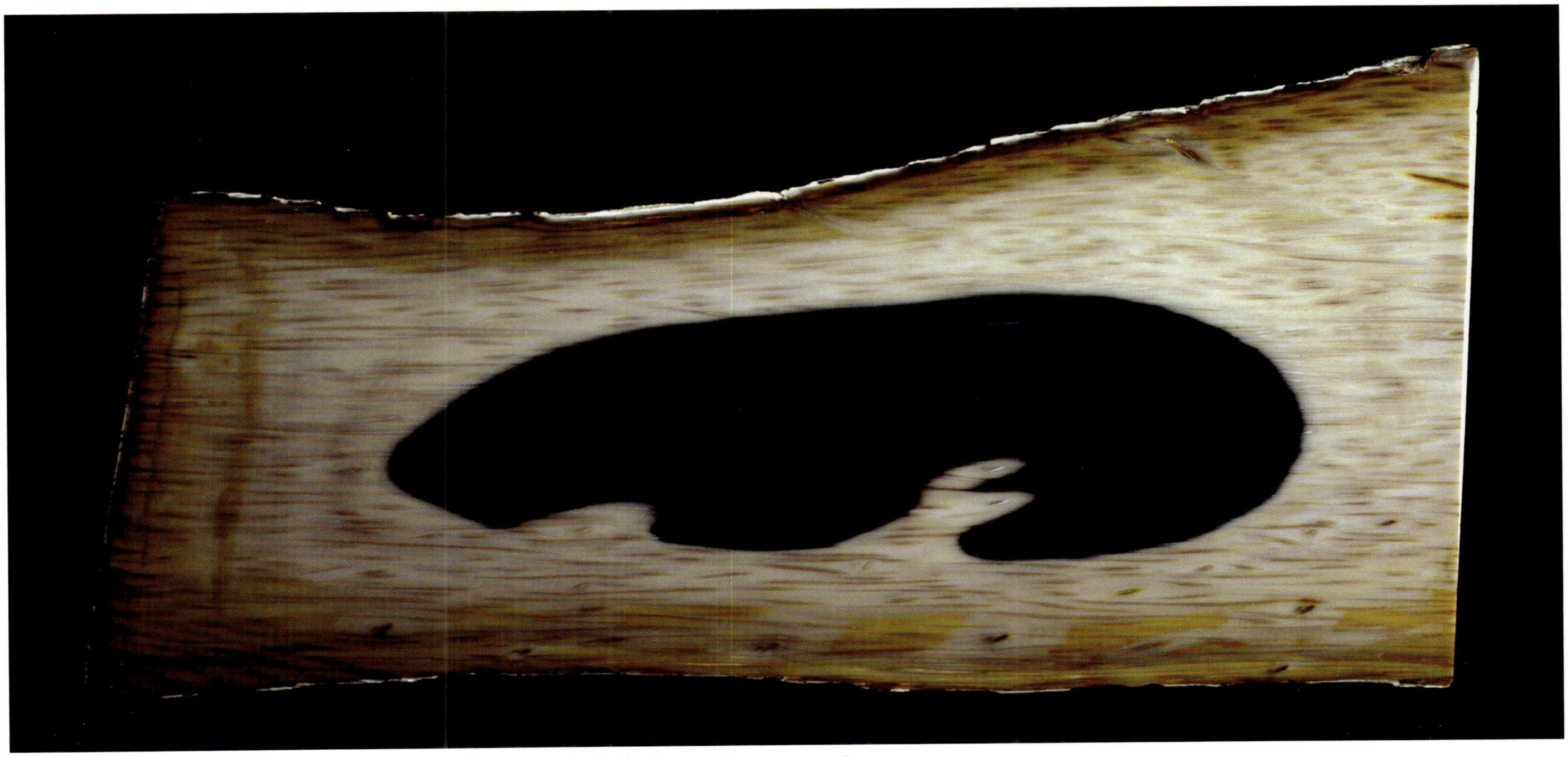

This longitudinally cut *Palmoxylon* specimen has a particularly interesting pattern in black that resembles a Native American bear fetish. The Shreveport Collection, 42 cm.

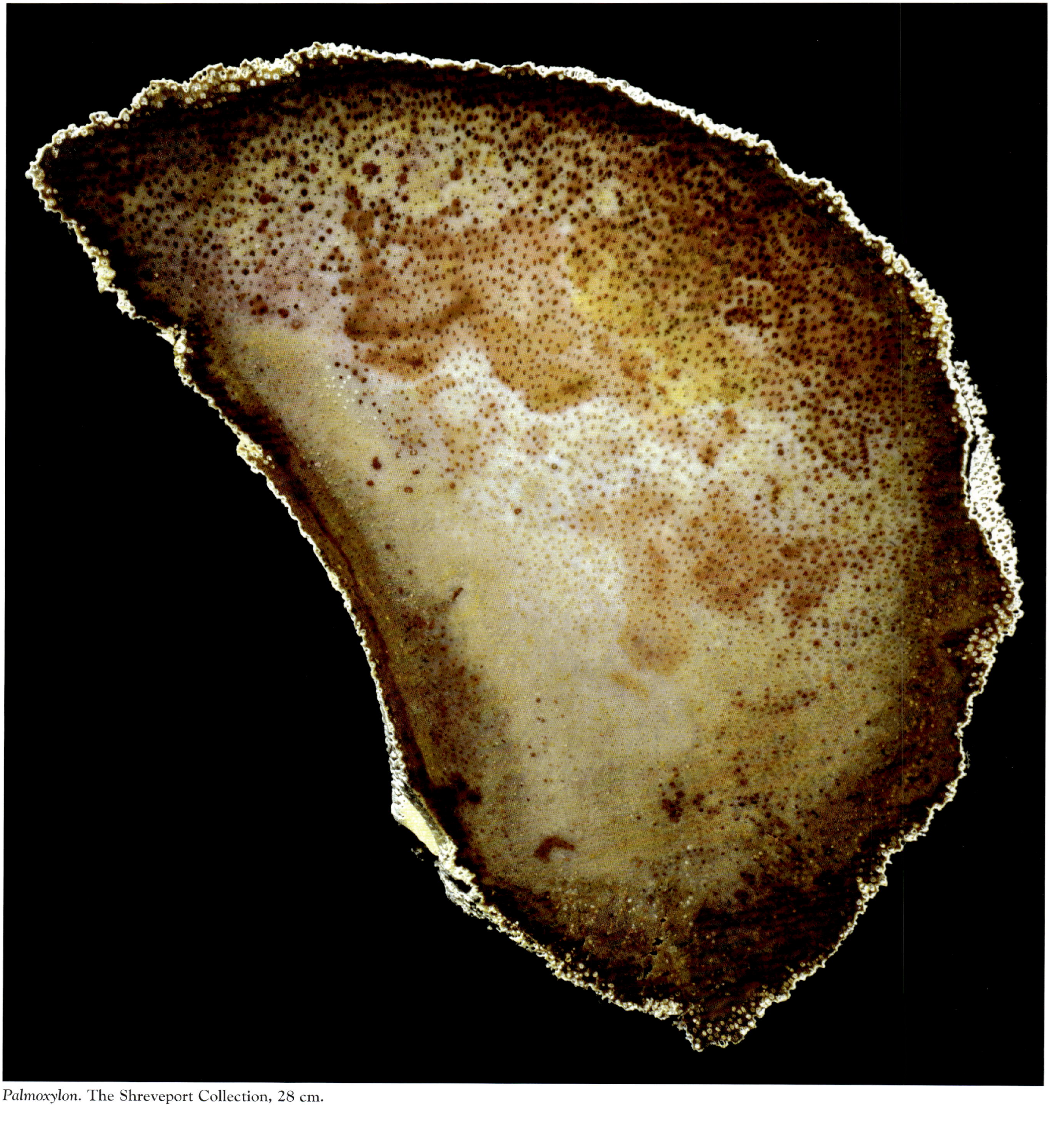

Palmoxylon. The Shreveport Collection, 28 cm.

Palmoxylon. The Shreveport Collection, 27 cm.

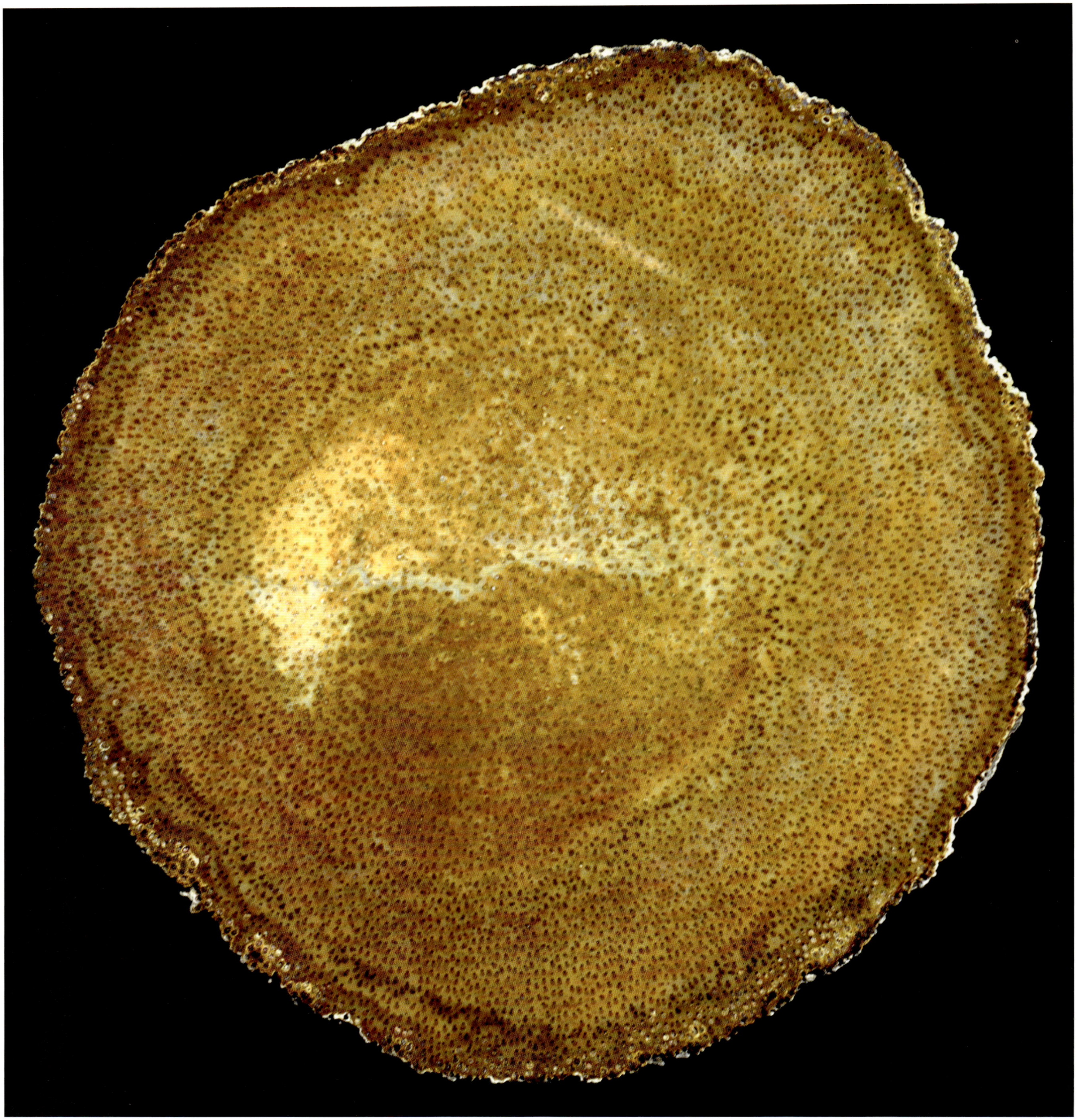

Palmoxylon. The Shreveport Collection, 19 cm.

Two slices from close proximity on the same *Palmoxylon* log, demonstrating how the black sections can vary in size and form. Other slices from this log have no black at all. The Shreveport Collection, 18 cm each.

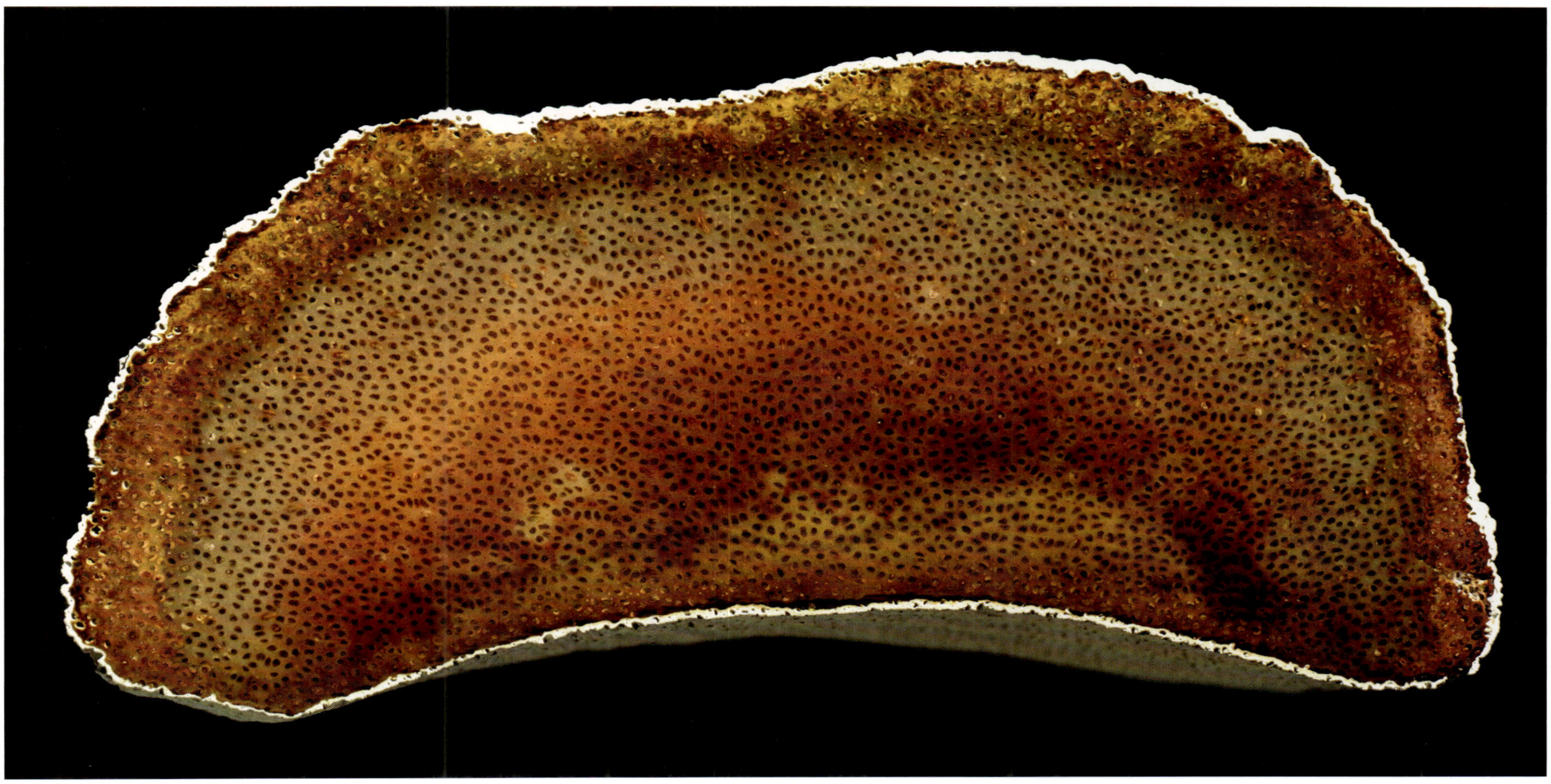

Palmoxylon. The Shreveport Collection, 12 cm.

Longitudinally cut *Palmoxylon* specimen with attractive display of vascular bundles. The Shreveport Collection, 13.5 cm.

Longitudinally cut *Palmoxylon*. The Shreveport Collection, 17 cm.

Palmoxylon. The Shreveport Collection, 22 cm.

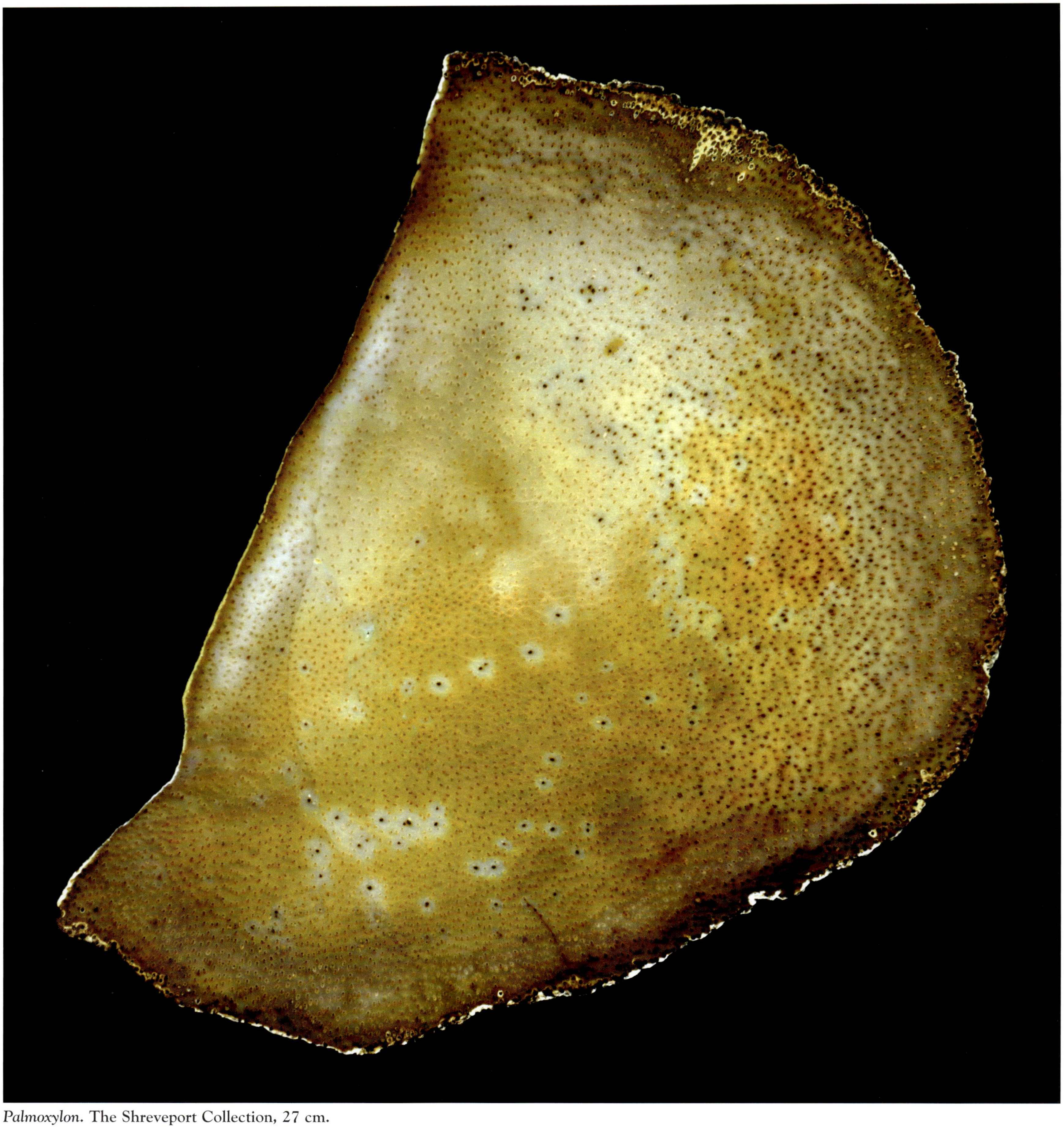

Palmoxylon. The Shreveport Collection, 27 cm.

Palmoxylon. The Shreveport Collection, 15 cm.

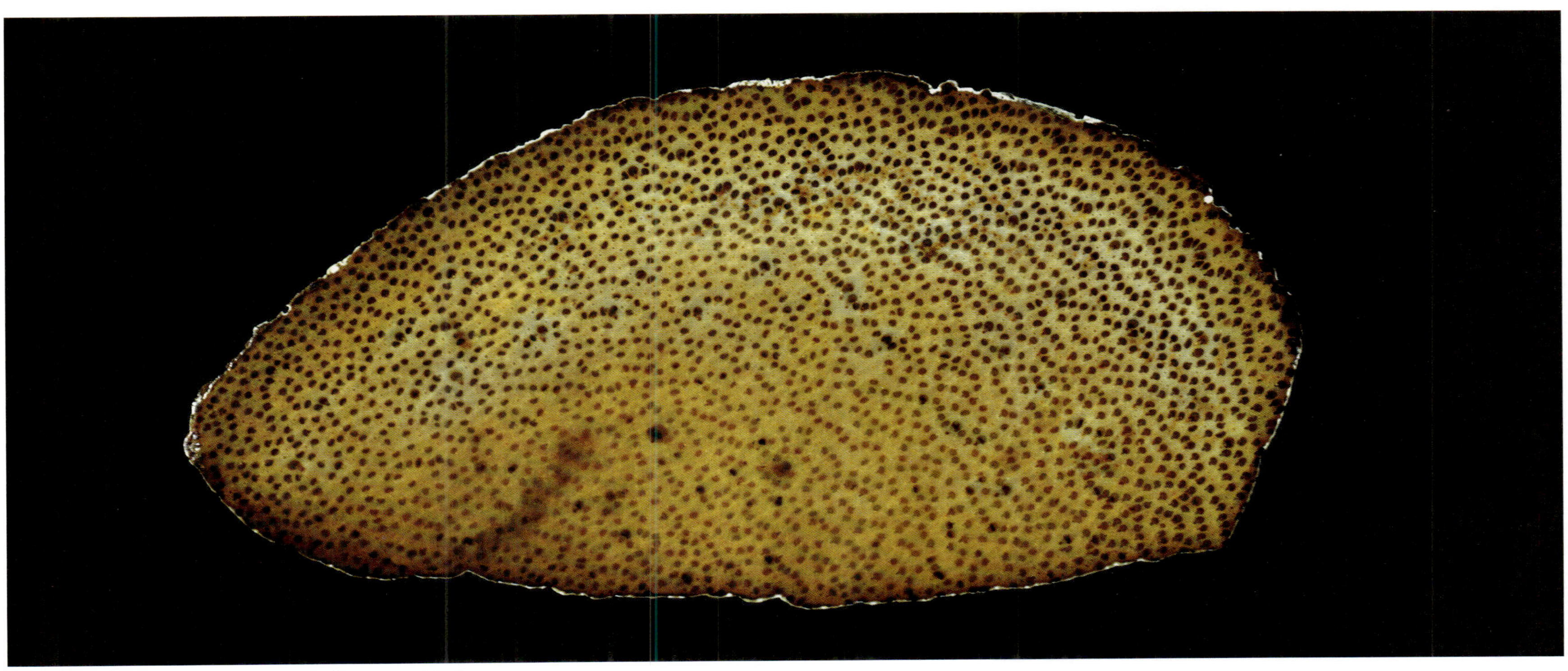

Palmoxylon. The Shreveport Collection, 13.5 cm.

This yellow and white combination is one of the more rare color combinations for Louisiana *Palmoxylon*. The Shreveport Collection, 16.5 cm.

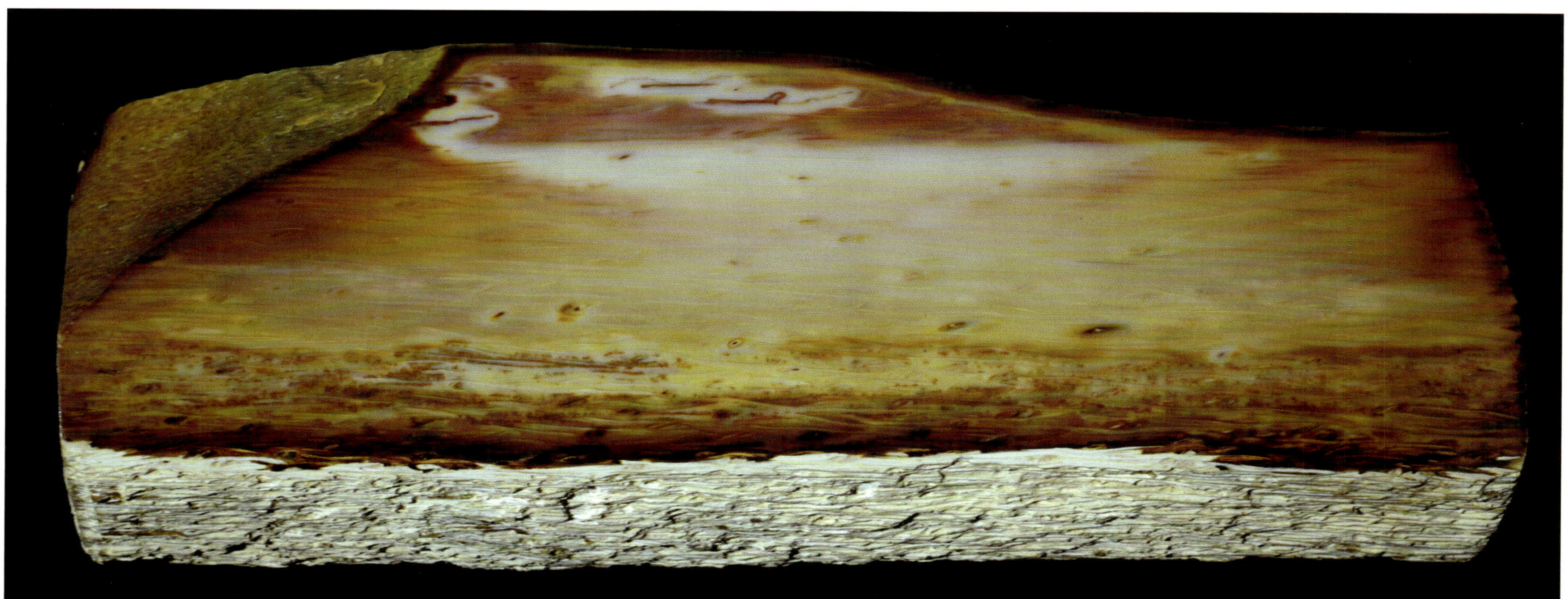

Longitudinally cut *Palmoxylon*. The Shreveport Collection, 29 cm.

Palmoxylon. The Shreveport Collection, 21 cm.

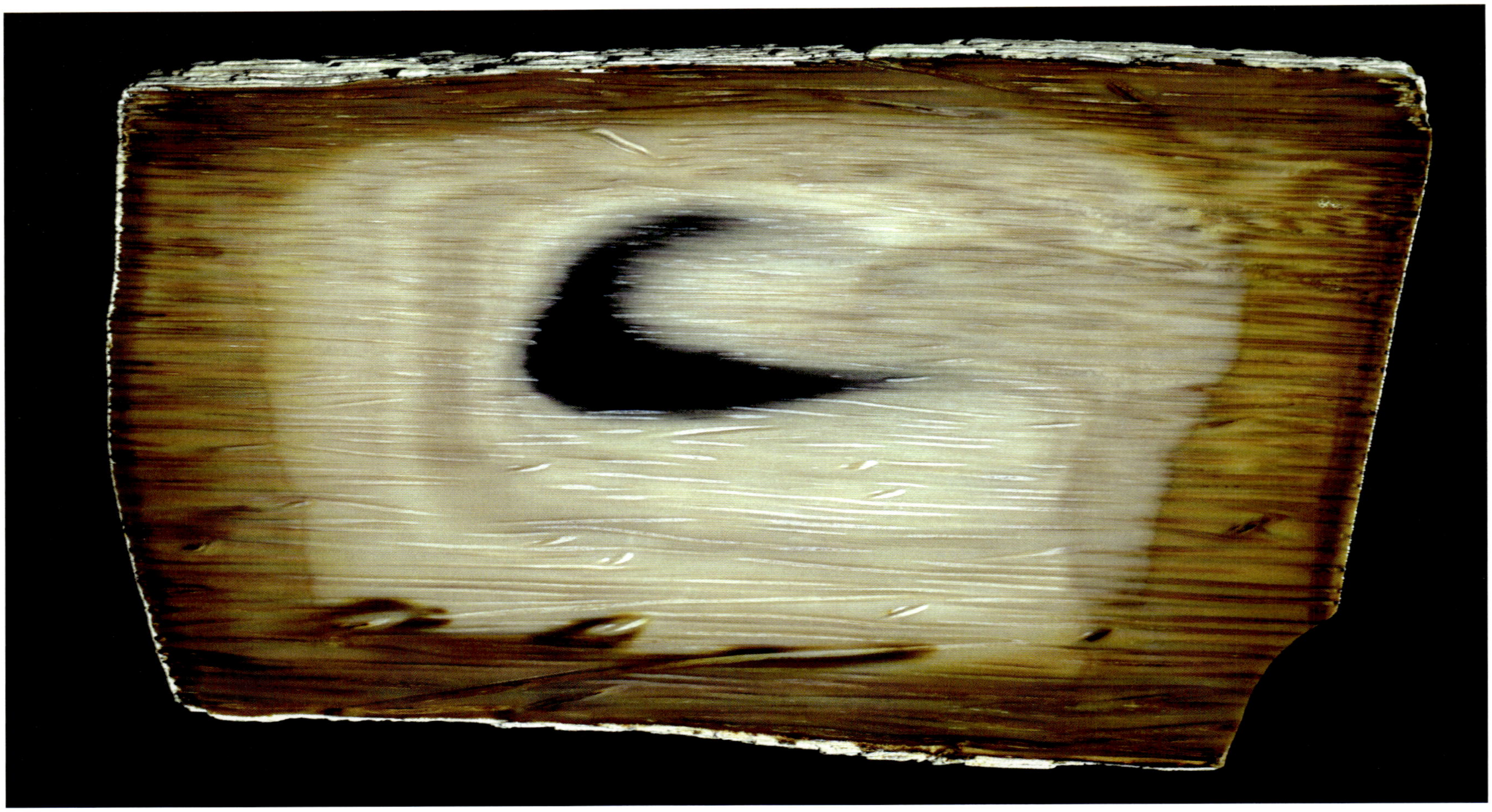

Longitudinally cut *Palmoxylon* with black "Nike swoosh." The Shreveport Collection, 24 cm.

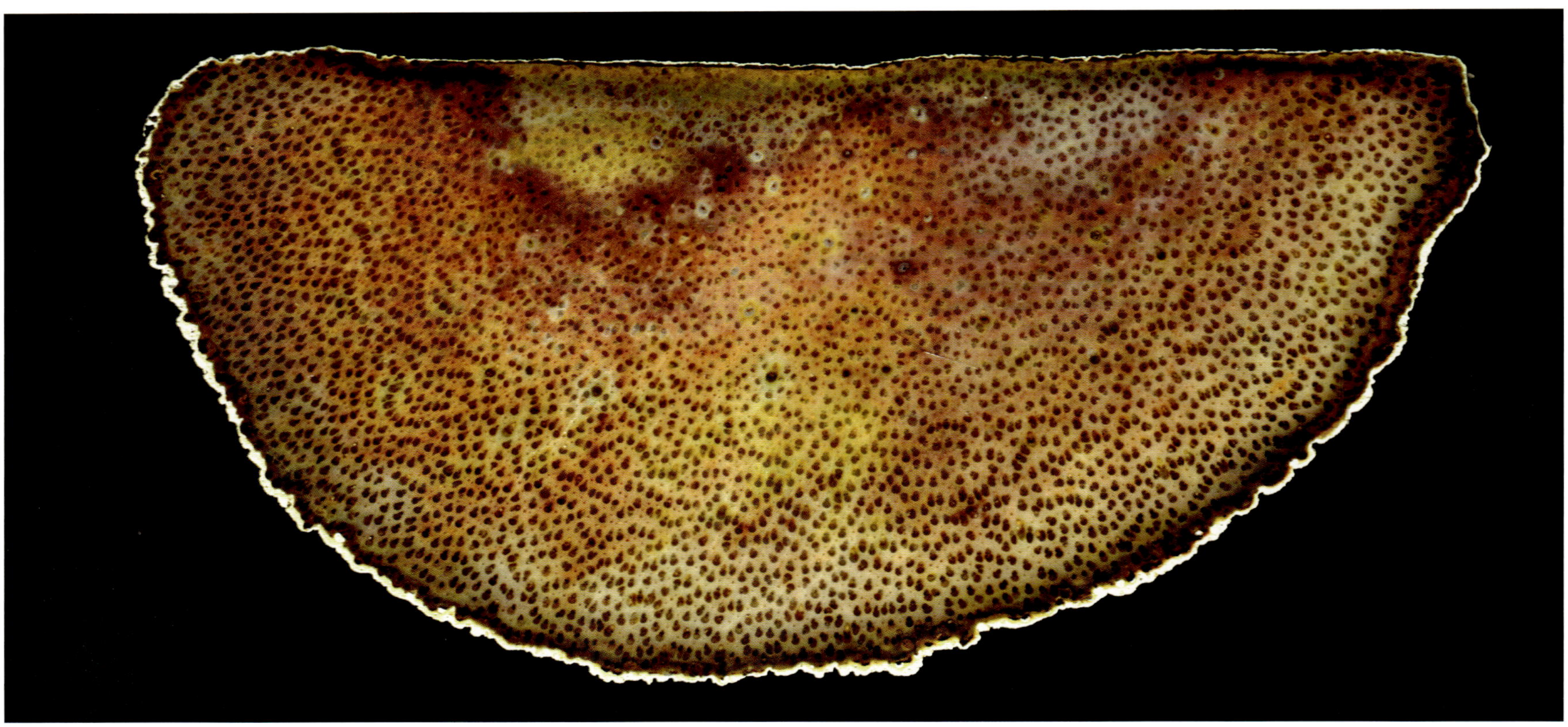

Palmoxylon. The Shreveport Collection, 15 cm.

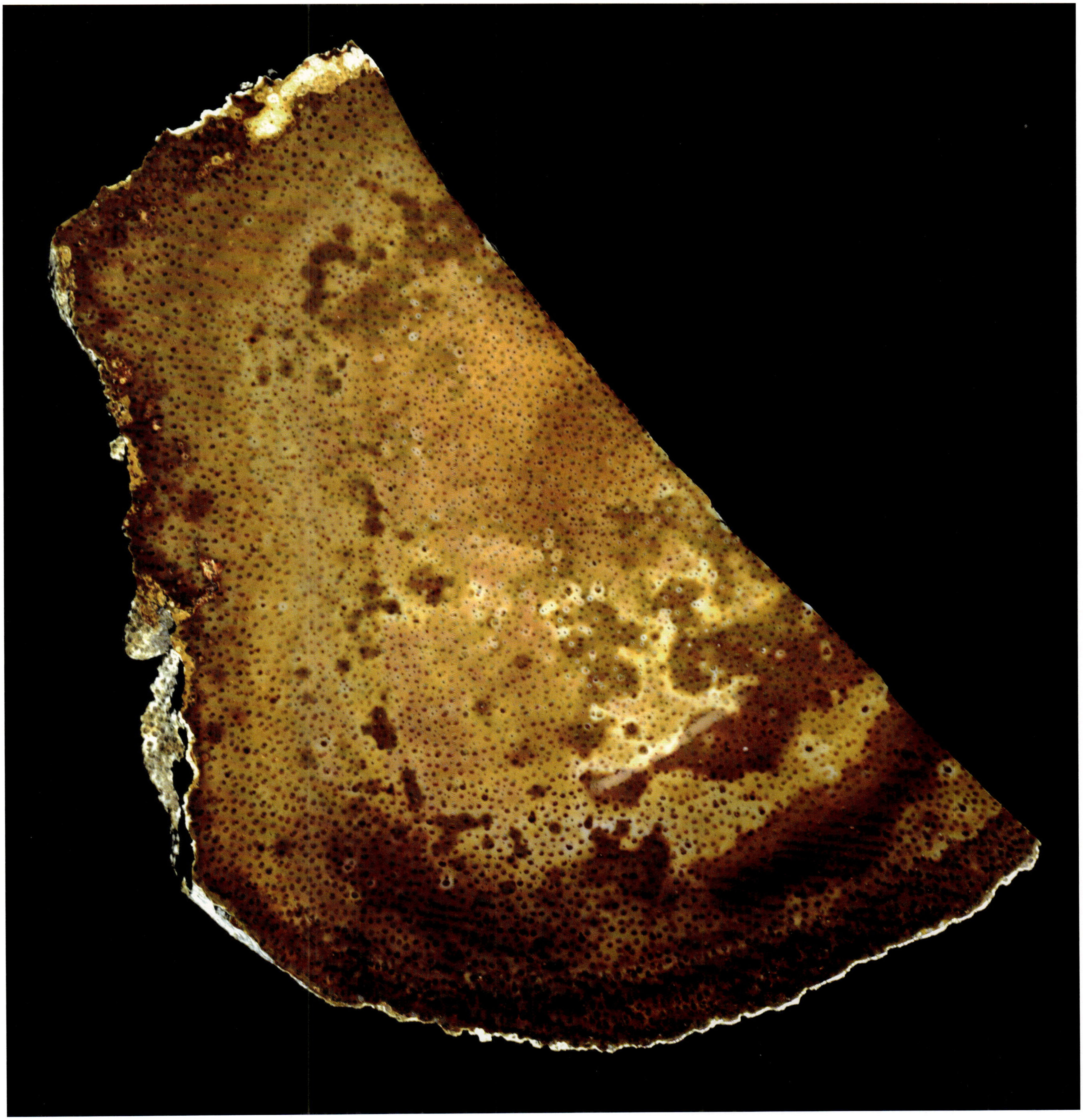

Palmoxylon. The Shreveport Collection, 20 cm.

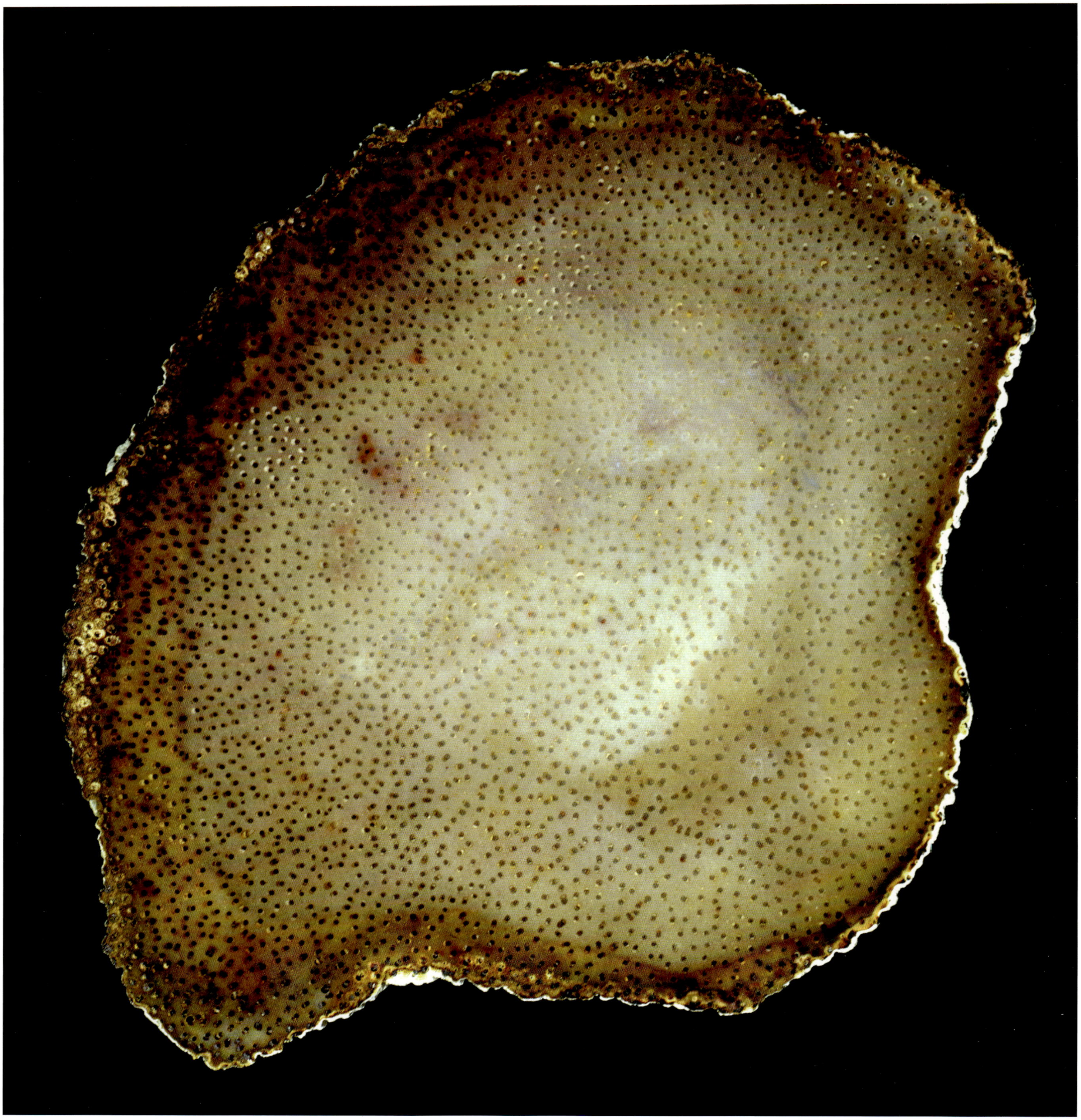

Palmoxylon. The Shreveport Collection, 15.5 cm.

Palmoxylon. The Shreveport Collection, 19.5 cm.

THE JOE TAUBR COLLECTION

A geologist by avocation, Joe Taubr has been a lifelong rockhound and student of the natural sciences. He has especially enjoyed the pursuit of agate, jasper, and fossil wood specimens, which he masterfully cut into perfectly matched cabochons.

Mr. Taubr's extensive knowledge of the geology of the western United States, including the locations of myriad mineral and fossil collecting localities, is unparalleled. Each cabochon measures 5.1 by 7.6 centimeters (2 by 3 inches).

Araucarioxylon. Triassic. Navajo County, Arizona, USA. The Joe Taubr Collection.

Araucarioxylon. Triassic. Navajo County, Arizona, USA. The Joe Taubr Collection.

Fossil conifer. Jurassic. Queensland, Australia. The Joe Taubr Collection.

Araucarioxylon. Triassic. Navajo County, Arizona, USA. The Joe Taubr Collection.

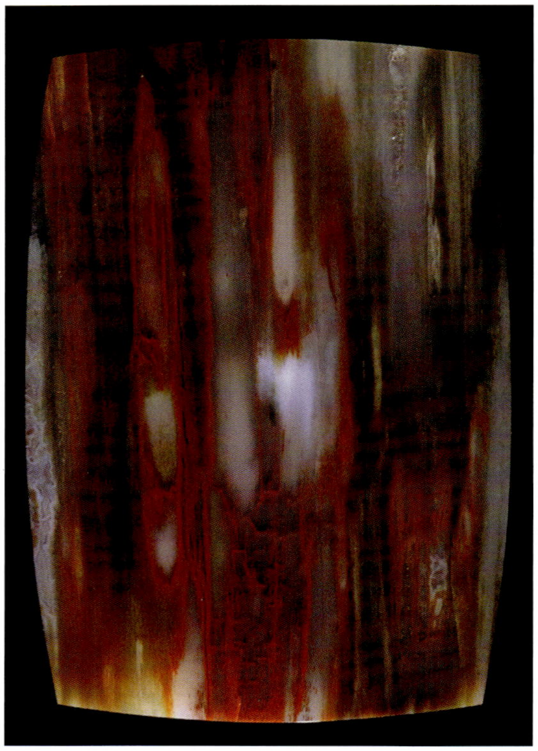

Araucarioxylon. Triassic. Navajo County, Arizona, USA. The Joe Taubr Collection.

Fossil conifer. Jurassic. Queensland, Australia. The Joe Taubr Collection.

Fossil wood. Tertiary. Arapahoe County, Colorado, USA. The Joe Taubr Collection.

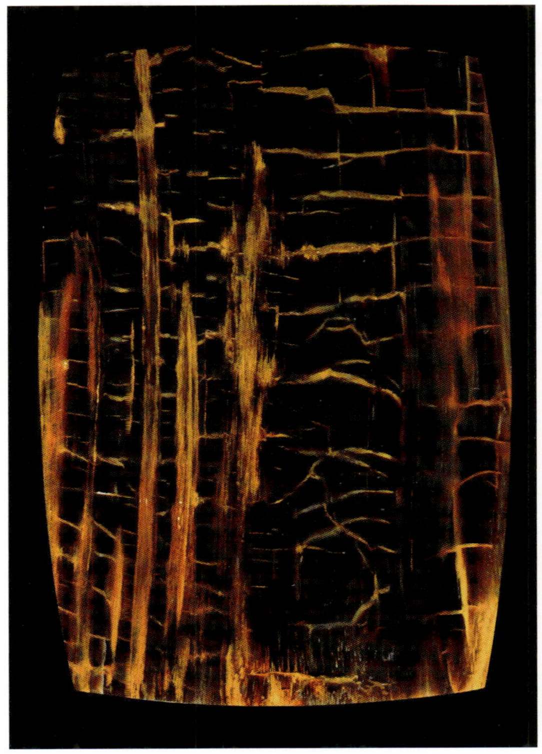

Sycamore-like fossil hardwood. Tertiary. Douglas County, Colorado, USA. The Joe Taubr Collection.

Sycamore-like fossil hardwood. Tertiary. Douglas County, Colorado, USA. The Joe Taubr Collection.

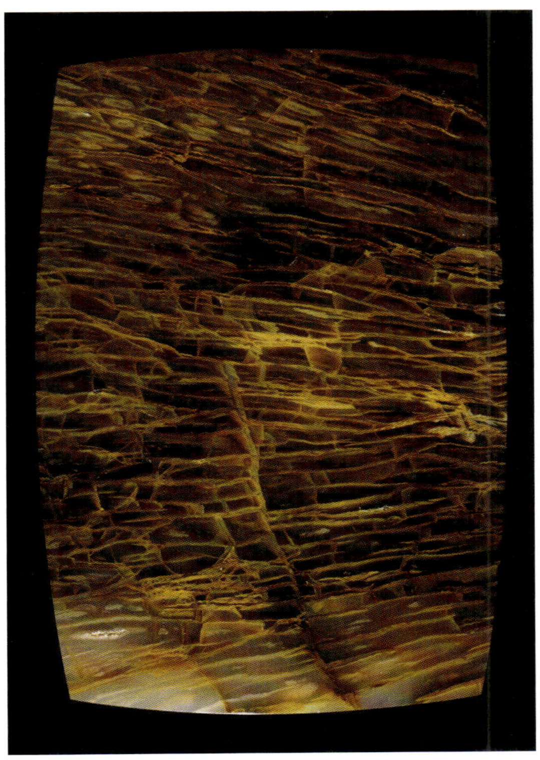

Fossil wood. Tertiary. Arapahoe County, Colorado, USA. The Joe Taubr Collection.

Sycamore-like fossil hardwood. Tertiary. Douglas County, Colorado, USA. The Joe Taubr Collection.

Sycamore-like fossil hardwood. Tertiary. Douglas County, Colorado, USA. The Joe Taubr Collection.

Fossil dinosaur bone. Jurassic. Colorado/ Utah, USA. The Joe Taubr Collection.

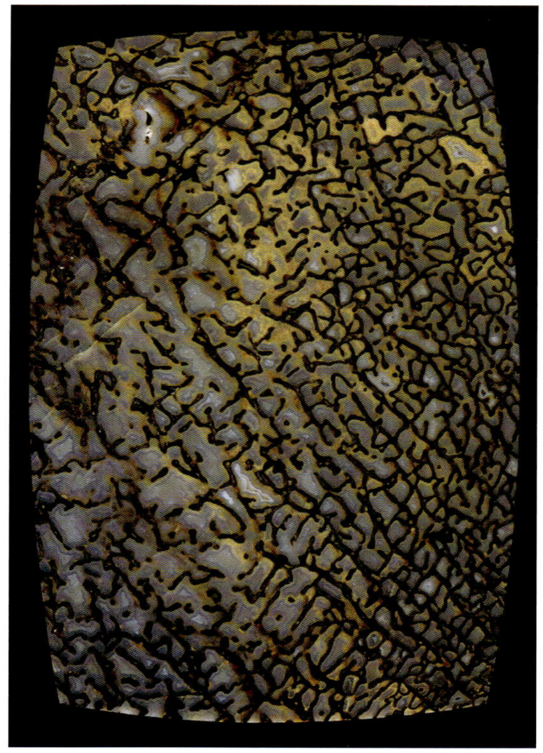
Fossil dinosaur bone. Jurassic. Colorado/ Utah, USA. The Joe Taubr Collection.

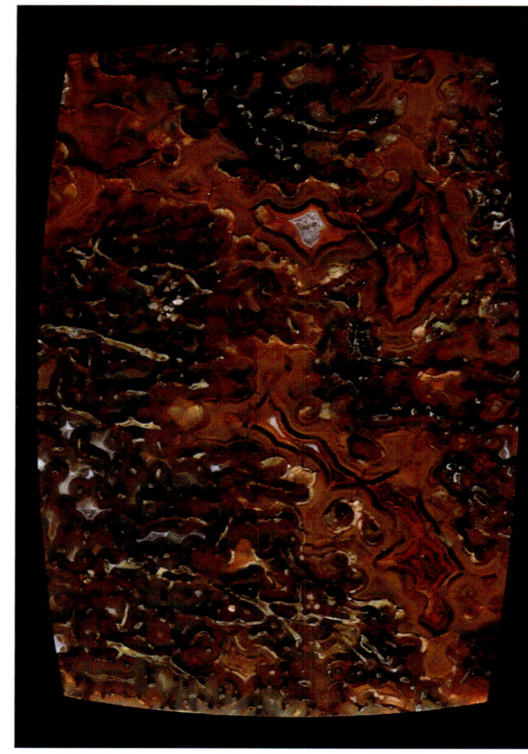
Fossil dinosaur bone. Jurassic. Colorado/ Utah, USA. The Joe Taubr Collection.

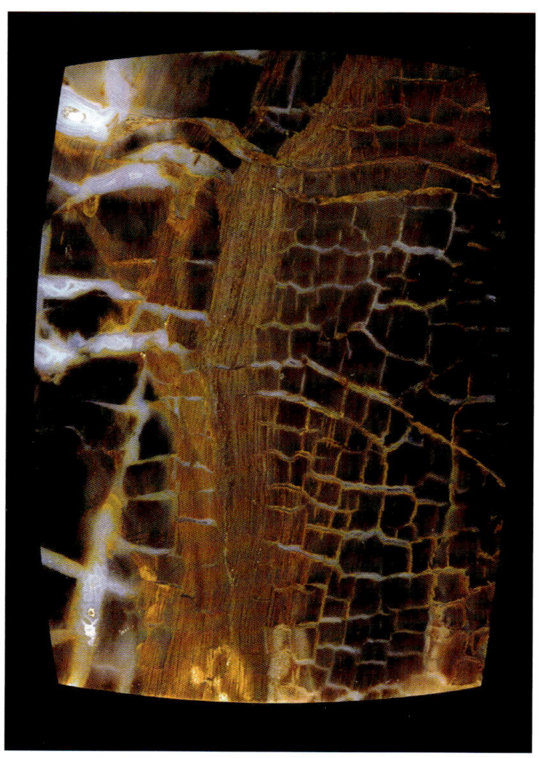
Fossil wood. Tertiary. Douglas County, Colorado, USA. The Joe Taubr Collection.

Fossil wood. Tertiary. Douglas County, Colorado, USA. The Joe Taubr Collection.

Fossil wood. Tertiary. Douglas County, Colorado, USA. The Joe Taubr Collection.

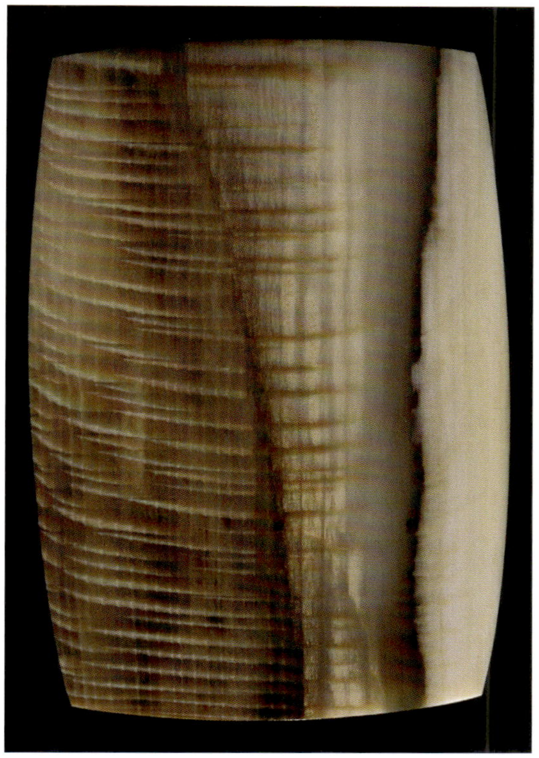

Fossil hardwood. Tertiary. Oregon, USA. The Joe Taubr Collection.

Fossil wood. Tertiary. Oregon, USA. The Joe Taubr Collection.

Fossil hardwood. Tertiary. Oregon, USA. The Joe Taubr Collection.

Palmoxylon. Oligocene. Louisiana, USA. The Joe Taubr Collection.

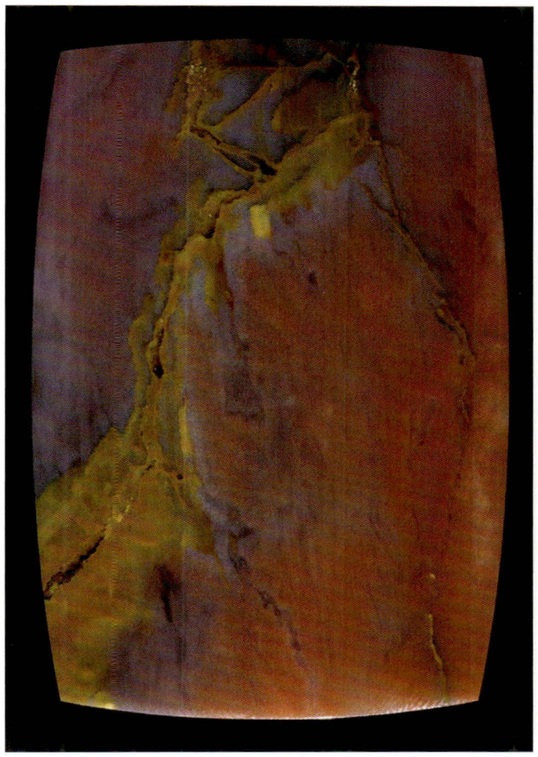

Agate. Nevada, USA. The Joe Taubr Collection.

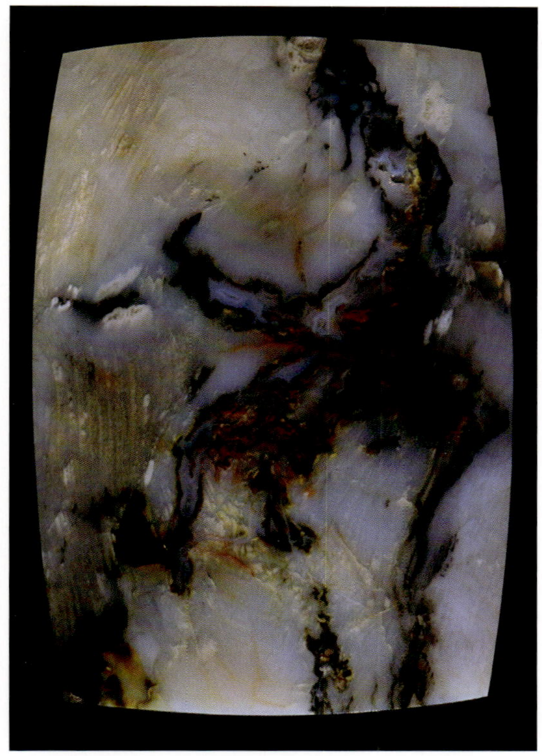

Taxodioxylon. Tertiary. Hubbard Basin, Nevada, USA. The Joe Taubr Collection.

Fossil conifer. Tertiary. Oregon, USA.
The Joe Taubr Collection.

Fossil wood. Tertiary. Oregon,
USA. The Joe Taubr Collection.

Fossil wood. Tertiary. Oregon, USA.
The Joe Taubr Collection.

Sycamore-like fossil hardwood. Tertiary.
Oregon, USA. The Joe Taubr Collection.

Fossil hardwood. Tertiary. Oregon, USA.
The Joe Taubr Collection.

Sycamore-like fossil hardwood. Tertiary.
Oregon, USA. The Joe Taubr Collection.

AGATES

As discussed in chapter two, agates are close cousins of fossil wood. The macro photograph and micro image essays in the introduction of this book depict the progression from well-preserved fossil wood structure to the absolute absence of wood structure and also show true agate structure. The following photographs show some of the textures and structures discussed regarding fossil wood development. Several show well-formed escape tubes.

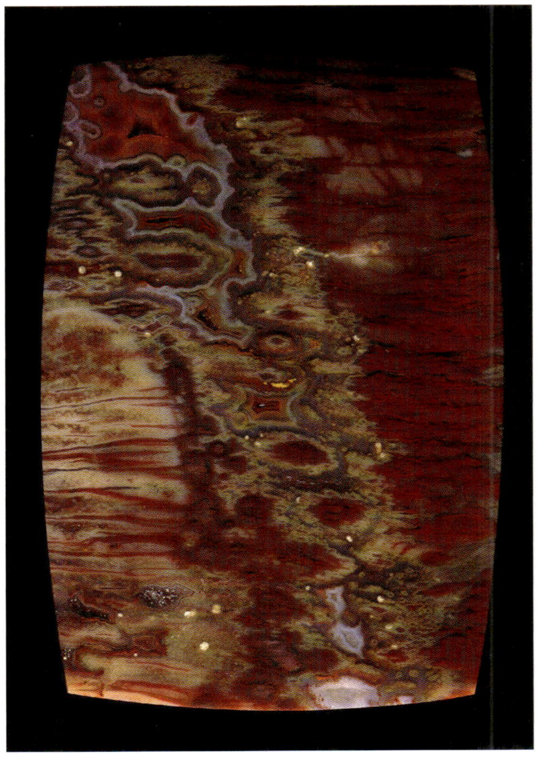

Fossil conifer. Triassic. Utah, USA.
The Joe Taubr Collection.

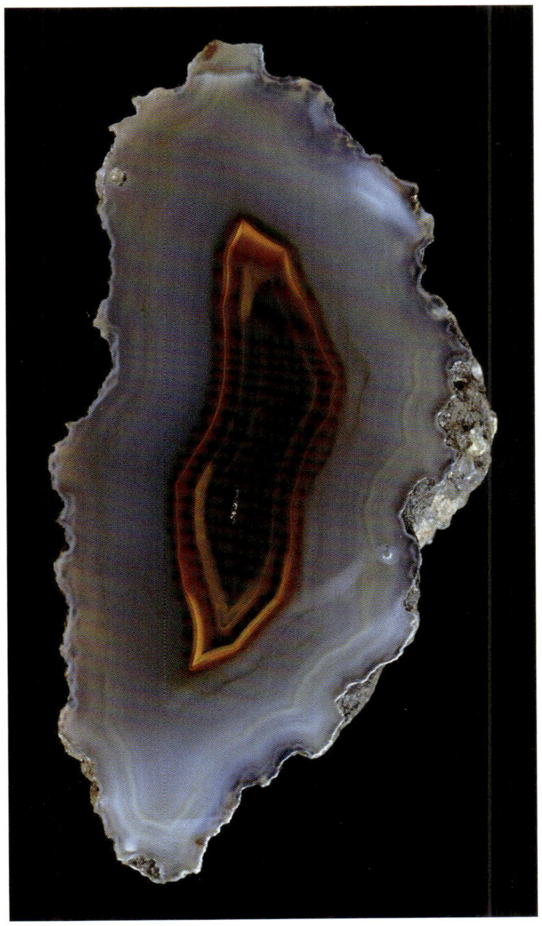

Condor agate. Patagonia, Argentina.
Daniels specimen, 9 cm.

Condor agate. Patagonia, Argentina.
Daniels specimen, 5 cm.

Condor agate. Patagonia, Argentina.
Daniels specimen, 5 cm.

Condor agate. Patagonia, Argentina.
Daniels specimen, 5 cm.

Condor agate. Patagonia, Argentina.
Daniels specimen, 6 cm.

Condor agate. Patagonia, Argentina.
Daniels specimen, 5.5 cm.

Condor agate. Patagonia, Argentina. Daniels specimen, 6 cm.

Condor agate. Patagonia, Argentina. Daniels specimen, 6.5 cm.

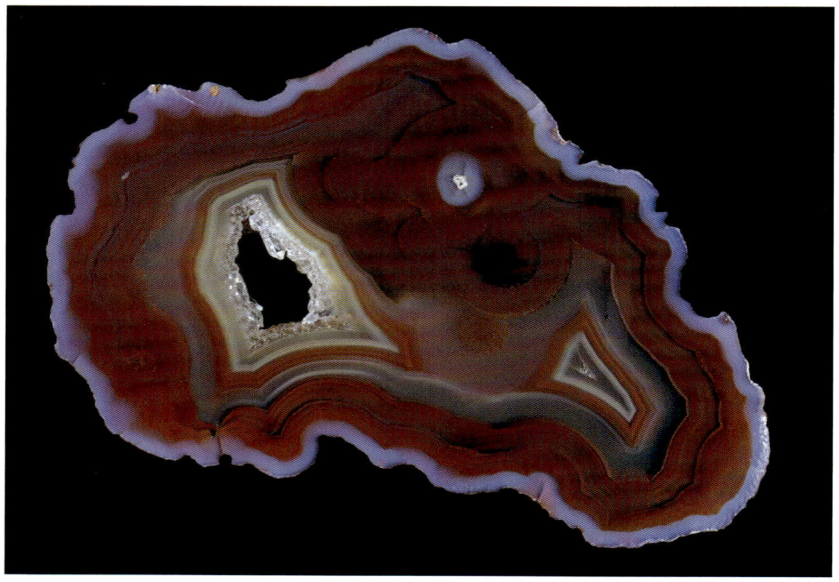

Condor agate. Patagonia, Argentina. Daniels specimen, 6 cm.

Condor agate. Patagonia, Argentina. Daniels specimen, 5 cm.

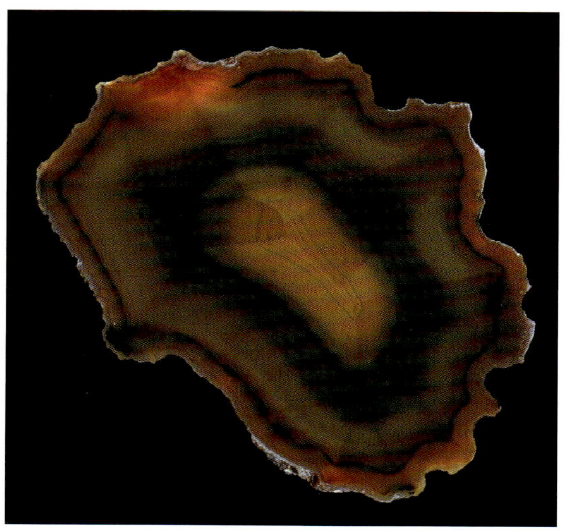

Condor agate. Patagonia, Argentina. Daniels specimen, 5.5 cm.

Condor agate. Patagonia, Argentina. Daniels specimen, 5.5 cm.

Condor agate. Patagonia, Argentina. Daniels specimen, 4.5 cm.

Condor agate. Patagonia, Argentina. Note the escape tube at the center right. Daniels specimen, 9 cm.

Laguna agate. Chihuahua, Mexico. Daniels specimen, 9.5 cm.

Brazilian agate. Brazil. Daniels specimen, 13.5 cm.

Laguna agate. Chihuahua, Mexico. Daniels specimen, 10 cm.

Condor agate. Patagonia, Argentina. Daniels specimen, 9 cm.

Condor agate. Patagonia, Argentina.
Daniels specimen, 7 cm.

Condor agate. Patagonia, Argentina. Daniels specimen, 7 cm.

Condor agate. Patagonia, Argentina.
Daniels specimen, 6.5 cm.

Condor agate. Patagonia, Argentina. Daniels specimen, 8.5 cm.

Condor agate. Patagonia, Argentina.
Daniels specimen, 8.5 cm.

Condor agate. Patagonia, Argentina. Daniels specimen, 6.5 cm.

Condor agate. Patagonia, Argentina. Daniels specimen, 8 cm.

Condor agate. Patagonia, Argentina. Daniels specimen, 7 cm.

Condor agate. Patagonia, Argentina. Daniels specimen, 12 cm.

Condor agate. Patagonia, Argentina.
Daniels specimen, 6.5 cm.

Ghost agate. Patagonia, Argentina.
Daniels specimen, 6.5 cm.

Ghost agate. Patagonia, Argentina.
Daniels specimen, 6 cm.

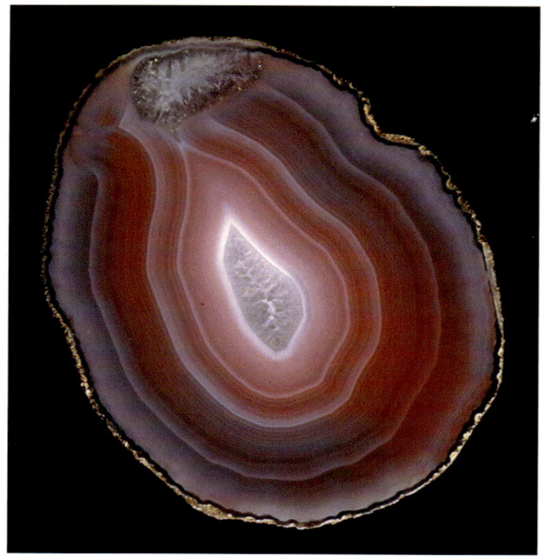

Queensland agate. Queensland, Australia.
Daniels specimen, 5 cm.

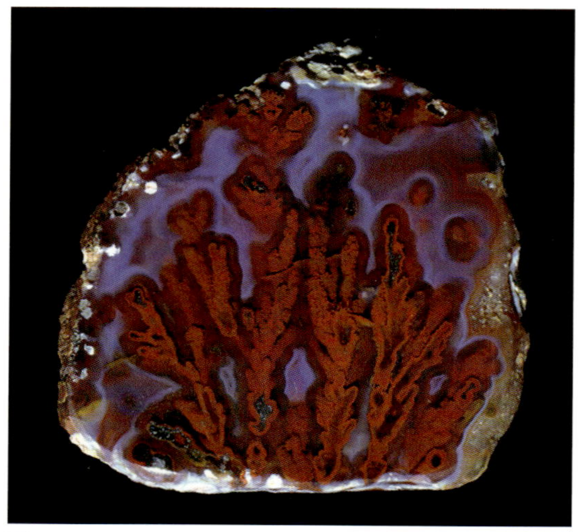

Plume agate. Davis Mountains, Texas,
USA. Daniels specimen, 5 cm.

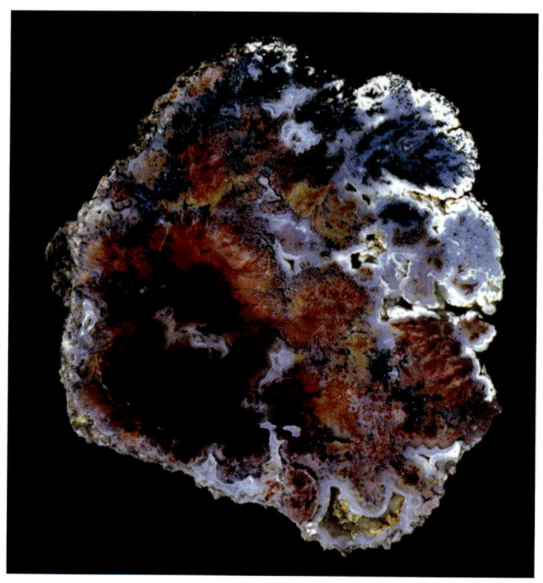

Plume agate. Davis Mountains, Texas,
USA. Daniels specimen, 8 cm.

Ghost agate. Patagonia, Argentina.
Daniels specimen, 6 cm.

Plume agate. Davis Mountains, Texas,
USA. Daniels specimen, 9 cm.

Dryhead agate. South-central Montana,
USA. Daniels specimen, 6 cm.

Moctezuma agate. Chihuahua, Mexico. Daniels specimen, 8.5 cm.

Laguna agate with shadow effect. Chihuahua, Mexico. Daniels specimen, 5.5 cm.

Laguna agate with shadow effect. Chihuahua, Mexico. Daniels specimen, 6 cm.

Laguna agate. Chihuahua, Mexico. Daniels specimen, 5.5 cm.

Coyomito agate. Chihuahua, Mexico. Daniels specimen, 6 cm.

Laguna agate with shadow effect. Chihuahua, Mexico. Daniels specimen, 6 cm.

Laguna agate with shadow effect. Chihuahua, Mexico. Daniels specimen, 8 cm.

Laguna agate with shadow effect. Chihuahua, Mexico. Daniels specimen, 7.5 cm.

Laguna agate with shadow effect. Chihuahua, Mexico. Daniels specimen, 6 cm.

APPENDIX A

HARDWOOD IDENTIFICATION AIDS

PORE ARRANGEMENT

Pore arrangement relative to one another in transverse section (in latewood for ring porous to semi ring porous; in entire growth ring for diffuse porous) is a feature used in hardwood identification.

- **Earlywood pores in one row**: American elm (usually in one row but can be in several, especially in juvenile wood) and inconsistently black cherry.
- **Predominately solitary pores**: beech, dogwood, live oak, maple, red oaks (latewood pores round, thick-walled, distinct with hand lens), sycamore, tanoak, white oaks (latewood pores, more numerous, thin-walled, angular, not sharp with hand lens); inconsistently in black ash, blackgum, California laurel (myrtle), magnolia, red alder, sassafras, white ash, and yellow poplar.
- **Pores in multiples**: aspen, basswood, beech, birch, black ash, black cherry, blackgum, black willow, butternut, California laurel (myrtle), cottonwood, dogwood, hickory, magnolia, maple, persimmon, red alder, sassafras, sweetgum, sycamore, white ash, yellow poplar; arranged in flame-shaped patches in American chestnut; inconsistently found in honey locust.
- **Pores in chains**: American holly and sweetgum.
- **Pores in wavy bands (Ulmiform)**: elm and hackberry; found inconsistently in black locust, catalpa, mulberry, Osage orange, and redbud.
- **Nested pores** (latewood pores in clusters): black cherry, black locust, catalpa, honey locust, Osage orange, red mulberry, and white oak; found inconsistently in red alder.

Note: *Earlywood pores of locust are large and filled with tyloses. Honey locust pores are large, typically solitary, yet sometimes banded in multiples of 2 to 5.*

PARENCHYMA GROUPINGS

Most hardwoods with longitudinal parenchyma contain more than one type of parenchyma cell arrangement. In red oaks, diffuse to banded apotracheal parenchyma occurs together with different types of paratracheal parenchyma. The wood of black locust and honey locust will contain paratracheal parenchyma, ranging from scanty to confluent, together with terminal parenchyma. Hickories typically have apotracheal and paratracheal banded parenchyma in addition to marginal parenchyma.

- **Nonexistent or rare**: black cherry.
- **Marginal**: aspen, basswood, black ash, black locust, black walnut, black willow, butternut, cottonwood, hard maple, hickory, honey locust, magnolia, persimmon, soft maple, white ash, yellow poplar (tuliptree); inconsistently present in birch and red alder; occurring as occasional cells in birches and maples; in essentially continuous bands in magnolia, poplars, and willows.
- **Paratracheal-scanty**: American chestnut, American holly, black ash, blackgum, black walnut, butternut, dogwood, hard maple, hickory, live oak, red alder, red oak, soft maple, sweetgum, sycamore, tanoak, white ash, and white oak; inconsistently present in black cherry and sassafras.

- **Paratracheal-vasicentric**: black ash, black locust, butternut, California laurel (myrtle), hickory, persimmon, sassafras, walnut, and white ash; abundant but not in complete sheaths around earlywood pores and subsumed in the wavy bands of latewood vessels and vascular tracheids in American elm, hackberry, and slippery elm.

- **Paratracheal-aliform**: honey locust, sassafras, and white ash.

- **Paratracheal-confluent**: honey locust and sassafras.

- **Apotracheal-diffuse**: American chestnut, American elm, American holly, beech, birch, blackgum, black walnut, butternut, hackberry, hard maple, honey locust, live oak, red alder, red oak, slippery elm, soft maple, sweetgum, sycamore, tanoak, and white oak; inconsistently present in black cherry and hickory.

- **Apotracheal-diffuse-in-aggregate**: basswood, beech, black walnut, birch, butternut, dogwood, hickory, live oak, persimmon, red oak, sycamore, tanoak, and white oak.

- **Banded**: basswood, beech, black walnut, butternut, dogwood, hickory, live oak, persimmon, red oak, tanoak, and white oak.

RAY GROUPINGS

Note: several species commonly have more than one type of ray and therefore appear in more than one grouping.

- **Uniseriate**: common in hard maple; sometimes in American chestnut, American holly (cells upright), aspen, basswood, black willow, cottonwood, dogwood (cells upright), live oak, maple, red alder, red oak, soft maple, sweetgum (cells upright), tanoak, and white oak. In soft maple, uniseriate rays are sometimes sparse and sometimes abundant.

- **Up to four seriate**: common in black ash, black cherry, black walnut, butternut, California laurel (myrtle), hickory, live oak, magnolia, persimmon, sassafras, soft maple (with exceptions), sweetgum, white ash, and yellow poplar (tulip tree).

- **Widest 5 to 10 seriate**: common in American elm, American holly, basswood, birch, black locust, dogwood, hackberry, hard maple, honey locust, and slippery elm; inconsistently present in black cherry, black walnut, catalpa, hickory, magnolia, soft maple, and yellow poplar (tuliptree).

- **Widest more than 10 seriate**: common in beech, hackberry, live oak, red oak, sycamore, tanoak, and white oak; inconsistently present in honey locust and slippery elm.

FACTORS THAT MAY AID FURTHER DISCRIMINATION BETWEEN DIFFUSE POROUS HARDWOODS

- Maple, dogwood, and holly are similar in structure. In maple, the pores are small and solitary or in radial multiples of two to several and uniform in size and distribution. Maple rays are distinct and, depending on the species, can range up to 8 seriate. Hard maple rays are two distinct sizes, whereas soft maple rays are the same size. The widest rays in maples are approximately as wide as the largest pores. Growth rings in maple display two or three rows of thick-walled fibers. Dogwood is similar to maple. Dogwood pores are evenly distributed and mostly solitary, sometimes in radial multiples of two or more. Dogwood rays are 2 to 8 seriate and about as wide as, or wider than, the largest pores. Holly is similar to maple, with distinct rays, small pores, some solitary but mostly in multiples arranged in a long radial set of pore chains. These pore chain sets often extend through growth ring boundaries. Holly rays range from 1 to 5 seriate, most being 5 seriate.

- In cherry, the first pores in the growth ring typically form a row along the earliest wood. These pores are not distinctly larger than the other pores throughout the growth ring. Pores are solitary and in multiples or small clusters. Rays range from 1 to 6 seriate but are most commonly 3 to 4 seriate. Cherry is similar to other fruitwoods such as apple and pear, but rays in apple and pear are normally biseriate, ranging from 1 to 3 seriate.

- Yellow poplar (tulip tree): parenchyma is marginal; rays are slightly wavy in heartwood and noded, 1 to 5 seriate, but mostly 2 to 3 seriate; pores are small, some solitary, most reside in irregular clusters or multiples.

- Magnolia is similar to yellow poplar (cannot be differentiated without microscopic analysis); rays are 1 to 5 seriate.

- Basswood has marginal parenchyma that is faint to distinct. Rays are small but visible to the unaided eye, noded, and are 1 to 6 seriate. Pores are small and quite evenly distributed and can be larger and more numerous at the early side of the growth ring; some pores are solitary, but most are in irregular multiples and clusters. The distinct marginal parenchyma and noded rays of basswood lend to confusion with yellow poplar.

- In birch, rays are difficult to see with the unaided eye but appear distinct with a hand lens. Pores are usually distinct and well separated, sometimes solitary but mainly in radial multiples; pore diameters are larger than the width of the rays, which is a way to distinguish birch from maple; rays are 1 to 5 seriate.

- California laurel (myrtle) has rays that are indistinct with a hand lens. Pores are small and solitary or in radial multiples of two to several, well spaced and evenly distributed. Some pore multiples extend across the growth-ring boundary. Each pore or multiple is surrounded by a layer of vasicentric parenchyma. Rays range from 1 to 4 seriate and are mostly biseriate.

- Most tropical woods are diffuse porous; an exception is teak, which is more often ring porous.

The following diffuse porous woods have rays that are barely visible or not visible to the unaided eye and barely visible with a hand lens.

- Willow is diffuse porous, although some examples show considerable diminution in pore size from earlywood to latewood. Earlywood pores are barely visible to the unaided eye and grade to invisible in outer latewood. Willow rays are fine and uniseriate. Willow is similar to poplar.

- Poplars — including aspen, cottonwood, and poplar — are all very similar. Pores are slightly larger in cottonwood, with the largest just visible to the unaided eye. Growth rings are usually distinct. Earlywood pores are larger, more crowded, and more numerous than latewood pores, and are close to a semi ring porous distribution. Some pores are solitary, but most are arranged in multiples—some radial, some tangential, some in small clusters. Rays are extremely fine, uniseriate, and barely visible even with a hand lens. Aspen and basswood are similar to each other, however rays in aspen are uniseriate and basswood rays are multiseriate.

- Sweetgum rays and growth rings are difficult to see even with a hand lens. The pores are extremely numerous and solitary, in multiples, in short chains, or in small irregular clusters. Rays are 2 to 3 seriate, but mostly biseriate.

APPENDIX B

TREE NAME INDEX

For selected family, common, and genus names used in this book.

FAMILY	COMMON	GENUS
Aceraceae	maple	*Acer*
Anacardiaceae	pepper tree	*Schinus*
Araceae (Plamae)	palm	numerous
Araucariaceae	araucaria	*Araucaria*
Aquifoliaceae	holly	*Ilex*
Betulaceae	alder	*Alnus*
	birch	*Betula*
Bombacaceae	ceiba	*Ceiba*
	balsa	*Ochroma*
Calamitaceae	*Calamites*	*Calamites*
Casaurinacaea	casaurina	*Casaurina*
Cornaceae	dogwood	*Cornus*
	blackgum	*Nyssa*
Cupressaceae	cedar	*Chamaecyparis*
	juniper	*Juniperus*
	red cedar	*Thuja*
	cypress	*Cupressus*
Ebenaceae	persimmon	*Diospyros*
Euphorbia	rubber tree	*Hevea*
Fabaceae (Leguminosae)	honey locust	*Gleditsia*
	black locust	*Robinia*
Fagaceae	beech	*Fagus*
	chestnut	*Castanea*
	tanoak	*Lithocarpus*
	oak, live oak	*Quercus*
Ginkgoaceae	ginkgo	*Ginkgo*
Hamamelidaceae	sweetgum	*Liquidambar*
Juglandaceae	pecan, hickory	*Carya*
	walnut, butternut	*Juglans*
Lauraceae	California laurel (Pacific myrtle)	*Umbellularia*
	sassafras	*Sassafras*
	stinkwood, imbuya	*Ocotea*
	camphor tree	*Cinnamomum*

FAMILY	COMMON	GENUS
Magnoliaceae	magnolia	*Magnolia*
	yellow poplar	*Liriodendron*
Moraceae	mulberry	*Morus*
	Osage orange	*Maclura*
Oleaceae	ash	*Fraxinus*
	olive	*Olea*
Osmundaceae	*Osmunda*	*Osmunda*
Pinaceae	fir	*Abies*
	larch	*Larix*
	spruce	*Picea*
	pine	*Pinus*
	Douglas fir	*Pseudotsuga*
	hemlock	*Tsuga*
Platanaceae	sycamore	*Platanus*
Podocarpaceae	various	*Podocarpus*
Protopinaceae	*Woodworthia*	*Woodworthia*
Rosaceae	mountain mahogany	*Cercocarpus*
	apple	*Malus*
	cherry	*Prunus*
	American mountain ash	*Sorbus*
Salicaceae	poplar, aspen, cottonwood	*Populus*
	willow	*Salix*
Simaroubaceae	ailanthus	*Ailanthus*
Taxaceae	yew	*Taxus*
Taxodiaceae	redwood	*Sequoia*
	giant sequoia	*Sequoiadendron*
	bald cypress	*Taxodium*
Tempskyaceae	*Tempskya*	*Tempskya*
Tilaceae	basswood	*Tilia*
Trochodendraceae	trochodendron	*Trochodendron*
Ulmaceae	hackberry	*Celtis*
	elm	*Ulmus*
Vitaceae	grape	*Vitus*

GLOSSARY

-aceae: plural suffix for names of orders or families of plants.

acicular: said of a crystal that is needlelike in form.

adventitious: said of a plant part that arises from an unusual place, such as a root that arises from a leaf or stem rather than from a primary root. Adventitious buds of some gymnosperms can be removed and rooted.

agate: a translucent cryptocrystalline variety of quartz, being a variegated chalcedony frequently mixed or alternating with opal and characterized by colors arranged in alternating strips or bands, in irregular clouds, or in moss-like forms.

alkaline: pertaining to substances that release hydroxyl (OH^-) ions in water; having a pH value greater than 7.

amethyst: a lavender variety of quartz.

ammonite: an extinct shelled mollusk, relative of modern chambered nautilus and squid.

anaerobic environment: a highly organic sediment characteristic of some basins where restricted circulation of the water results in the absence of oxygen at the sediment surface and bottom water is often rich in hydrogen sulfide.

andesite: an extrusive igneous rock that is usually tan to gray and of intermediate chemical composition between silica-rich rhyolite and silica-poor basalt.

angiosperm: flowering plant with seeds that form within an ovary.

anticline: a fold, generally convex upward, whose core contains stratigraphically older rocks.

apex: the tip of a pinnule, leaf, or trunk where new growth is occurring.

arcuate: said of a fold whose axial trace is curved or bent.

ash: particles of pulverized magma less than 2 mm in diameter.

authogenic: formed or generated in place such as a mineral that crystallized in the place where it is now found.

basalt: dark gray to black volcanic rock rich in iron and magnesium and poor in silica.

bentonite: a soft, plastic, porous, light-colored rock composed essentially of clay minerals of the smectite (montmorillonite) group that has characteristic swelling properties upon wetting.

biogenic: pertaining to or produced by life. Biogenic rock: a rock produced directly by the physiological activities of organisms, either plant or animal.

bleb: a small usually rounded inclusion.

bordered pit: a pit in which the margin projects over a chamber separated from the thin pit-closing membrane, as in tracheids of coniferous wood.

botryodial: having the form of a bunch of grapes; also said of a crystalline aggregrate in which the spherical shapes are composed of radiating crystals.

breccia: rock consisting of coarse, angular, broken rock fragments imbedded in a finer material such as volcanic ash.

broadleaf: hardwood tree.

calcrete: a conglomerate of surficial gravel and sand cemented by calcium carbonate.

carnelian: a red, orange, or brownish-red form of chalcedony.

cellulose: a polymerized carbohydrate composed of glucose units, of which the permanent cell membranes of plants are formed.

chalcedony: a cryptocrystalline variety of silicon dioxide or quartz with a composition $SiO_2 \cdot nH_2O$. It is usually microscopically fibrous, translucent to semi-transparent, and has a nearly waxy luster.

clastic: pertaining to a rock or sediment composed principally of broken fragments that are derived from preexisting rocks or minerals and that have been transported some distance from their places of origin; also said of the texture of such a rock.

clay: very fine grain sediment with particle sizes of less than 4 microns (1/256 inch) in diameter.

claystone: a rock with more than 67% of clay-sized minerals or fragments but lacking a fine lamination.

compression fossils: these flat fossils form in freshwater and marine environments and any environment where a sedimentary layer buries plant parts. As the layers build, water expresses from the plant; as the layer becomes more compact, the plant material flattens.

compression wood: see reaction wood.

conchoidal: smoothly curving fracture surface of certain minerals and rock in three-dimensional form, either concave or convex, and is in contrast to planar and angular cleavage of most samples.

contact: the boundary between two geologic rock units.

cortex: tissue external to the phloem including bark.

cotyledon: the earliest leaf arising from a seed.

craze: minute cracks or patterns of cracks in a stone (especially opal) when it is being ground or polished.

cristobalite: a mineral of composition SiO_2, like quartz, but of a different crystal form.

deciduous: plants that shed their leaves annually.

devitrification: conversion of glass to crystalline material.

diagenesis: all the chemical, physical, and biologic changes undergone by a sediment after its initial deposition and during and after its lithification, exclusive of surficial alteration and metamorphism.

diapir: a dome or anticlinal fold in which the overlying rocks have been ruptured by the squeezing out of plastic core material (often salt).

dicotyledon: an angiosperm in which the seed germinates to two leaves and the leaves have complex, net-like venation; abbreviated as dicot.

discontinuous rings: these result where the cambium remains dormant at one or more places along its circumference during a growth season while functioning normally in other places. The appearance is as though a new growth layer runs into the late wood of the earlier growth ring.

double-hearted, triple-hearted, and so on: a wood specimen revealing a multiple set of annual rings, representing a transverse section of a dividing trunk or branch, as in the fork of a branch.

double ring: an annual ring containing one false ring.

false rings: in some trees there appear to be more annual rings than truly exist. This is due to false rings, where a band of ostensible latewood preceded what appears to be early-wood. The cells composing the false late-wood grade to the inside and to the outside into more porous tissue. In true rings, the transition from the latewood of one year to the earlywood of the following year is abrupt.

fold: a curve or bend of a planar structure such as rock strata, bedding planes, foliation, or cleavage.

formation: an igneous, sedimentary, or metamorphic rock unit or series of rock units sufficiently homogeneous or distinctive to be mapped at a scale of 1:25,000.

fossil: remains or traces of plant or animal in rock.

fructification: a reproductive organ or fruiting structure of a plant.

full round: fossil wood specimen that is cut across the transverse aspect [cross section] and encompasses a near complete set of annual rings, or if a cast, has the near full circumference of the original branch or bole. The term also applies to an uncut fossil wood specimen that appears to be a fossil of the near complete branch or bole.

genus: a classification of plants or animals with common distinguishing characteristics. The name is always capitalized and italicized (e.g., *Homo sapiens*).

grating: in optics, equidistant and parallel lines or textures that are used in producing spectra by diffraction.

gymnosperm: a seed plant with seeds borne naked in the cone.

hardwood: deciduous angiospermous wood.

helical: a regular spiral such as the curve of a screw.

hemicellulose: a polysaccharide component of plant tissue composed of linked hexose and pentose sugars.

hogback: a shape-crested ridge formed by the outcropping of steeply inclined resistant rocks, and produced by differential erosion. The feature resembles in outline the back of a hog.

hydrate: to cause the incorporation of water into the chemical composition of a material.

hydrolysis: decomposition in which a compound is split into other compounds by taking up elements of water.

hydroxyl ion: an atom of oxygen and hydrogen that combine and produce a negative charge, shown as OH^-.

integument: a natural covering such as skin, shell, rind; any covering, coating, or enclosure.

jasper: an opaque, often red or otherwise colorful, cryptocrystalline quartz that is colored by iron oxide.

laccolith: a dome formed by forceful injection of magma between sedimentary layers.

Laramide Orogeny: mountain building of Late Cretaceous and Early Tertiary periods that created the Rocky Mountains.

lava: molten magma that reached the Earth's surface and the rock formed when it cooled.

lava flow: outpouring of molten lava and the rock mass formed thereby.

leaf scar: a scar left on a twig where the leaf had been attached.

lepisphere: micrometer-sized spheroidal body, usually composed of a silica mineral, with radiating crystal orientation and scaly crystal terminations of the outer surface.

lignin: a complex amorphous polysaccharide containing methoxyl and phenyl propane units always occurring with cellulose and forming an important constituent of wood.

lignotuber: plant structures that arise in the axils of leaves, possibly in response to plant trauma, that can fall off into the soil and regenerate a new plant.

longitudinal section: a section formed by cutting parallel to the rays or along the radius of the stem.

lumen: the cavity that the cell walls enclose (plural lumina).

magma: naturally occurring molten or partially molten rock, generated within the Earth.

mantle: the thick, partially molten zone between the Earth's core and crust.

matrix: the natural rock or earthy matter in which a fossil is embedded.

meristem: formative tissue consisting of cells actively growing and dividing, as at the apex of young stems.

micrite: microcrystalline limestone with grain size less than 4 microns; the crystalline matrix of limestone usually formed from carbonate mud.

microfibril: one of the submicroscopic elongated bundles of cellulose of the plant cell wall.

micron: short for micrometer; a length of one-millionth of a meter, one-thousandth of a millimeter, abbreviated as mm.

monocotyledon: an angiosperm in which the seed germinates to one leaf (as in grasses, lilies, and palms) and the leaves have parallel veins; abbreviated as monocot.

mudstone: a poorly bedded sedimentary rock formed from mud and composed of particles smaller than silt.

multiple ring: an annual ring containing two or more false rings.

mybp: millions of years before present.

nanometer: a length one-billionth of a meter.

oblate: flattened or depressed at the poles.

opal: $SiO_2 \bullet n\ H_2O$; a glassy hydrated form of silicon dioxide, softer than quartz, less tough than chalcedony.

opal-A: amorphous opal with a gel-like structure Most precious opal and common (or potch opal) are this variety.

opal-CT: opal that contains an internal structure similar to cristobalite and tridymite. It represents an opal variety that is beginning to crystallize and may have a platy habit under high magnification.

outcrop: bedrock that appears at the Earth's surface.

oxidizing environment: a depositional environment characterized by having abundant oxygen.

paleoecology: the study of life in past geologic time, based on fossil plants and animals, environments, and the chronology of the Earth.

Pangea: a supercontinent that existed from about 300 to about 200 million years ago and included most of the continental crust of the Earth from which the present continents were derived.

periderm: the outer bark and the layer of growing tissue between the bark and the wood.

permineralization: the process whereby soluble minerals enter cells or pore spaces and also fill the spaces between cells or pore spaces, resulting in a hardened mineral that contains and supports original tissues.

petrifaction: the general term for a process of fossilization whereby organic matter is converted into a stony substance by the infiltration of water containing dissolved inorganic matter that replaces the original organic materials, sometimes retaining the structure.

petrify: to convert to stone or a stony substance.

petrographic microscope: a microscope that can use polarized light on a revolving stage for analyses of thin slices of rock (about 30 microns thick) mounted on glass slides.

phloem: food-conducting tissues of vascular plants, consisting of various types of cells.

-phyte: a combining form suffix meaning a plant growing in a specified way.

pinnate: having simple leaflets arranged on both sides of a common stem in a featherlike arrangement.

pit chamber: the space in a plant cell wall between the a bordered pit and the open area.

pith: the spongy central tissue of a stem, consisting chiefly of parenchyma.

polymorphism: the characteristic of a chemical substance to crystallize in more than one form (e.g., hexagonal and isometric for SiO_2).

polysaccharide: a carbohydrate such as cellulose containing more than three monosaccharide units per molecule and being capable of hydrolysis by acids or enzymes.

potch opal: common opal, translucent to subtransparent exhibiting a waxy luster.

precious opal: opal that exhibits flashes of color.

prolate: extended or elongated in the direction of a line joining the poles.

quartz: SiO_2 (silicon dioxide), a hexagonal form of silica.

ramentum: epidermal hairs or scales and leaf bases covering the outside of a stem, typically in ferns and cycadophytes (ramenta is plural).

reaction wood: Not all growth rings are evenly concentric. In some specimens, the pith is considerably to one side, and the rings are uneven, often wider on one side than on the other. One explanation for this is **reaction wood**. Reaction wood can complicate wood identification. In conifers, reaction wood forms toward the underside of a leaning stem and is known as **compression wood**. The pith is offset to the upper side of the stem, and the growth rings in the compression wood are wider than normal. The latewood tends to be wider and denser. In hardwoods, the reaction wood forms on the upper side of the leaning stem. The reaction wood in hardwoods is under stress, or tension, and is called **tension wood**. Tension wood does not normally complicate identification.

redbed: red sedimentary layer, usually of siltstone or mudstone.

red wood: colloquialism for red agate limb casts from eastern Utah.

refractive index: the ratio of the velocity of light in a vacuum compared to the velocity in a crystal.

round: term for a specimen of fossil wood in transverse section that includes the near complete branch or trunk circumference; **full round** will be very nearly 360 degrees complete. A **specimen round** is a section of fossil wood that is cut and normally polished and includes the full circumference of the original fossil wood specimen as found but does not include a complete transverse section of the original bole or branch.

sand: rock fragments between 0.74 mm and 4.76 mm in diameter.

sandstone: sedimentary rock formed predominately from sand with particles larger than silt.

scanning electron microscope (SEM): a microscope in which a finely focused beam of electrons is moved rapidly across a specimen to be examined and the reflected and emitted electron intensity is measured and displayed to form an image.

sedimentary rock: rock composed of particles of other rock transported and deposited by water, wind, or ice.

shadow effect: the appearance of movement and perception of depth in a polished agate or fossil wood specimen when rocked slightly from side to side. Caused in agate and agate-replaced fossil wood specimens by tightly packed alternating translucent and opaque banding. Seen rarely in fossil wood with well-preserved cell structure infilled with translucent chalcedony.

shipworm wood: fossil wood that contains silicified bore tunnels of a bivalve mollusk such as *Teredo* or *Bankia*, commonly known as shipworms.

short shoot: short branches usually bearing leaves or sometimes fruiting structures; also called spur shoots.

silcrete: a conglomerate consisting of surficial sand and gravel cemented into a hard mass by silica.

silica: silicon dioxide (SiO_2); common at the Earth's surface as crystalline, microcrystalline, cyptocrystalline, and impure forms, frequently with water or hydroxyl included or attached.

siltstone: sedimentary rock formed from silt, with particles smaller than sand grains and larger than clay particles.

specific gravity: the mass (weight on Earth) of a substance compared to an equal mass of water.

specimen round: fossil wood specimen that has been cut and polished, commonly across the transverse aspect, and has a perimeter that is the natural rock as found; the polished face reveals the color and detail, although the rock as found was not a complete trunk or branch, sometimes having broken and tumbled millennia ago.

stratigraphy: the study of time as revealed in the succession of rocks.

syncline: a fold in which the core contains the stratigraphically younger rocks; it is generally concave upward.

tangential section: a cut made a right angle to the ray of a limb.

taxonomy: the branch of biology concerned with the naming and classifying of plants and animals.

template: in the case of plant silicification, the detailed replication of plant physiology by polymerized silica.

Tertiary Period: the geologic period that includes the Pliocene through Paleocene epochs in the United State Geological Survey time scale.

tracheid: an elongated, tapering plant cell with secondary walls and perforations or pits on its surface; used for support and water conduction.

transverse section: across the long axis of a limb; a cross section.

tridymite: a vitreous colorless mineral with a composition of SiO_2, usually occurring as minute tabular white crystals or scales.

volcaniclastic: a general term for all clastic volcanic materials formed by any process of fragmentation, dispersed by any kind of transporting agent, deposited in any environment, or mixed in any significant portion with non-volcanic fragments.

xylem: in vascular plants, a complex tissue that is involved with water conduction, food storage, and strengthening. Types of cells commonly found in xylem include tracheids for conduction and strengthening, vessels for conduction, fibers for strengthening, and parenchyma for storage.

BIBLIOGRAPHY

Akahane, H., Furuno, T., Miyajima, H., Yoshikawa, T., and Yamamoto, S. (2004) Rapid wood silicification in hot spring water: an explanation of silicification during earth's history. *Sedimentary Geology*, vol. 169, pp. 219-228.

Allen, E. T., and Day, A. L. (1935) *Hot Springs of the Yellowstone National Park, Publication No. 446*. Washington, D.C.: Carnegie Institute of Washington Publication.

Anderson, O. J., and S. G. Lucas. (1997) The Upper Jurassic Morrison Formation in the Four Corners region: Mesozoic geology and paleontology of the Four Corners region. In *New Mexico Geological Society Forty-eighth Annual Field Conference*, New Mexico Geological Society, pp. 139-155.

Arnold, C. A., (1953) Origin and relationships of the cycads. *Phytomorphology*, vol. 3, pp. 51-65.

Ash, S. R., and Creber, G. T. (2000) The Late Triassic *Araucarioxylon arizonicum* trees of the Petrified Forest National Park, Arizona, USA. *Paleontology*, vol. 43, part 1, pp. 15-28. The Paleontological Association.

Aubrey, W. M. (1996) Stratigraphic architecture and deformational history of Early Cretaceous Foreland Basin, eastern Utah and southwestern Colorado, pp. 211-220. In *Geology and Resources of the Paradox Basin: Utah Geological Association Guidebook 25*.

Baker, A. A., Dane, C. H., and Reeside, J. B., Jr. (1936) *Correlation of Jurassic Formations of Parts of Utah, Arizona, New Mexico, and Colorado*. U.S. Geological Survey Professional Paper 183.

Bass, C. (1964) Significant new plant locality in Utah. *Contributions to Geology*. Laramie, Wyoming: University of Wyoming, vol. 3, no. 2, pp. 94-95.

Beck, G. F. (1943) *Squaw Creek Mascall Type Petrified Forest of Central Washington*. Ellensburg, Washington: Science Department of Central Washington College.

Bestland, E. A., Hammond, P. E., Blackwell, D. L. S., Retallack, G. J., and Stimac, J. (1999) Geologic framework of the Clarno Unit, John Day Fossil Beds National Monument, central Oregon. *Oregon Geology*, vol. 61, no. 1. Portland, Oregon: Oregon Department of Geology and Mineral Industries.

Brookins, D. G. (1980) *Geochronologic Studies in the Grants Mineral Belt*. New Mexico Bureau of Mineral and Mineral Resources Memoir 38.

Buckland, W. C. (1828) On the cycadeoidea, a family of fossil plants found in the oolite quarries of Isle of Portland. *Transactions of the Geological Society of London*, 2nd series, vol. II, pp. 395-401.

Buurman, P. (1972) Mineralization of fossil wood. Leiden, Deutschland: *Scripta Geologica, vol. 12, Rijksmuseum van Geologie en Mineralogie*.

Calder, M. G. (1953) A coniferous petrified forest in Patagonia. *Bulletin of the British Museum of Natural History Geology*, vol. 2, pp. 99-138.

Cannon, H. L. (1964) *Geochemistry of Rocks and Related Soils and Vegetation in the Yellow Cat Area Grand County, Utah*: U.S. Geological Survey Bulletin 1176.

Carlson, M. R. (2002) *The Beauty of Banded Agates*. Edina, Minnesota: Fortification Press.

Carruthers, W. (1870) On fossil cycadean stems from the secondary rocks of Britain. *Transactions of the Linnean Society*, vol. 26, pp. 675-708.

Castor, S. B., and Ferdock, G. C. (2004) *Minerals of Nevada, Nevada Bureau of Mines and Geology, Special Publication 31*. Reno, Nevada: University of Nevada Press.

Chandler, M. E. J. (1966) Fruiting organs from the Morrison Formation of Utah, U.S.A. *Bulletin of the British Museum of Natural History Geology*, vol. 12, no. 4, pp. 10-151.

Chaney, R. W. (1948) *The Ancient Forests of Oregon*. Condon Lecture Series. Eugene, Oregon: Oregon State System of Higher Education.

Chronic, H. (1988) *Pages of Stone: Grand Canyon and the Plateau Country*. Seattle, Washington: The Mountaineers.

Cleal, C. J., and Thomas, B. A. (1999) *Plant Fossils: The History of Land Vegetation*. Woodbridge, Suffolk, England: The Boydell Press.

Core, H. A., Cote, W. A., and Day, A. C. (1976) *Wood Structure and Identification*. Syracuse, New York: Syracuse University Press.

Cornell, R. M., and Schwertmann, U. (1996) *The Iron Oxides,* New York: VCH Publishers.

Craig, L. C. (1955) *Stratigraphy of the Morrison Formation and Related Formations, Colorado Plateau Region, a Preliminary Report*. U.S. Geological Survey Bulletin 1009-E, pp. 125-168.

Crepet, W. L. (1974) Investigations of North American cycadeoids: the reproductive biology of Cycadeoidea. *Palaeontolographica,* vol.148, pp. 144-169.

Daniels, F. J. (1998) *Petrified Wood: The World of Fossilized Wood, Cones, Ferns, and Cycads*. Grand Junction, Colorado: Western Colorado Publishing Company.

Dare, W. L. (1961) *Uranium Mining Methods and Costs at Several Mines on the Colorado Plateau*. U.S. Bureau of Mines, Information Circular 8015.

Darragh, P. J., Gaskin, A. J., and Sanders, J. (1976) Opals. *Scientific American*, vol. 24, pp. 84-95.

Dayvault, R. D. (2000) "Red Wood" from the Yellow Cat area of Grand County, Utah. *Rocks & Minerals*, vol. 75, no. 4, pp. 258-267.

Dayvault, R. D., and Hatch, H. S. (2003) Short shoots from the Late Jurassic Morrison Formation of southeastern Utah. *Rocks & Minerals*, vol. 78, no. 4, pp. 232-247.

Dayvault, R. D., and Hatch, H. S. (2005) Cycads from the Upper Jurassic and Lower Cretaceous rocks of southeastern Utah. *Rocks & Minerals*, vol. 80, no. 6, pp. 412-432.

Delevoryas, T. (1962) *Morphology and Evolution of Fossil Plants*. New York: Holt, Rinehart, and Winston.

Demko, T. M. (1995) Taphonomy of fossil plants in Petrified Forest National Park, Arizona. *Fossils of Arizona*, vol. 3. Southwest Paleontological Society and Mesa Southwest Museum.

Dibble, H. L. (2002) *Quartz: An Introduction to Crystalline Quartz*. Angola, New York: Dibble Trust Fund.

Doelling, H. H. (2000) Geology of Arches National Park, Grand County, Utah, pp. 11-36 in *Geology of Utah's parks and monuments, 2000 Utah Geological Association Publication 28*. Salt Lake City, Utah: Publishers Press.

Dorf, E. (1964) The petrified forests of Yellowstone Park. *Scientific American*, vol. 210, pp. 107-114.

Drum, R. W. (1968) Petrification of plant tissue in the laboratory. *Nature*, vol. 218, pp. 784-785.

Eubanks, W. (1960) Fossil woods of Oregon. *The Ore Bin*. Portland, Oregon: State of Oregon Department of Geology and Mineral Industries.

Eubanks, W. (1966) Fossil woods supplement knowledge of the Succor Creek fossil flora. *The Ore Bin*. Portland, Oregon: State of Oregon Department of Geology and Mineral Industries.

extraLapis No. 7 (1994) *Versteinertes Holz*. Munich, Germany: Christian Weise Verlag.

Fergusen, L. (1988) *The Fossil Cliffs of Joggins*. Halifax, Nova Scotia, Canada: Nova Scotia Museum.

Flynn, J. H., Jr., and Holder, C. D., eds. (2001) *A Guide to Useful Woods of the World*. Madison, Wisconsin: Forest Products Society.

Folk, R. L. (2002) *The potential role of nannobacteria (dwarf forms, 30-100 nm) in the precipitation of chert and its precursors*. Geological Society of America Paper No. 243-2.

Folk, R. L. (1997) *Nanobacteria: surely not figments, but what under heaven are they?* naturalSCIENCE.com, vol. 1, article 3.

Frondel, C. (1978) Characters of quartz fibers. *American Mineralogist*, vol. 63, pp. 17-27.

Gains, R. V., Skinner, H. C. W., Foord, E. E., Mason, B., and Roxenzweig, A. (1997) *Dana's New Mineralogy* (8th edition). New York: John Wiley and Sons, Inc.

Gilbert, G. K. (1877) *Report of the Geology of the Henry Mountains*. U.S. Geological and Geographical Survey, Rocky Mountains region. pp.1-160, Washington, D.C.

Graham, A. (1999) *Late Cretaceous and Cenozoic History of North American Vegetation, North of Mexico*. New York/Oxford: Oxford University Press.

Gregory, I. (1968) The fossil woods near Holley in the Sweet Home Petrified Forest, Linn County, Oregon. *The Ore Bin*. Portland, Oregon: State of Oregon Department of Geology and Mineral Industries.

Gregory, I. (1969) Fossilized palm wood in Oregon. *The Ore Bin*. Portland, Oregon: State of Oregon Department of Geology and Mineral Industries.

Hergert, H. L. (1961) Plant fossils in the Clarno Formation, Oregon. *The Ore Bin*, vol. 23, pp. 55-62. Portland, Oregon: State of Oregon Department of Geology and Mineral Industries.

Hintze, L. F. (1988) *Geologic History of Utah*. Provo, Utah: Brigham Young University.

Hoadley, R. B. (1990) *Identifying Wood: Accurate Results with Simple Tools*. Newton, Connecticut: The Taunton Press.

Hunt, C. A. (1953) *Geology and geography of the Henry Mountains region, Utah*. U.S. Geological Survey professional paper 228.

Iler, R. K. (1979) *The Chemistry of Silica*. New York: John Wiley & Sons.

Jackson, J. A., ed. (1997) *Glossary of Geology* (4th ed.). Alexandria, Virginia: American Geological Institute.

Jefferson, T. H. (1987) The preservation of conifer wood: examples from the Lower Cretaceous of Antarctica. *Paleontology*, vol. 30, part 2, pp. 233-249.

Jones, D. L. (2002) *Cycads of the World*. Washington, D. C.: Smithsonian Institution Press.

Jones, J. B., and Segnit, E. R. (1971) The nature of opal. *Journal of the Geological Society of Australia*, vol. 18, pp. 57-68.

Keith, H. D., and Paddon, J. F. (1963) A phenomenological theory of spherulite crystallization. *Journal of Applied Physics*, vol. 38, no. 8, pp. 2409-2421.

Kemner, K., and Boyanov, B. (2005) Advanced Photon Source Laboratory unpublished data, Argonne National Laboratory.

Kenrick, P. and Davis, P. (2004) *Fossil Plants*. Washington, D. C: Smithsonian Books.

Kirkland, J. I., Cifelli, R. L., Britt, B. B., Burge, D. L., DeCourten, F. L., Eaton, J. G., and Parrish, J. M. (1999) Distribution of vertebrate fauna in the Cedar Mountain Formation, East-Central, Utah, pp. 201-217. In *Vertebrate paleontology in Utah,* D. D. Gillette, ed., Miscellaneous Publication 99-1. Utah Geological Survey.

Kirkland, J. I., Lucas, S. G., and Estep, J. W. (1998) Cretaceous dinosaurs of the Colorado Plateau, pp. 79-89. In *Lower and Middle Cretaceous Terrestrial Ecosystems*. Lucas, S. G., Kirkland, J. I., and Estep, J. W,. eds. New Mexico Museum of Natural History and Science Bulletin No. 14.

Knowlton, F. H. (1928) *The Fossil Forests of the Yellowstone National Park*. United States Geological Survey.

Kowallis, B. J., and Heaton, J. S. (1987) Fission-track dating of bentonites and bentonitic mudstones from the Morrison Formation in central Utah. *Geology* vol.15, pp. 1138-1142.

Kuczumow, A., Vekemans, B., Schalm, O., Dorrine, W., Chevallier, P., Dillmann, P., Ro, Chul-Un, Janssens, K., and van Greiken, R. (1998) Analyses of petrified wood by electron, X-ray, and optical microprobes. *Journal of Analytical Atomic Spectrometry*, vol. 14, pp. 435-446.

Landmesser, M. (1984) Das Problem der Achatgenese, Mitteilungen *POLLICHIA*, vol. 72, pp. 5-137.

Landmesser, M. (1988) Structural characteristics of agates and their genetic significance. *Neues Jachbuch fur Mineralogie, Abandlungen*, pp. 223-235.

Leo, R. F. and Barghoorn, E. S. (1976) *Silicification of Wood*. Botanical Museum Leaflets, Harvard University, vol. 25, no. 1.

Litwin, R. C., Turner, C. E., and Peterson, F. (1998) Palynological evidence on the age of the Morrison Formation, Western Interior, U.S. In *Modern Geology*. vol. 22, *The Upper Jurassic Morrison Formation. Part 1: An interdisciplinary study*. pp. 279-319. Denver, Colorado: Denver Museum of Natural History and Gordon and Breach Science Publishers.

Longyear, B. (1941) Time for petrifaction of wood. *The Mineralogist*, vol. 9, pp. 218-220.

Mamay, S. H. (1976) *Paleozoic Origin of the Cycads*. U.S. Geological Survey Professional Paper 934.

Manchester, S. R. (1987) Oligocene fossil plants of the John Day Formation, Fossil, Oregon. *Oregon Geology*, vol. 49. no. 10. Portland, Oregon: Oregon Department of Geology and Mineral Industries.

Manchester, S. R. (2000) Late Eocene fossil plants of the John Day Formation, Wheeler County, Oregon. Oregon Geology, vol. 62, no. 3. Portland, Oregon: Oregon Department of Geology and Mineral Industries.

Meyer, F. A. A. (1791) *Briefe uber Einige Mineralogische Gegenstande an Herrn Peter Camper*. Gottingen: Johann Christian Dieterich, Deutchland.

Meyer, H. (1973) The Oligocene Lyons Flora of northwestern Oregon. *The Ore Bin*, vol. 35, no. 3. Portland, Oregon: State of Oregon Department of Geology.

Meyer, H. W. (2003) *The Fossils of Florissant*. Washington and London: Smithsonian Books.

Meylan, B. A., and Butterfield, B. G. (1972) *Three-dimensional Structure of Wood: A Scanning Electron Microscope Study*. Syracuse, New York: Syracuse University Press.

Moore, P. R., and Wallace, R. (2000) Petrified wood from the Miocene volcanic sequence of Coromandel Peninsula, northern New Zealand. *Journal of the Royal Society of New Zealand*, vol. 30, no. 2, pp. 115-130.

Moxon, T. (1996) *Agate, Microstructure and Possible Origin*. South Yorks, England: Terra Publications.

Murata, K. (1940) Volcanic ash as a source of silica for the silicification of wood. *American Journal Science*, vol. 238, pp. 586-596.

Mustoe, G. E. (2001) Washington's fossil forests. *Washington Geology*, vol. 29, no. 1/2, pp. 10-20. Olympia, Washington: Washington State Department of Natural Resources.

Mustoe, G. E. (2003) Microscopy of silicified wood. *Microscope Today*, vol. 11, no. 6, pp. 34-37.

Nicklas, K. J. (1997) *The Evolutionary Biology of Plants*. Chicago, Illinois: The University of Chicago Press.

Norstog, K. J., and Nicholls, T. J. (1997) *The Biology of the Cycads*. Ithaca, New York: Cornell University Press.

Oehler, J. H. (1976) Hydrothermal crystallization of silica gel. *Geological Society of America Bulletin*, vol. 87, no. 10, pp. 1143-1152.

Orsen, M. (1998) *Ginkgo Petrified Forest*. Vantage, Washington: Ginkgo Gem Shop.

Pabian, R. K., and Zahrins, A. (1994) *Banded Agates Origins and Inclusions: Educational Circular No. 12*. Lincoln, Nebraska: University of Nebraska-Lincoln Press.

Panshin, A. J., and de Zeeuw, C. (1970) *Textbook of Wood Technology*. New York: McGraw-Hill Book Company.

Peterson, F. (1988) *Stratigraphy and Nomenclature of Middle Jurassic Rocks, Western Colorado Plateau, Utah and Arizona*. U.S. Geological Survey Bulletin 1633-B.

Peterson, F., and Turner-Peterson, C. E. (1987) The Morrison Formation of the Colorado Plateau: Recent advances in sedimentology, stratigraphy, and paleotectonics. *Hunteria,* vol. 2 no. 1, pp. 1-18.

Ransom, J. E. (1955) *Petrified Forest Trails: Guide to the Petrified Forests of America*. Portland, Oregon: Mineralogist Publishing Company.

Robler, R. (2001) *The Petrified Forest of Chemnitz*. Chemnitz, Germany: The Museum of Natural History of Chemnitz.

Scott, R. A., Barghoorn, E. S., and Prakash, U. (1962) Wood of *Ginkgo* in the Tertiary of Western North America. *American Journal of Botany*, vol. 49., no. 10, pp. 1095-1101.

Scott, R. B, Carrara, P. E., Hood, W. D, and Murray, K. E. (2002) *Geologic map of the Grand Junction quadrangle, Mesa County, Colorado.* U.S. Geological Survey Miscellaneous Field Studies Map, MF-2362.

Scurfield, G., and Segnit, E. R. (1984) Petrification of wood by silica minerals. *Sedimentary Geology*, vol. 39, pp. 149-167.

Siever, R. (1957) The silica budget in the sedimentary cycle. *American Mineralogist*, vol. 42, pp. 821-842.

Siever, R. (1972) Silicon: in Wedepohl, K. (ed.), *Handbook of Geochemistry*, vol. II/3. New York: Springer-Verlag, pp. 241-265.

Smith, E. (1999) *Black Opal Fossils of Lightning Ridge*. East Roseville, New South Wales, Australia: Kangaroo Press.

St. John, R. (1927) Replacement vs. impregnation in petrified wood. *Journal of Geology*, vol. 35, pp. 729-739.

Stein, C. L. (1982) Silica recrystallization in petrified wood. *Journal of Sedimentary Petrology*, vol. 52, no. 4, pp. 1277-1282.

Stevenson, D. W. 1998. *A 300-million-year twosome: The cycad and its pollinator*, New York Botanical Garden, http://www.nybg.org.

Stewart, W. N. (1983) *Paleobotany and the evolution of plants*. Oakleigh, Victoria, Australia: Cambridge University Press, London.

Stewart, W. N., and Rothwell, G. W. (1993) *Paleobotany and the Evolution of Plants*. Oakleigh, Victoria, Australia: Cambridge University Press.

Stockey, R. A. (2004) A reinterpretation of the Cerro Cuadrado fossil "seedlings," Argentina, pp. 164-171, in *Secrets of Petrified Plants: Fascination from Millions of Years*. Heppenheim, Germany: D'Oro Publishers.

Stockey, R. A., and Taylor T. N. (1978) On the structure and evolutionary relationships of the Cerro Cuadrado fossil conifer seedlings. *Botanical Journal of Linnean Society*, vol. 76, pp.161-176.

Stokes, W. L. (1944) Morrison Formation and related deposits in and adjacent to the Colorado Plateau. *Geological Society of America Bulletin* vol. 55, pp. 951-992.

Stokes, W. L. (1980) Stratigraphic interpretations of Triassic and Jurassic beds of Henry Mountains area. In *Henry Mountains Symposium*, Utah Geological Association publication 8, pp.113-122. Salt Lake City: Utah Geological Association.

Sullivan, K. R. (1998) Isotopic ages of igneous intrusions in southeastern Utah. In *Laccolith complexes of the southeastern Utah: Time of emplacement and tectonic setting – workshop proceedings*. U.S. Geological Survey Bulletin 2158, pp. 33-35.

Tidwell, W. D. (1998) *Common Fossil Plants of Western North America*. 2nd ed. Washington, D. C.: Smithsonian Institution Press.

Tidwell, W. D., and Ash, S. R. (1990) On the Upper Jurassic stem *Hermanophyton* and its species from Colorado and Utah, USA. *Palaeontographica*. Stuttgart, Germany: E. Schweizbart'sche Verlagsbuchhandlung.

Tidwell, W. D., Britt, B. B., and Ash, S. R. (1998) Preliminary flora analysis of the Mygatt-Moore quarry in the Jurassic Morrison Formation, west-central Colorado. In *Modern Geology. vol 22, The Upper Jurassic Morrison Formation. Part 1: An interdisciplinary study*, pp. 341-378. Denver, Colorado: Denver Museum of Natural History and Gordon and Breach Science Publishers.

Tidwell, W. D., and Medlyn, D. A. (1992) Short shoots from the Upper Jurassic Morrison Formation, Utah, Wyoming, and Colorado, USA. *Review of Paleobotany and Palynology*, vol. 71, pp. 219-238. Amsterdam, Netherlands: Elsevier Science Publishers.

Tidwell, W. D., and Parker, L. R. (1989) *Protoyucca shadishii* gen. et sp. novol., an arborescent monocotyledon with secondary growth from the Middle Miocene of northwestern Nevada, USA. *Review of Paleobotany and Palynology*, vol. 62, pp. 79-95. Amsterdam: Elsevier Science Publishers.

Turner, C. E., and Fishman, N. S. (1991) *Jurassic Lake T-oo-dichi': A large alkaline, saline lake, Morrison Formation, Eastern Colorado Plateau.* Geological Survey of America Bulletin, vol. 103, pp. 538-558.

Vail, J. G. (1928) *Soluble Silicates in Industry*. New York: Chemical Catalog Company.

Walton, J. (1956) *Rhexoxylon* and *Dadoxylon* from the Lower Shire Region of Nyasaland and Portuguese East Africa. London: Her Majesty's Stationery Office.

Wang, Y. (2001) *Formation of Petrified Woods – An Organic Molecule-templated Silica Mineral Precipitation in Nature*, Paper No. 99-C. Geological Society of America.

Wang, Y., and Merino, E. (1995) Origin of fibrosity and banding in agates from flood basalts. *American Journal of Science*, vol. 295, pp. 49-77.

Ward, L. F. 1894. Recent discoveries of cycadean trunks in the Potomac Formation of Maryland. *Bulletin of the Torrey Botanical Club*, vol. 21, pp. 291-299.

Ward, L. F. (1900) Descriptions of a new genus and twenty new species of fossil cycadean trunks from the Jurassic of Wyoming. *Proceedings of the Washington Academy of Science*, vol. 1, pp. 253-300.

Wardrop, A. B., and Bland, D. E. (1959) Process of Lignification in Woody Plants. *Proceedings of the 4th International Congress on Biochemistry*, vol. 2, pp. 92-116.

Watson, J., and Sincock, C. A. (1992) Bennettitales of the English Wealden. *Monograph of the Palaeontographical Society*. London.

Wehrfeld, X. X. (1935) Patagonia el gran acervo de fosiles de la Argentina. *Revol. Geogr. Am.*, vol. 4, pp. 117-130.

Wheeler, E. A., and Manchester, S. R. (2002) Woods of the Middle Eocene Nut Beds Flora, Clarno Formation, Oregon, USA. *IAWA Journal*, supplement 3. The Netherlands: International Association of Wood Anatomists.

Wheeler, E. A., and Michalski, T. C. (2003) Paleocene and Early Eocene woods of the Denver Basin, Colorado. *Rocky Mountain Geology*, vol. 38, no. 1, Spring 2003.

White, M. E. (1990) The *Flowering of Gondwana: The 400 Million Year Story of Australia's Plants*. Princeton, New Jersey: Princeton University Press.

Wieland, G. R. (1906) *American Fossil Cycads, Vol. 1*. Washington, D.C.: Carnegie Institute.

Wieland, G.R. (1916) *American Fossil Cycads, Vol. 2*. Washington, D.C.: Carnegie Institute.

Wieland, G. R. (1934) Fossil cycads, with special reference to *Raumeria Reichenbachiana Goeppert sp.* of the Zwinger of Dresden. *Paleontohraphica*. Stuttgart: E. Schweizerbart'sche Verlagsbuchhandlung.

Willis, K. J., and McElwain, J. C. (2002) *The Evolution of Plants*. Oxford/New York: Oxford University Press.

Woodward, L. A., Kaufman, W. H., Schumacher, O. L., and Talbott, L. W. (1974) Strata-bound copper deposits in the Triassic sandstone of the Sierra Nacimiento, New Mexico. *Economic Geology*, vol. 69, pp. 108-120.

Xeros, N. (1954) Occurrence of an invertebrate iridescent-like virus. *Nature*, vol. 174, p. 562.

Zahrins, A. (1977) *Origins and Geologic History of Siliceous Metacolloidal Deposits, Cathedral Mountain Quadrangle, Brewster County, Texas*, Unpublished M.S. Thesis. Lincoln, Nebraska: University of Nebraska.

INDEX

bold number indicates reference in photograph or micro image caption, possibly in addition to appearing in the text

Q

R

S